Proteins
Analysis and Design

Proteins
Analysis and Design

Edited by

Ruth Hogue Angeletti

Albert Einstein College of Medicine of Yeshiva University
Bronx, New York

Academic Press

San Diego London Boston New York Sydney Tokyo Toronto

Front cover photograph: Alpha carbon trace of trimeric tetrahydrodipicolinate *N*-succinyltransferase. The triangular domains are the left-handed parallel beta helices formed by tandem-repeated copies of hexapeptide amino acid sequence motif. Courtesy of Todd W. Beaman, David A. Binder, John S. Blanchard, and Steven L. Roderick, Albert Einstein College of Medicine, Bronx, New York.

This book is printed on acid-free paper. ♾

Academic Press
a division of Harcourt Brace & Company
525 B Street, Suite 1900, San Diego, CA 92101-4495, USA
http://www.academicpress.com

Academic Press
24–28 Oval Road, London NW1 7DX, UK
http://www.hbuk.co.uk/ap/

Proteins : analysis and design / edited by Ruth Hogue Angeletti.
p. cm.
Includes bibliographical references and index.
ISBN 0-12-058785-8 (hc : alk. paper)
1. Proteins. 2. Proteins--Analysis. 3. Protein engineering.
I. Angeletti, Ruth Hogue.
QP551.P697774 1998
572′.6--dc21 98-4746
CIP

Printed in the United States of America
98 99 00 01 02 MM 9 8 7 6 5 4 3 2 1

Finn Wold

1928–1997

This volume is dedicated to the memory of Finn Wold, who devoted his career to the study of protein biology and to the nurturing of protein science and protein scientists.

Contents

Contributors

Numbers in parentheses indicate the pages on which the authors' contributions begin.

Ruedi Aebersold (3) Department of Molecular Biotechnology, University of Washington, Seattle, Washington 98195

Ruth Hogue Angeletti (1) Albert Einstein College of Medicine of Yeshiva University, Bronx, New York 10461

Gary S. Coombs (259) Howard Hughes Medical Institute, Department of Pharmacology, University of Texas Southwestern Medical Center at Dallas, Dallas, Texas 75235

David R. Corey (259) Howard Hughes Medical Institute, Department of Pharmacology, University of Texas Southwestern Medical Center at Dallas, Dallas, Texas 75235

Gregg B. Fields (207) Department of Chemistry and Biochemistry, Center for Molecular Biology and Biotechnology, Florida Atlantic University, Boca Raton, Florida 33431

Michael W. Klemba (313) Department of Molecular Biology, Princeton University, Princeton, New Jersey 08544

Radha Gudepu Krishna (121) Diagnostic Systems Laboratories, Inc., Webster, Texas 77598

Janelle L. Lauer (207) Department of Chemistry and Biochemistry, Center for Molecular Biology and Biotechnology, Florida Atlantic University, Boca Raton, Florida 33431

Mary Munson (313) Department of Microbiology, GBF-National Research Center for Biotechnology, D-38124 Braunschweig, Germany

Lynne Regan (313) Department of Molecular Biophysics and Biochemistry, Yale University, New Haven, Connecticut 06520

Scott D. Patterson (3) Amgen Inc., Protein Structure, Amgen Center, Thousand Oaks, California 91320

Finn Wold* (121) Department of Biochemistry and Molecular Biology, University of Texas Medical School, Houston, Texas 77225

*Deceased.

Preface

It is now possible to perform our dream experiments with proteins: to analyze details of their structure–function relationships and their life cycles within the cell. Many of us need to embrace these technologies to achieve full understanding of the systems we have chosen to study. Although the experimental approaches are within the reach of all of us, we have not all been trained in their implementation. This book was designed to provide guides, both philosophical and practical, from experts in protein analysis and design.

Ruth Hogue Angeletti

Introduction

Ruth Hogue Angeletti

Protein biology is the study of protein structure revealing details of the function and life cycle of individual polypeptides. Completion of the sequences of animal and plant genomes will not be the consummation of modern biology, but a new beginning. The experimental and philosophical challenge is not just to identify polypeptide gene products for all coding genes, but to provide a rationale for how they work. As part of a cellular response to external stimuli, proteins may be altered by chemical modification or change subcellular localization, processing, degradation, or concentration. How do proteins work? How do they work together? How do they work over time and space?

The development of technologies and experimental approaches that were required to answer the questions of protein biology accelerated during the early period of genomic analysis. These technologies are now essential tools of experimental biology. All of the strategies aim toward the highest sensitivity analysis possible. Chromatographic and electrophoretic separation and detection, Edman sequence analysis, amino acid analysis, and mass spectrometry are now all performed routinely with a few picomoles of protein or peptide. Subpicomole and low femtomole levels of analysis are reported in the literature more frequently, and experiments detecting attomoles of protein or peptide have been described. Pushing these barriers of sensitivity is of intense interest because of the desire and need to understand changes that take place in a few molecules within a few cells or a single cell, to study not the population of molecules, but those few relevant to the physiological events at hand.

Covalent modifications of proteins are an essential part of the language of intracellular and intercellular communication. They may be reversible or irreversible. They may be required for biological activity, or simply modulate it, functioning as "molecular switches." They are important for signaling and molecular and cellular recognition. These modifications may impart structural stability or unique structural features, facilitate protein folding, or promote intermolecular interactions. They may anchor a protein in a membrane or determine intracellular or extracellular position. Modifications can alter the biological lifetime of a protein, and define the process by which it is degraded. Whereas some modifications are present on an entire population of a polypeptide, others are present on only a small subset of the protein at any one moment. Quantification of the extent and range of modifications present in this population is not routinely analyzed, but could have important biological implications. Thus, knowledge of the possible modifications of a particular protein, as well as modifications induced by external stimuli, are of fundamental importance to research in structural biology and cell biology alike.

The core technology for analyzing protein covalent structure is mass spectrometry. Once the province of specialists, mass spectrometry must now be in the repertoire of all biologists. The ability to measure accurately the molecular weights of both large and small biological molecules means that mass measurements can be used to unambiguously identify a polypeptide or gene product, to establish the presence of a covalent modification and localize its position within the polypeptide chain, and to sequence or verify structures by patterns of molecular bond breakage. The strategies of mass spectrometry provide insight into biological processes and are a key element in the discovery process. Establishing lines of communication between analytical protein chemists and cell biologists will be needed to release the full power of detailed genomic and cellular information.

Protein biology also includes the design of new proteins to test biological and physiological hypotheses. One can now design and prepare novel polypeptides with new functions that cannot be isolated from nature. How can one manipulate proteins to understand their original function or to make them perform new ones? On what principles can these designs be based? Protein analysis and protein design are linked disciplines. In order to understand how a protein functions, one must understand its complete covalent structure, including all cross-linking, processing, and modification events. To design a novel polypeptide, one must know what those modifications are in order to incorporate them into the structure, or to know when they can be ignored. The most creatively designed protein model cannot be used to prove the starting hypothesis if the final structure has not been rigorously analyzed and verified.

Both molecular biological and chemical synthetic procedures have made it possible to alter polypeptide structure to probe details of function. Strategies can be developed to alter a single amino acid, to substitute a nonnatural residue, or to mix and match functional units to create new biological entities. Pragmatic concerns, bioassays, and considerations of secondary structural elements can all be used to build on knowledge and function of known molecules. More challenging still is *de novo* design of new protein activities from first principles, a less developed field. Progress is being made in this direction, and should proceed rapidly as the related fields of protein design mature.

We can no longer be practitioners of a single art or technology, but must take a problem solving, or systems approach, to the biological questions that we ask. These protein strategies are only a part of modern experimental biology, yet are a core mechanism to identify the most important questions and find their solutions.

Chapter 1

Current Problems and Technical Solutions in Protein Biochemistry

Ruedi Aebersold and Scott D. Patterson

I. Introduction

The vast majority of biological processes and pathways are tightly controlled. This applies equally to well-understood, relatively simple processes such as oxygen transport and storage and to as yet molecularly poorly understood phenomena of extreme complexity such as development, cell differentiation, and signal transduction pathways that serve to elicit the appropriate intracellular responses to extracellular stimuli.

Such biological processes differ significantly in their complexity as well as in the level and mechanism of control. They have in common that they involve multiple components, that the activities of at least some components are regulated, and that multiple components structurally and functionally interact. Frequently, key components are proteins and regulatory control is mediated by a variety of mechanisms, including reversible protein modification, most notably protein phosphorylation, formation of protein : protein complexes, proteolytic protein processing, *de novo* synthesis, and intracellular translocation. For the experimental biologist this means that a comprehensive analysis of such a process requires the identification of each system component, the characterization of the identified species with respect to their chemical composition and functional control, and the investigation of the interaction between system components. Finally, the results obtained at these levels should be sufficient to establish a model of the process studied that satisfies all the experimental observations.

The traditional approach to investigating complex systems has been reductionist in nature. Proteins representing a single step in a process were purified and analyzed in great detail. This step was repeated until seemingly all the elements of the pathway were identified, isolated, and studied. Reconstitution of the system from the isolated elements was used as a criterion to indicate that the system had been analyzed in its full complexity or at least to a sufficient degree to support a model of the process. Contraction of striated and smooth muscle are illustrative examples for the success of this approach. Insights from light microscopic, electron microscopic, physiologic, and biochemical studies culminated in the formulation of the sliding filament model for contraction of striated muscle (1). Further biochemical analyses resulted in the identification and structural and functional analysis of the components involved, as well as in the reconstitution of contractile elements from purified proteins *in vitro* (2,3).

Significantly, more than 30 years after formulation of the sliding filament model, substantial structural and regulatory features of muscle contraction remain to be discovered [e.g., (4)]. This suggests that the reductionist approach, although very successful in describing essential events in a complex system, is less suited for a comprehensive systems analysis.

Dramatic advances in several independent research areas indicate that an integrated biotechnology capable of successfully analyzing biological systems in their whole complexity is rapidly emerging. The most dramatic developments include the following:

1. *Initiation of systematic, genome analysis programs.* Large-scale DNA mapping and sequencing efforts, aimed at deciphering the complete genomic sequences of a number of species (5–8) or the expressed sequences represented by cDNAs of differentiated tissues and cells (9), are the most widely publicized and discussed global programs in biology. It is projected that the genome sequence of yeast (*Saccharomyces cerevisiae*) will be completed before this chapter appears in print (6) and that the sequence of the complete human genome may be determined within the next 10 years (8). Although these programs promise to generate a unique resource for research into biology and medicine, the frequent implication that knowledge of genome sequences will be sufficient to achieve understanding of biological systems is simplistic. An array of expressed gene sequences does not describe a biological system, mainly because proteins, the most important biological effector molecules, are not simply the linear translation of gene sequences. Proteins should be considered the products of mRNA translation *and* of diverse posttranslational processing and modification events.
2. *Development of powerful new analytical technology.* In particular, the introduction of mass spectrometric techniques and instruments compatible with the routine analysis of proteins and peptides at high sensitivity now makes investigation of the dynamics of biological systems, including induced protein modifications, reversible protein : protein associations, protein translocations, targeted destruction and *de novo* synthesis experimentally accessible.
3. *Rapid advances in data analysis, data storage, and data distribution.* Centralized, continuously updated, and interconnected databases that are remotely accessible are a necessary prerequisite for integrated systems analysis, and new algorithms for the analysis of data derived from large-scale programs are being developed.
4. *The capability for targeted genetic manipulation of cells and organisms.* In a number of experimental organisms, transgene expression, including the expression of dominant negative mutants, targeted gene knockout, and gene replacement, are now relatively standard procedures. These techniques are indispensable for testing hypotheses related to the structure and function of genes and proteins.

In this chapter we describe recently developed protein analytical techniques and concepts and attempt to demonstrate how these tools converge toward a comprehensive technology for the analysis of complex biological systems. We focus on the analysis of proteins because proteins are the most important effector molecules in biology. To highlight the stunning technological progress during the past few years in this area, we will begin each section with a case study, describing the experimental approach and the techniques used to successfully analyze a complex, regulated system. As the case study we chose the signal transduction pathway induced by the virus-fighting protein interferon from the cell membrane to the nucleus, which represents one of the best-understood regulatory pathways in cells. Following the case study we will discuss novel technology and approaches and we will illustrate the impact of these new tools on biological research. The sections will, therefore, predominantly focus on discussing challenges and promising approaches to protein analytical problems, rather than on detailed treatment of individual techniques. Each builds on chapters elsewhere in this volume describing mass spectrometry of proteins and peptides, separations technologies, and posttranslational modifications, respectively.

II. Early Events in Signaling by Interferons: From Membrane Receptor to Induced Genes

It is well established that binding of polypeptides such as lymphokines, hormones, neurotransmitters or growth factors to specific cell surface receptors causes changes in the transcription rate of specific sets of genes by transmitting a signal from the cell membrane to the nucleus. At least the early events (transcriptional activation within minutes of membrane receptor stimulation) are based on the activation of preexisting transcription factors by a variety of mechanisms, including protein phosphorylation, protein : protein complex formation, intracellular translocation, and proteolysis. The involvement of multiple protein components and the tight regulation of function define such signal transduction pathways as complex biological systems.

The interferons are a family of secreted polypeptides (cytokines) that act as mediators in the host defense against viruses and parasites. As with other cytokines, the interaction of interferons with their specific membrane receptor induces rapid transcription of a set of immediate-early genes in the absence of protein synthesis (10). Over the past 10 years, studies from a number of groups converged on establishing the complete sequence of events required to transduce the signal from the membrane to the nucleus, making this the most completely understood signal transduction pathway in mammalian cells (11).

The work began with the identification of several genes that were transcriptionally upregulated in response to interferon stimulation of cells (12,13). Detailed analysis of the 5′ upstream genomic sequence of these genes identified the interferon-stimulated response element (ISRE) (14–18) as the binding site occupied after interferon α (IFNα) stimulation, presumably by a transcriptional activator (17). One of the factors binding to the regulatory regions was termed interferon-stimulated gene factor 3 (ISGF3). This factor was the only one to show several prerequisite characteristics as an interferon α activated transcription factor, including absolute dependence on interferon treatment (13,19), rapid increase in activity without protein synthesis (16,17) and specific binding to the nucleotide sequence required for interferon α dependent gene activation (18,20). Subsequent work led to the isolation of the ISGF3 complex (21), sequence analysis of the respective proteins and coding genes (22–24), and elucidation of the sequence of events leading to the interferon-induced activation of ISGF3

(25,26), and finally, to the establishment of a molecular and functional connection between interferon receptor stimulation and transcriptional activation of interferon-induced genes (27).

The model now accepted suggests that binding of interferon to its cognate receptor activates a nonreceptor tyrosine kinase of the family of Janus kinases (JAKs), which is noncovalently associated with the cytoplasmic tail of the receptor (see Fig. 1). The tyrosine kinase, activated through ligand–receptor interaction, phosphorylates on tyrosine residues one or more of three large polypeptides of 113 kDa, 91 kDa, and 84 kDa that preexist in the cytoplasm, causing them to aggregate and translocate to the nucleus, where they interact with a smaller protein of 48 kDa. The now complete ISGF3 protein complex forms the transcription factor that activates immediate-early gene transcription. By convention, the latent cytoplasmic transcription factors have been termed STATs (signal transducers and activators of transcription), whereby the 91- and 84-kDa proteins are named Stat 1α and Stat 1β, respectively, and the 113-kDa polypeptide is named Stat 2. In this chapter we have chosen this pathway (shown in Fig. 1), which has been described in more detail in several excellent reviews (27–29), as a prototypical example for a complex biological system for the following reasons: (i) Transduction of the interferon signal is currently the most completely understood signal transduction pathway. There is a direct and complete link from the membrane receptor to the nuclear factors transactivating immediate-early genes. (ii) The type of regulatory events controlling this pathway, including protein phosphorylation, translocation, and protein : protein complex formation are common to the regulation of many, if not most, complex biological systems. (iii) Most of the elements involved in this pathway, as well as the mechanisms controlling the activity, were determined using protein

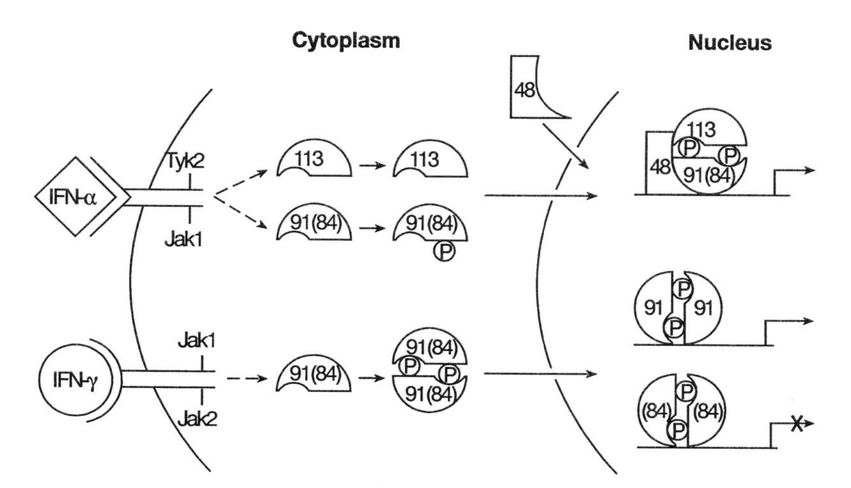

Fig. 1 Overview of the proteins identified in interferon α- and γ-dependent signal transduction and gene activation. The Jak kinases are phosphorylated on tyrosine in response to ligand, but the sites and the requirement for such modification are not yet established. The circled P's on the Stat proteins are tyrosine phosphates, and the indentations symbolize SH2 domains. Reprinted with permission from Darnell, J. E., Jr., I. M. Kerr, and G. R. Stark. 1994. Jak-STAT pathways and transcriptional activation in response to IFNs and other extracellular signaling proteins. *Science* 264:1415–1421. (27). Copyright 1994 American Association for the Advancement of Science.

analytical technology, thus providing a framework to discuss novel technology in comparison with accepted, standard techniques. Finally, (iv) it has become apparent that at least some of the elements involved in the interferon signal transduction pathway and proteins structurally and functionally related to the Jak and Stat proteins constitute a network of intracellular signaling molecules that are collectively involved in transmitting the signal from at least 25 different membrane receptors (29). This suggests that signal transduction pathways initiated by different receptors (in the same cell) are not independently controlled, but are interwoven to an extent that is only slowly becoming apparent. This illustrates the premise of this chapter that comprehensive investigation of complex biological systems requires new analytical technologies, which are better suited to achieving a global perspective.

III. Detection of Systems Components

A. Overview

Depending on the complexity of the organism, a cell contains thousands to tens of thousands of different types of proteins. Collectively, these proteins constitute the structure and perform and control numerous cellular functions. Establishment of a complete inventory of all those proteins that are involved in a particular biological process is complicated by the fact that usually neither the number, nor the identities, nor the activities of the elements of the process are known. A credible model for a process cannot be formulated, however, as long as the components and their functions are not known. In this section we describe experimental approaches for the detection of those proteins that are part of a particular process. The discussion is limited to approaches for the detection of proteins with known or unknown function that are part of a process. DNA-based approaches such as subtractive hybridization and cDNA profiling using ordered DNA arrays are not further discussed (30,31).

B. ISGF3 Case Study: Detection of Proteins Involved in Interferon Signal Transduction

Detection of the components of the ISGF3 complex was essentially based on the observation that an interferon induced transcription factor was binding to specific DNA sequences in the promoter regions of interferon-regulated genes (Fig. 2A). This provided a bioassay for the detection and purification of the factor. Comparative gel-shift assays from interferon stimulated and unstimulated control cells showed that two of the three observed protein : oligonucleotide complexes were specifically induced by interferon stimulation of the cells (17). The more rapidly induced of the two complexes, which did not require protein synthesis, was furthermore shown to constitute, or at least to contain, the transactivating factor ISGF3 (16,17). The gel-shift assay (or electrophoretic mobility shift assay, EMSA), like many other bioassays, conclusively demonstrates the presence of a specific activity in a sample. It does not, however, yield any molecular information on the activity. To further investigate the activity, ISGF3 factor was purified by column chromatography, incubated with the [^{32}P]phosphate-labeled oligonucleotides containing the specific recognition

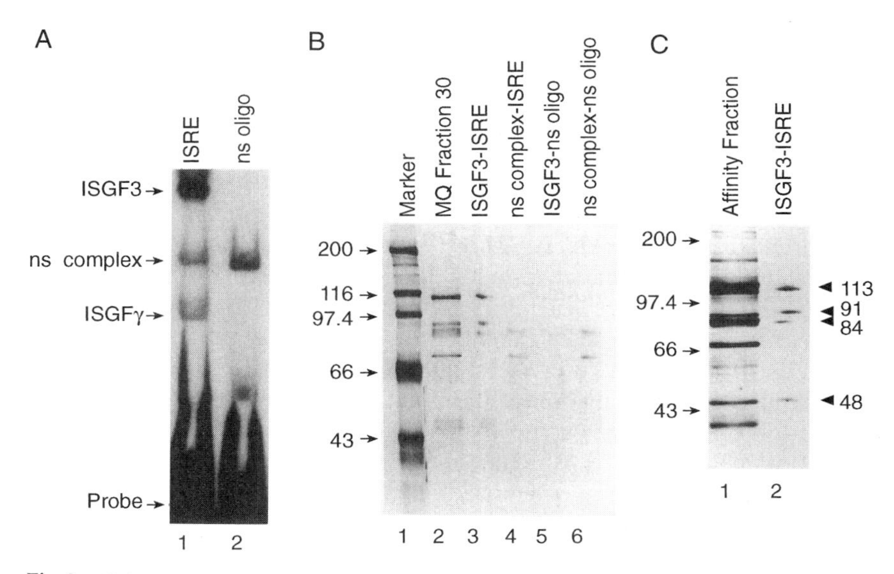

Fig. 2 ISGF3 is composed of four distinct polypeptides. (A) Partially purified ISGF3 was incubated with a radiolabeled ISRE (lane 1) or unrelated oligonucleotide [nonspecific (ns) oligo; lane 2], and protein : DNA complexes were fractionated by EMSA. ISGF3 and ISGF3γ complexes were formed with the ISRE, and a contaminant nonspecific (ns) complex was formed with both probes as indicated at left. After autoradiography, regions of the preparative gel corresponding to all three complexes were excised from both lanes. (B) Proteins present in the excised slices were electroeluted onto SDS–PAGE (8% gel) for electrophoretic separation and stained with silver. (C) An experiment similar to that described in A and B, but using a less pure affinity fraction of ISGF3 (lane 1). Four proteins (48, 84, 91, and 113 kDa; lane 2) were specifically recovered from the ISGF3–ISRE complex. Reproduced with permission from Fu, X. Y., D. S. Kessler, S. A. Veals, D. E. Levy, and J. E. Darnell, Jr. (1990). ISGF3, the transcriptional activator induced by interferon alpha, consists of multiple interacting polypeptide chains. *Proc. Natl. Acad. Sci. USA* **87**:8555–8559. Copyright 1990 American Association for the Advancement of Science.

sequence, and separated from the contaminating proteins by nondenaturing gel electrophoresis. The ISGF3 protein was excised from gels and subjected to denaturing sodium dodecyl sulfate–polyacrylamide gel electrophoresis (SDS–PAGE). These and complementing experiments established that ISGF3 consisted of at least four polypeptides of apparent molecular mass 48 kDa, 84 kDa (Stat 1β), 91 kDa (Stat 1α), and 113 kDa (Stat 2), respectively, of which the 48-kDa polypeptide and Stat 1 appeared to directly interact with DNA (21,32) (Fig. 2 B,C).

In the work described previously, the proteins were identified as components of the ISGF3 complex using a biochemical assay. The proteins were isolated and assayed based on their ability to bind to a specific nucleotide sequence. The example illustrates the success of the approach and also highlights some of the limitations of the biochemical "purify-and-assay" technique for the analysis of multicomponent systems. The limitations include the following: (i) Components participating in one activity, in the present case specific transcriptional activation of genes carrying the ISGF3 binding element, were successfully detected. Elucidation of the whole signaling pathway from the membrane to the nucleus required the combined use of genetics, biochemistry, and cell biology approaches (29). (ii) Analysis of the ISGF3 factor did not necessarily suggest further experiments (biochemical, genetic) toward the isolation of upstream or downstream segments of the pathway. (iii) Isolation of the ISGF3 complex and demonstration of its transactivating activity did not rule out the presence of additional components associated with the ISGF3 factor *in vivo,* nor did it rule out the presence of other protein complexes with an overlapping activity. The Stat proteins participating in the ISGF3 complex and related proteins participate in signal transducing and transactivating complexes induced by at least 25 cytokines (29). (iv) Specific functions could not be associated with individual elements without further experiments. In fact, the protein pattern obtained from isolated ISGF3 complexes represented an average picture of the composition of all the complexes in the sample and did not reveal any information concerning the stoichiometry and the composition of individual complexes. (v) Detection of the polypeptides associated with the ISGF3 transactivating activity does not provide any mechanistic information. Insight into the mechanism or the function of a protein can, however, frequently be deduced from the gene or amino acid sequence (see Section IV). It is, therefore, advantageous that the same analytical techniques that are used for the detection of system components also serve as micro preparative techniques for the isolation of the proteins for further analysis.

These and additional limitations could be overcome if a profile of all the expressed proteins could be established from the tissue or cell source under investigation. If such a profile could be established as a function of time (e.g., time course of changes in protein profile induced by cell stimulation), substantial amounts of critical information could be obtained. Profiles obtained at different time points or profiles obtained under different experiment conditions could be analyzed by subtractive pattern analysis. Changes in the protein profile filtered out from the total protein contents of the sample would be interpreted as relevant to the process under investigation, and all the invariant elements would be interpreted as unrelated to the process under investigation. Unfortunately, at present, there is no technology capable of establishing a complete quantitative protein profile of a cell. Among the established techniques, quantitative high-resolution two-dimensional polyacrylamide gel electrophoresis (2-DE) comes closest to fulfilling these objectives.

C. Quantitative Two-Dimensional Gel Electrophoresis: Capabilities and Limitations

Two-dimensional (2-D) gel electrophoresis was first shown to be a powerful technique for separating thousands of proteins in 1975, when O'Farrell and Klose independently developed a system that combined isoelectric focusing gel electrophoresis in the first dimension with SDS gel electrophoresis in the second dimension, to create a p*I* vs molecular weight separation matrix (33,34). With the use of a large gel format, more than 10,000 protein spots could be separated (35–37). It was soon recognized that quantitation of separated spots would allow construction of a protein profile of a sample. Subsequent subtractive analysis of such profiles from samples generated under different experimental conditions would result in the detection of those proteins that were altered in a particular process within the background of proteins unaffected by the conditions applied. Furthermore, many such protein profiles could be archived in a computer for later comparative analyses. Figure 3 shows an overview of the approach, which employes 2-DE for quantitative analyses.

For the detection of those proteins in a complex protein mixture that are involved in a particular process, subtractive pattern analysis has a number of unique advantages: (i) The technique provides for a simultaneous survey of a large number of proteins. For a relatively simple cell such as yeast or *Escherichia coli*, the majority of proteins can be surveyed in a single operation, whereas for the more complex cells of a higher organism only a fraction of the proteins present are detectable. (ii) No functional assay is required to follow the fate of a protein. This is of particular importance in cases in which the function of the proteins involved in a process is not known or in which multiple proteins participate in a particular function. Also, in cellular regulatory systems, activities can change in very subtle ways, such as through translocation or association with other proteins. Such changes are difficult to follow biochemically *in vitro*. (iii) The technique is micro preparative. Section IV shows that in principle every protein spot identified by the 2-D gel approach can be identified at the sequence level. (iv) The technique is multidimensional. Even though the approach is based on two-dimensional protein separation, the results obtained

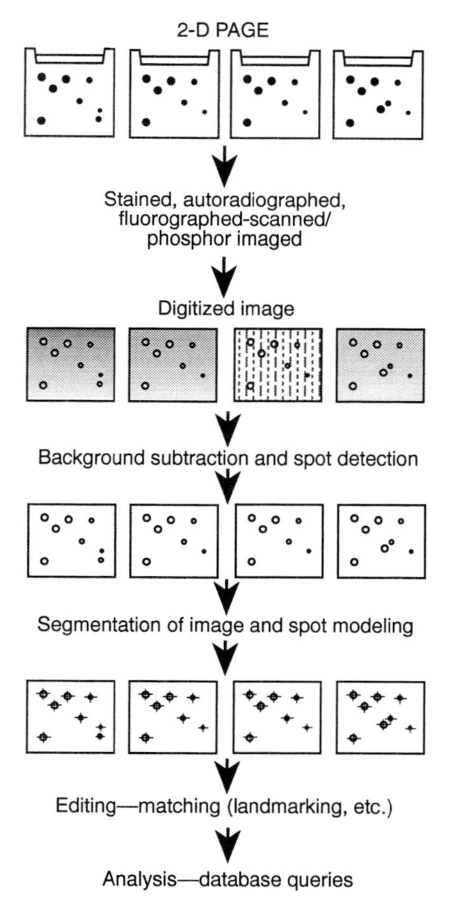

Fig. 3 Flow chart of the steps involved in quantitative 2-DE analysis. Following digitization of the protein image by either direct means (e.g., phosphor imaging) or indirect means (scanning of fluorographs, autoradiographs, or stained gels), the computer program detects these spots and performs cleanup functions such as streak removal if necessary. In some cases, multiple exposures of the same gel may have to be performed. The image also needs to be calibrated. A background removal function is then performed. The protein spots are now modeled using formulas such as Gaussian curves. The mathematically described spots are now ready to be matched between gels. This is usually accomplished by manually assigning a few protein spot matches (landmarks) and allowing the computer program to match the remaining spots. following verification of these matches, the database of linked spot information can be queried to determine which protein spots display coordinate regulation. This scheme is based on that of Monardo, P. J., T. Boutell, J. I. Garrels, and G. I. Latter. (1994). A distributed system for two-dimensional gel analysis. *Comput. Appl. Biosci.* **10:**137–143.

by the technique are multidimensional. Additional data dimensions that can be derived by relatively simple experiments include protein quantities in spots, protein translocation as detected by following the fate of a protein through subcellular fractions, protein turnover and half-life as detected by pulse-chase metabolic labeling experiments, protein associations as detected by coprecipitation experiments, and posttranslational modifications as detected by specific labeling with radioactive metabolic precursors.

In spite of these conceptual advantages, to date the approach has contributed relatively little to solving fundamental biological problems. A number of technical limitations have prevented the widespread use of quantitative 2-DE. The most significant limitations are as follows: (i) Limitations in protein detection. Only a fraction of the proteins present in a cell are usually detected. Size and p*I* limitations of 2-DE and poor solubility preclude some proteins from being represented in the 2-D pattern. More importantly, proteins of low abundance, which frequently include regulatory proteins of intense interest, are not detectable if total cell lysates are separated. Applying larger amounts of sample to the gel is generally not an option, since electrophoretic separation breaks down if the gels are overloaded. (ii) Reproducibility in gel patterns. In spite of intense efforts to standardize the gel patterns obtained from the same sample in different laboratories or at different time points, direct comparison of such patterns has remained difficult. (iii) Accurate spot quantitation. Spots containing radiolabeled proteins are quite accurately quantified, whereas proteins detected by staining are more difficult to quantify. This is particularly true for proteins detected by silver staining.

D. Recent Technology Developments in Quantitative Two-Dimensional Gel Electrophoresis

In the past few years incremental improvements have made 2-D gel technology easier to use, more reproducible, and more robust. Technology development has focused on eliminating the limitations described previously. Work has primarily focused on improving capabilities for computerized pattern analysis, spot quantitation, and spot detection and on enhancing the reproducibility of 2-D gel patterns.

1. Spot Quantitation and Pattern Acquisition

The first attempts at quantitative analysis involved cutting spots from radiolabeled samples from the gel and counting radioactivity. The disadvantage of this technique is that it can be applied only to a few hundred well-separated spots out of the thousands on the gel. Partially or completely overlapping spots cannot accurately be quantified. Early attempts at computer analysis of films to quantify the relative amounts of protein in 2-D gel spots aimed at alleviating this limitation (38,39). Later, several powerful software systems with the ability to detect protein spots, to quantify the protein contents in each spot, to build a quantitative 2-D database, and to perform subtractive protein pattern analysis were developed. The characteristics and relative performance of a number of such systems were reviewed in a recent comprehensive article (40) and a detailed comparison of two popular pattern analysis packages was recently undertaken (41). The systems available include a commercial software package from Bio Image (Bio Image Products, Ann Arbor, MI); the Tycho system initially developed at the Argonne National Laboratory (42) and further developed commercially by Large Scale Biology Inc. (Rockville, MD); the Melanie system, a derivative of the Elsie system initiated by Miller and Olson (43) now being developed at Geneva University Hospital (44); the GELLAB (45); HERMeS, a system developed by Vincens and colleagues (46); GalTool, developed by Solomon and Harrington (47); and the QUEST database system (48,49), as well as the Discovery Series, a commercial derivative of an earlier version of the Quest software (PDI Inc., Huntington Station, NY). Considerable effort also went into refining the methods of spot detection and quantitation. To quantify more than four orders of magnitude of spot intensity, phosphor imaging has reduced the recording time required over fluorography from over 1 month (for multiple exposures) to 3–5 days (50). To better quantify nonradioactive samples, Rodriguez *et al.* (51) explored silver staining and concluded that real-time monitoring during silver-stain development is necessary, but not necessarily sufficient, for quantitation of silver-stained spots. Neumann *et al.* (52) attempted to elute dye absorbed to stained proteins for quantitation of protein spots. Although the procedure was not very sensitive, with a detection limit of approximately 20 ng, it was simple to apply and linear at least in the range of 50 ng to 10 μg of protein present in the spot.

2. *Reproducibility of Two-Dimensional Polyacrylamide Gel Electrophoresis Protein Patterns*

Several groups attempted to make 2-D gel protein patterns more reproducible and easier to correlate by improving the gel running equipment, by improving the gel matrix, or by introducing improved p*I* and molecular weight (MW) standards. Using carrier ampholyte-based isoelectric focusing, Harrington *et al.* (53) and Nokihara *et al.* (54) demonstrated that experimental gel electrophoresis equipment employing carefully controlled electrophoretic parameters and exact positional alignment of the first dimension gels onto the second dimension gels provided superior pattern reproducibility. Patton *et al.* (55) introduced a commercial gel system that provided reproducible gel patterns through computer control of electrophoresis parameters, including temperature and electric power profile of a run. Studies comparing 2-D protein pattern reproducibility using immobilized pH gradient (IPG) gels (56) as a first dimension within a laboratory (57), or between laboratories (58,59), concluded that using this technology, highly reproducible 2-D patterns could be obtained with as little as 1 mm positional variability between spots observed in standardized samples run in different laboratories. Gersten *et al.* (60) used tritiated bacteriophage ghost proteins producing 48 protein spots as a standardized sample to aid in pattern correlation between different laboratories. Bjellqvist and co-workers (61) used a panel of 41 well-characterized proteins common to most cell types as landmarks to correlate gel images, even in cases in which IEF dimensions with different pH gradients were used. To verify the identity of protein spots observed in different gels and different gel systems, Dean *et al.* (62) used a procedure they termed "spot transfer," which involves elution and comigration of proteins. To enhance stability of the gel matrices, in particular under very basic conditions, and to enhance compatibility of the matrix with organic solvents, Righetti and colleagues and Harrington and colleagues have explored alternative matrices for electrophoresis (63–66).

With the exception of the IPG first dimension gels, to date none of the described advances has found widespread application. It is anticipated, however, that reproducibility and quality of gel patterns will incrementally increase over the next few years. Furthermore, the need for absolute pattern reproducibility might actually decrease. Section IV will show that the methods for identifying gel separated protein by their amino acid sequence have improved dramatically. The identity of spots in different gels can, therefore, be verified by more conclusive criteria than their electrophoretic mobility, thus relaxing the need for exact pattern reproducibility.

3. *Enhanced Detectability of Separated Proteins*

The development of IPGs with a wider pH range (67) and a nonlinear wide-range gradient shape for optimized resolution and pattern reproducibility (68) significantly enhanced the usable p*I* range for the separation of proteins compared to that of IEF gels with carrier ampholytes. Focusing of very basic proteins with a p*I* range of 10–12 was successfully achieved in such a gel system (69). To detect proteins of lesser abundance, Bjellqvist and colleagues (70) used IPGs in the first dimension with a modification of the method allowing higher total protein load. Although the protein load achieved with these gels is impressive, it has to be pointed out that the significant limitation of detecting only the more abundant proteins was not solved using this approach.

E. Use of Two-Dimensional Gel Electrophoresis for Analysis of Complex Systems: Subtractive Analyses and Protein Databases

Protein profiles established by quantitative 2-D gel electrophoresis have been used in two principal ways for approaching biological problems. The first is straightforward subtractive pattern analysis of two or a few gels representing protein samples obtained under different experimental conditions, with the aim of detecting those proteins whose expression levels or location in the 2-D pattern have changed. The second is the establishment of a data structure containing protein gel patterns of a particular cell type or tissue obtained under a variety of experimental conditions, with identification of selected proteins and possible additional information concerning the system. Such structures are commonly referred to as 2-D protein databases. In the following we describe some representative work and discuss the limitations of the subtractive approach and the 2-D databases.

Subtractive pattern analysis was successfully used to identify proteins that are linked to a variety of complex biological or pathological processes, including development of mouse embryos (71–74), meal worms (75), nematodes (76), and sea urchin (77); aging (78); inherited and induced disorders (79–83); cancer (84–95); experimentally induced conditions (96–98); cell biology research (99–101); metabolism (102); infectious disease (103); and apoptosis (or programmed cell death) both in cultured cells (104–109) and in tissues (110–113). In addition, several groups developed tools to enhance the data analysis capabilities of subtractive 2-D gel analysis. Caudill *et al.* (114) discuss strategies for extraction of data contained in 2-D protein profiles, Rovensky and Lefkovits developed data management software (115), and Nugues (116) introduced artificial intelligence techniques to build an image interpretation. An example of the results of using this approach to study alterations in protein turnover in apoptotic

Jurkat T lymphoblasts compared with control Jurkat T lymphoblasts is shown in Fig. 4.

In these studies, typically two to a few gel patterns were compared. It was soon recognized that it would be extremely valuable if the pattern from a large number of experiments could be archived for later detailed analysis. Researchers trying to find marker proteins, which correlate with a specific disease, would benefit from a large number of protein patterns from healthy and affected individuals. Pattern differences based on protein polymorphisms and other disease-unrelated differences could be filtered out by multiple comparisons, and those differences that correlated with the disease would be recognized at a higher level of confidence if multiple gel patterns were available. Similarly, observations made in experimental cell research could be substantiated if the observed event could be followed over a defined time course. Therefore, several groups added database capabilities to their software for quantitative protein pattern analysis [(42–45,48,49); reviewed in (40)]. These programs were used to construct 2-D databases from a variety of tissues, cell types, organisms, and organelles, including human plasma (117), breast (94), cerebrospinal fluid (118), heart (119,120), liver (121), plasma and red blood cells (122), lymphocytes (123,124), transformed amnion cells (125), keratinocytes

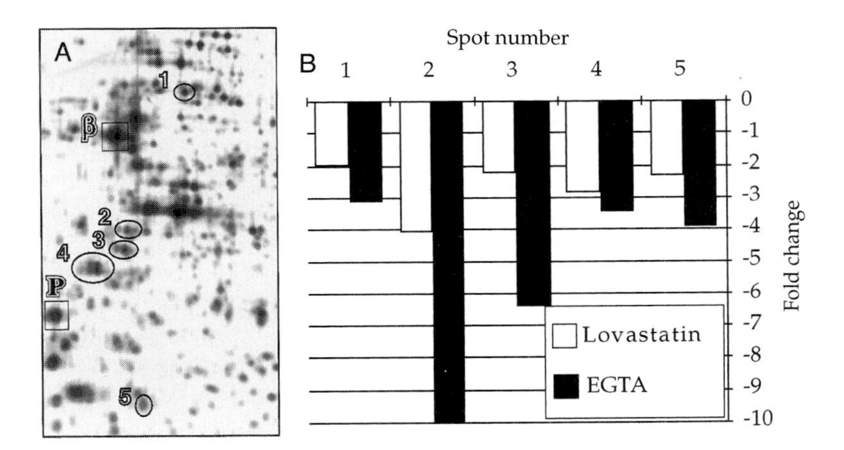

Fig. 4 Quantitative 2-DE analysis reveals coordinate changes in protein turnover associated with the induction of apoptosis. Two different treatments (lovastatin 100 μM for 48 hr, and EGTA 5 mM for 24 hr) were used to induce Jurkat T lymphoblasts to undergo apoptosis. Following induction, the apoptotic cells were enriched by Percoll gradient density centrifugation to yield a population of cells of increased buoyant density compared to control cells that were clearly apoptotic by FACS, DNA laddering, and morphological appearance. These cells were labeled for 3 h with [^{35}S]methionine/cysteine and the cells lysed and separated by 2-DE, as were control (untreated) cells. Interestingly, very few changes in protein turnover were observed in the 2-DE patterns. Quantitative analysis of the resulting 2-DE patterns employing the Quest II software (49) revealed five proteins that displayed a coordinated decrease in turnover in replicate experiments. Those five proteins are labeled in panel A (together with β-tubulin, β, and proliferating cell nuclear antigen, P, which were unchanged in these experiments). The fold change in amount of label is shown for each of the five proteins in panel B for the two treatments. The protein spot #4 was identified as the nucleolar phosphoprotein, numatrin/B23/nucleophosmin (108). Adapted from Patterson, S. D., J. S. Grossman, P. D'Andrea, and G. I. Latter. (1995). Reduced Numatrin/B23/Nucleophosmin labeling in apoptotic Jurkat T-lymphoblasts. *J. Biol. Chem.* **270:**9429–9436.

(126,127), common human proteins (128), rat and mouse liver (129–131), rat embryo fibroblasts (132,133), yeast (134,135), *E. coli* (136), *Drosophila* (137), the hyperthermophile *Pyrococcus furiosus* (138), focal adhesion plates (139), rat liver epithelial cell nuclear proteins (91), and human sperm membrane proteins (140). Although some of these databases are very extensive and contain hundreds of annotations (126,127), others are in very preliminary versions. To make these databases really useful to biologists, database structure, accessibility, and portability need to be addressed, and continued updating must be assured.

F. Conclusions and Perspective

The impact of 2-D protein databases and 2-D protein gel technology in general on research areas currently in the focus of interest, including development, signal transduction, cell-cycle control, and molecular immunology and neurobiology, has been relatively small. Deficiencies in detecting low and very low abundance proteins by 2-DE and difficulties in correlating protein patterns obtained by different gel systems, different laboratories, or at different time points have left some promises of the technology unfulfilled. The most significant benefits anticipated from 2-D databases and 2-D gel technology, namely, the extraction of significant information from a global survey of the proteins present in biological samples, a reduction of duplicated work (e.g., repeated identification of the same protein by different groups), and the synergistic data accumulation and data transfer between laboratories, remain to be realized. In fact, several of the 2-D gel databases described in the early 1990s either have been abandoned or have not been forcefully updated.

What is the right strategy to take advantage of the powerful protein 2-D gel technology? Several groups propose making the information in 2-D gels more accessible by publishing 2-D gel protein patterns from selected sources on the World Wide Web (141,142). A workshop on 2-D protein databases concluded in 1992 that construction of a relational 2-D database, which would catalog pertinent information such as sample source, experimental protocols, loci of identified proteins in 2-D gels, relevant biological information, and other annotations to protein spots, would be desirable (143). The fact that this recommendation was never implemented attests to conceptual and technical difficulties of this approach. Furthermore, recent advances in protein identification capabilities (see Section IV) suggest that the need for large centralized 2-D protein databases has been reduced because protein patterns are now much easier to correlate.

We suggest that a realistic and feasible strategy for identifying components of complex systems should not be based on 2-DE alone, but rather should use 2-DE as one of a collection of complementary tools. Such complementary techniques include biochemical fractionation of cell lysates, an area explored by Jungblut *et al.* (144) and Ramsby *et al.* (145); integration of 2-D gel data into genomic sequence databases (146); protein coprecipitation to find proteins interacting under certain experimental or physiological conditions [e.g., (147)]; the yeast 2 hybrid system for detecting weakly interacting proteins (148); and DNA-based techniques for establishing quantitative profiles of expressed messages. It can be expected that the optimal use of DNA sequence databases for the analysis of biological systems will be a topic of intense debate. The high resolving power, along with the ability to generate quantitative data and to provide a global overview over a complex protein sample, virtually assures a prominent role for protein 2-DE in that discussion.

IV. Identification of Components of Complex Systems

A. Overview

Following detection of the components of interest for a specific biological system, the next step is their identification. A schematic representation of the various approaches that can be applied is shown in Fig. 5. In the past, the most utilized and most definitive method to identify proteins has been partial N-terminal or internal amino acid sequencing. N-Terminal sequencing of intact protein is the simplest method, but it only produces limited amounts of sequence information and fails if the protein is N-terminally blocked, that is, if the protein has no free α-amino group. Internal amino acid sequencing overcomes the problem of N-terminal blocking, and usually several peptide se-

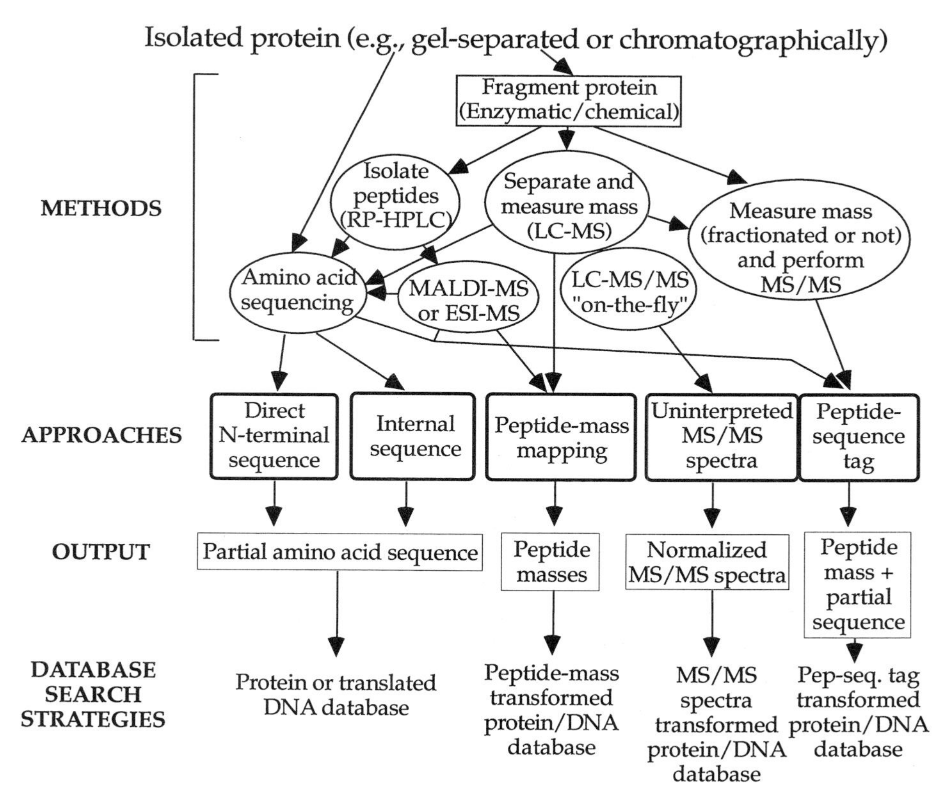

Fig. 5 Schematic view of protein identification methods. Identification of proteins generally follows two approaches, either direct N-terminal sequencing or analysis of protein fragments by either mass spectrometric or sequencing approaches. The data derived from these approaches is then used to search various protein sequence databases in one form or another. See text for details.

quences from a protein are generated. Obtaining internal sequences requires, however, additional sample-handling steps, as peptides obtained from a protein digest have to be fractionated [usually by reversed-phase high-performance liquid chromatography (RP-HPLC)] before being subjected to sequence analysis. The major limitations with these techniques are their limited sensitivity and sample throughput. Approximately 10–50 pmol of isolated protein is generally required for successful sequencing, and only one to two peptide sequences can be determined per day per automated protein sequencer. In spite of these limitations, protein and peptide sequencing continues to be used successfully in numerous projects. For most applications at the sensitivity limit of the technique, proteins are isolated by one-dimensional (1-D) or 2-D gel electrophoresis. Western blotting (or immunoblotting) is also used extensively to identify proteins based on recognition of an epitope by an antibody. If the antibodies are well characterized, this provides a rapid means of detecting known, or structurally related, proteins in a mixture. The use of antibodies will be described in Section VI. This section will deal with the identification of proteins whose identity is not known.

There is a pressing need for methods that can routinely and quickly identify proteins even if less than 10 pmol of a specific protein is present in a biological sample. Isolation, detection, and analysis of such small sample amounts is technically challenging and requires integration of individual experimental steps into a streamlined process covering the path of the sample from the biological source to the analytical instrument. Since the early 1990s there have been tremendous advances in protein chemistry, mainly due to the introduction of mass spectrometers into the hands of a greater number of protein biochemists. Mass spectrometry (MS) has enhanced many analytical techniques in protein chemistry, either by substituting detectors for established techniques or by enabling novel approaches to protein identification and characterization.

Approaches to protein identification are also dramatically influenced by the enormous amount of DNA-derived sequence information being generated by genome projects. The availability of large-scale DNA sequence databases enhances and accelerates protein identification and has led to the concepts of protein-based gene expression studies and the "proteome" (*prote*in complement expressed by a gen*ome*) (149–151). The ease with which information can be accessed and distributed through the Internet has greatly facilitated this work.

This section outlines the advances being made to increase the sensitivity, reliability, and throughput of protein identification. Refinements of traditional methods of protein identification, such as amino acid composition and sequencing, the introduction of new mass spectrometry-based techniques, and the development of computational tools for the combination of data from multiple identification methods, which individually may not provide conclusive identification, are now routinely used to identify extremely small quantities of proteins. Sections IV,C,D, and F were adapted from a previous review of ours (152), with permission of VCH Weinheim.

B. ISGF3 Case Study: Characterization of Proteins Involved in Interferon Signal Transduction

The analytical experiments described in Section III.B suggested that four polypeptides with apparent molecular masses of 48 kDa, 84 kDa, 91 kDa, and 113 kDa, respectively, constituted the ISGF3 complex. To further characterize these polypeptides, large-scale sample preparation was required. Cell lysates from approximately 200 liters of cultured cells were processed to purify sufficient amounts of the polypeptides for sequencing. Almost the same amount of sample was required for internal sequencing as for N-terminal sequencing. To avoid the disappointment of a blocked N terminus of one or several proteins, we opted for internal sequencing. The ISGF3 proteins were purified from cell lysates by repeated passage over oligonucleotide affinity chromatography columns (153). Synthetic oligonucleotides containing the known ISGF3 binding sequence or a related sequence unable to bind the transcriptional activator were immobilized on a resin and constituted the affinity matrix. Proteins specifically binding to the relevant matrix but not to the control matrix were eluted, concentrated, and separated by SDS–PAGE (see Fig. 6). The proteins were electrophoretically transferred to nitrocellulose and detected by staining, and individual protein bands were excised and subjected to tryptic cleavage (154). The resulting peptide fragments were released from the membrane, separated by RP-HPLC, and manually collected. Selected peptides were chemically sequenced using automated Edman degradation. The peptide sequences were used to design degenerate oligonucleotide probes in forward and reverse orientation. PCR (polymerase chain reaction) amplification with such oligonucleotide probes generated extended DNA products, which were shown to contain the determined peptide sequences. These PCR products were used to screen cDNA libraries to finally isolate full-length clones of the genes coding for the four polypeptides [sequencing of the 84- and 91-kDa proteins (22), sequence of the 113-kDa protein (23), sequence of the 48-kDa protein (24)].

Careful analysis of the obtained complete gene sequences revealed several

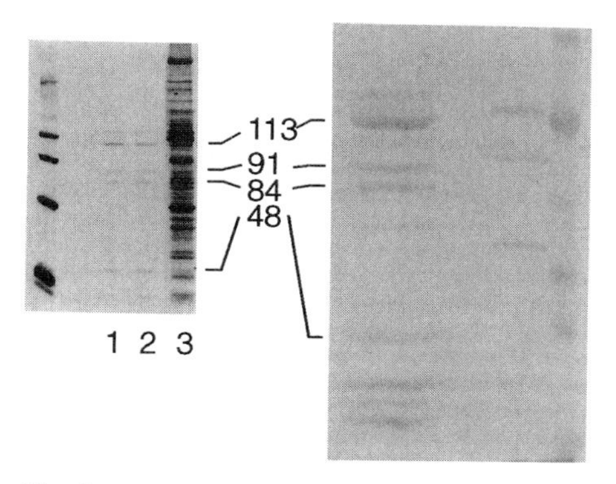

Fig. 6 Purification of ISGF3. (Left) Purification of ISGF3 showing the proteins present after the first oligonucleotide affinity column (lane 3) and two preparations after the final chromatography step (lanes 1 and 2). The leftmost lane contains protein molecular weight markers (high-molecular-weight, Sigma Chemical Co., St. Louis, MO). ISGF3 component proteins are indicated at 113, 91, 84, and 48 kDa. (Right) Purified ISGF3 from 2–3 $\times$ 10^{11} cells was electroblotted to nitrocellulose after preparations from 1 and 2 (left, lanes 1 and 2) had been pooled and separated on by SDS–PAGE (75% gel) (lane to the left). The blot was stained with Amido Black. ISGF3 component proteins are indicated. The two lanes to the right contain protein markers (high-molecular-weight and prestained markers, Sigma Chemical Co., St. Louis, MO). Reproduced with permission from Schindler, C., X. Y. Fu, T. Improta, R. Aebersold, and J. E. Darnell, Jr. (1992). Proteins of transcription factor ISGF-3: One gene encodes the 91- and 84-kDa ISGF-3 proteins that are activated by interferon alpha. *Proc. Natl. Acad. Sci. USA* **89:**7836–7839.

important features that explained previous observations and suggested further experimentation. We found that the sequences of the 84- and the 91-kDa proteins were identical, with the exception of a C-terminal extension of the 91-kDa protein, which explained its slower migration in the gel. The two proteins were splicing isoforms of the same gene. Their relatedness is expressed in the now generally accepted nomenclature Stat 1α (91-kDa polypeptide) and Stat 1β (84-kDa polypeptide). It was also found that the sequence of the 113-kDa polypeptide (termed Stat 2) was related but different from the Stat 1 protein sequences, and the Stat 1 and Stat 2 sequences contained SH2 domains. The presence of this extended sequence motif, the function of which is specific binding to tyrosine-phosphorylated peptide sequences (155), not only suggested a mechanism for the assembly of the ISGF3 complex in interferon-activated cells, but also indicated that further experimentation should concentrate on the detection and localization of tyrosine-phosphate residues in the Stat proteins and on the isolation of the putative protein tyrosine kinase(s).

This case study illustrates several strengths and weaknesses of the approach used. Direct sequencing of the gel-separated proteins was successful in this case and led to the determination of the complete sequence of all the components involved in the ISGF3 activity. The internal sequencing approach is almost always successful, provided that sufficient amounts of the protein can be isolated. Final separation of enriched protein preparations by gel electrophoresis has the significant advantage that many of the idiosyncrasies of proteins (e.g., poor solubility, sticking to surfaces) are eliminated by the presence of strongly solubilizing detergents. Furthermore, gel-separated proteins are generally completely cleaved by proteolytic enzymes because cleavage sites are accessible in the denatured polypeptides. In the present case, which involved the analysis of relatively nonabundant intracellular regulatory proteins, the production of the microgram amounts of protein required for sequencing was a massive and expensive effort. Furthermore, the limited amount of protein sequence information generated failed to identify the structural relationship between the 84-kDa and the 91-kDa polypeptides. The major limitations of this approach are, therefore, the limited sensitivity and the limited amount of structural information that is obtained prior to gene cloning. The major strengths of the approach are its generality and the unambiguous nature of the information obtained.

Since this work was performed in the years 1991 and 1992, the technology for the identification of polypeptides and for the generation of partial protein sequences has changed dramatically. Traditional techniques and novel approaches for the identification of proteins are described in the following sections (see Fig. 5).

C. Generation of Contiguous Amino Acid Sequence

The most direct method to obtain amino acid sequence information of a protein is to subject the intact polypeptide to N-terminal sequence analysis. There are few instances where the protein of interest will not have been separated by gel electrophoresis as the last step in the purification, particularly if only small amounts of the protein are available. The standard approach to sequencing gel-separated proteins is by electroblotting onto a polyvinylidene difluoride

(PVDF) membrane (156,157). Transfer of gel-separated proteins to PVDF has also been achieved by passive elution and centrifugation with microconcentrators containing PVDF-based filters (158,159). Electroblotting of proteins onto inert membranes as a technique for isolating proteins for N-terminal sequencing provided the opportunity for the first time to directly insert membrane-immobilized, gel-separated proteins into the amino acid sequencer without the need for electroelution of the protein from the gel matrix (160,161). The introduction of the first PVDF membrane for that purpose (157) was a major advance, since this membrane could be mass manufactured at a consistent quality. Improvements on the initial PVDF membrane led to membranes with a smaller pore size allowing for higher protein capacity per unit area (see Table I). The use of membranes consisting of other polymers, including Teflon, which was found to be suitable for direct N-terminal (and C-terminal) sequencing at the 50 pmol level, and polypropylene, was also investigated (162). Several reports compared the characteristics of different membranes (163,164). Although the sensitivity of direct N-terminal sequencing is higher than that of internal sequencing because losses incurred during the digestion/elution/fractionation process are avoided, N-terminal sequencing fails with blocked proteins. Since a substantial fraction of cellular proteins are refractory to the Edman chemistry (165,166), internal sequencing is the preferred method unless there is a reasonable expectation for a free α-amino group. This could be expected if the protein was secreted, as these proteins are often cleaved from proforms near the N terminus during translocation from the cytosol to the extracellular space, or if the protein was membrane bound and is released from the cell surface by proteolytic cleavage.

In the following sections, advances in chemical sequencing and mass spectrometry-based sequencing will be outlined (Sections IV,C,1 and 2) and methods for the isolation and analysis of peptides from gel-separated proteins

Table I
Protein Blotting Membranes Suitable for Amino Acid Sequence Analysis[a]

Name	Membrane type (pore size)[b]	Manufacturer/distributor	Ref.
Immobilon-P	PVDF (0.45 μm)	Millipore (Bedford, MA)	(156)
Immobilon-P[SQ]	PVDF (0.1 μm)	Millipore (Bedford, MA)	
Immobilon-CD	Charge mod. PVDF (0.1–0.22 μm)	Millipore (Bedford, MA)	(289)
ProBlott	PVDF (0.2 μm)	ABI (Foster City, CA)	(569)
Trans-Blot	PVDF (0.2 μm)	Bio-Rad (Hercules, CA)	
Westran	PVDF (0.45 μm)	Schleicher & Schuell (Keene, NH)	
FluoroTrans	PVDF (0.20 μm)	Pall Corp (Port Washington, NY)	
Selex 20 (PP 20)	Polypropylene (0.2 μm)	Schleicher & Schuell (Keene, NH)	(331)
SM 17507	Polypropylene (0.2 μm)	Sartorius, Germany	
SM 17558	Polypropylene (0.1 μm)	Sartorius, Germany	
SM 17506	Polypropylene (0.45 μm)	Sartorius, Germany	
Teflon tape	Teflon (mil. spec. T-27730A)	3M (St. Paul, MN)	(162)
PCGM-1[c]	Glass fiber (polybase treated)	Janssen Life Sci., Belgium	(160)
Glassybond[c]	Siliconized glass	Biometra, Germany	(570)

[a] Direct and in some cases, following elution. Data compiled from catalogs and references (163,164,571).
[b] Data from manufacturers or Eckerskorn and Lottspeich (164).
[c] Can also be modified by investigator using GF/C, according to references cited.

will be addressed (Section IV,D). Significant advancements in chemical C-terminal amino acid sequencing have been achieved (167–170). As chemical C-terminal sequencing is currently being performed in the low hundreds of picomoles range, and is therefore unlikely to be used in studies of unknown proteins present in low quantities, this technique will not be covered further in this section.

1. Chemical Sequencing Approaches

Numerous incremental improvements have increased the sensitivity of chemical protein sequencing to the stage where a number of laboratories can routinely sequence very low picomole quantities of peptides/proteins applied to the sequencer. The improvements include the following: (i) optimization of the sequencing conditions (171–174); (ii) introduction of a microcartridge (175); and (iii) the use of microbore columns for the phenylthiohydantoin (PTH)-amino acid separations (176–178).

Detection of many proteins of intense biological interest that are present only in very small quantities in polyacrylamide gels illustrates the necessity for methods to obtain amino acid sequence information from even lower amounts of peptide/protein. Advances in protein purification techniques now allow the isolation of subpicomole quantities of sample (179). Further improvements in the overall sensitivity of chemical protein sequencing, therefore, depends on a more sensitive degradation chemistry. It is unlikely that the sensitivity of detection for PTHs by UV absorbance will be dramatically increased. Beer's law describes the physical boundaries limiting the sensitivity of UV absorbance detection as well as the technical tradeoffs (e.g., sample concentration vs the length of the light path), which prevent the construction of a UV detector with highly superior sensitivity. Furthermore, PTHs in a chromatography–UV absorbance detection system are identified by comparing the elution time of the PTH to be identified with the retention time of PTH standards. Therefore, only a few PTH derivatives of posttranslationally modified amino acids are conclusively identified using standard or even modified chromatography (180,181). Consequently, a number of laboratories have investigated the use modified Edman reagents that are able to generate amino acid derivatives detectable at high sensitivity by mass spectrometry. Such reagents should provide not only increased sensitivity, but also the ability to determine the nature of posttranslational modifications stable to the sequencing chemistry, and should allow faster sequencing cycles. Figure 7 shows the structures and a performance assessment of a panel of reagents tested. Others have investigated the use of Edman reagents that can be detected by spectrofluorometry (182–191), by light absorbance in the visible range of the spectrum (192), or by liquid scintillation counting (193,194).

We synthesized and evaluated a panel of reagents that are characterized by a phenyl isothiocyanate (PITC) group for efficient stepwise peptide degradation, a base for optimal electrospray (ES) ionization, and an inert chemical spacer separating the other groups and providing desirable solubility (195,196). We showed that several of these reagents, including the most promising, 4-(3-pyridinylmethylaminocarboxypropyl)phenyl isothiocyanate (referred to as PITC 311) (see Fig. 7), yielded amino acid derivatives that could be detected in the low femtomole to high attomole range (195,196). Using PITC 311, 10

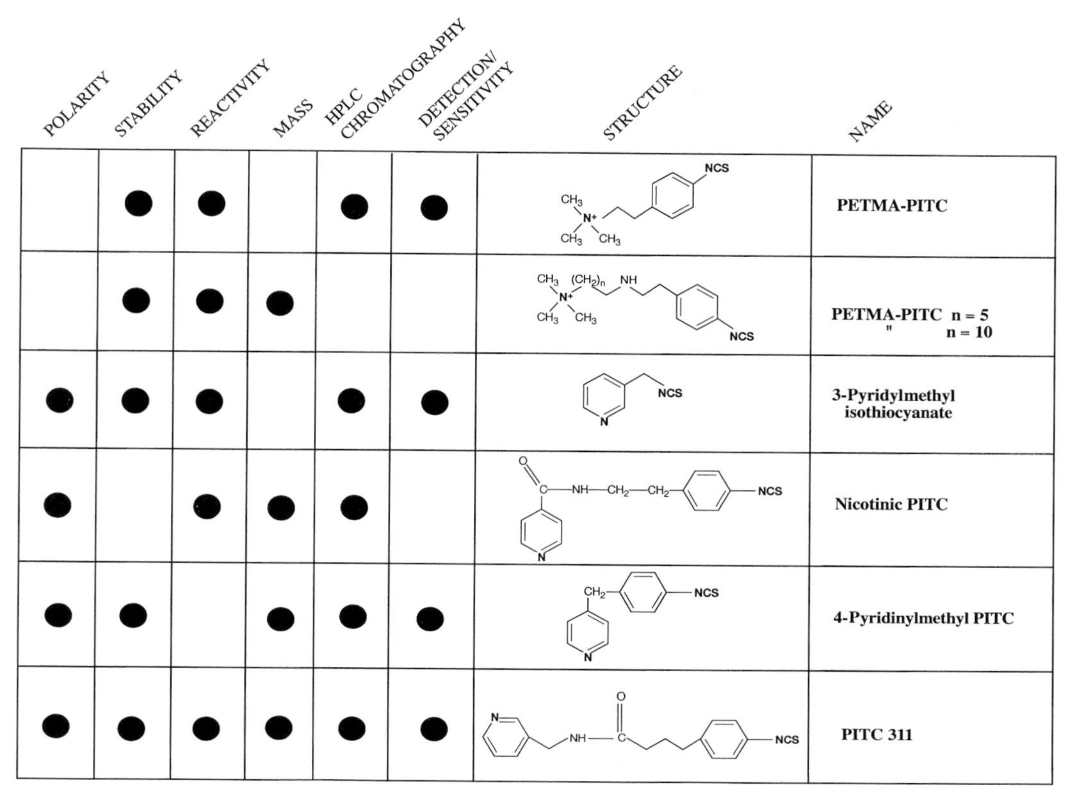

Fig. 7 Evaluation of panel of sequencing reagents yielding amino acid derivatives that are detectable by ESI-MS. The structure of the reagents is shown to the right, and some essential parameters are listed on the left-hand side. A dot in the matrix indicates that a reagent behaves favorably with respect to the parameter indicated.

amino acid long peptides were successfully sequenced when between 0.5 and 1 pmol of sample was loaded to the sequencer (197), and we have succeeded in implementing automated absorptive sequencing protocols of a sensitivity of 250 fmol of peptide applied to the sequencer (Bill Fisher, PE-SCIEX, Ontario and RA, unpublished, 1995). As the 311 PTHs are detected by on-line liquid chromatography–mass spectrometry (LC-MS) separation, both, retention time and compound mass are used to assign the released residues, thus enhancing the confidence level of sequence assignments and permitting the characterization of modified residues (198,199). Finally, 311 PITC chemistry was implemented on a sequencer and LC-MS system without modification to either instrument and protocols were developed allowing a 20-min injection to injection analysis cycle (198).

Bailey and colleagues (200,201) have reported work in progress with the conceptually similar reagent dimethylaminopropyl isothiocyanate, and Stolowitz *et al.* (202) used a different approach by employing gas chromatography/ion-trap mass spectrometry (GC-MS) for detection of 2-benzoyl-5-*O*-(4′-nitrobenzene)sulfonylthiazoles. Although the GC-MS detector has demonstrated sensitivity in the low femtomole range, it requires the samples to be delivered in extremely small volumes not compatible with currently available protein sequencers.

Therefore, it appears that new protein sequencing chemistries based on the detection of the cleaved amino acid derivatives by MS can achieve very high sensitivities, even on widely used commercial sequencing instrumentation. At present the high cost of the mass analyzers limits the wide application of these methods.

In Edman-type sequencing, the peptide sequence is determined by the analysis of the amino acid derivatives released in successive sequencing cycles. Since the chemistry is not quantitative, a series of truncated peptides is left behind after extraction of the cleaved derivatives. These peptides differ in mass by the mass of the amino acid released in each step. In principle, the peptide sequence could be determined by accurate mass measurement of these truncated peptides. Chait *et al.* (203) implemented this concept into a practical technique termed ''protein ladder sequencing'' (see Fig. 8). The technique uses matrix-assisted laser desorption/ionization (MALDI)-MS for the analysis of a ''ladder'' of peptides, truncated by chemical degradation.

In the first implementation of the technique, the peptide ladder was generated by Edman degradation using the standard reagent phenyl isothiocyanate. In each cycle a small amount (5%) of the reagent phenyl isocyanate was used to enhance the level of each truncated peptide species in the mixture (203). Subsequent experiments showed that suitable peptide ladders could be generated by incomplete Edman degradation performed with 6-min cycle times on an automated sequencer (204).

Although to date no novel sequences have been reported by ladder sequencing, the method is very appealing. Analysis of the peptide ladder by MALDI-MS is very fast and sensitive. Potentially, ''peptide degraders'' could be multiplexed to a single mass analyzer, providing a peptide sequencing system with high throughput. Provided sufficient mass accuracy can be obtained, only the isobaric amino acids Leu and Ile cannot be distinguished. Gln (128.13 Da) and Lys (128.17 Da) can be distinguished because the ε-amino group of lysine will become modified with the coupling reagent. The ability of this method to determine the site of posttranslational modifications was demonstrated by sequencing through a site of serine phosphorylation (203). Wang *et al.* (204) also showed that ladder sequencing, at least in some instances, can cope with a sample consisting of two peptides.

Bartlet-Jones *et al.* (205) used a slightly different chemical approach for the generation of a series of truncated peptides using the new degradation reagent, trifluoroethyl isothiocyanate (TFEITC). They attempted to drive the coupling and cyclization/cleavage reactions to completion. A ''nested set'' of fragments was generated by the addition of a new aliquot of sample at the beginning of each degradation cycle (see Fig. 8). The peptide sequence was then determined by analyzing the resulting sample by MALDI-MS and by calculating the mass difference between the detected peptide ions.

This method resembles that of the ''ladder sequencing'' approach of Chait *et al.* (203), except that there is no addition of a terminating reagent to the free α-amino groups of the truncated peptides, and that the peptide degradation can be implemented in a very simple multiplexed instrument, thanks to the fact that TFEITC is volatile and excess reagent can therefore be removed simply by applying reduced pressure to the reaction chamber.

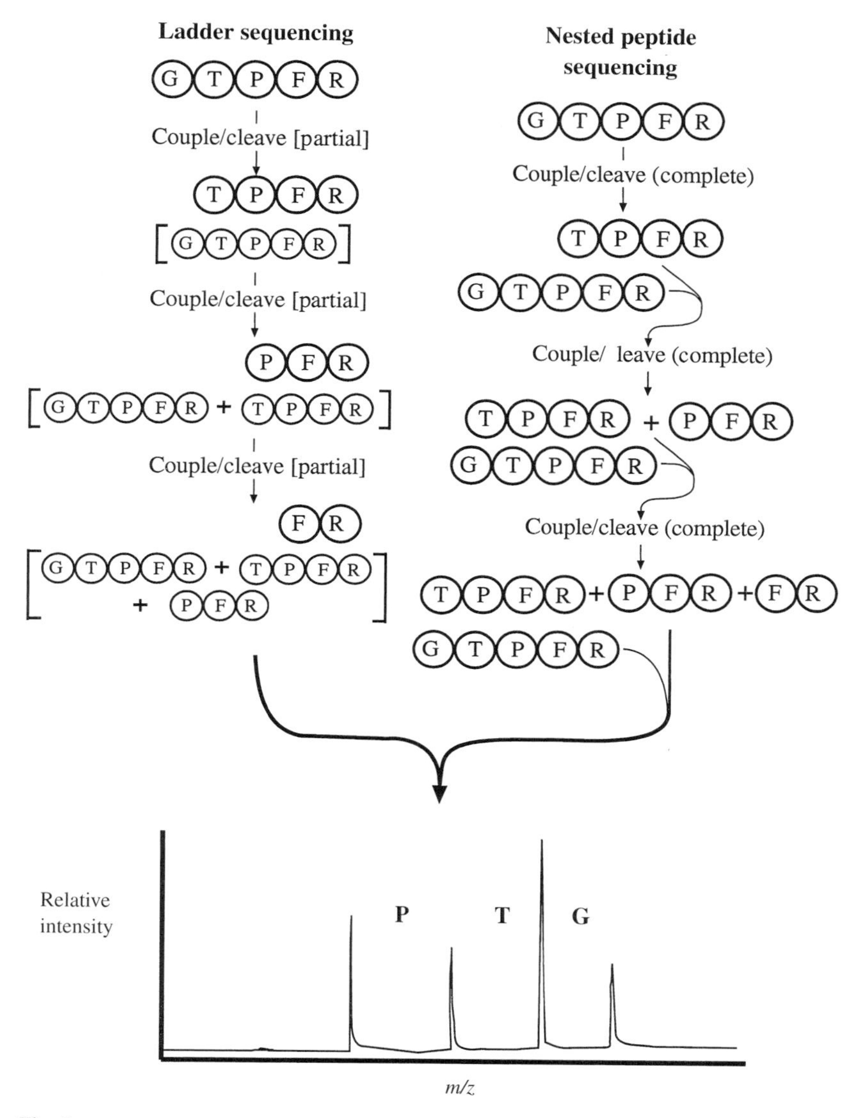

Fig. 8 Ladder and nested peptide sequencing with MALDI-MS analysis. Both of these methods use the conventional Edman chemistry approach with analysis of the remaining peptide rather than the released amino acid. However, with the ladder sequencing approach of Chait *et al.* (203) (shown at the left), there is only one addition of peptide and the chemical cleavage at each step is partial. In the nested peptide sequencing approach of Bartlet-Jones *et al.* (205) (shown at the right), it is attempted to take the coupling/cleavage steps to completion and add new intact peptide at each step. In both methods the sample is analyzed by MALDI-MS and the mass differences between the ions reveals the amino acid identity. Adapted from Patterson, S. D., and R. Aebersold. (1995). Mass spectrometric approaches for the identification of gel-separated proteins. *Electrophoresis* **16:**1791–1814, with permission.

Both methods share the same limitations. Isobaric amino acids are not distinguished, and the lengths of the peptides conducive to sequencing with this method are limited by the mass resolution of the mass spectrometer. Both methods also suffer from the conceptual disadvantage that the total available signal is distributed among the species in the peptide ladder. At constant detector sensitivity, this means that the longer the peptide sequence, the more

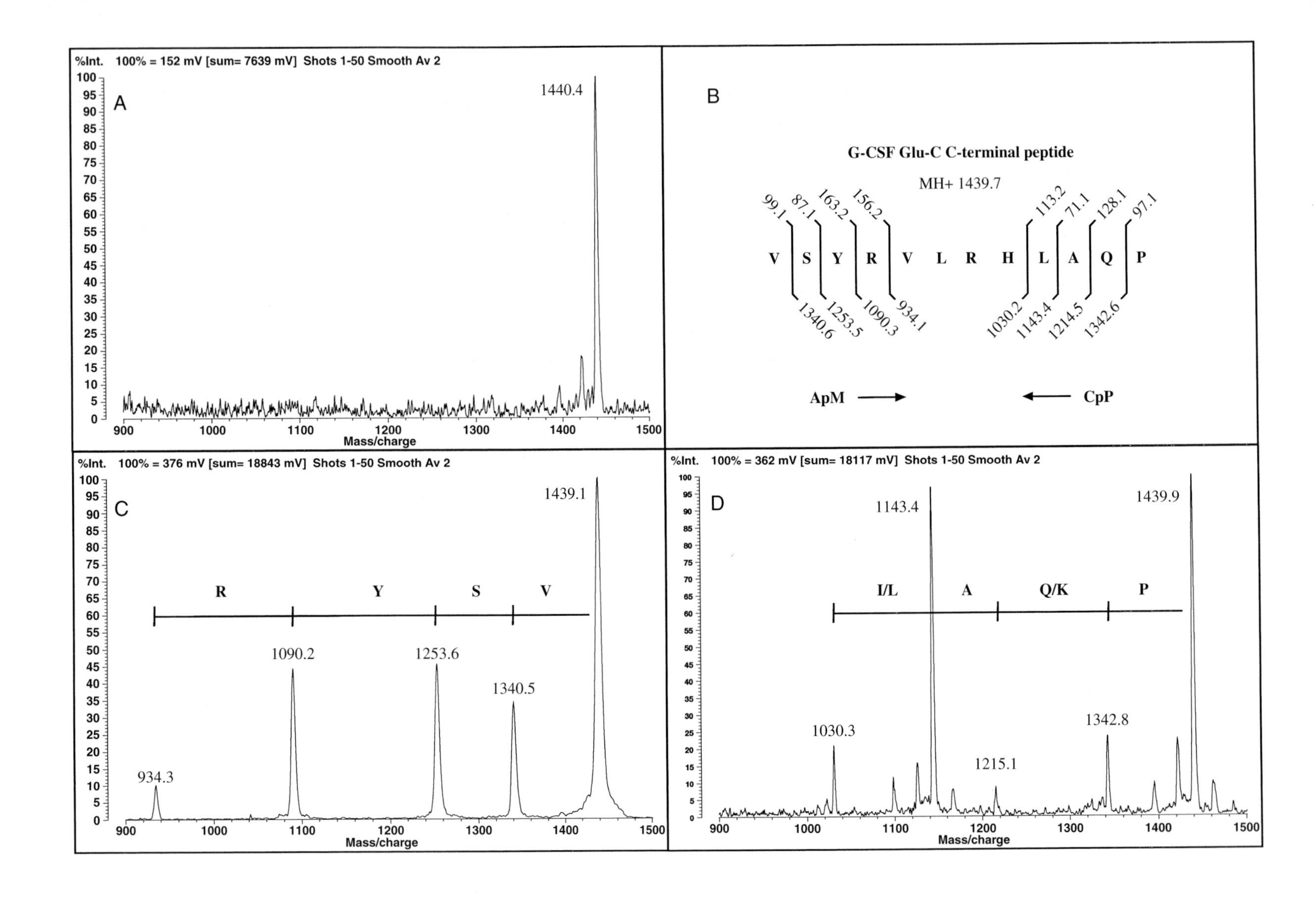
%Int. 100% = 152 mV [sum= 7639 mV] Shots 1-50 Smooth Av 2
A
1440.4
Mass/charge
B
G-CSF Glu-C C-terminal peptide
MH+ 1439.7
V S Y R V L R H L A Q P
99.1 87.1 163.2 156.2
1340.6 1253.5 1090.3 934.1
113.2 71.1 128.1 97.1
1030.2 1143.4 1214.5 1342.6
ApM
CpP
%Int. 100% = 376 mV [sum= 18843 mV] Shots 1-50 Smooth Av 2
C
R Y S V
934.3
1090.2
1253.6
1340.5
1439.1
Mass/charge
%Int. 100% = 362 mV [sum= 18117 mV] Shots 1-50 Smooth Av 2
D
I/L A Q/K P
1030.3
1143.4
1215.1
1342.8
1439.9
Mass/charge

sample has to be available. An attempt to address this was made by Bartlet-Jones *et al.* (205), who developed a "quaternary tagging procedure" aimed at increasing the sensitivity of detection of peptides by MALDI-MS by selectively modifying the N terminus with a quaternary ammonium alkyl N-hydroxy succinimide (NHS) ester. This limitation, potentially restricting the maximal sensitivity achievable by this method is, however, compensated by the relatively high tolerance of MALDI-MS for the presence of buffer salts and other contaminants.

The two methods just described generate a series of peptides that differ in their N termini; thus, they generate N terminal peptide sequences. Methods for the generation of a series of peptides that share the same N termini but are truncated at their C-terminal end, in conjunction with mass analysis of the peptide mixture, would provide a method for C-terminal sequencing, a notoriously difficult task in protein chemistry (206,207). Tsugita *et al.* (208) used pentafluoropropionic acid (and heptafluorobutyric acid) as a means of generating sequentially truncated C-terminal fragments that were analyzed by electrospray/pneumatically assisted/thermospray ionization (ESI)- or fast-atom bombardment (FAB)-MS. The demonstrated sensitivity level was in excess of 100 pmol.

Earlier, Klarskov *et al.* (209) used plasma desorption (PD)-MS and FAB-MS to demonstrate that peptides truncated at their N or C termini by degradation with amino- and carboxypeptidases, respectively, could be used to determine partial N- and C-terminal sequences, although the method lacked sensitivity. Several groups have used the more sensitive MALDI-MS technique to determine C-terminal sequences after carboxypeptidase digestion of peptides.

An early example is the report by Schar *et al.* (210), who used solution phase digestion of 500 pmol of peptide using carboxypeptidases Y and B. Thiede *et al.* (211), using carboxypeptidases P and Y, showed successful degradation for a wider range of peptides and smaller proteins at lower concentrations. Woods *et al.* (212) have demonstrated the use of carboxypeptidases to generate sequence information from low to subpicomole quantities of peptides by a simple method of mixing the peptide and enzymes directly on the sample probe (or slide), and successfully obtained sequence information from major histocompatability complex (MHC)-associated peptides (213). The same authors also demonstrated that other enzymes, including endoproteases and aminopeptidases, are compatible with digestion directly on the sample slide, thus eliminating any potential sample losses incurred through the sample transfer to the probe, following digestion in solution (212). A separate study suggested that the use of a series of increasing concentrations of carboxypeptidase Y provided increased confi-

---

Fig. 9 Exopeptidase digestions on the sample slide with subsequent MALDI-MS analysis to generate amino acid sequence "ladders." The C-terminal endoproteinase Glu-C derived peptide of rHu G-CSF (A) was subjected to on-probe digestions with either aminopeptidase M (ApM) for 3 min (C) or carboxypeptidase P (CpP) for 20 sec (D). Amino acid assignments shown are based upon the observed mass differences in panels C and D. The sequence of the peptide is shown in (B), together with the residue masses for the amino acids removed by the treatments, and the mass of the peptide remaining. All masses are listed as the protonated average mass. The analysis of the data yielded 8 residues of the 12-residue peptide with only two ambiguities due to isobaric amino acids. The spectra in (A) and (D) were obtained through the reflectron of the Kompact MALDI III (Shimadzu/Kratos, Columbia, MD) and the spectrum in (C) was obtained in linear mode; all were calibrated externally (Patterson, unpublished data, 1995).

dence, but also required increased sample quantity (214). An example of this approach is shown in Fig. 9, where a peptide has been digested with aminopeptidase M and a separate aliquot with carboxypeptidase P on the sample slide.

Hydrochloric acid has long been used for the hydrolysis of peptide bonds for amino acid composition analysis (215–217). Two reports (218,219) describe partial peptide hydrolysis by hydrochloric acid to generate fragments that potentially could be reassembled into the initial peptide after their masses were determined by MALDI-MS. Because acid hydrolysis has no preference for N- or C-terminal truncation or for hydrolysis of internal peptide bonds, mass analysis of the resulting peptide mixture was complex and complicated by the fact that it was impossible to deduce the orientation of the sequence (N- or C-terminal). Clearly, the N-terminal-specific tagging procedure described earlier (205) could prove extremely useful for this approach.

With any of the methods just described involving the analysis of sets of truncated peptides by MALDI-MS, the uniformity of ionization efficiency for peptides with different physicochemical characteristics is a concern. It has been found empirically that the truncated peptides usually are most successfully ionized if a basic residue remains on the truncated peptide. Woods *et al.* (212) investigated other parameters affecting ionization, such as peptide hydrophobicity, and showed that incorporation of ammonium-containing buffers prior to matrix addition assisted with ionization of hydrophobic peptides, and in one instance ionization of a peptide without a basic residue.

2. *Mass-Spectrometric Sequencing*

In the techniques described previously, the samples to be analyzed by MS consisted of either sequentially removed amino acid derivatives or ordered sets of truncated peptides generated by chemical or enzymatic methods. In these cases, the mass spectrometer was used as a single-stage mass analyzer, essentially acting as a very sensitive and accurate balance to weigh the molecule. With two-stage or tandem mass spectrometry (MS/MS), the whole peptide sequencing process, namely, selecting a molecular species for sequencing, fragmentation of the selected ion, and analysis of the fragment ions, can be performed in a single operation, provided that the molecular ion of the peptide can initially be formed under conditions that minimize fragmentation (220,221). The two most popular methods for forming peptide ions, ESI and MALDI, are both considered "soft" ionization methods, that is, intact, desolvated ions are formed, allowing mass analysis in the gas phase with minimal ionization-induced fragmentation. Also, both ionization methods have been implemented in two-stage mass spectrometers with the possibility of inducing fragmentation of the peptide bonds. A description of tandem mass spectrometry for sequencing and characterization of posttranslational modifications can be found in Patterson and Aebersold (152) and is not repeated here. In this section we briefly cover some recently developed methods for transferring small quantities of peptides into the mass spectrometer without incurring significant sample loss.

For ESI-MS, these technical improvements include the introduction of microelectrospray sources (222–229). Such sources simply provide smaller diameter capillaries from which the peptide solutions are sprayed. Smaller diameter capillaries allow smaller drops to be formed during spraying, which are desolvated more rapidly than larger droplets. Microsprayers can, therefore, be positioned much closer to the ESI-MS orifice (inlet to the mass spectrometer) than conventional electrospray sources. This means that a greater percentage of the ionized peptides can be introduced to the ESI-MS, thereby greatly increasing the sensitivity of the analysis. In particular, the NanoES source developed by Wilm and Mann uses a pulled glass capillary to produce a needle with an inner diameter of 1–2 μm that allows spraying times of more than 30 min from 1 μl of sample (223). Such long spraying times mean that multiple peptides can be sequenced from a mixture without prior separation (230). Applications of this method to the analysis of gel-separated proteins have demonstrated sensitivities in the low femtomole range for recovered peptides (230,231).

MALDI-MS is often used for the mass analysis of complex mixtures. It has been observed that the intact molecular ions generated by the relatively soft MALDI ionization technique undergo significant metastable fragmentation, termed "postsource decay" or PSD (232,233). It is thought that PSD stems from multiple early collisions between the analyte and matrix ions during plume expansion and ion acceleration, and from collision events in the field free drift region (234). As fragmentation is often sequence-specific (232,233) and generates the same types of ions observed by ESI-MS/MS, PSD spectra generated by MALDI-MS can be used to determine peptide sequences, even though time-of-flight mass spectrometers lack a designated collision cell. A description of PSD-MALDI-MS can be found in Patterson and Aebersold (152).

The ability of PSD-MALDI-MS to provide sequence information for low picomole to subpicomole quantities of peptides has been demonstrated (235), and metastable fragmentation of glycosidic linkages in glycopeptides (236), disulfide bonds (237), Asp-X cleavages in small proteins (238), and sulfate and phosphate elimination from peptides carrying these modifications (239) have also been observed. However, the fragment ion spectra obtained by PSD-MALDI-MS are not always easily interpreted and seem, in some cases, to differ from low-energy collision-induced (CID) spectra obtained by ESI-MS/MS. Before the method can be considered routine for peptide sequence determination, the PSD spectra of a wide range of peptides need to be reconciled. To enhance data interpretability, Spengler *et al.* (240) used on-probe deuteration after initial PSD-MALDI-MS, and James *et al.* (241) described the use of methyl esterification after initial PSD-MALDI-MS to increase the mass of carboxylic acid containing peptides by 14 Da/COOH and so assist in determination of C-terminal derived fragments. Such strategies greatly assist with the interpretation of PSD-MALDI-MS spectra of unknown peptides. Kauffman *et al.* (242) have also described in detail an approach to assist in the interpretation of PSD-MALDI-MS spectra by starting with interpretation of the small fragment ion (immonium ion) region of the spectrum, in a manner similar to that applied to the interpretation of low-energy ESI-MS/MS spectra (243). Although parent ion selection in PSD-MALDI-MS is not as discriminating as in ESI-MS/MS (e.g., an empirical standard is the resolution of a protonated species of about 1000 m/z from its sodiated adduct) (244), it is sufficient to analyze individual peptides from many peptide mixtures. Postsource decay analysis by MALDI-MS has potentially been made even more rapid and powerful with the introduction of a "curved-field reflector" (CFR-MALDI-MS), which focuses all of the metastable ions in a single spectrum (245), eliminating the need to scan the reflector voltage and therefore concatenate the obtained spectra. With parent ion selection, this provides the opportunity to rapidly generate sequence specific fragment ions from a number of peptides in a mixture. Using the CFR-MALDI-MS, sequence-specific fragment ions have been obtained from *in situ* digests of proteins electroblotted to PVDF (246). This digestion approach was described by Vestling and Fenselau (247,248), who demonstrated that MALDI-MS spectra could be obtained directly from the PVDF membrane inserted into the instrument following *in situ* proteolysis without prior elution of the peptides.

D. Identification of Proteins by Peptide-Mass Mapping and Related Methods

1. *Generation of Peptides*

The generation of peptides from intact polypeptides is useful for protein identification for two main reasons. First, sequences of isolated peptide fragments, derived by chemical degradation or tandem mass spectrometry, unambiguously identify a protein. Second, accurate peptide masses, frequently used in conjunction with additional information, are suitable for the identification of proteins by correlating the isolated protein with a sequence in a sequence database. This section covers methods for obtaining peptide fragments that are suitable for protein identification by either approach. Since most proteins are separated by gel electrophoresis as a last purification or concentration step, the discussion is restricted to the treatment of gel-separated proteins.

The following three strategies have been used to generate peptide fragments from gel-separated proteins: (i) electroelution of the intact protein with subsequent chemical or enzymatic fragmentation in solution, (ii) direct digestion of the protein in the gel matrix (in-gel digestion), or (iii) electrotransfer of the proteins to a membrane and fragmentation on the membrane. Generally, the methods developed earlier for the generation of peptides from gel-separated proteins for subsequent RP-HPLC separation and sequencing could be easily adapted to be compatible with LC-MS or MALDI-MS detection of the peptide fragments. The most important point to consider is the complete removal of residual SDS or other detergents from the peptide sample, because this detergent strongly interferes with ESI-MS and MALDI-MS (249). Even in cases in which the peptide sample is separated by RP-HPLC prior to MS analysis, removal of SDS is essential since it negatively affects chromatography. It should also be kept in mind that high-grade solvents used in protein biochemistry usually are optimized for UV transparency and not for minimal contamination detectable by MS. For high-sensitivity analytical work involving MS detection, careful evaluation of the solvents used is therefore advised. Finally, the objective of the fragmentation experiment should dictate the method used. For protein identification at high sensitivity, the priority is the recovery of some peptides at high yield and in a very clean condition. In contrast, the characterization and localization within the polypeptide chain of rare posttranslational modifications requires recovery of peptides spanning the whole protein sequence.

a. Electroelution. Electroelution and passive elution are the oldest methods for the recovery of gel-separated proteins. Both methods, electroelution (250,251) and passive elution (252), were initially developed for recovery of proteins for N-terminal sequencing. Protein yields with these methods were variable, especially when small quantities of protein were present in the band. This was most likely due to difficulties in resolubilizing the denatured, SDS-depleted polypeptide. Although the use of 2 *M* urea to solubilize the protein during digestion at least in part alleviated the solubility problem, Clauser *et al.* (253) observed N-terminal carbamylation of some peptides. Several groups have successfully used electroelution to isolate proteins for MS analysis (253–255). In these studies (253,254), SDS was removed either by cold acetone precipitation according to Konigsberg and Henderson (256), or by decreasing the SDS concentration by eliminating the detergent from the lower buffer reservoir during the run (255). In an alternative approach to obtaining mass information by ESI-MS analysis on intact proteins separated by preparative IPG-IEF gels for ESI-MS analysis, Breme *et al.* (257) eluted intact proteins separated by preparative IPG-isoelectric focusing (IEF) gels using organic extraction, passive elution, or centrifugation through a 0.22-μm membrane. This method has so far only been demonstrated in the low microgram range, but will no doubt be extremely useful for the characterization of charge variants, especially as some of the material could also be peptide mapped.

b. In-Gel Digestions. Most protocols for in-gel digestion of gel-separated proteins call for the use of detergents such as SDS (258,259) and Tween 20 (260,261) to enhance recovery of peptides. These conditions have proven successful for the fractionation of peptides by RP-HPLC with UV detection, where the samples are being prepared for N-terminal sequence analysis. However, the presence of these detergents suppresses MALD ionization of the unfractionated peptide sample, most likely by interfering with crystallization of the sample matrix. Even using a modified Rosenfeld *et al.* (260) protocol, together with the sample preparation method of Xiang and Beavis (262), MALDI-MS spectra were still dominated by Tween 20 signal in one study of in-gel digested proteins (263). Kirchner *et al.* (264) evaluated 16 different detergents for compatibility with digestion of PVDF-bound protein and subsequent MALDI-MS analysis, revealing that decyl- and octylglucopyranoside and decyl- and dodecylmaltopyranoside detergents were the best choices. The presence of detergents is equally detrimental for LC-MS. SDS is essentially incompatible with RP-HPLC, and even small amounts of neutral detergent pose a significant problem. Most neutral detergents used, including Tween 20, Titron X-100, and Nonidet P-40 (NP-40), are complex mixtures of polymeric molecules that are separated, at least in part, by RP-HPLC, are efficiently ionized by ESI, and readily undergo collision-induced fragmentation. Although these compounds exhibit very low UV absorbance, their strong MS signal precludes them from LC-MS analysis because a significant fraction of the LC-MS trace is obscured by polymer signals that make detection and mass analysis of small amounts of peptides difficult (265).

Hence, when MS analysis is the method of choice, detergents need to be eliminated. Several adaptations of the protocols aimed at eliminating the detergents in the peptide samples have been published. The simplest protocol eliminated the use of detergents for peptide recovery. Interestingly, recoveries in the absence of detergents were apparently still high, as long as the extraction solution contained acetonitrile (266). Moritz *et al.* extracted tryptic peptides from as little as 20 pmol of phosphorylase *b* ($M_r \sim 97,000$), which had been separated by SDS–PAGE, detected by protein staining, and digested in the gel (267). If the peptides were extracted in 0.1% (v/v) trifluoroacetic acid (TFA) and 0.1% (v/v) TFA/60% (v/v) acetonitrile, apparent recoveries of 80% compared to a solution digest of the same protein were reported (267). Quantitative analysis of peptides recovered from more hydrophobic proteins will be useful to determine whether these excellent yields can be achieved generally. Two reports presented chromatography-based solutions for removing detergents from peptide samples. The samples were run through small detergent trapping columns for either capillary LC-MS (268) or microbore RP-HPLC (269).

Highly disulfide-linked proteins may not be amenable to standard digestion protocols if they are electrophoresed under nonreducing conditions. Jeno *et al.* (270) have shown that standard reduction and carboxymethylation protocols are compatible with in-gel digestion, although for subsequent mass spectrometric analyses, SDS has to be removed as the final step.

Re-electrophoresis has been used to concentrate small quantities of gel-separated sample from a number of parallel runs (271–278). Following electrophoretic concentration, proteins were either blotted to a membrane for subsequent digestion or digested in-gel. Rider *et al.* (277) altered their earlier method to reelectrophorese pooled samples (in gel pieces) into an agarose gel rather than another SDS-PA gel. The concentrated sample could then be melted out of the agarose, providing substrates for digestion in essentially the same form as if the purified protein were to be digested in solution. When the peptide digest obtained from this method was analyzed by LC-MS, *N*-octyl-β-glucoside was substituted for RTX-100. However, it was noted that there were some ions present throughout the LC-MS run that were derived from the agarose, and that some brands of agarose gave an unacceptably high number of contaminants (277). This method has been demonstrated effective at the 50-pmol level.

*c. **On-Membrane Digestions.*** Peptide samples generated by the original method for peptide recovery from blotted proteins for sequencing (154) and adaptations of this method are compatible with mass spectrometric peptide mapping (279). This is supported by a series of reports in which proteins separated by 1-D or 2-D gels and transferred to PVDF, Immobilon-CD, or nitrocellulose membranes were fragmented on the support and the recovered peptide mixtures were analyzed by either MALDI-MS (263,266,280–282) or LC-MS (278,283). The presence of SDS in these samples is considered a minor problem (281), and trace amounts left on the membrane can be easily removed (280,282).

Enzymatic digestion of proteins on membranes requires that permanent absorption of the proteolytic enzyme be prevented, and that peptide fragments be released into solution. Enzyme absorption was initially quenched by treating the protein band with PVP-40 prior to addition of the protease, and peptide recovery was shown to be enhanced by the addition of organic solvent such as acetonitrile, which provides good peptide solubility (154,284). Samples generated using these conditions are directly compatible with MALDI-MS as well as with LC-MS.

More recently, several modifications to the method were described, all aimed at increasing peptide recovery. Reduced Triton X-100 (RTX-100) as a substitute for PVP-40 was reported to increase the peptide yield from digests conducted on either nitrocellulose or high-retention PVDF (285,286). Although RTX-100 exhibits low UV absorbance at 214 nm and is, therefore, well suited for peptide detection by UV absorbance (285,286), the detergent caused significant problems during LC-MS, including signal suppression and obscuring of peptide signals coeluting with RTX-100 species (249,265). Similar problems were observed during MALDI-MS of RTX-100-containing peptide samples. Total signal loss, as observed by Fernandez *et al.* (287), could be overcome when the protein-doped thin polycrystalline film method (262) was employed for sample preparation (263).

Other modifications to the method include substitution of PVP-40 by PVP-360, which only obscured the LC-MS signals of very hydrophobic peptides (265), inclusion of a reduction and alkylation step prior to digestion for mass analysis of PVDF-blotted proteins from 2-DE (280), and inclusion of octyl-β-glucoside as a wetting agent during tryptic digests of proteins on PVDF membranes (288). The presence of octyl-β-glucoside did not significantly suppress the signals in MALDI-MS. Octyl-β-glucoside, because of its small and polar structure, is also compatible with LC-MS, although MS signals from characteristic contaminants can obscure low-abundance peptides (R.A., unpublished data, 1994).

A further solution for avoiding detergent contamination of peptide samples is to perform protein digestion on charge-modified membranes, which release peptides either in a pH-dependent manner or in high salt (289,290). Two protocols using the charge-modified PVDF membrane Immobilon-CD have been published. Hess *et al.* (283) used high salt concentrations in the digest buffer to recover peptides for LC-MS, and Zhang *et al.* (282) described conditions for the recovery of peptides from Immobilon-CD in acidic organic solvents prior to MALDI-MS. Both methods have been demonstrated to successfully generate peptide mass data from less than 10 pmol of protein loaded on the gel. A slight modification of the method of Zhang *et al.* (282) produced MALDI-MS peptide maps from 5 pmol of protein loaded on the gel (291).

Two studies that compare peptide yields obtained after in-gel digestion or digestion of proteins on membranes have arrived at conflicting conclusions. Moritz *et al.* (266) reported that on-membrane digestion resulted in higher peptide recoveries, whereas Merewether *et al.* (263) provided amino acid sequence data in support of slightly higher recoveries when using a modified in-gel protocol. These results suggest either that the optimal protocol has not yet been found, or, more likely, that because of the idiotypic behavior of proteins, there is no single best method for all proteins. It is interesting to speculate that the protein available for digestion following blotting (if it is essentially quantitative) may be the same as that available for in-gel digestion, that is, there may be a portion of protein irreversibly trapped in the gel matrix and not available for either transfer or digestion.

2. *Peptide-Mass Mapping*

The size of the contents of genomic and cDNA sequence databases is growing exponentially. The significance of *de novo* amino acid sequence determination for the purpose of cloning the coding gene is therefore diminishing, at least for those species that are the primary focus of genome sequencing projects. Concurrently, the need for rapid, sensitive, and conclusive methods of identifying the gene coding for a protein by correlating the protein with an entry in a sequence database is increasing. A number of methods have been developed that, either by themselves or in combination with other techniques, are able to rapidly identify the proteins in a gel band or gel spot. Sections IV,D,2,a and b summarize approaches that use peptide mass information and partial MS/MS derived sequence information, respectively, for the identification of proteins.

a. Protein Identification by Peptide-Mass Mapping. Peptide mapping is one of the most commonly employed approaches to identifying and characterizing proteins. Isolated proteins, including proteins separated by gel electrophoresis, are chemically or enzymatically cleaved, and the resulting peptides are separated and isolated. For the isolation of relatively small peptides, RP-HPLC is the preferred method, since individual fractions are collected in a liquid phase. However, electrophoretic separations of cleaved proteins, particularly in gel systems resolving low molecular weight polypeptides (292), have also been used successfully. Traditionally, isolated peptides were subjected to amino acid sequencing. The advent of MALDI- and ESI-MS has added an attractive dimension to the characterization of peptides in protein digests. As described earlier, MALDI-MS has been used to analyze unfractionated peptide digests, or to systematically screen the contents in each HPLC fraction. ESI-MS is most effectively conducted on-line with the HPLC separation by splitting the column effluent. This allows simultaneous mass measurement and fraction collection (267,283,293). ESI-MS is equally suited for off-line mass measurement of HPLC fractions, but is less suited for the analysis of unfractionated peptide mixtures (294).

Knowledge of the accurate mass of a peptide is very useful in conjunction with peptide sequencing. It provides an assessment of sample purity prior to application of the peptide to the sequencer, and it confirms the determined peptide sequence. The mass of a single peptide by itself is, however, not sufficient to identify the protein. A few groups independently have realized that the collective peptide mass information obtained by MS analysis of a protein digest provides an idiotypic fingerprint for that protein (295,296). This insight was implemented into a series of algorithms, all of which match sets of experimentally derived protein fragment masses against a database of peptide masses derived by calculating all the possible peptide masses that can be generated by fragmenting all the sequences in a sequence database under certain, deliberately chosen parameters [(280,281,297–299), and reviewed in (300,301)]. All of these programs are available as commercial packages, from the authors, across the Internet as an e-mail server [mowse@dl.ac.uk, (281)], or by accessing the Darwin suite of programs on the World Wide Web (URL http://cbrg.inf.ethz.ch/) (241,297). Other programs that offer similar packages are available on the World

Wide Web from the groups of Burlingame (URL http://rafael.ucsf.edu/) and Chait (URL http://chait-sgi.rockefeller.edu/), and the list continues to grow.

The original approach is schematically outlined in Fig. 10. Accurate masses of several peptides from a specifically cleaved protein are experimentally determined. These masses are then used to search peptide-mass databases, which include the fragment masses of all sequences in amino acid sequence databases, including those deduced from DNA sequences. Importantly, this can include the six-way translations of EST (expressed sequence tag) databases. These fragment masses are computed by using specific cleavage rules, including the substrate specificity of proteases, preferred cleavage sites and enzyme-resistant peptide bonds, and provisions for dealing with incomplete proteolysis and unspecific cleavage. The search results are displayed in a ranked list of candidate proteins, with the highest ranked being the most likely candidate.

Of course, these programs can only identify proteins for which the exact sequence, or the sequence of close homologs, is in the database. It is important to note that the programs themselves cannot determine whether an identification is correct, even though the results of some programs are annotated with scores, which are useful for the scientist in distinguishing between true and false positives. To make protein identification by peptide-mass database searching conclusive, secondary criteria are frequently required. This is supported by a number of studies, which have evaluated the reliability of protein identifications by peptide-mass mapping. Clauser *et al.* (253) used high-energy MS/MS-derived sequence information of selected peptides to either confirm or reject the protein identification obtained using the MOWSE peptide mass searching program (281). They found that for 6 of the investigated 11 proteins that were isolated from 2-D gels, and for which the sequences were in the database, the MOWSE search correctly identified the protein. In the other 5 cases the MOWSE search was unable to predict the subsequently identified protein. A study by Sutton *et al.* (288) yielded similar results. From a set of 16 proteins, which had previously been identified by Western blotting, by protein sequencing, or in one instance by database comparison, 11 were correctly identified using the MOWSE program with masses obtained by MALDI-MS. One of the 5 mismatches was explained as a probable previous misidentification due to comigration of the protein under investigation with a contaminant. In two other mismatches the MOWSE search was in contradiction either to previous protein sequence information or to identification by Western blotting. In other reported studies the identification by peptide-mass searching was confirmed by amino acid sequencing (84,278,302). These studies suggest that one has to be very careful when interpreting the results of peptide-mass searches. Although the correct answer is often obtained, elimination of false positives requires conclusive secondary criteria.

Some reports have tested the sensitivity limit of protein identification by peptide-mass mapping. Various quantities of known proteins were subjected to SDS-PAGE and peptides were obtained by enzymatic fragmentation either in-gel or on Immobilon-CD membrane. Peptide masses were determined by MS, and these results were used to estimate the lowest amount of protein required for positive identification by peptide mass searches (266,291). Moritz *et al.* (266) showed data to 60 pmol, and the present authors (291) showed data to 5 pmol, of protein loaded on the gel, with both studies concluding that low

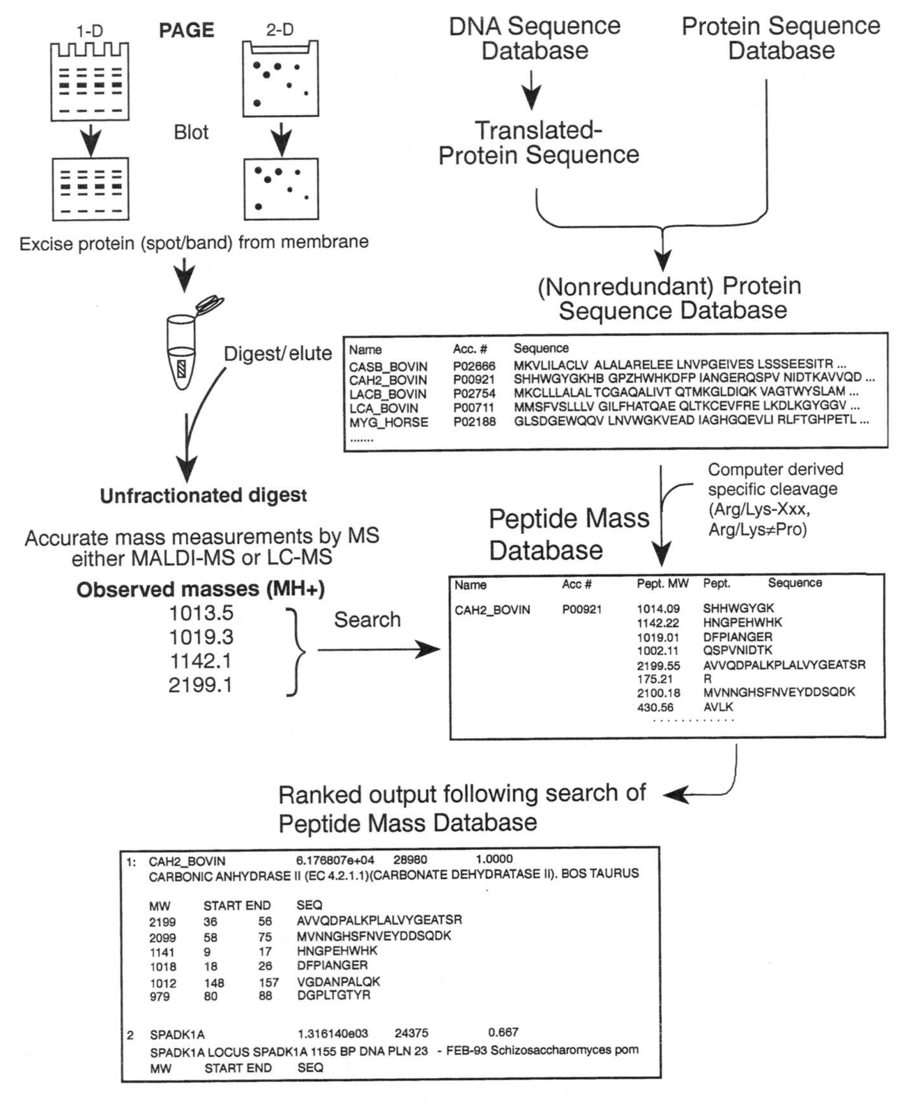

Fig. 10 Flow diagram for the identification of gel-separated proteins by peptide-mass searching. Either 1-D or 2-D PAGE gels are blotted (or not) to a membrane, and the protein spot/band of interest is excised from the membrane for blotted samples or from the stained gel for in-gel digests, and then cleaved chemically or enzymatically. The mass of the released peptides is determined accurately by MALDI-TOF-MS and LC-MS. Protein databases, and DNA sequence databases translated into protein sequences, are assembled into a nonredundant sequence database. Peptide fragment masses (using specific cleavage rules, i.e., for trypsin, after Arg or Lys but not when followed by Pro) are computed from this database to generate a peptide mass database. The peptide mass database can then be searched using the experimentally derived masses. A ranked output of the most likely candidates, based on the number of matches and the frequency of such matches occurring randomly in that mass range, is generated. This scheme is based on that of Pappin *et al.* (281), available on the Internet, but the other versions mentioned in the text work on the same principle. Reproduced with permission from Patterson, S. D., and R. Aebersold. (1995). Mass spectrometric approaches for the identification of gel-separated proteins. *Electrophoresis* **16**:1791–1814.

picomole amounts of protein are sufficient for positive identification, although the latter report showed that even with 74% of β-casein sequence mapped by four peptides, the protein was not identified. This was because only one of the four peptides was not modified. The other three peptides were oxidized, were phosphorylated, or were a putative sequence variant (291). These studies also illustrated that it is predominantly the quality of the peptide mass data that determines the quality of the search results. The more accurate the data, the more conclusive the matches. As Clauser *et al.* indicated (253), the chance that peptides of similar masses are present in different sequence database entries increases with increasing size of the sequence databases. Therefore, higher mass accuracy is desirable to minimize the size of the mass window during the peptide mass search. ESI-MS generally yields data of higher mass accuracy than MALDI-MS. MALDI-MS data typically are associated with an error of ± 2–3 Da, which can be somewhat reduced by the inclusion of an internal calibrant. High-resolution mass spectrometric approaches such as Fourier transform mass spectrometry (FTMS) coupled with ESI and MALDI ionization will no doubt prove useful in addressing this problem.

If the data from peptide-mass database searches are carefully analyzed, it is almost always observed that not all of the experimentally determined peptide masses match a list of the possible peptide masses from the protein identified by the computer as the best match. Several plausible explanations for these nonassigned peptide masses should be considered, not least of which is that neither the protein or any related protein is in the database. Some are easily testable by careful analysis of the data, and others require additional experiments.

1. The peptide-mass database search identified the correct protein, but some of the peptides are posttranslationally or artifactually modified, or the protein has undergone posttranslational processing. By taking into account plausible protein modifications (see the chapter by Wold in this volume), some of the initially unassignable masses may be rationalized. It is important to note, however, that without experimental data, such assignments should be considered tentative.

2. The peptide-mass database search identified the correct protein, but some of the peptides were derived by unspecific proteolytic cleavage or by specific proteolysis by a contaminating protease. This is easily tested by probing whether the tentatively identified protein, in the form of an unprocessed and unmodified polypeptide, can produce the observed masses if no assumptions about the specificity of the cleaving agent are made.

3. The peptide-mass database search identified the correct protein, but the protein was contaminated with one or several other proteins. If sufficient peptide masses from the putative contaminating protein can be determined, and if the sequence of that protein is in a database, a secondary database search with a subtracted dataset consisting of the all the determined masses subtracted by the peptide masses matching the tentatively identified protein can explain the origin of the unassigned masses, as well as identify the contaminating protein (278).

4. The peptide-mass database search identified a homolog, from the same or a different species, to the protein that was actually isolated, or a processing

variant (303,304). Sequence homologs from other species are easily eliminated by species-specific database searching. The identification of possible homologs from different species, if verified by confirmatory data, can, however, be very useful, especially for scientists working with genetically poorly characterized organisms (149,150).

5. The peptide-mass database search identified a false positive. This potentially disastrous result is difficult to verify or disprove, although in those programs that provide a ranking score, the difference between the highest and the second-highest score, compared to the difference between the second-highest and subsequent scores, can be used to spot false positives. This result could also be due to the fact that the sequence of the protein has yet to be deposited in the database, or it may be truly novel. If there is any doubt about the conclusiveness of peptide mass search results, confirmatory criteria should be used. The following section describes some approaches that have been used to complement protein identification by peptide-mass mapping.

b. Strategies to Enhance Protein Identification by Peptide-Mass Mapping. The previous sections documented the need for strategies to enhance protein identification by peptide-mass mapping. Several groups have published useful solutions on the software level by enabling the inclusion of secondary search criteria, and on the sample preparation level by site-specific chemical modification or by data integration allowing the synergistic use of data from unrelated types of experiments.

James *et al.* (241) and Pappin *et al.* (305) recognized that the selectivity of peptide-mass map searches is dramatically enhanced if more than one cleavage reagent is used on the same substrate, and the search results from both datasets are combined. This option is now realized and available in the Darwin, and in the near future, on the MOWSE search program.

A number of procedures have been described that target specific groups in peptides for chemical modification, and provisions to use the induced mass differences as confirmatory criteria in peptide mass database searches have been implemented at least by two groups (241,306).

The amino acid composition of a peptide is reflected by the number of exchangeable hydrogens present and can be measured by mass spectrometry if the protons are exchanged by heavy isotopes. Hydrogen exchange with an excess of a deuterated solvent is essentially instantaneous, no acid, base, or catalyst is required, and there is generally no exchange of carbon-bound hydrogen (307). Significant back exchange can occur, however, if precautions are not taken. In a peptide the following groups undergo hydrogen exchange: hydroxyl, carboxyl, nitrogen bound hydrogen, and sulfhydryl groups. The number of exchangeable hydrogens present on each amino acid residue in a peptide, including the amide hydrogen, is therefore as follows: Ala, Phe, Gly, Ile, Leu, Met, and Val all exchange one hydrogen; Cys, Asp, Glu, His, Ser, Thr, Trp, and Tyr exchange two; Lys, Asn, carboxymethyl-Cys, and Gln exchange three; and Arg exchanges five hydrogens (241). For MALDI-MS, deuteration can be conducted on the sample probe by repeated mixing of the dried crystallized sample with a deuterated solution ($D_2O:C2H_5OD$, 1:1, v/v) under a nitrogen countercurrent (240). The number of exchangeable hydrogens is determined simply by subtracting the peptide mass measured prior to deuterium exchange from the mass of the heavy-isotope labeled peptide. Using this approach, Spengler *et al.* (240) reported 98.5% hydrogen exchange for the peptide substance P. The measured mass increase, by 23 Da from MH^+ of 1347.7 to MD^+ of 1370.7, matched the expected exchange of 22 hydrogens plus one deuteron to make MD^+. Deuteration can also be performed prior to ESI-MS analysis (308). James *et al.* (241) also reported the use of deuterium exchange, specifically for the purpose of enhancing the selectivity of peptide-mass database searching. Their program (Darwin) incorporates such additional data into the search algorithm. Provided that high deuterium exchange rates can be achieved and stabilized, hydrogen exchange is a simple, fast, and extremely useful technique.

The addition of methyl groups by methyl esterification of free carboxyl groups in the side chains of aspartic acid and glutamic acid residues and the terminal α-carboxyl group is another selective chemical modification of peptides. The reaction, which has been used frequently for the derivatization of peptides prior to MS/MS sequencing (221), can be performed with essentially quantitative yields on very small quantities of sample (306). This reaction also has the advantage that addition of a single methyl group increases the peptide mass by 14 Da, and is therefore easier to detect than hydrogen exchange, especially in relatively low-resolution MALDI-MS. There are a number of protocols for methyl esterification [e.g., (221,309)]. Craig *et al.* (310) described a simple procedure for modification of tyrosine residues by iodination on the sample probe prior to MALDI-MS analysis. This protocol is reported to be specific for tyrosines and adds 126 Da per residue. In particular, derivatization of histidines was not observed.

Finally, programs for protein identification by peptide mass database searches have been modified to incorporate data from unrelated analytical techniques such as partial peptide sequencing or amino acid composition analysis [e.g., (305,306)], thus dramatically enhancing the level of confidence in the protein identification results.

3. *Peptide Sequence Tags and Automated MS/MS Spectra Interpretation*

Complete or partial peptide sequences, either by themselves or in conjunction with peptide mass database searches, are the most discriminating criteria for the identification of proteins. In the following we describe two approaches, which identify proteins by taking advantage of reproducible peptide fragmentation patterns in the mass spectrometer, without explicitly interpreting the complete peptide sequence.

Mann and Wilm (311) described a mass spectrometry-based search approach that uses sequence-specific fragmentation information in combination with the mass of the intact peptide. This approach can be used in addition to the peptide-mass search methods described previously (or on its own), and as such it provides a significant amount of additional information for the search. The sequence-specific fragmentation data can be obtained by either ESI-MS/MS or PSD-MALDI-MS. However, the approach can also use Edman-derived sequence data in combination with the mass of the peptide sequenced even if only partial, noncontiguous sequence information is obtained. The theory outlined in Fig. 11 is based on interpretation of a portion of the ESI-MS/MS

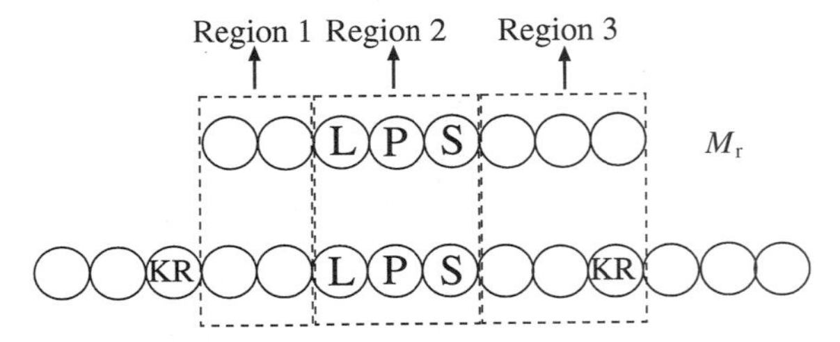

Fig. 11 The peptide sequence tag approach. A peptide is subjected to either PSD-MALDI-MS or ESI-MS/MS and the resulting spectrum is partially interpreted. In the example given, this resulted in the finding of a fragment ion series, the ions of which differed in mass by 113, 97, and 87, which yields a partial sequence of L, P, S (although there is no way to distinguish between I or L, and if the mass resolution was not sufficient Q may not be distinguishable from K). The partial sequence LPS, therefore, represents the sequence "tag." The mass of region 1 (m_1) is defined as the mass of the smallest ion in the sequence defining fragment ion series. The mass of region 3 (m_3) is defined as the mass difference between the total mass of the peptide and the largest sequence defining fragment ion (from the LPS series). The five criteria that can be used in the search include these three (m_1, partial sequence, and m_3) and the N-terminal and C-terminal cleavage specificity (i.e., whether the expected cleavage is tryptic, as in this example). This complete peptide sequence tag can be used to search an appropriately computed protein sequence database, resulting in an output that ranks any protein, which contains a sequence that corresponds to the peptide sequence tag. Adapted with permission from Patterson, S. D., and R. Aebersold. (1995). Mass spectrometric approaches for the identification of gel-separated proteins. *Electrophoresis* **16:**1791–1814.

(or CID) or PSD-MALDI-MS fragmentation data. Burlingame's group (University of California San Francisco) also has a similar peptide-sequence tag program available on their World Wide Web server at URL http://rafael.ucsf.edu/.

For this search, it is not a requirement that the whole spectrum be interpreted, only a portion that yields mass differences consistent with amino acid residues. Therefore, the partial sequence information is derived from fragment ions, the mass difference of which is consistent with an amino acid residue. This partial sequence information might only be two residues (derived from three fragment ions), which on its own is not sufficient to be useful in a standard sequence-based search. The partial sequence information or "tag," shown schematically in Fig. 11 and represented by "LPS," has a mass that is represented as m_2 (and is referred to later as region 2). However, there are two other pieces of information that can be derived from this fragmentation spectrum in addition to the sequence itself: the mass difference between region 2 and the N-terminus (known as region 1, with mass m_1), and the mass difference between region 2 and the C-terminus (known as region 3, with mass m_3). The sum of regions 1 and 3 and the mass of the amino acid residues of the sequence (m_2) equals the molecular mass of the intact peptide, that is $m_1 + m_2 + m_3 = M$. If the fragment ions are N-terminal-derived (b series; see Section V,C,2), then m_1 is the lowest mass of the ion series from region 2, and m_3 can be derived from the mass difference of the largest interpreted fragment in the ion series and the intact mass of the peptide, M. The converse is true if the ion series is C-terminal derived (y series), that is, m_1 is the difference between the largest region 2 defining ion and the intact mass plus the mass of two hydrogens, and m_3 is equal to the lowest mass in the interpreted ion series (region 2) minus two hydrogens—and the sequence obtained has to be reversed in orientation (that is, SPL in Fig. 11) because it is being read from the C terminus. These three pieces of information—two masses and a portion of sequence—are referred to as the peptide-sequence tag or PST (311). The specificity of the cleavage reagent used can also be incorporated into the search, adding a further two pieces of information that can be used as search critiera (i.e., the cleavage specificity at the N and C termini). As shown in Fig. 11 these are the five criteria that can be used in the PST approach. However, use of all five criteria is not an absolute requirement. It is also not always necessary to know which series of ions one is observing (b or y series), that is, one conducts the search assuming they are either N- or C-terminal derived. Of course, as with any of the mass search programs, the presence of posttranslational modifications on the peptides will negatively influence the result.

A peptide modified posttranslationally with a moiety that could not be predicted and included in the peptide mass database (i.e., unexpected modification of cysteines) would not be found using the complete peptide-sequence tag. However, another important feature of this search routine is that one can define what is referred to as "N-terminal" or "C-terminal" specificity, that is, if there is a modification (and therefore the mass of the peptide has been increased over that predicted), then only data for two of the regions (1 and 2, or 2 and 3) need be included in the search. Of course, this lowers the specificity of the search, but in some cases it can still be very discriminating. This is a powerful search approach and is extremely useful either on its own or—as it will more likely be used—in conjunction with the peptide-mass searches described previously.

The PST approach can also be used with partial amino acid sequence data derived using chemical approaches, either traditional Edman sequencing or one of the newer methodologies described previously. In this case, the intact mass of the analyzed peptide is used together with the sequence, which may have any number of unknown residues labeled "X." In the case of Edman-derived PTH data, the Q = K and I = L options can be removed (311). This is also a powerful tool for identification of proteins whose gene sequence has already been determined, whether completely or partially as an EST.

Yates and colleagues (312) developed a protein identification approach that is based on automated interpretation of ESI-MS/MS fragmentation data in a format that can be used to directly search sequence databases. The method is schematically illustrated in Fig. 12 and is referred to as SEQUEST. In a first step, the program identifies all those peptides as candidate peptides that can be generated by a protein in the sequence database and whose mass matches the measured mass of the peptide ion. In a second step, the program predicts the fragment ions expected for each of the candidate peptides if they were fragmented under the experimental conditions, used, and then compares the experimentally determined MS/MS spectrum with the predicted spectra using cluster analysis algorithms. Each of the comparisons receives a score, and the highest-scoring peptide sequences are then reported. The protein is therefore identified based on peptide MS/MS spectra, even though the peptide sequence is never explicitly determined. This method has several extremely attractive features:

1. The method can effectively identify proteins in protein mixtures. Because every peptide in a peptide mixture provides the data for an independent

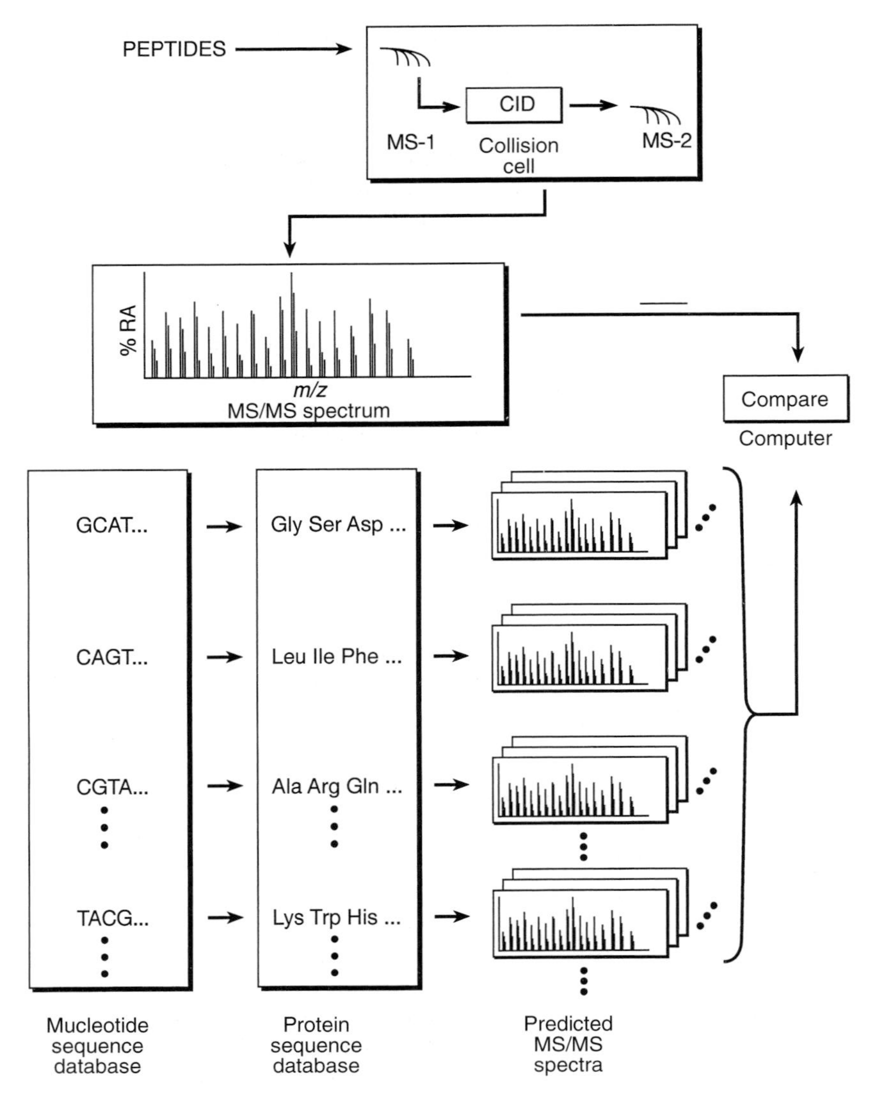

Fig. 12 The SEQUEST system for identification of proteins through analysis of uninterpreted tandem MS spectra. In this approach, low-energy ESI-MS/MS spectra from individual peptides are obtained and normalized. These MS/MS spectra are then compared with predicted MS/MS spectra computed from protein sequence databases. All possible peptides corresponding to the mass of the experimented peptide, irrespective of the cleavage specificity of the experimental enzyme, are used in the search. A ranked output of possible matches is then produced. Reproduced with permission from Patterson, S. D., and R. Aebersold. (1995). Mass spectrometric approaches for the identification of gel-separated proteins. *Electrophoresis* **16:**1791–1814.

database search, the program is uniquely suited to cope with complex peptide mixtures, even if they were derived by digesting a protein sample containing several proteins.

2. The method provides data that are inherently autoconfirmatory. Typically, the method is sequentially applied to several peptides from a single protein digest, each peptide providing an independent database search. Top-

ranked scoring of the same protein by several peptide searches essentially confirms the protein identity.

3. The method is easily automatable. It has been adapted for automated identification of peptide digests analyzed by LC-ESI-MS/MS where the ions subjected to collision-induced fragmentation are automatically selected during the run (313) and the data are automatically analyzed.

4. The method is adaptable to locate peptides carrying specific posttranslational modifications (313) and to identify the protein from which the peptide originated. For these reasons this approach is very appealing for research projects that require a large throughput of data in an automated fashion, for example, identification of many spots from a 2-DE separation, or when proteins in protein mixtures are to be identified.

Automated execution of the method is currently limited to triple quadrupole ESI-MS instruments with a data system capable of selecting ions for fragmentation on-the-fly. The SEQUEST system is available from Finnigan MAT (San Jose, CA).

4. *Gel Electrophoresis-Induced Protein Modifications*

High-resolution protein electrophoresis is usually conducted in a matrix of polyacrylamide (PA), which is formed by polymerization of reactive acrylamide monomers. It has long been considered a problem that residual unreacted acrylamide monomers could form covalent adducts with proteins during gel electrophoresis (314,315). Because polymerization of gels is rarely greater than 90%, leaving potentially 30 m*M* of free acrylamide in the matrix, this is a valid concern (315). The failure to obtain N-terminal sequences of gel-separated proteins was often thought to be caused by artifactual blocking, but because of the high proportion of endogenously blocked proteins and the potential for protein blocking prior to application to the gel, the magnitude of the effect was difficult to assess. Furthermore, the advent of methods for acquiring internal sequence from peptide digests provided a general solution to the problem of obtaining sequence from N-terminally blocked proteins.

The mass spectrometry-based methods just described have revived interest in the question. For the first time, acrylamide adducts arising from gel electrophoresis can be quantitatively and qualitatively characterized in a comprehensive manner. A detailed overview of the approaches used to study the effect was included in a review (300). The following is a summary of the subject, updated with relevant data.

A series of extensive studies involving nuclear magnetic resonance (NMR), mass spectrometry, amino acid composition analysis, and protein sequencing showed that the amino acid residue most at risk of acrylamide adduction was cysteine, resulting in cysteinyl-*S*-β-propionamide (see Fig. 13 for the structures of gel-induced protein adducts). Using ^{1}H NMR, GC-MS, and chemical synthesis, Chiari and colleagues verified the structure of the predicted adduct (315,316) and confirmed that this modification could occur in native proteins. Interestingly, in this case only one of five Cys residues of β-lactoglobulin (Cys-160) was modified (315). Ploug *et al.* (317) found modified cysteines following amino acid composition analysis of gel-separated proteins, and Brune demonstrated the presence of cysteinyl-*S*-β-propionamide in gel-separated proteins by se-

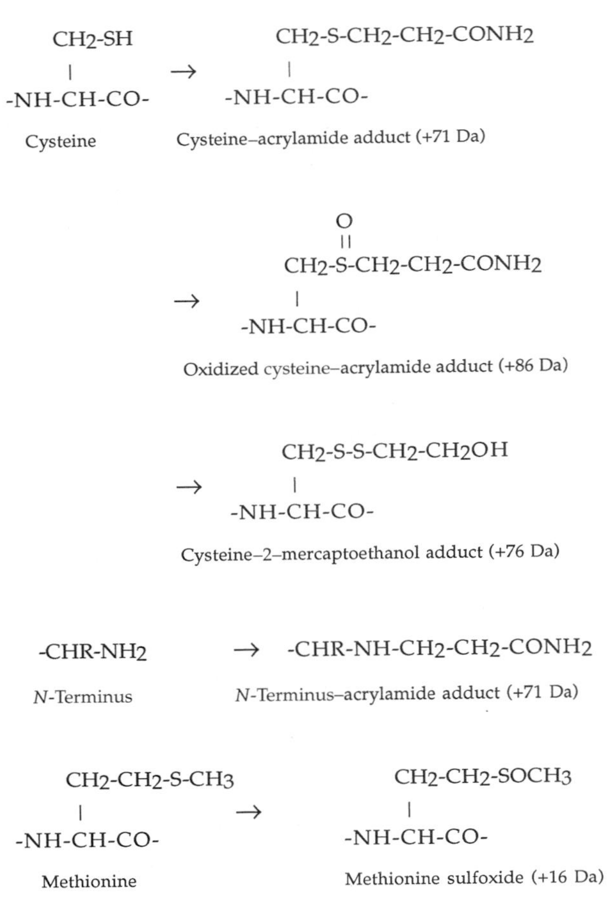

Fig. 13 Gel electrophoresis-induced protein modifications. Chemical formulas of gel electrophoresis-induced protein modifications together with the mass difference from the unmodified residue in brackets. The cysteine acrylamide adduct is referred to in the text as cysteinyl-*S*-β-propionamide. Data obtained from references cited in the text, and for the oxidized acrylamide adduct, from K. Clauser ((253) and personal communication). Reproduced with permission from Patterson, S. D., and R. Aebersold. (1995). Mass spectrometric approaches for the identification of gel-separated proteins. *Electrophoresis* **16:**1791–1814.

quence analysis (271). The same author also developed a method to stably modify cysteinyl residues with acrylamide prior to sequence analysis (271). Using ESI-MS to analyze electroeluted proteins and peptides, le Maire *et al.* (318) consistently found cysteinyl-containing peptides with masses 71 Da higher than expected for the bare peptide chain. This mass increase is consistent with acrylamide-adducted cysteine (Fig. 2). Similar results have been obtained by other groups using either ESI-MS or MALDI-MS for the analysis of peptide maps generated from gel-separated proteins (253,255,266,288), and Hall *et al.* (254) unequivocally showed the acrylamide adduct on a cysteine residue using high-energy MS/MS analysis of a tryptic peptide isolated by HPLC from an electroeluted, 2-DE-separated protein spot. Results from Clauser *et al.* (253), who reported that all cysteine-containing tryptic peptides derived from 2-DE-separated proteins revealed the presence of either Cys-acrylamide or, in one case, an oxidized Cys-acrylamide (Fig. 13) (revealed by high-energy MS/MS analysis), suggest that acrylamide adduction may be quite frequent. Bonaventura *et al.* (319) reported that for unexplained reasons, hemoglobin is susceptible to acrylamide adduction, as revealed by ESI-MS analysis of gel-separated material, even with storage of the gels for 26 hr postpolymerization.

There have only been two reports of mass increases consistent with acrylamide adduction in peptides that do not contain cysteine residues. Klarskov *et al.* (320) isolated an internal peptide with the sequence GSTGK from an in-gel digested protein. LC-MS analysis showed a peptide mass that was consistent with an acrylamide adduct. Although the nature of this modification was not confirmed, the suggestion was offered that other nucleophiles, including amino groups, may be alkylated under the conditions of in-gel digestion (320). Haebel *et al.* (255) have provided evidence that the N-terminal α-amino group of an electroeluted protein was modified with an acrylamide adduct (Fig. 13). This modification, which was observed in both the intact protein and the N-terminal tryptic fragment, rendered the protein refractive to sequencing by the Edman degradation. However, a decrease in N-terminal blocking from 55% to 7% was achieved by inclusion of 0.2% (v/v) mercaptoacetic acid in the second dimension electrophoresis buffer of their 2-DE separation. The authors postulated that the presence of an N-terminal glycine may favor the formation of this adduct because of low steric hindrance, or alternatively, the high pK_a of the α-amino group (255). Although the conditions favoring N-terminal modification in polyacrylamide gels have not been exhaustively investigated, it appears beneficial to use reagents to quench the reactive double bond of free acrylamide and to use gel ingredients of the highest purity. In addition to the mercaptoacetic acid described previously, the following scavengers and their uses have been published: 3-mercaptopropanoic acid for Tris-Tricine SDS gells (317,321), glutathione or sodium thioglycolate for modified SDS-PAGE gels (322,323), and free cysteine for IPG gels (315).

In addition to the formation of acrylamide adducts, a number of different substrate modifications can occur in gels, which are now detectable by mass spectrometric analyses of gel-separated proteins. An increase in mass of a methionine-containing peptide by 16 Da can be interpreted as partial oxidation to methionine sulfoxide (Fig. 13). This oxidation may be due to the presence of residual persulfate in the gel (324). Klarskov *et al.* (320) reported a mass increase consistent with the addition of a single 2-mercaptoethanol moiety to a peptide containing a free Cys residue (Fig. 13). The same group had previously observed mercaptoethanol modifications following reduction of a recombinant human immunodeficiency virus type-1 (HIV-1) p25 protein (325).

Therefore, in addition to the more than 200 endogenously produced posttranslational modifications [(326) and chapter by Krishna and Wold elsewhere in this volume], detailed analyses of the covalent structure of gel-separated proteins have to consider artifactual modifications, at least some of which can occur in the polyacrylamide gel. The extensive summary of the mass differences induced by posttranslational modifications by Krishna and Wold (327) is an invaluable resource for such studies. For an Internet-accessible version see the World Wide Web site of Ken Mitchelhill (St. Vincents Hospital, Victoria, Australia) set up at URL http://www.medstv.unimelb.edu.au/WWWDOCS/SVIMRDocs/MassSpec/deltamassV2.html.

5. Limitations of Peptide-Mass Search Approaches

The peptide-mass search programs described previously provide a rapid means of identifying proteins whose sequence has already been determined and is contained within the database being searched. The major technical limitations and pitfalls of the methods have been discussed. They have to be seriously considered if the number of false positive protein identifications is to be minimized. Generally, the level of confidence in a protein identification is dramatically enhanced if two or more independent sets of data converge on identifying the same protein. Such datasets can be derived from independent identifications using the same method or by combining the results from different types of experiments.

Although the precision of the method clearly relies on the specificity of the proteolytic agent, the measured mass accuracy, the purity of the protein substrate (except for the MS/MS pattern analysis approach), and other experimental parameters, it is also limited by the quality of the sequence databases and the ability of the search programs to accept qualifying criteria such as information on posttranslational modifications. Aside from being able to specify the modification of cysteine, as yet no other modifications have been incorporated into the peptide-mass database search, although work is in progress to allow others. In addition, the contents of database annotation tables, which include information on signal peptides, as well as determined and *putative* sites of protein modification, are not considered in the peptide mass database searches. Of course, the modifications listed may be specific for a particular physiological state, and therefore not relevant to the system being studied.

E. Identification of Proteins by Amino Acid Composition

The amino acid composition of a protein, if used in conjunction with an estimate of the polypeptide molecular weight, *may* provide sufficient information to identify a protein or at least to determine sequence relationships with a high level of confidence (328,329). Therefore, the amino acid composition may be a useful means to identify protein isoforms, including charge isoforms separated by 2-DE. Chemical analysis and differential radiolabeling are the two methods that are most commonly used to determine the amino acid composition of a protein.

1. Chemical Methods

Determination of the amino acid composition by chemical methods involves the hydrolysis of all the peptide bonds in a protein, followed by separation, detection, and quantitation of the resulting amino acids. Polypeptides are most frequently hydrolyzed by exposure to strong acids (215), although enzymatic methods have also been described (330). Determination of the amino acid composition of microgram amounts of protein to a level of accuracy that is consistent with protein identification is difficult to achieve (331). Several investigators have demonstrated the ability to establish amino acid compositions of gel-separated proteins that were electroblotted to PVDF membranes prior to hydrolysis. Ploug *et al.* (317) and Nakagawa and Fukuda (332) showed data in the 100-pmol range of protein. Tous *et al.* (333) were able to obtain composition data from as little as 10 pmol of protein. These data were generated using postcolumn derivatization with *o*-phthaldialdehyde (OPA) derivatizing reagent. Lottspeich *et al.* (334) described the combined use of OPA and 9-fluoromethoxycarbonyl (FMOC) for analysis of submicrogram quantities of PVDF blotted gel-separated protein with high accuracy. Tous *et al.* (333) noted that staining of the bands prior to hydrolysis increased artifactual peaks and baseline noise. This could be avoided if the protein bands on the wet PVDF membrane were detected by transillumination. Other observed artifacts include polyacrylamide gel-induced Cys modification (317), which could be minimized by preelectrophoresis of the gel. In extensive studies Gharahdaghi *et al.* (335) probed the limits of the method. Analyzing precolumn labeled PTC amino acids, they demonstrated an average accuracy of 85%, an average precision of 5%, and average recovery of 88% for a range of proteins with molecular masses between 14.3 and 77 kDa. For sensitive analysis of any PVDF-blotted and Amido Black-stainable protein, Cordwell *et al.* and Wasinger *et al.* (149,150) employed a rapid 1-h hydrolysis combined with a single extraction, automated derivatization, and chromatography to achieve very high throughput. This was performed in a batch mode (with replicates also run) to increase reliability. This method was developed for high-throughput identification of 2-DE separated proteins (149,150).

2. *Radiolabeling Methods*

An earlier approach for the simultaneous identification of many 2-DE separated proteins was based on differential metabolic labeling of proteins with radioactive amino acids (336,337). Cultured cells were labeled separately, each one with one of 20 ^{14}C-labeled amino acids. The 20 samples were separated individually by 2-DE and the patterns were recorded by autoradiography followed by densitometry of the films. The patterns were merged and analyzed in a computer. The known composition of selected proteins in the 2-DE pattern served to calibrate the system. Storage phosphor imaging, which is more quantitatively accurate, more sensitive, faster, and of a larger dynamic range than autoradiography (50,338,339), may render this approach more practical. Garrels *et al.* (340) proposed a similar radiolabeling method for the global identification of proteins separated by 2-DE. They used storage phosphor imaging and double-labeling with a short-lived (^{35}S) and a long-lived (^{14}C) isotope. Quantitative data from multiple exposures over a period of 128 days, related to the known half-life of the two isotopes, were used to derive partial amino acid compositions of the separated proteins. Application of this methodology to the study of the yeast *Saccharomyces cerevisiae* revealed that the double-labeling method was not as predictive as the single-label method, i.e., more proteins were in the match list (134). However, the single-label method, which requires only one set of images per analysis, may be somewhat more susceptible to experimental errors (134). The advantage of these approaches to protein "identification" is that they are truly parallel, i.e., hundreds of proteins can be analyzed in a single experiment. However, low-abundance proteins cannot be analyzed without pre-enrichment.

3. *Computer Programs for Protein Identification by Their Amino Acid Composition*

The programs described earlier for the analysis of radiolabeled, 2-DE separated proteins contain provisions for the cross-correlation of spots between different gels or for the identification of proteins by comparing the composition with the calculated composition of proteins in sequence databases (340). Several groups also wrote compute programs with the intention of searching sequence databases with experimental data derived by chemical composition analysis of gel-separated proteins. In these programs the amino acid composition is either used by itself, or more frequently, in combination with estimated p*I* and M_r data of the proteins to be analyzed. FINDER (329), PROPSEARCH (341), ASA (342), and SWISS2D-PAGE Amino Acid Composition (149) (accessible on the World Wide Web at URL http://expasy.hcuge.ch/ch2d/aacompi.html) and algorithms by Sibbald *et al.* (343) and Eckerskorn *et al.* (344) are examples of such systems. Some of these programs contain provisions for correcting systematic errors in the experimentally determined data and for weighing the relative reliability of the determination for each amino acid. In an effort to increase the reliability and throughput of amino acid analyses for large-scale protein identification projects, simplified and automated protocols for amino acid composition analyses were worked out. As described earlier, Cordwell *et al.* and Wasinger *et al.* (149,150), developed a rapid batch method for amino acid analysis in which one operator with one HPLC unit was reported to be able to process 100 samples per week (151). The sensitivity of these analyses was high. The authors claim that any protein spot detectable on PVDF by Amido Black staining could be reliably analyzed (149,150).

F. Synergy of Peptide Mass, Amino Acid Composition, and Amino Acid Sequence Data for Protein Identification

The use of MS in peptide-mapping experiments provides the accurate mass of isolated peptides or of peptides in inhomogeneous peptide fractions. This simple statement has striking consequences for the identification of proteins by peptide sequencing. (i) The peptide mass determines the number of sequencing cycles required to complete the peptide sequence, thus minimizing reagent costs and sequencer time. (ii) The peptide mass confirms the determined peptide sequence and indicates whether the sequence is complete. For a correctly and completely interpreted peptide sequence the sum of the mass of the identified amino acids must match the peptide mass. (iii) Mass analysis indicates the purity of peptide fractions and the relative abundance of peptides in inhomogeneous samples, and thus prevents sequencing of mixed peptides, which usually results in uninterpretable data. (iv) The molecular mass suggests the origin of the fragment in proteolytic digests. Proteolytic digestion of small amounts of peptides requires relatively high enzyme-to-substrate ratios. Therefore, some of the isolated peptides may be derived from autolysis of the protease. Comparison of the determined peptide masses with the peptide masses predicted from autolysis identifies such peptides. (v) The peptide mass can be used to complete a peptide sequence in cases in which a residue could not be conclusively identified. (vi) The peptide mass is a useful parameter for pooling appropriate fractions for sequencing. Frequently, peptide fractions from different digests need to be pooled to achieve the amounts of peptide required for successful sequencing. This is particularly the case if differentially modified protein isoforms are separated by high-resolution gel electrophoresis (304).

Various strategies have been proposed and successfully employed to integrate ESI-MS or MALDI-MS data with results from chemical peptide sequencing. They have in common that a fraction of the peptide sample is used for determination of the peptide mass and the remainder of the sample is subjected to sequencing. As peptide mass determination can be performed more sensitively than peptide sequencing, typically a small fraction of the peptide sample (5–10%) is used for peptide mass determination.

Tempst *et al.* (173,345) used MALDI-MS data of aliquot peptide fractions collected from RP-HPLC columns to confirm sequences determined by chemical degradation, and to assign missed amino acid residues by determining the mass difference between the observed residues and the total peptide mass. This work clearly documented the practical aspects and caveats of such an approach when the aim is to obtain sufficient sequence information for gene cloning. It was noted that not all peptides in an HPLC fraction were necessarily ionized by MALDI-MS. Therefore, mixed peptide sequences were observed even when mass data indicated that only a single peptide was present (345). Similarly, LC-MS was used to determine the mass of peptides eluting from RP-HPLC columns. A flow-splitting device was used between the column and the mass spectrometer, which introduced approximately 10% of the eluting peptide into the mass spectrometer while the remainder of the sample was collected. This configuration allowed sequential monitoring on-line of the UV absorbance and peptide mass without any additional manipulations (283,346). To achieve mass analysis at higher sensitivity by ESI-MS, Wong *et al.* (347) and Moritz *et al.* (348) used capillary column HPLC for either ESI-MS/MS or sample collection for Edman sequencing.

These reports represent early attempts to synergistically use peptide mass and chemical sequencing data to more reliably determine peptide sequences. In these cases the complementary data are manually analyzed individually and then jointly interpreted. Several groups have attempted automation of integration of different data types for peptide sequencing. Johnson and Walsh (349) described a computer algorithm that utilizes chemical sequencing data, as well as MS and MS/MS data, in a complementary manner. The procedure has the great advantage that sequences in peptide mixtures can be determined without further fractionation. First a peptide mixture is N-terminally sequenced. Then the masses of peptides in the mixture are determined and partial CID spectra obtained. The algorithm then determines all hypothetical sequences with a calculated mass equal to the observed mass of one of the peptides present in the mixture, taking into account the chemical sequencing data. These sequences are then ranked according to how well each of them accounts for the fragment ions in the MS/MS spectrum of that peptide, and the process is then repeated for all peptides in the mixture (349). Kellner *et al.* (235) used a combination of PSD-MALDI-MS and Edman sequencing as complementary methods for sequence determination in the low picomole range by applying ~90% of the sample to the sequencer, with the remainder being used for PSD-MALDI-MS. The present authors have successfully employed the combined approaches of peptide-mass mapping and either PSD-MALDI-MS or Edman sequencing to generate peptide-sequence tags, to identify gel-separated proteins enriched on a GST-fusion protein affinity column (350,351).

Although none of these strategies increases the sensitivity of the peptide sequence determination per se, they are extremely useful for increasing the sample throughput and the reliability of sequence data. However, there are other combinations of data that are also being explored to increase the confidence of various search strategies: amino acid composition and amino acid sequencing or mass spectrometric peptide mapping. Cordwell *et al.* (149) and Wasinger *et al.* (150) have used the approach of combined amino acid composition and MALDI-MS peptide mapping (albeit conducted separately) for cross-species identification in the small organisms *Spiroplasma melliferum* and *Mycoplasma genitalium*. This approach is especially suited to cross-species identification where proteins from the species of interest have not already been sequenced and are unlikely to be as part of large-scale genome sequencing initiatives. This group has also coupled amino acid composition analysis with a few cycles of amino acid sequencing to generate what they refer to as a "sequence tag" [cited in (151)]. Following three or four cycles of N-terminal sequencing, the blotted protein band is then subjected to amino acid analysis. The amino acid composition is then used in the search, and all of the ranked candidates are then searched using the few N-terminal residues [cited in (151)]. As described earlier for the "peptide-sequence tag" approach of Mann and Wilm (311), the combination of approaches is extremely powerful and greatly increases the confidence of protein identifications in the absence of sufficient contiguous amino acid sequence information.

G. Summary

In this chapter we have described techniques for the characterization of proteins by their amino sequence. We attempted to illustrate (Fig. 5) the technical revolution that has dramatically changed this field. The ISGF3 case study indicates that internal protein sequencing was the only general technique to determine the primary structure of a protein as recently as a few years ago. Since then new instrumentation, improved techniques, and novel strategies have enhanced the speed, the sensitivity, and the accuracy with which proteins can be identified by their amino acid sequence. Given the very high performance of current analytical instrumentation, the technical challenges are once again in the area of preparing minuscule amounts of samples in a form compatible with analytical instruments, avoiding losses during sample workup, and integrating powerful techniques into an optimized analytical process. Of equal importance is an effective technology transfer from specialized laboratories into the general research community in order for the novel technology to be used to solve fundamental biological and biomedical questions.

V. Analysis of Regulatory Modifications of Components of Complex Systems

A. Overview

Regulated biological systems and processes are frequently characterized by a rapid response to altered external conditions. As an example, the earliest events induced by receptor stimulation of cells such as protein tyrosine phosphorylation (352), enzyme activation (353–356), and production of second messengers (357–364) are detectable within seconds to minutes after initiation of the signal.

Such rapid responses are incompatible with transcriptional or translational control and are frequently based on processing and modification of preexisting proteins. The functional consequences of posttranslational modifications are diverse and dramatic. They can target proteins to specific sub- or extracellular locations or to destruction by proteolysis, induce specific protein : protein, protein : DNA, or protein : ligand interaction, or activate or deactivate enzymatic activities. The characterization of posttranslational modifications is, therefore, an essential step toward the complete understanding of regulated biological processes.

The investigation of posttranslational modifications poses specific technical problems at several levels. First, those proteins that are modified must be detected in complex biological samples. Second, the type(s) of protein modification(s) present need to be determined. Third, the modified residues within the polypeptide chain remain to be localized. Detection of modified proteins in protein mixtures has typically been accomplished by a combination of gel electrophoresis, in particular 2-D gell electrophoresis, and detection techniques specific for a modification. Radiolabeling of modified residues either using radiolabeled metabolic precursors (365) or through the addition of radiolabels by enzymatic or chemical reactions specific for the modified residue [e.g., (366,367)] and specific "staining" protocols of gel-separated proteins are the most common techniques for the detection of modified proteins. Although group specific staining methods such as the detection of dihydroxyphenylalanine by a modified silver stain (368) are relatively rare, detection of modified proteins by attachment of a specific reagent (lectin, antibody) followed by a color reaction is more common. The major limitation of these methods is that only those modifications that are specifically probed for are found.

Localization of the modified residues within the polypeptide chain and detection of rare or novel modifications is very difficult. The standard method involves digestion of the protein, separation and purification of the fragments carrying the modification, and detailed chemical, enzymatic, or mass spectrometric analysis of the modified peptide (see Fig. 14). A general, integrated strategy for the detection, characterization, and localization of modified residues at high sensitivity is an important objective in analytical biochemistry.

Hundreds of types of posttranslational modifications have been described (326,369) and the analysis of posttranslational modifications has been concisely reviewed (370), as well as in the chapter by Krishna and Wold in this volume. To date, only a few of those modifications have been recognized as inducible or reversible. To continue the focus of this chapter on approaches for the analysis of regulated biological systems, reversible and/or inducible modifications are primarily discussed. The technologies described for investigating induced modifications are, however, equally applicable for investigating permanent modifications.

This section outlines technical advances that have been made to increase the generality, sensitivity, and reliability of the analysis of posttranslational modifications. Like protein identification, the analysis of posttranslational modifications has been revolutionized by the development of mass spectrometric techniques, which are compatible with the analysis of gel-separated proteins.

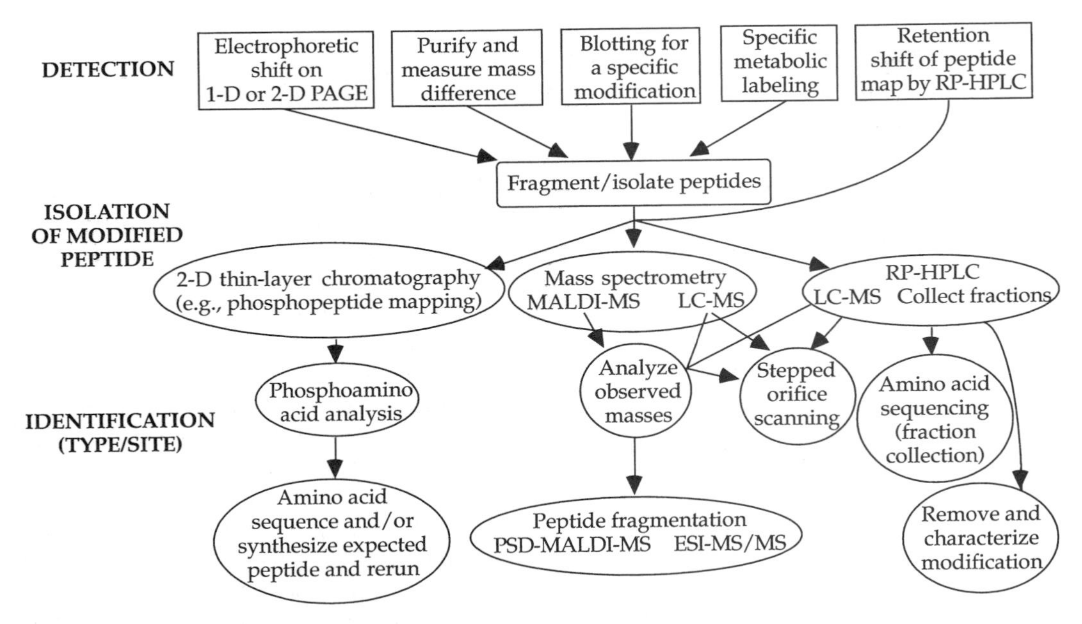

Fig. 14 Overview of some of the approaches available for the analysis of posttranslational modifications. Following detection of a postranslationally modified protein, the protein is generally fragmented and the resulting peptides isolated prior to analysis to determine upon which peptide the modification lies using one of the methods listed. Additional methods are then employed to determine the site of modification on the peptide. Approaches used for some specific classes of modifications are detailed in the text.

B. Activation-Induced Modifications on ISGF3 and Their Function: A Case Study

The rapid time course of interferon-induced gene activation suggested that at least the initial response was posttranslationally controlled. Careful analysis of the Stat 1 and Stat 2 protein sequences with respect to identifiable sequence motifs revealed the presence of SH2 and SH3 domains in all three (Stat 1α and 1β, and Stat 2) proteins, suggesting the involvement of protein tyrosine kinase(s) and tyrosine phosphorylation of signaling proteins. Fu (26) and Schindler and colleagues (371) independently demonstrated that the Stat 1α and 1β and Stat 2 proteins were inducibly tyrosine phosphorylated after stimulation of HeLa cells with interferon α, and that tyrosine phosphorylation coincided with the formation of the hetero-oligomeric ISGF3 complex and was necessary for the translocation of the complex from the cytoplasm to the nucleus. Schindler *et al.* (371) furthermore demonstrated that Stat 1α, but not the two other polypeptides, was also tyrosine phosphorylated after interferon γ (IFNγ) stimulation of fibroblasts. The two groups used similar methods, which were essentially based on immunoprecipitation of the interferon-induced ISGF3 complex with antibodies raised against individual Stat proteins. Tyrosine phosphorylation was shown by Western blotting of the immunoprecipitated and gel-separated Stat proteins using phosphotyrosine-specific antibodies. Alternatively, protein phosphorylation was detected by autoradiography of gel-separated Stat proteins immunoprecipitated from cells that were metabolically labeled with [^{32}P]phosphate (see Fig. 15). The presence of tyrosine phosphate in the labeled proteins was then confirmed by phosphoamino acid analysis.

Phosphopeptide mapping suggested that in Stat 1, as well as in Stat 2, protein tyrosine phosphorylation of a single tyrosine residue was induced after interferon α stimulation (371) (Fig. 15). The location of the tyrosine-phosphate residue in the Stat 1α protein was determined by radiosequencing using Edman degradation. The single radioactive tryptic peptide derived from immunoprecipitated Stat 1α was prepared by 2-D (electrophoresis/thin layer chromatography) phosphopeptide mapping on a silica thin layer chromatography plate (372,373), recovered from the plate, and subjected to automated stepwise chemical degradation (180,374). The phenylthiohydantoins recovered in each sequencing cycle were separated by TLC and the radiolabeled residue, PTH phosphotyrosine, was detected in cycle 4 of the sequencing experiment (25). Tyrosine residue 701 was the only tyrosine residue in Stat 1α that was four residues downstream of a predicted tryptic cleavage site, and was therefore tentatively identified as the phosphorylated residue. This assignment was confirmed by comigration in a 2-D phosphopeptide mapping experiment of the *in vivo* phosphorylated Stat 1α phosphopeptide with a synthetic phosphopeptide with the amino acid sequence predicted around residue 701 (25). The phosphorylation sites on Stat 1β and Stat 2 were determined in analogous fashion. Knowledge of the site and type of induced Stat phosphorylation not only suggested the mechanisms that led to the assembly and stabilization of the ISGF3 complex, but also led to further experimentation toward biochemical dissection of the signaling pathway and identification of potential pharmacologic targets for modulating interferon action.

This case study illustrates several technical issues involved in the identification and localization within polypeptides of posttranslational modifications. (i) Detection of induced sites of protein modification in a cell is frequently based on metabolic radiolabeling and, therefore, not general. Only those modifications that are targeted by the labeled precursor will be detected. (ii) Although the presence of multiple protein modifications can be detected by high-resolution 2-DE, the number of electrophoretic isoforms does not reveal the number of modified residues in a polypeptide. By themselves, the results just discussed therefore do not rule out Stat modifications other than phosphate. (iii) The sensitivity of analytical techniques is frequently not sufficient for the direct analysis of the modified polypeptides isolated from the biological source. In the case of the Stat proteins described earlier, the site of protein phosphorylation was confirmed indirectly by comigration of a synthetic phosphopeptide with the phosphopeptide isolated from interferon-stimulated cells. It is essential that indirect or tentative assignments be confirmed by conclusive experiments. The most conclusive confirmation of the functional relevance of protein modifications frequently involves expression and functional analysis of proteins mutated at the postulated modified site. (iv) Sequence motifs provide immensely useful hints for the investigation of protein modifications. In the present case study, the detection of SH2 domains in the Stat proteins strongly suggested the involvement of tyrosine phosphate in interferon signaling. Other useful sequence patterns include the consensus recognition sequences of enzymes known to modify proteins. It should be stressed, however, that any modification

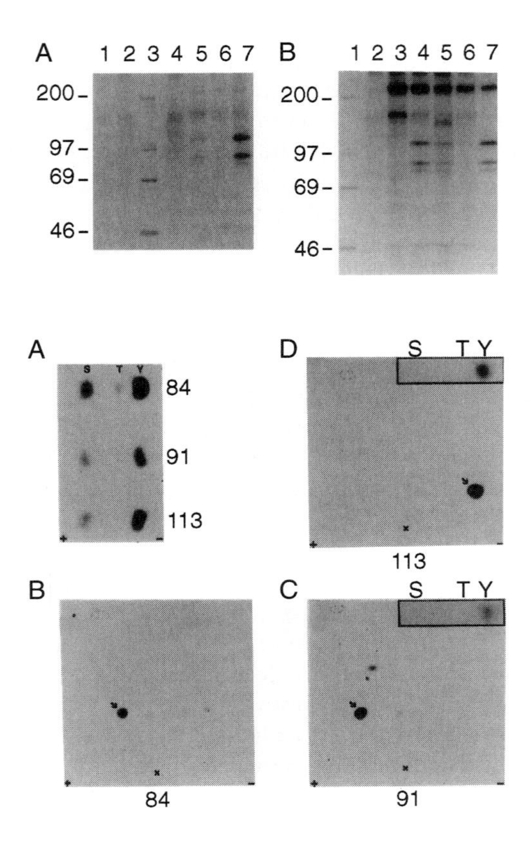

Fig. 15 Interferon α-dependent phosphorylation (*upper*), and phosphoamino acid and phosphopeptide analysis (*lower*) of 113-, 91-, and 84-kDa proteins. (*Upper*) HeLa cells (1×10^7) were labeled with phosphate-free DME (supplemented with dialyzed serum) containing ortho[^{32}P]phosphate (400 μCi / ml) for 1 hr. After various treatments, proteins from cell extracts were immunoprecipitated with anti-113 or preimmune serum. Phosphatase inhibitors (50 m*M* NaF, 30 m*M* pyrophosphate, and 0.1 m*M* Na_3VO_4) were added during extract preparation. (A) Interferon α-dependent phosphorylation of ISGF3 proteins. Proteins were immunoprecipitated with preimmune serum (lanes 1 and 2) or anti-113 (lanes 4 to 7). Interferon treatments were: lanes 1 and 6, interferon γ alone (16 h); lane 5, interferon α alone (7 min); lanes 2 and 7, interferon γ followed by interferon α; lane 4, proteins from untreated cells; and lane 3, protein molecular weight markers. (B) Inhibition by staurosporine of phosphorylation of ISGF3α proteins. Lane 1, protein molecular weight markers. Proteins were precipitated with preimmune serum (lane 2) or anti-113 (lanes 3 to 7). Cells were left untreated (lane 3) or treated with interferon γ for 16 hr and interferon α for 7 min (lanes 4 to 7). Cells were treated with the protein kinase inhibitors 9 m*M* 2-aminopurine (lane 5), 500 μ*M* staurosporine (Biomol) (lane 6), and 50 m*M* H7 (Biomol) (lane 7) for 20 min. (*Lower*) (A) ^{32}P-labeled, immunoprecipitated 84-, 91-, and 113-kDa protein bands obtained as in upper panel were eluted, acid hydrolyzed, and analyzed for the presence of phosphoserine (S), phosphothreonine (T), and phosphotyrosine (Y). (B to D) Two-dimensional phosphopeptide analysis of ^{32}P-labeled immunoprecipitated (B) 84-, (C) 91-, and (D) 113-kDa proteins digested with thermolysin. Phosphopeptide analysis was then performed. The polarity of electrophoresis was as indicated; "X" marks the origin, and the dashed oval shows the migration of bromophenol red marker. The predominant phosphopeptide is indicated by an arrow, and phosphoamino acid determinations for these recovered proteins are shown in the insets. Reprinted with permission from Schindler, C., K. Shuai, V. R. Prezioso, and J. E. Darnell, Jr. 1992. Interferon-dependent tyrosine phosphorylation of a latent cytoplasmic transcription factor. *Science* **257**:809–813. (371). Copyright 1992 American Association for the Advancement of Science.

suspected as a result of an amino acid sequence pattern needs to be experimentally verified.

Since the work just described was published, novel, powerful analytical techniques and experimental approaches have been developed that represent significant progress. Some of these techniques are described later. It should be pointed out, however, that because of the frequently low stoichiometry of induced modifications *in vivo,* and the inherent low abundance of many regulatory proteins, direct biochemical analysis of sites of protein modification remains a difficult task.

C. Activation-Induced Modifications: Biological Significance

Among the hundreds of types of modifications described to date, only a few have been shown to be reversible or to be of regulatory significance in biological processes. The best-studied among them include protein phosphorylation; some types of protein glycosylation, in particular attachment of *N*-acetylglucosamine to serine residues in intracellular proteins; and attachment of lipid groups to proteins. The following sections are limited to discussing these modifications. A more comprehensive summary, focusing on the structures of posttranslational modifications, is presented in the chapter by Krishna and Wold elsewhere in this volume.

1. *Protein Phosphorylation*

Because of the importance of reversible protein phosphorylation in cell physiology (352,355,375) the enzymology, structure, and function of this modification has been extensively studied. The most common form of protein phosphorylation involves the formation of phosphate esters of the hydroxyamino acids serine, threonine, and tyrosine, of which serine and threonine phosphorylation are much more prevalent. The formation of phosphoramidates with arginine, histidine and lysine, and acyl phosphates with aspartic acid and glutamic acid are considered less frequent and their respective functions are less well understood. Since at least some of these modifications (e.g., phosphohistidine) are chemically labile, it has been proposed that they are more frequent than generally assumed (376). Methods for their analysis have been described (377).

Two counteracting enzyme systems, the protein kinases and protein phosphatases, modulate the level of protein phosphorylation. Their structures, specificity, and regulation have been extensively studied and reviewed (352,355,375). It is assumed that there are hundreds of protein kinases with different activities and regulatory characteristics and possibly a similar number of protein phosphatases (355). Although a more detailed description of protein phosphorylation systems is beyond the scope of this chapter, it should be apparent that knowledge of the phosphorylation state of proteins is important for the understanding of many regulated biological systems.

The functional consequences of protein phosphorylation are dramatic and diverse and provide fascinating insights into how biological function is controlled. Glycogen phosphorylase, one of the best-studied allosteric enzymes, illustrates at atomic resolution how a single phosphorylation event regulates enzyme activity (378). Analysis of the protein tyrosine kinase p56lck has demonstrated that protein phosphorylation at a single site, Ser-59, possibly by mitogen-activated protein (MAP) kinase (379), can alter the substrate specificity of the enzyme (380). As discussed in the ISGF3 case study in this chapter, protein tyrosine phosphorylation can induce the formation of functional enzyme complexes through the association of a tyrosine phosphorylated peptide sequence with a complementary SH2 domain. Such associations by themselves can be sufficient for the activation of complexed enzymes (25). Alternatively, tyrosine phosphate-induced protein complexes have been shown to provide a scaffold required for stabilizing functionally interacting proteins in close proximity (381,382). Protein phosphorylation, therefore, controls biological function by altering the catalytic activity and substrate specificity of enzymes and by stabilizing functional enzyme complexes.

2. *Protein Glycosylation*

Protein glycosylation is the most complex posttranslational modification. The number of different components, some of which may be themselves modified, the different anomeric forms (α or β), chain length, and position of linkages and branching points translates into an enormous number of possible structures that can occupy a single site of glycosylation (for a recent review of protein glycosylation, see Lis and Sharon (383)). Multiple glycosylation sites in a single protein further compound the structural diversity of glycoproteins. For example, analysis of the carbohydrates released from the hamster scrapie prion protein (PrP 27-30), with a mass of 27–30 kDa and two *N*-linked glycosylation sites, revealed the presence of variable numbers of branched carbohydrate structures. It was estimated that more than 400 different glycoforms of PrP 27-30 could exist (384).

In a detailed study of recombinant human erythropoietin, Rush *et al.* (385) revealed extensive heterogeneity associated with the *N*-linked and *O*-linked carbohydrate side chains on this Chinese hamster ovary (CHO) cell-derived protein (see Fig. 16). In addition to the known variables, sialic acid, lactosamine, and *N*-linked branching structures, heterogeneity of *O*-acetylation of the sialic acids was also observed, suggesting that the availability of improved analytical techniques will uncover even more extensive variation in carbohydrate structures. Generally, protein glycosylation appears to be permanent and little is known about the functional relevance of the diversity in carbohydrate structure. The occurrence of species- and cell-specific glycosylation patterns (386) suggests, however, that the activities that perform protein glycosylation are controlled. Furthermore, there are examples that indicate that protein glycosylation is enzymatically modulated with distinct functional consequences. The trafficking of blood-borne lymphocytes to lymph nodes is mediated by adhesive receptors on lymphocytes known as homing receptors (387,388). The homing receptor, L-selectin, is a cell-surface C-type lectin that directs lymphocyte traffic to lymph nodes and contributes to lymphocyte homing to Peyer's patches and to leukocyte interactions with inflamed venules. MAdCAM-1 is a facultative ligand for L-selectin, which resides on the surface of mesenteric lymph nodes (389). The interaction of the L-selectin with MAdCAM-1 is dependent on the sialylation of MAdCAM-1 (389), as are other homing receptor interactions (e.g., (390)). Furthermore, Kearse and Hart (391) showed that glycosylation of nuclear proteins at serine side chains by *N*-acetylglucosamine (*O*-GlcNAc) is induced rapidly in DO-11.10 T-cells by either PMA/Ionomycin treatment (activation)

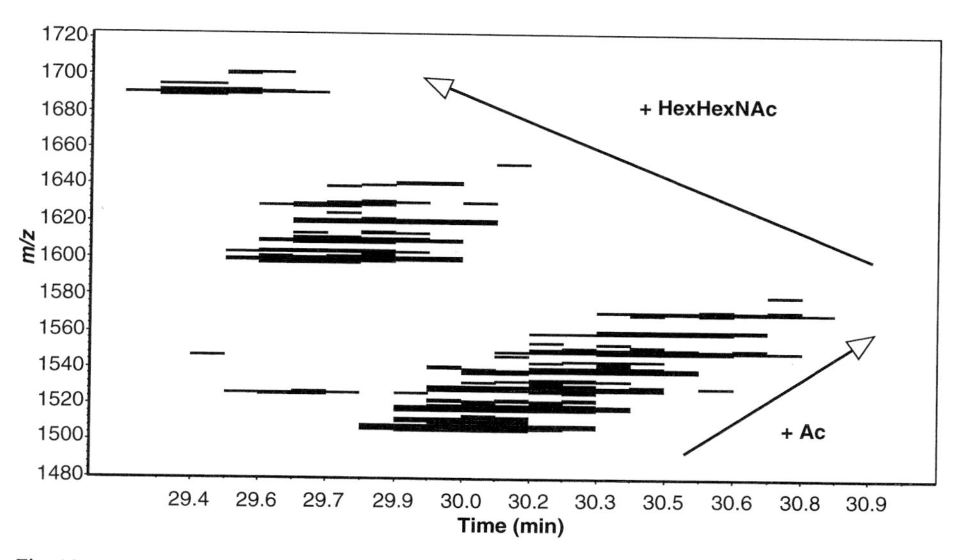

Fig. 16 Portion of an LC-MS contour plot of a tryptic digest of a rHu-EPO showing the extensive heterogeneity associated with the N-83 glycopeptide. The region of the contour plot shown corresponds to the 29–31 min RP-HPLC column eluate, and the ESI-MS spectra obtained in the m/z range of 1480–1720 (6+ charge state). The heterogeneity labeled +Ac results from increasing numbers of acetyl groups on the terminal sialic acid residues, which results in increased retention time on the RP column. Additional heterogeneity revealed in this plot is a result of lactosamine (+Hex-HexNAc) extensions on the complex side chains of the N-83 glycopeptide, which result in decreased retention on the RP column. For a detailed description of the analysis of the carbohydrate heterogeneity of rHu-EPO, see Rush *et al.* (385). The contour plot was kindly provided by Dr. M. Rohde, Amgen Inc. (Thousand Oaks).

or cycloheximide treatment (protein synthesis inhibition) and that there is a return to control levels of modification over a few hours. The reverse is true for cytoplasmic proteins in this system, which show a rapid decrease in modification followed by an increase to control levels over a few hours (391). There are a large number of proteins that have been found to possess *O*-GlcNAc modifications, and these, together with the possible role these modifications play in cellular physiology, have been reviewed (367,392).

Carbohydrate analysis is an enormous field in its own right that cannot be covered in this chapter. The aim of this section is to briefly cover some of the more common approaches used and to provide insight into some recent developments in high-sensitivity glycoprotein analysis.

3. *Protein Acylation*

Originally, permanent targeting of proteins to membranes within the cell was thought to be the predominant function of lipid attachment to proteins. More recently, structural diversity of lipid modifications (393), different modes of lipid attachment, and the discovery of reversible lipid attachment (394) suggested that this type of posttranslational modification might be of more diverse functional significance.

Structural diversity of protein-attached lipids was discovered by careful analysis of the N-terminally acylated proteins recoverin (395) and transducin

(396), the catalytic subunit of cAMP-dependent protein kinase (397), and the palmitoylated G-protein α subunits (398). Variation in the type of lipid attachment was mainly investigated on myristoylated and palmitoylated proteins. The first elucidated mode of myristate attachment was an amide linkage to N-terminal glycine (399). More recently it has been shown that myristoylation on platelet proteins predominantly occurs through thioester linkages to the side chains of serine or threonine residues (400). Palmitate initially was shown to be linked to cysteine residues through thioester linkages (399), but attachment of palmitate through ester linkages to serine or threonine residues has also been found (401). Taken together, these data suggest unanticipated complexity of protein acylation.

Diverse functional significance of protein acylation has recently been experimentally confirmed. Several studies concluded that small lipid modifications mediate protein : protein interactions (402,403). At least for the calcium-sensing protein recoverin, *N*-acylation with myristate appears to play an essential role in Ca^{2+}-dependent membrane targeting (404). These functional studies are consistent with heteronuclear NMR data, which shows that in the absence of calcium the myristoyl group is sequestered within a hydrophobic helical pocket. An increase in calcium concentration results in extrusion of the acyl group and insertion into the lipid bilayer (404). It has been proposed that the myristoyl : membrane interaction is further stabilized by basic residues close to the site of myristoylation that are involved in electrostatic interactions with acidic phospholipids in the bilayer (405). The 3D structure of the catalytic domain of cAMP-dependent kinase at atomic resolution (406) showed that the *N*-myristate group was embedded in the protein and significantly contributed to the stability of that protein. Finally, lipid attachments have been implicated in enzyme catalysis. Using a myristoyl-CoA analog, Berthiaume *et al.* (407) demonstrated reversible acylation of an active site cysteine in the mitochondrial enzyme methylmalonate-semialdehyde dehydrogenase. Since this finding was repeated in alcohol dehydrogenase from yeast and a glutamate dehydrogenase, this regulatory modification might be more prevalent than initially realized (407). Clearly, further structural characterization of lipids and their linkages to proteins is necessary to elucidate the biological relevance of these modifications.

D. Detection of Inducibly Modified Proteins

1. *Detection of Phosphorylated Proteins*

Classical methods for the detection and characterization of phosphorylated proteins have been published in two volumes of *Methods in Enzymology* (407a,b) and are not repeated here. Generally, phosphoproteins separated by gel electrophoresis are detected either by autoradiography after ^{32}P radiolabeling or by immunoblotting. Proteins are metabolically radiolabeled by equilibrating the cellular ATP pool with $^{32}PO_4$, by *in vitro* kinase reactions in crude or fractionated cell lysates using [γ-^{32}P]ATP, or by *in vitro* kinase reactions using purified kinases. Although *in vitro* kinase reactions provide higher incorporation of the radiolabel, not all of the *in vitro* generated phosphorylation events can be considered physiological. Conversely, protein phosphorylation sites detected by metabolic radiolabeling can be considered physiological. Because of the frequently low stoichiometry of phosphorylation and the relatively low specific activity of the equilibrated [^{32}P]ATP pool, such sites are more difficult to detect. The type of amino acid that is labeled can be determined by phosphoamino acid analysis. This technique employs gas- or liquid-phase hydrolysis of the peptide bonds under conditions that leave at least a fraction of the phosphate ester bonds intact. The hydrolyzates of ^{32}P-labeled phosphoproteins are added with unlabeled phosphoamino acid standards and separated by thin-layer chromatography. The unlabeled standards are detected by ninhydrin staining. The type of phosphoamino acid present in the protein is then determined by correlation of the radiolabeled species with the stained standard samples. LeGendre and Matsudaira (408) detailed both acid and base hydrolysis conditions based on previous protocols (409,410), which can be employed to distinguish phosphoserine, phosphothreonine, and phosphotyrosine. If picomole amounts of the phosphoamino acids are available, nonradioactive methods for phosphoamino acid analysis have been described. Murthy and Iqbal (411), used RP-HPLC and precolumn derivatization of an amino acid hydrolyzate with PITC to detect phosphoamino acids at a level of 5 pmol in 100-fold excess of nonphosphorylated amino acids, and Meyer and colleagues identified the type of phosphoamino acid by capillary electrophoresis of dabsylated residues (412).

The simplest approach for identifying tyrosine phosphorylated proteins is immunoblotting of gel-separated proteins with an antiserum specific for tyrosine phosphate. Several such (monoclonal) antibodies are commercially available. This provides a rapid and sensitive method for monitoring the tyrosine phosphorylation status of proteins separated by 1-D or 2-D gel electrophoresis. It should be noted, however, that lower molecular weight tyrosine phosphorylated proteins may not be detected as efficiently (413,414) as proteins with larger molecular weight. Numerous groups have attempted to produce antisera with specificity for phosphothreonine or phosphoserine. A polyclonal antiserum with specificity for phosphothreonine was reported (415), and Sigma Chemical Company (St. Louis, MO) offers monoclonal antibodies against phosphoserine and phosphothreonine that are said to react with some but not all appropriately phosphorylated proteins. The difficulties in raising these antisera are possibly due to the presence of high levels of serine and threonine phosphatases in serum, which dephosphorylated the phosphoamino acid antigens before a humoral immune response could be induced.

2. *Protein Glycosylation*

Determination of glycosylated proteins in biological samples has been achieved by metabolic labeling of cells with radioactive precursors, by direct monosaccharide analysis of released sugars, by the introduction of reporter groups by carbohydrate-specific chemical or enzymatic reactions, by carbohydrate-specific staining and by affinity detection with lectins. The common aim of these strategies is the introduction of an easily detectable group specific for glycoproteins. Typically, the labeled proteins are separated by gel electrophoresis and detected either by the gel matrix or on electroblots. For metabolic radiolabeling of glycoproteins in cultured cells, a wide variety of suitable precursors is commercially available. These reagents used either individually or in combination provide limited structural information about the carbohydrates present. For example, incorporation of radiolabeled mannose is a specific indicator for proteins that contain *N*-linked carbohydrates (132,365). Specific glycosylation inhibitors, which prevent protein glycosylation in cultured cells (416), are frequently used to complement or confirm results obtained by metabolic radiolabeling.

To detect *O*-linked *N*-acetylglucosamine-modified serine residues (*O*-GlcNAc) the specificity of galactosyltransferase to catalyze the addition of galactose to terminal GlcNAc residues can be used (366). This major form of protein glycosylation, first described by Torres and Hart (366), is found in all eukaryotes from yeast to man, has been shown to be reversible, and has been implicated as a regulator of enzyme activity similar to protein phosphorylation (for reviews, see (367,392)). In this method, proteins containing terminal GlcNAc are labeled using radiolabeled UDP-galactose and autogalactosylated galactosyltransferase, either in solution prior to gel electrophoresis (366) or following blotting on a membrane (417).

A number of methods have been developed that take advantage of the relative specificity of periodate to form aldehyde groups from the vicinal hydroxyl groups on sugars, which can then be detected with hydrazide probes containing different reporter groups (418–420). One particularly sensitive, nonradioactive method employs commercially available digoxygenin hydrazide, which can be detected using an enzyme-conjugated anti-digoxygenin Fab fragment (421). This method can be performed either on glycoproteins or glycopeptides in solution or, following electroblotting, on membranes. Detection sensitivities to the level of tens of nanograms of glycoprotein have been achieved. Based on similar principles, the periodic acid–Schiff stain has been used for many years to detect glycoproteins after SDS–PAGE. More recently, the method has been modified to detect glycoproteins in nondenaturing PA gels (422) and on PVDF membranes (423).

The simplest approach for the identification of glycoproteins is based on the specificity of lectin : carbohydrate interactions. Using lectins, glycoproteins separated by gel electrophoresis have been detected in the gel and, more recently, after electroblotting on membranes. Lectins with a broad specificity are used to detect most glycoproteins, whereas lectins with a narrower specificity can be used to gain limited information on the composition and linkage of the carbohydrate side chains (383,424). In a similar, but more specific approach, Valenzano *et al.* (425) used labeled mannose 6-phosphate receptor to specifically identify mannose 6-phosphate on glycoproteins. This principle was validated in both a ligand blot assay for gel-separated proteins, and a microtiter plate-based format for glycoproteins and glycopeptides in solution (425). A variety of methods for detecting the glycoprotein-bound lectins have been applied. They include direct radiolabeling of the lectin (426), lectin biotinylation for detection with avidin–enzyme complexes (427), digoxygenin labeling for detection with enzyme conjugated anti-digoxygenin antibodies (428), or lectin–enzyme conjugates (429). The sensitivity of detection of glycoproteins by enzyme-based colorimetric development has been enhanced by about 10-fold by the introduction of chemiluminescent methods (430,431).

3. *Lipids*

The identification and characterization of lipid modifications is dominated by the use of radiolabeled precursors that either are directly incorporated into the target proteins or are metabolized into structures that in turn are incorporated into proteins (e.g., (365)).

Metabolic labeling for protein myristoylation and palmitoylation are typically accomplished using [^{3}H]myristic acid or [^{3}H]palmitic acid, respectively, as metabolic precursors (365,432). However, it has been reported that myristate can be metabolically converted into palmitate (433). To determine whether incorporated palmitate was metabolically converted to myristate or whether myristate was incorporated, inhibitors of myristoyl-CoA : protein *N*-myristoyltransferase can be used (434). A study on *in vitro* translation of pp60^{src} using iodinated fatty acids (435) showed that myristoylation was successfully performed in rabbit reticulocyte lysates.

For prenylated proteins, most notably members of the small GTP binding family of *ras*-related proteins, G proteins, and the lamins A and B (for reviews see (436,437)), the labeling is accomplished by blocking the rate-limiting enzyme of the cholesterolgenesis pathway (3-hydroxy-3-methylglutaryl-CoA reductase, HMGR) with a specific inhibitor such as lovastatin or compactin. The product of this enzyme is then added in radiolabeled form, i.e., [^{3}H]mevalonolactone to cultured cells which convert it into mevalonate. Mevalonate is then incorporated into proteins as either a farnesyl (C_{15}) or a geranylgeranyl (C_{20}) group. The gene for the mevalonate transporter, MEV, has been cloned (438), and transfection of this cDNA into cells overcomes the need for blockage of HMGR. Reticulocyte lysates also contain a prenyltransferase, and therefore at least some labelings can be performed in this readily available *in vitro* system (439,440). Radiolabeled proteins can then be detected by autoradiography of gels. Provided that specific antibodies are available, the modified protein can simply be identified by immunoblotting.

Therefore, the detection and identification of proteins carrying specific lipid modifications is generally quite straightforward as long as a number of precautions are taken. (i) The length of the labeling period should be optimized for maximal incorporation and maximal specificity. Frequently, metabolism of the radiolabeled precursor during extensive labeling periods result in the accumulation of radiolabeled products that themselves can be incorporated into proteins. (ii) The specific activity of the radiolabeled precursor pool has to be considered. Frequently, the size of the endogenous unlabeled pool of the lipid precursor of interest will substantially dilute the radiolabel so that only very low levels of incorporation can be achieved. (iii) The stability of the lipid : protein conjugate needs to be evaluated. In particular, it is important that the conditions for sample preparation and gel electrophoresis will not result in release of the label from the protein.

E. Determination of Sites of Protein Modification

1. *Protein Phosphorylation*

The methods for determining the site of protein phosphorylation have in common that the phosphoprotein is purified, preferably to homogeneity, and cleaved by specific chemical or enzymatic reactions to produce a peptide mixture in which one or a few peptides are phosphorylated (see Fig. 14). The methods differ in the strategies used to isolate the phosphopeptide(s), to determine their amino acid sequence(s), and to localize the phosphorylated residue(s) within the phosphopeptide. Peptide separations are most frequently performed by 2-D phosphopeptide mapping (electrophoresis / thin-layer chromatography) or by high-performance liquid chromatography. These protocols have been described in great detail (372,373,407a,b) and are not repeated here. Analysis of phosphopeptides until recently was essentially performed by chemical stepwise degradation. As for peptide sequence determination, chemical techniques are increasingly complemented or replaced by mass spectrometric approaches. Several groups have attempted to develop systems integrating phosphopeptide separation and analysis for the determination of the sites of phosphorylation at high sensitivity. For the generation of the peptide mixture for phosphopeptide analysis, essentially the same considerations apply as to the generation of peptides for sequencing described in Section IV.

a. Chemical Phosphospeptide Analysis. Determination of the site of peptide phosphorylation by the Edman degradation is characterized by the different chemical stabilities of aliphatic and aromatic phosphate ester bonds. Although chemical sequencing eliminates the phosphate from serine and threonine residues, phenylthiohydantoyl (PTH) phosphotyrosine is stable under the conditions used and can be isolated as such (180,441). Therefore, phosphoamino acid-specific nonradioactive and radioactive sequencing strategies were developed.

i. Nonradioactive Sequencing Strategies This work has been pioneered and reviewed by Meyer and colleagues (412). In their protocols phosphoserine is detected following β-elimination and reaction with ethanethiol to form the stable *S*-ethylcysteine. Phosphothreonine is detected using the same reaction forming β-methyl-*S*-ethylcysteine. These compounds are easily extracted from automated sequencers in the form of the corresponding PTHs and analyzed by RP-HPLC, even though some types of automated amino acid sequencers require adaptation of the extraction conditions. The method has been used successfully in a variety of studies, but is of limited sensitivity. Meyer and colleagues found that PTH phosphotyrosine, although extracted from the sequencer, was difficult to analyze by chromatography, and therefore introduced by off-line capillary electrophoresis separation for the positive identification of PTH-phosphotyrosine (412). In contrast, we found that PTH phosphotyrosine could be recovered quantitatively and as a sharp peak from RP-HPLC columns if the chromatographic conditions were optimized (180). This resulted in development of a method for the determination of phosphotyrosine residues during automated solid-phase amino acid sequencing with on-line PTH detection (180). The method is equally compatible with radiolabeled and nonradiolabeled phosphopeptides. For nonradiolabeled peptides, positive identification of PTH phosphotyrosine by on-line UV absorbance detection requires approximately 1 pmol of the compound. Using the complementary approach of analyzing the peptides left behind in the sequencer cartridge after stepwise peptide degradation, rather than the extracted derivative ("ladder sequencing"), Chait *et al.* (203,204) demonstrated localization of phosphoserine within a peptide. Interestingly, this result suggested that phosphate elimination observed by chemical sequencing occurred during the conversion reaction rather than the actual chemical stepwise peptide degradation.

ii. Radioactive Sequencing Strategies Determination of the site(s) of phosphorylation of ^{32}P-radiolabeled peptides by chemical sequencing involves the extraction of the cleavage product in each sequencing cycle and determination of the level of released radioactivity, typically by liquid scintillation or Cerenkov counting. The extracted radiolabeled compounds eliminated are inorganic phosphate for phosphoserine and phosphothreonine and PTH-phosphotyrosine for a phosphotyrosine residue. Although inorganic phosphate is not retained by RP-HPLC columns, PTH-phosphotyrosine is less polar and elutes with a retention time similar to that of PTH-glutamic acid (180). The retention time of the radiolabeled compound on the chromatography column can, therefore, discriminate between serine/threonine phosphorylation and tyrosine phosphorylation. Sites of phosphorylation in a peptide are characterized by a burst of radioactivity in a particular sequencing cycle. If more than approximately 5 pmol of the phosphopeptide is available, the sequence can be followed by PTH analysis and the burst of radioactivity can be correlated with the amino acid sequence. If, as in the ISGF3 case study described in Section V,B, much smaller amounts of phosphopeptides are available and the PTH's cannot be detected, only the position of the phosphorylated residue within the phosphopeptide, but not the surrounding amino acid sequence, can be determined. Since in this situation the amino acid sequence of the phosphopeptide cannot be directly determined, this method is error prone, and tentative assignments of sites of phosphorylation must be verified by complementary experiments (see Section V,B, or (25)).

To achieve maximum sensitivity of detection, the cleaved radiolabeled products of each sequencing cycle can be directly collected from the cartridge of the protein sequencer for subsequent Cerenkov counting (374,442). Although this method avoids the losses associated with conversion and transfer to an on-line chromatography system, it cannot distinguish between serine, threonine phosphorylation on one hand, and tyrosine phosphorylation on the other. Eliminated inorganic phosphate is most efficiently extracted from the sequencer cartridge by solvents such as methanol or methanol : water mixtures, which are more polar than the extraction solvents (*n*-butyl chloride or ethyl acetate) commonly used for sequencing. The use of polar solvents bears the danger of also extracting the phosphopeptide substrate from the sequencer. Phosphopeptide sequencing is, therefore, most successful with covalently immobilized peptides (374,443,444), although phosphopeptide sequencing on the positively charged Immobilon-N PVDF membrane using pulsed-liquid phase protocols has also been reported (445). These sequencing results showed considerable lag, possibly due to poor solubility of inorganic phosphate in the organic solvents used, and identification of phosphotyrosine was not demonstrated.

b. Mass Spectrometric Approaches to Phosphorylation Analysis. Numerous mass spectrometry-based methods for the determination of sites of peptide phosphorylation have been published. They are variations of two basic themes. The first relies on the observation that phosphate ester bonds in phosphopeptides can be dissociated either at the source or in a collision cell of an ESI-MS, or during post-source decay for MALDI-MS, under conditions that leave most of the peptide bonds intact. Among the peptides applied to the mass spectrometer, phosphopeptides are therefore identified by characteristic reporter ions that indicate fragmentation of a phosphate ester bond. The second uses the mass increment of 80 Da that is added to the mass of a peptide by each phosphate group. Among the peptides applied to the mass spectrometer, phosphopeptides are therefore identified by a difference in experimentally determined peptide mass and the peptide mass predicted from the amino acid sequence. Although both of these approaches do not strictly determine the site of protein phosphorylation, they locate the phosphorylated residues within a specific peptide fragment. If that peptide only contains one residue that can be phosphorylated, the assignment can be considered conclusive. If there are several hydroxylamino acids in the peptide, enzymatic or chemical cleavage between the putative phosphorylation sites followed by mass analysis of the fragments or sequencing of the phosphopeptide can result in conclusive assignment of the phosphorylated residue.

Through the use of high collisional excitation potentials during analysis of a peptide mixture by negative-ion ESI-LC-MS, Huddleston *et al.* (446,447) have demonstrated the ability to identify phosphopeptides in peptide mixtures based on dissociation at the ion source. The method is compatible with serine, threonine, or tyrosine phosphorylated peptides. This is, therefore, a more universal approach than a previous tandem mass spectrometric method based on neutral-loss scanning, which was limited to the analysis of peptides phosphorylated at serine and threonine residues (448). In the new method phosphopeptides are detected by characteristic low-mass signature ions. Phosphopeptide dissociation is achieved by stepping (or scanning) the orifice potential of the ES-ion source from a high potential (up to -350 V) over the low-mass region of the mass spectrum to a lower (normal) potential over the higher mass region of the spectrum during each mass scan. Diagnostic low-mass fragment ions of m/z 97, 79, and 63 for phosphoserine and phosphothreonine, and m/z 79 and 63 for phosphotyrosine, as well as the mass of the intact precursor ion, conclusively identify phosphopeptides. The method is a very useful adjunct to LC-MS peptide mapping, as it provides the ability to detect the presence and the mass of phosphopeptides in a single RP-HPLC run without any additional manipulations or loss of sample. Huddleston *et al.* (446) suggest that the practical sensitivity limit of the method is 50 pmol if a 320-μm i.d. C_{18} packed fused-silica column is used. Independently, Ding *et al.* (449) demonstrated that stepping of the orifice potential from high (-250 V) to low potential generates the diagnostic 79 m/z ion. They claim that the detection of this reporter ion is sufficient for phosphopeptide identification. It should be noted, however, that MS analysis of phosphorylated peptides is often complicated by the neutral loss of H_3PO_4 during collisional activation and that the reporter ions are detected in the low-mass region of the mass spectrum, which frequently suffers from high levels of background signals. Furthermore, the sensitivity of these methods currently is not sufficient for the analysis of sites of phosphorylation induced *in vivo*.

MALDI-MS has also been used to determine sites of phosphorylation in peptide digests of proteins of known sequence. Phosphopeptides were identified by a mass increment of 80 Da per phosphate group to the expected mass of an unmodified peptide. The high sensitivity of MALDI-MS makes this technique particularly attractive. Yip and Hutchens (450,451) analyzed synthetic peptides phosphorylated by *in vitro* kinase reactions as well as an unfractionated digest of human β-casein. As expected, the sensitivity of MALDI-MS was at the subpicomole level. Zhang *et al.* (282) demonstrated the ability of MALDI-MS to identify sites of phosphorylation on two isoforms of synapsin separated by 2-D (NEPHGE/SDS) PAGE. The separated proteins were electroblotted to Immobilon-CD and cleaved on the membrane. Interestingly, the present authors observed in that study that for basic proteins such as synapsin, differentially phosphorylated forms could not be resolved by gel electrophoresis (NEPHGE). It was, however, possible to distinguish differentially phosphorylated species in the same protein spot by MALDI-MS. Using CNBr and endoprotease Lys-C cleavage of consecutive slices of a single protein spot followed by MALDI-MS analysis of the resulting peptide mixture, a profile of unphosphorylated, single phosphorylated, and doubly phosphorylated synapsins was established across the spot (282). A similar approach had been used by Wang *et al.* (452), who employed MALDI-MS to analyze the phosphorylation sites of phorbol 12-myristate 13-acetate induced phosphorylation of Op18 (or stathmin, p19) that was purified from cell lysates by prerparative 2-D PAGE. PSD-MALDI-MS has also been demonstrated to be useful for the analysis of phosphorylated peptides in the negative ion mode. Talbo and Mann (239) showed that the site of modification was able to be determined for threonine-, serine-, and tyrosine-phosphorylated peptides.

The obvious advantages of MALDI-MS are the exquisite sensitivity of the method and the relative tolerance to many buffers, solvents, and additives that are preferred by biochemists. It was noted, however, that MALDI-MS cannot be used for quantitative analysis of phosphorylation reactions because of an increase in specific signal size of the dephosphorylated peptide compared to a phosphopeptide of the same sequence (453).

The recently described matrix-assisted laser desorption quadrupole ion trap mass spectrometer with external ion source (454–456) potentially combines the high sensitivity of MALDI-MS with the ability to perform collision-induced fragmentation experiments for detailed structural analysis of peptides. Early results obtained with this type of instrument demonstrated efficient collision-induced dissociation of HPO_3 ($m/z = 98$) from phosphopeptides in the ion trap (457,458), suggesting that this might develop into a tool with the sensitivity and selectivity required for the direct structural analysis of *in vivo* phosphorylation sites in low-abundance regulatory proteins.

c. Integrated Approaches to Phosphopeptide Analysis. To enhance the sensitivity and reliability of phosphopeptide analysis and to attempt to interface the analytical techniques with the biological source of the phosphopepteins, several groups have reported integrated strategies that typically combine chemical, enzymatic, and mass spectrometric elements. Comparative analysis of peptide masses before and after chemical or enzymatic dephosphorylation and the use of affinity-based methods to selectively enrich for phosphorylated peptides prior to mass analysis represent two successful implementations of integrated approaches.

Amankwa *et al.* (459) employed a "phosphatase enzyme microreactor" (ER) connected on-line to a capillary electrophoresis (CE)-MS or capillary LC-MS system for phosphopeptide analysis. Phosphopeptides were detected in peptide mixtures by comparative examination of the total ion current of samples exposed to the phosphatase reactor and control samples that were not exposed to the phosphatase. The instrument setup for the ER/LC/MS approach is schematically shown in Fig. 17. Enzymatic dephosphorylation of the phosphopeptides resulted in a decrease in peptide mass by 80 Da per cleaved phosphate ester bond and an increased retention time on RP-HPLC or decreased electrophoretic mobility on CE. This method therefore provides two means, peptide mass and residence time in the separation system, for the identification of the phosphorylated peptides within a complex mixture. Furthermore, it determines the number of phosphorylated residues in a peptide and can distinguish between tyrosine and serine/threonine phosphorylation if phosphatases with the respective specificities are used (459). Using on-column UV absorbance detection the authors showed that the ER-CE implementation of the method allowed the analysis of phosphopeptides below the 50 fmol level. Because of limitations in the ESI-MS interface, the practical sensitivity limit achieved was in the low picomole range: 2.6 pmol of tyrosine phosphorylated peptide within a 13.6 pmol tryptic digest of carboxymethylated α-lactalbumin could be positively

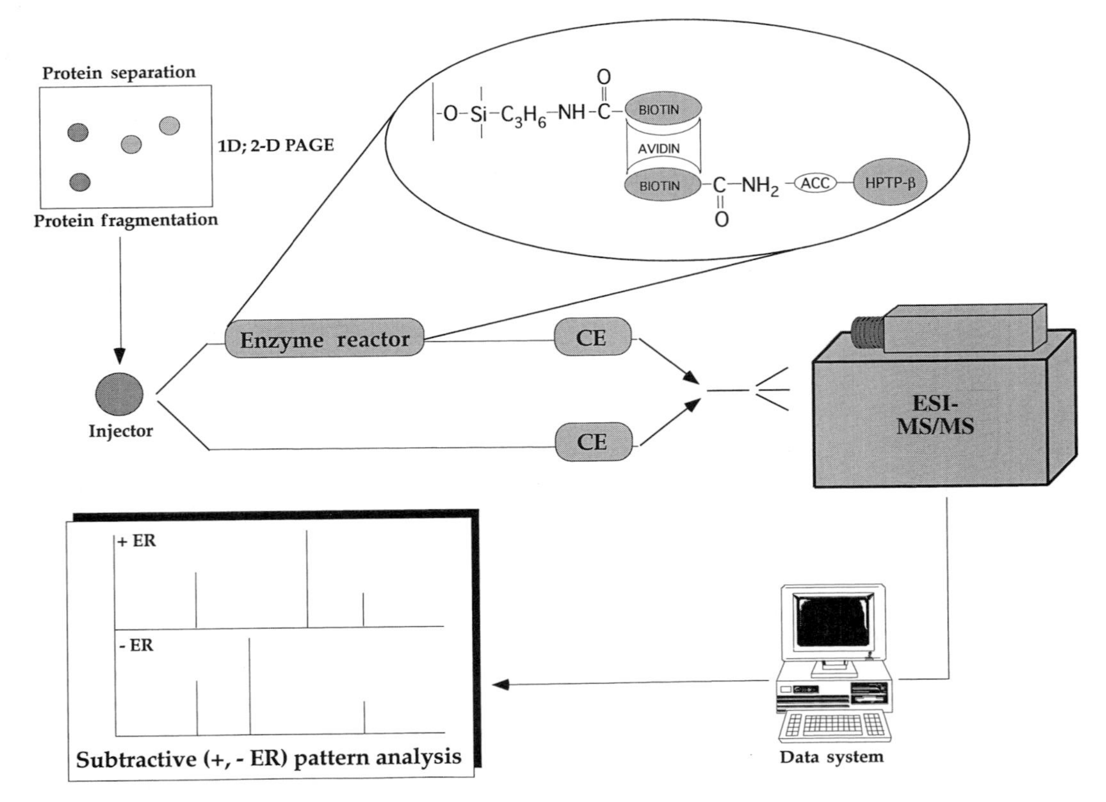

Fig. 17 Schematic representation of an integrated system for the determination of posttranslational modifications. Proteolytic digests of proteins separated by 2-DE are exposed (or not exposed for control samples) to a microreactor containing an immobilized enzyme specific for the removal of one type of posttranslational modification. Peptide samples are then comparatively analyzed by CE-ESI-MS. The modified peptide is identified in the peptide mixture by a change in electrophoretic mobility and a corresponding shift in molecular mass. The insert represents one effective way to immobilize the enzymes in the enzyme reactor [see (459)].

identified. With the development of microsprayer technology for ESI-MS (see Section IV and (222–229)), it is anticipated that the whole sensitivity potential of the method will be realized.

To construct the ER, the surface of a fused-silica capillary column (250 μm i.d.) was first chemically modified by covalent attachment of biotin according to Amankwa and Kuhr (460), followed by coupling of avidin to the biotin. This column was then ready to accept any biotinylated enzyme, allowing a range of specificities to easily be incorporated into the basic design. In the work just described, the authors flushed the column with a lysate prepared from *E. coli* expressing a metabolically biotinylated broad-specificity tyrosine phosphatase (human protein tyrosine phosphatase β). The expression system used (PinPoint by Promega, Madison, WI) results in a fusion protein consisting of a quantitatively, metabolically biotinylated 12.5-kDa domain of the transcarboxylase complex from *Propionibacterium shermanii* and the protein of interest (461), thus avoiding the need for chemical enzyme biotinylation, which frequently eliminates enzymatic activity.

Resing *et al.* (462) applied a conceptually similar strategy to the analysis of serine and threonine phosphorylation within an internal repeat of rat profilaggrin. They incubated peptide mixtures with $Ba(OH)_2$ causing a β-elimination of H_3PO_4 and generation of dehydrated serine or threonine at the site of phosphorylation, thus making the interpretation of multiple sites of phosphorylation considerably easier. Peptides modified in this manner were stable for at least a few hours. From the peptides analyzed it appeared that the treatment resulted in some conversion of Gln to Glu (462) and that collision-induced fragmentation in a tandem MS was more pronounced at both the N-terminal and C-terminal peptide bonds of the dehydroamino acids, thereby assisting the search for phosphopeptides in peptide mixtures (462).

A combination of enzymatic and MS techniques for the characterization of phosphopeptides was also implemented for MALDI-MS. Two groups demonstrated *in situ* (on probe) dephosphorylation reactions using calf alkaline intestinal phosphatase prior to analysis of the peptide mixture by MALDI-MS (453,463). Both groups conducted the reactions on the probe, which was placed in a humidified chamber for time periods between 1 min and 4 hr (453,463). The method is attractive because the same peptide sample can be repeatedly analyzed and the dephosphorylation reaction can be monitored as a function of time.

Rather than detecting phosphopeptides in peptide mixtures by phosphopeptide specific chemical or enzymatic reactions, Nuwaysir and Stults (464) decided to specifically enrich peptide mixtures for phosphopeptides prior to analysis by ESI-MS. They used an immobilized Fe^{3+} metal-ion affinity chromatography (IMAC) column to selectively retain phosphopeptides, taking advantage of the known affinity of phosphate for ferric ions (465). Retained phosphopeptides were subsequently eluted under basic conditions and analyzed by ESI-MS (464). A sensitivity in the 10- to 20-pmol range was reported. Interestingly, in demonstrating their respective methods Nuwaysir and Stults (464), Huddleston *et al.* (447), and Ding *et al.* (449) all examined β-casein as one of their model compounds. Using RP-HPLC with positive-ion ESI-MS separation, Nuwaysir and Stults failed to detect a quadruply phosphorylated β-casein tryptic peptide. They attributed this to low solubility of this fragment. Both

Huddleston *et al.* and Ding *et al.* (447,449), using their respective methods for negative ion ESI-MS and phosphopeptide dissociation at the ion source, were able to detect a quadruply phosphorylated Lys-C peptide, which is three residues longer than a putative tryptic peptide. Ding *et al.* (449) suggest that this observed difference is due to the use of different proteolytic enzymes.

Kassel *et al.* (466) developed a similar but more selective method for specifically enriching peptides phosphorylated at tyrosine residues. They prepared a phosphotyrosine-specific immunoaffinity column upstream of a RP-HPLC for the capture of tyrosine-phosphorylated peptides from a peptide mixture. Non-tyrosine-phosphorylated peptides passed through the resin directly onto a RP-HPLC column, from which they were eluted by organic solvent and analyzed by on-line ESI-MS. After reequilibration of the RP-HPLC column in the polar mobile phase, the immobilized phosphopeptides were acid eluted, transferred to the RP-HPLC column, and analyzed by LC-MS. The data presented were limited to the analysis of a synthetic peptide mixture at the 300 pmol per peptide level. The group did, however, present a potentially even more selective enrichment technique by immobilizing a recombinant SH2 domain for trapping of tyrosine-phosphorylated peptides containing the specific sequence motif that is recognized by the SH2 domain (466).

d. Determination of Sites of in Vivo Phosphorylation. Currently, techniques for the analysis of *in vivo* protein phosphorylation sites rely largely on 2-D phosphopeptide mapping of isolated proteins that have been phosphorylated *in vivo* following metabolic labeling of cells with radiolabeled phosphate. Because of the small quantities of such material recoverable from typical cell lysates, and the often low stoichiometry of phosphorylation observed on proteins of interest, direct biochemical determination of phosphorylation sites inducible *in vivo* is almost impossible, even with the improved techniques described previously. To develop a general strategy for the mapping of physiological sites of phosphorylation, the authors combined the adapted current techniques in column chromatography and ESI-MS. We connected microbore IMAC and narrow-bore HPLC columns in series, with on-line analysis of column eluates by ESI-MS (467). With this system we were able to determine both the sites of autophosphorylation of the T-cell specific tyrosine kinase ZAP-70 and those induced on ZAP-70 by the tyrosine kinase $p56^{lck}$ *in vitro*, by using recombinantly expressed and isolated ZAP-70 and $p56^{lck}$ and the recovery of tryptic phosphopeptides from 2-D phosphopeptide maps. By comparison of these data with tryptic phosphopeptide maps of *in vivo* phosphorylated ZAP-70, isolated from stimulated Jurkat T lymphoblasts following ^{32}P-metabolic labeling, we were able to determine the sites of tyrosine phosphorylation induced on ZAP-70 following TCR engagement (382,468). This methodological approach is generally applicable and is summarized in Fig. 18.

Since many of the important phosphoprotein components of cell signaling pathways have now been identified, the availability of expressed forms of these proteins means that comparison of *in vitro* and *in vivo* generated phosphopeptide maps and subsequent analysis by ESI-MS of the *in vitro* derived phosphopeptides should be a more rapid approach than mutagenic analysis of all potential phosphorylation sites. This approach thus represents a direct biochemical bioassay for the determination of the exact phosphorylation state of any

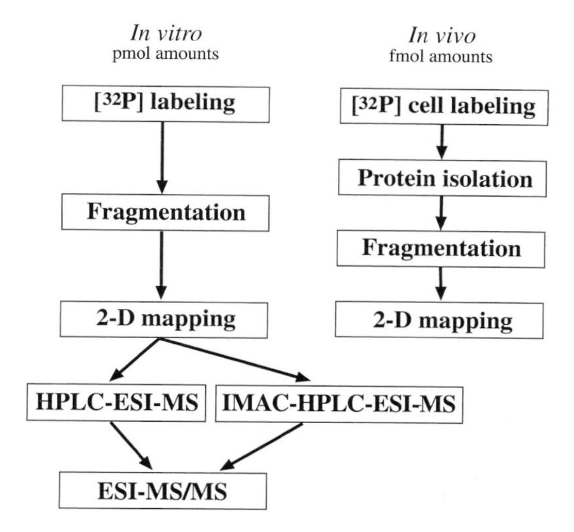

Fig. 18 Schematic representation of the experimental approach, utilizing both *in vivo* and *in vitro* derived material for the determination of protein phosphorylation sites. Because of the low-femtomole amounts of phosphoproteins of interest available from *in vivo* sources, the *in vivo* derived samples are correlated at the level of 2-D phosphopeptide mapping, whereas larger (picomole) amounts of *in vitro* prepared samples are analyzed via ESI-MS.

phosphoprotein of interest. Indeed, by its very nature, direct biochemical determination of phosphorylation sites is more reliable than indirect analysis via mutagenesis. The availability of rapid, sensitive, and direct approaches for the determination of inducible phosphorylation sites on proteins of interest involved in signal transduction pathways should thus greatly facilitate investigation of the specific roles played by these modifications via site-directed mutagenesis and subsequent expression of these mutant proteins in appropriate transgenic or cell culture systems.

2. *Glycoproteins*

The sensitivity of current methods for the complete structural analysis of the carbohydrate moiety of glycoproteins is incompatible with the amounts of glycoprotein that can typically isolated from cellular sources. In an extensive study on the glycoprotein erythropoietin, Rush *et al.* (385) demonstrated convincingly that even in cases in which an abundant amount of the purified glycoprotein is available, detailed structural analysis of the carbohydrate chains and localization of carbohydrate attachment sites is a major research project. Therefore, most projects involving the analysis of glycoproteins are limited to relatively crude structural analyses or the determination of the glycosylated amino acid residues in the polypeptide chain.

a. Structural Analysis of Gel-Separated Glycoproteins. Description of the technology for complete structural analysis of carbohydrates is beyond the scope of this chapter. The following is therefore limited to methods that are suitable for a relatively crude structural analysis of glycoproteins and for the determination of the average monosaccharide composition of complex, protein-linked carbohydrates.

Two alternative strategies have been described for obtaining limited structural information on protein-attached carbohydrates. They have in common that they use gel-separated proteins as substrates and are, therefore, compatible with relatively small amounts of glycoprotein. The first method is based on chemical and/or enzymatic reactions that are performed in sequence on proteins electroblotted onto inert membranes such as PVDF. Periodate oxidation and reduction (426), degradation by glycosidases (469,470), and sequential chemical reactions including a complete "Smith degradation" (427) are examples of reactions that have been successfully performed on electroblotted proteins. Frequently, if total cell lysates are separated by gel electrophoresis, the amount of glycoprotein in a gel band is not sufficient for obtaining conclusive results. This limitation has been overcome by enriching samples for glycoproteins or glycopeptides prior to gel electrophoresis by lectin affinity chromatography (421,471). In an interesting application of this procedure, Hayes *et al.* (472) demonstrated the selective absorption of *O*-GlcNAc containing glycopeptides by the lectin RCA I prior to RP-HPLC.

The second method is based on fragmentation of the electroblotted glycoprotein on the membrane and structural analysis of the recovered glycopeptide(s). The work of Shao and Chin (473) describes an elegant and sensitive implementation of this approach. Peptides in small aliquots of fractions collected from RP-HPLC were biotinylated at the primary amino groups and immobilized in microtiter plates as peptide/glycopeptide–streptavidin complexes. Each well in the microtiter plate was then incubated with specific enzyme-conjugated lectins, and fractions containing glycopeptides were detected colorimetrically at sensitivities below 1 pmol. Additional structural information could be obtained if the immobilized glycopeptides were sequentially probed with lectins of different specificities or if the carbohydrate chains were subjected to digestion with specific glycosidases (473).

The monosaccharide composition of complex carbohydrates provides alternative or complementary structural information to the methods described above. To determine the sugar composition, carbohydrate side chains are released from the glycoprotein, isolated, and hydrolyzed. The monosaccharide composition of the resulting mixture can then be determined by a variety of established methods based on HPAEC-PAD (high pH anion-exchange chromatography with pulsed-amperometric detection) (474,475), HPLC (476), or high resolution electrophoretic separation of fluorescently labeled sugars (fluorophore-assisted carbohydrate electrophoresis, FACE; for an extensive review, see (477)). A successful strategy for carbohydrate analysis of electroblotted glycoproteins described by Weitzhandler *et al.* (474) is shown in Fig. 19. To release the carbohydrate side chains from glycoproteins on PVDF membranes, in polyacrylamide gels, or from proteins in solution, chemical or enzymatic methods have been used. Kawashima *et al.* (478) used hydrazinolysis to remove *N*-linked carbohydrates from proteins in gel bands and on membranes. The released products were isolated by gel permeation chromatography. The procedure had an overall yield on the order of 15 to 25%. Alternatively, Weitzhandler *et al.* (474,475) used endoglycosidases (H, F_2, or endo-β-galactosidase) to remove the intact carbohydrate chains from gel-separated glycoproteins and HPAEC-PAD for the isolation of the released products.

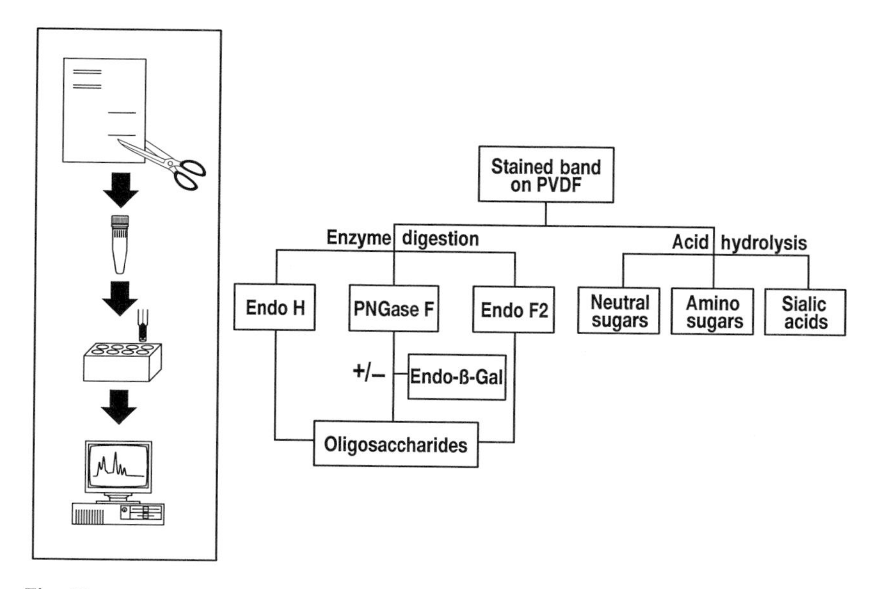

Fig. 19 Carbohydrate analysis of glycoproteins following SDS–PAGE, blotting to PVDF, and protein staining. The left panel shows the general scheme of glycoprotein separation by SDS–PAGE, blotting to PVDF, excision of the band of interest, treatment, and analysis by HPAEC-PAD. As shown in the right-hand panel, the membrane is subjected to a range of either chemical or enzymatic treatments to release either sugars or oligosaccharides for HPAEC-PAD analysis. The figure was courtesy of Dr. M. Weitzhandler and P. Henson, Dionex Corporation, and adapted from Weitzhandler, M., D. Kadlecek, N. Avdalovic, J. G. Forte, D. Chow, and R. R. Townsend. (1993). Monosaccharide and oligosaccharide analysis of proteins transferred to polyvinylidene fluoride membranes after sodium dodecyl sulfate-polyacrylamide gel electrophoresis. *J. Biol. Chem.* **268:**5121–5130; and Weitzhandler, M., D. Kadlecek, N. Avdalovic, and R. R. Townsend. (1993). Monosaccharide and oligosaccharide analysis of recombinant erythropoietin electro-transferred onto polyvinylidene fluoride membrane. *In* "Techniques in Protein Chemistry IV," pp. 135–142. Academic Press, San Diego.

b. Determination of Site of Protein Glycosylation. In addition to the composition and structure, the site(s) of attachment and number of carbohydrate groups are of particular interest. Whereas attachment sites for *N*-linked carbohydrates conform to the consensus sequence Asn-Xxx-Ser/Thr (Xxx = any amino acid) (383), the sequence motif around *O*-linked carbohydrates is much less defined (479,480). It is important, however, that even the attachment of *N*-linked carbohydrates to a particular site is experimentally verified. Traditionally, the glycopeptides were sequenced by chemical stepwise degradation. More recently, MS-based techniques have been developed that simultaneously determine the site of carbohydrate attachment and provide structural information. For an excellent review of MS analysis of carbohydrates, see Reinhold *et al.* (481).

Using the standard gas-liquid phase or pulsed-liquid phase chemical sequencing protocols, glycosylated Asn or Ser residues are not conclusively identified because under the extraction conditions used the PTH derivatives of glycosylated amino acids are not recovered [e.g., see (482)]. One possible interpretation for the lack of a specific signal in a sequencing cycle is, therefore, protein glycosylation. Since there are numerous other reasons for the absence of a specific signal in a peptide sequence, it is essential that the presence of a glycosylated amino acid be positively identified even if the blank cycle conforms the Asn-Xxx-Ser/Thr sequence motif. Resequencing of an enzymatically or chemically deglycosylated aliquot of the putative glycopeptide is one, albeit tedious, method of confirmation. Peptide sequencing using solid-phase protocols provides a more direct method to confirm the site of peptide glycosylation. Covalent glycopeptide attachment for solid-phase sequencing the more stringent extraction conditions required for quantitative extraction of PTH-(glyco) amino acids. Solid-phase sequencing through a glycosylation site was implemented in a simple and low-cost manual method to determine the site of attachment of *O*-GlcNAc (483). The *O*-GlcNAc group was radiolabeled with [^{14}C]galactose, and the presence of the modification was confirmed by liquid scintillation counting of the extracted product of each sequencing cycle. Automation of this approach was demonstrated in a series of publications (484–486) in which extraction of anilinothiazilinone (ATZ) (glyco)amino acids with anhydrous TFA and separation of both PTH-glyco- and PTH-amino acids by RP-HPLC was achieved. The same group also demonstrated that extracted PTH (glyco)amino acids remain intact and can be further structurally characterized by either HPAEC-PAD or ESI-MS (486,487). It was noted, however, that large oligosaccharide chains degraded slowly during the Edman degradation. This increased the observed heterogeneity of recovered PTHs and thereby made interpretation of the data more difficult (487). Furthermore, detection sensitivity of PTHs by ESI-MS is limited and varies with the composition of the amino acid side chain. The recent introduction of sequencing chemistries designed for detection of the resulting amino acid derivatives by ESI-MS (195,196,198) appears better suited for the detection at high sensitivity and structural characterization of (glyco)PTH-amino acid derivatives, at least for simple carbohydrate structures. Stability will still be an issue for the larger side-chain structures, such as *N*-linked moieties. This was demonstrated by Tull *et al.*, who used an LC-ESI-MS/MS system connected on-line to a solid-phase protein sequencer to structurally characterize a glycosylated glutamate in the active site of a cellobiosidase (199).

MALDI-MS and ESI-MS methods have been developed for the analysis of glycoproteins and glycopeptides by mass spectrometry. In addition to the high sensitivity, mass spectral analysis of glycopeptides offers the significant advantage that the site of glycosylation and limited structural information are obtained in the same operation. The approaches applied to glycopeptides are conceptually similar to the ones applied to the analysis of phosphopeptides (see earlier discussion).

Carr *et al.* (488) and Huddleston *et al.* (489) described ESI-MS methods for analysis of both *N*- and *O*-linked carbohydrates. Huddleston *et al.* (489) used collisional fragmentation at the ion source of a single quadrupole mass spectrometer to dissociate positive glycopeptide ions. The orifice voltage was alternated between a high (dissociating) voltage and the lower, nondissociating voltage normally used for the ionization of peptides. While at the high orifice potential, the mass spectrometer was scanning a low mass range (m/z 150–500) to detect characteristic fragments ions (sugar oxonium ion fragments) such as m/z 204 (HexNAc$^+$, *N*-acetylhexosamine), m/z 366 (Hex-HexNAc$^+$), m/z 163 (Hex$^+$, hexosamine), and m/z 292 (NeuAc$^+$, *N*-acetylneuraminic acid or sialic acid). When the orifice potential was at the normal value, the mass spectrometer scanned a higher mass range (m/z 501–2100) for determining the masses of intact glyco(peptides). In the same manner as for phosphopeptide analysis, this method allows simultaneous determination of the modified peptide mass and confirmation of the presence of a particular modification. If sufficient material is available and the experiment is performed on a triple quadrupole instrument, MS/MS analysis can determine the sequence of the peptide and the carbohydrate (489). This method should be applicable to gel-separated proteins, as it has been successfully used at the 25-pmol level (488,489).

The high sensitivity and simplicity of use make MALDI-MS a desirable method for the analysis of glycopeptides and carbohydrates. In a study of model oligosaccharides, Harvey (490) showed that these structures can be analyzed quantitatively over a range of 100 fmol to about 30 pmol using 2,5-dihydroxybenzoic acid as the matrix. Huberty *et al.* (236) demonstrated glycopeptide analysis by MALDI-MS at the 150-fmol level. Operation of the instruments in the reflectron mode showed that NeuAc residues were lost due to metastable decay (post-source decay) and that under specific experimental conditions significant fragment ions representing loss of Hex and HexNAc residues could be observed (236). Generally, however, the structural information that can be obtained by MALDI-MS of glycopeptides is limited. Results from Huberty *et al.* suggest that treatment of the glycopeptides with specific glycosidases is a more practical approach to characterize the side-chain structure (236). This is particularly promising since MALDI-MS is quite tolerant to salt contamination at the probe, and analyses could be performed without removal of the digest buffers (236). The combination of enzymatic reactions and MALDI-MS for the analysis of glycopeptides was further refined by Sutton *et al.* (491). They adapted the "reagent-array" method of Edge *et al.* (492) by miniaturizing the reaction volume to 1 μl and using MALDI-MS for detection of the reaction products rather than the usual Bio-Gel P4 (BioRad, Richmond, CA) column chromatography. The reagent-array method is based on eight different exo- or endoglycosidases that, used sequentially or in combination, generate an array of products that permit the interpretation of the carbohydrate sequence based on the mass differences of specific substrate and product pairs. The method is currently limited by the long reaction times, typically 18–24 hr, which are incompatible with enzyme reactions on-probe, and by difficulties in quantitation of the observed species. Extensive sialic acid modification inhibited ionization in MALDI-MS, as demonstrated by an increase in signal intensity following sialidase treatment (491).

In a detailed study of recombinant human erythropoietin using ESI-MS of RP-HPLC derived fractions and LC-MS, Rush *et al.* (385) revealed the extensive heterogeneity that can be associated with the *N*-linked and *O*-linked carbohydrate side chains on a Chinese hamster ovary cell-derived protein. In addition to the known structural variables such as sialic acid, lactosamine, and *N*-linked branching structures, heterogeneity of *O*-acetylation of the sialic acids was also observed (see Fig. 16). Masses consistent with no, mono-, or di-*O*-acetyl forms were observed. The permutations and combinations of these structures made the interpretation of mass spectra and the carbohydrate structures extremely complex. In the context of this discussion, the most important results of the study relate to sample handling. Rush *et al.* showed convincingly that variation of the pH maintained during carbohydrate analysis affected the sugar structure (385). Specifically, they reported that pH values outside the pH 4 to 6 range resulted in loss of sialic acid and/or acetyl groups on the sialic acid, and that exposure to relatively mild base conditions (0.2*N* NaOH for 2 hr at 0°C) was sufficient to remove the *O*-acetyl groups from the sialic acid (385).

3. *Lipid Localization*

Although the detection of proteins thought to carry a specific lipid modification can be relatively simple (see Section V,D,3), confirmation of the nature of the modification requires removal, extraction, and analysis of the lipid group. The following sections briefly cover traditional approaches used to identify and characterize specific lipid modifications and newer methodologies that have been applied to these analyses. For a more in-depth coverage of techniques related to analysis of covalently bound lipids, the reader is referred to an issue of *Methods: A Companion to Methods in Enzymology* (492a,b) and for a review of the wide variety of rare lipid linkages to (393).

*a. **Prenylation.*** To verify the presence and structure of isoprenyl groups on a protein, the lipid is released, extracted, and analyzed. Release of the thioether-linked prenyl group is achieved by treatment with either methyl iodide (493) or Raney nickel, and the lipid group is extracted with pentane or chloroform : methanol (494). Extracted isoprenoid products have been separated by either RP-HPLC (after methyl iodide release) or radiometric gas chromatography (GC)-MS analysis (after Raney nickel release), and the spectra compared with spectra obtained with authentic standards to distinguish between farnesyl or geranylgeranyl groups. More conclusive structural proof of extracted isoprenoid groups was achieved by mass spectrometric analysis of the separated products, that is, by GC-MS in either full-spectrum ionization mode or in selected ion monitoring mode for qualitative analysis (494).

These methods identify the type of modification at a relatively high sensitivity and are compatible with proteins separated by gel electrophoresis (495). However, they do not yield any information on the site of modification. To determine the site, the prenylated peptide needs to be isolated and identified. Any one of the methods described in the previous sections for peptide mapping can be used for the generation of the prenylated peptide. Further analysis is usually performed by amino acid sequence analysis. During chemical amino acid sequencing the site of prenylation is usually found to be a blank cycle. More definitive information can be obtained through tandem MS experiments on isolated modified peptides.

*b. **Carboxymethylation.*** It has been suggested that most, if not all prenylated proteins are coincidentally carboxymethylated at their C-terminal prenylated cysteine and that this modification is reversible (437). Analysis of carboxymethylation is accomplished either through the use of membrane preparations containing the carboxymethyltransferase activity for *in vitro* labeling with the methyl group donor, *S*-adenosyl[*methyl*-^{3}H]methionine, or by metabolic labeling using the precursor [*methyl*-^{3}H]methionine (496). However, unlike the metabolic labeling of prenyl groups described earlier, it is highly unlikely that the label will be transferred to other precursors and not solely reside on the C-terminal prenylated cysteine. Therefore, this analysis often only determines whether or not the protein could be carboxylmethylated. The simplest approach to confirm carboxylmethylation is to subject the putative carboxylmethylated protein to base treatment. The protein sample is placed into an uncapped microfuge tube containing NaOH and put into a plastic scintillation fluid, capped, and reacted for 12 to 24 hr at 37°C, at which time the amount of [^{3}H]methanol is determined by liquid scintillation counting (496,497). Xie *et al.* (496) also described methods for confirming the C-terminal nature of the prenylation and carboxylmethylation, which consisted of digestion of the protein with multiple proteases followed by performic acid oxidation to remove the prenyl group.

c. Myristoylation and Palmitoylation. The analysis of myristoylated and palmitylated proteins follows the same principles as that of prenylated proteins, except that there is no necessity to block an endogenous enzyme to allow efficient incorporation of the radiolabel into the target proteins. As for prenylated proteins, after release from the polypeptide, the lipids can be extracted directly from the gel or, following electroblotting from nitrocellulose (498) or PVDF (499,500), analyzed by cochromatography with authentic lipid myristate and palmitate standards in a RP-HPLC system (400,501). Cleavage conditions have been worked out that are specific for either type of myristate and palmitate attachment known. Kunz *et al.* (500) used acid methanolysis of PVDF-blotted proteins to remove amide linked myristate. Callahan *et al.* (498) used alkaline methanolysis of nitrocellulose-blotted proteins to remove thioester-linked palmitate. Sequential application of alkaline methanolysis to disrupt thioester linkages followed by an acid methanolysis to disrupt amide linkages, therefore, permits the determination of the nature of the attached lipid as well as the type of attachment from the same sample.

These methods do not provide any information on the site of lipid attachment. Chemical sequencing by the Edman degradation, the traditional method for localizing modified residues, does not yield conclusive results on myristylated and palmitoylated proteins. *N*-Terminal myristoylation renders peptides refractory to the Edman degradation (502), and palmitoylated sites will give a blank cycle (503). Therefore, it is obvious that mass spectrometry either by itself or in combination with chemical or enzymatic degradation is the method of choice for the localization of myristate and palmitate groups within polypeptides. Stults *et al.* have combined chemical methods with mass spectral analysis to investigate palmitoylation of the lung surfactant SP-C protein (504). Recognizing that some thioester bonds are sensitive to reduction with 2-mercaptoethanol or dithiothreitol, they used mass spectrometric analysis of pre- and post-cleavage peptide fragments to demonstrate the mass shift corresponding to palmitate in specific peptides (504). Analysis of myristoylated and palmitoylated peptides by MS and LC-MS has been successful and has revealed striking structural heterogeneity in *N*-acylations (395–397,505).

The stepped-orifice LC-MS approach described earlier for phosphorylation and glycosylation site analysis was shown to be compatible with the analysis of lipidated peptides as well, although detection of protein acylation required higher orifice voltage than fragmentation of glycosyl or phosphate ester bonds (506). Using a tripalmitylated cysteinyl peptide, Bean *et al.* detected signature ions at m/z 114, m/z 239, m/z 256, and m/z 257 that collectively provide reasonable discrimination between different lipid structures (506).

F. Summary

In Section V we have described the biological significance of some of the most frequent inducible and/or reversible posttranslational modifications, methods for the detection of specifically modified proteins in a biological sample, and methods for the localization of the modified sites in the polypeptide. The ISGF3 case study described how it was determined that the Stat proteins were inducibly phosphorylated and the methods that were used to suggest and verify the type and the location of phosphorylated residues.

Comparing the methods used in the interferon signaling project with the methods available today makes it apparent that in spite of intense research efforts, the analysis of protein modifications remains a challenging field. Detection of a modification in a protein *in vivo* essentially relies on selective radiolabeling or on the availability of group-specific reagents. Some recently developed mass spectrometric techniques either lack the sensitivity for direct analysis of modifications induced *in vivo*, or remain to be optimally interfaced with the biological source of the sample. The problem of localizing the modified residues within a polypeptide chain has benefitted more from recent technical developments. It has to be stressed, however, that even the methods with the highest sensitivity are not yet, or are only barely, compatible with the analysis of physiological modifications. Until the sensitivity of detection can be further increased, indirect but conclusive strategies, such as the one outlined for the determination of sites of phosphorylation, appear to be a promising way to obtain the required results.

As mass spectrometric techniques are able to analyze increasingly smaller quantities of gel-separated protein, their use in posttranslational modification analysis is rapidly increasing. The next few years will probably see optimization of protocols, integration of diverse analytical techniques, and the development of software that will make the detection and analysis of modified proteins easier. As with new techniques for the identification of proteins, a significant challenge for the technology developers will be to find optimal, general, and simple application protocols so that the improved techniques will find rapid application in biological research projects.

VI. Analysis of Protein: Ligand Complexes

A. Overview

The analysis of protein : ligand, in particular protein : protein and protein : DNA, complexes is one of the most important emerging themes in the analysis of complex biological systems. The example cited throughout this chapter was that of a cytokine-induced protein : protein complex binding to a specific DNA sequence. The proteins were biochemically analyzed after coprecipitation and separation by SDS/PAGE. In addition to the analysis of coprecipitated components by gel electrophoresis, a range of other technologies have also been used to examine protein : ligand complexes. They include ESI-MS, CZE, biosensors, ligand blotting, the genetic approach of two-hybrid analysis, and chemical cross-linking. This section will describe selected approaches and a limited number of examples (see Fig. 20). A comprehensive review provides an excellent treatment of this topic (507).

B. ISGF3 Case Study: Induced Protein : Protein and Protein : DNA Complexes

The investigation of interferon-α induced signaling illustrates the significance of protein : ligand interactions for the investigation and the understanding of biological processes, in particular signal transduction pathways. The interaction of the induced transactivator with a specific DNA sequence in the promoter region of interferon-induced genes provided the basis for the isolation and subsequent biochemical and structural analysis of the gene factor. The recognition that tyrosine phosphorylation induced the assembly of the ISGF3 complex

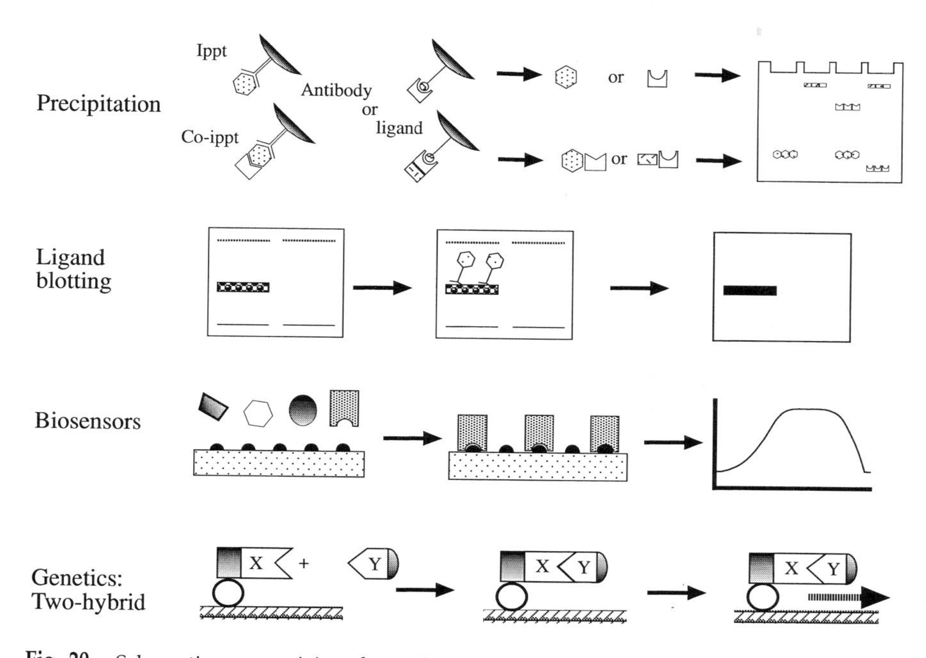

Fig. 20 Schematic summarizing the various approaches employed to analyze protein : protein interactions. In the first approach, precipitation, proteins and or protein complexes are bound by a ligand (e.g., antibody) and precipitated from the solution in which they are present because (in this case) the ligand is attached to an insoluble particle. The protein(s) can then be eluted and separated by SDS–PAGE. Ligand blotting relies on the ability of proteins to be able to at least partially renature following PAGE and blotting. Specific ligands can then be used to probe the blot to detect proteins that have a sufficiently high affinity for the ligand. Western blotting, where the ligand is an antibody, is one example of this approach. When biosensors are employed to analyze protein : protein interactions, a protein is immobilized to a preactivated surface, and a solution containing the ligand of interest is passed over the surface. The binding of a ligand (protein or even small molecule) is then detected because of a change in surface plasmon resonance (see text) shown as an increase in the signal of the "sensorgram." Elution of the ligand will result in a return of the signal to its prebound state. In the last approach shown here, the genetic approach of yeast two-hybrid analysis is employed, where a fusion protien is constructed that consisted of the "bait" (labeled X) and the DNA-binding domain of the GAL4 protein. A fusion protein between a protein (or library of proteins) Y is constructed with the activation domain of the GAL4 proteins. When the two GAL4 domains are associated, transactivation of a reporter gene results, and this can be detected.

from preexisting, nonphosphorylated proteins explained the rapid induction of the transactivating activity and provided a direct mechanistic link to the membrane receptor and the noncovalently associated tyrosine kinase(s).

To assay for the presence of the ISGF3 complex, Levy *et al.* (17) incubated ^{32}P-labeled oligonucleotides with nuclear lysates of interferon-stimulated cells. The oligonucleotide sequence represented the sequence in the promoter region of interferon-induced genes that was protected in footprinting assays. The sample was then subjected to nondenaturing gel electrophoresis. Whereas the majority of labeled oligonucleotides migrated rapidly through the gel, a small fraction was retarded by binding to DNA binding factors. Gel bands detected by autoradiography of the gel, which moved more slowly than the free oligonucleotides, indicated the presence of a DNA binding activity (see Fig. 21). This popular assay is commonly referred to as gel shift or electrophoretic motility shift assay (EMSA). To further investigate the factor(s) interacting with the DNA sequence, Fu *et al.* (21) partially purified the activity by ion-exchange

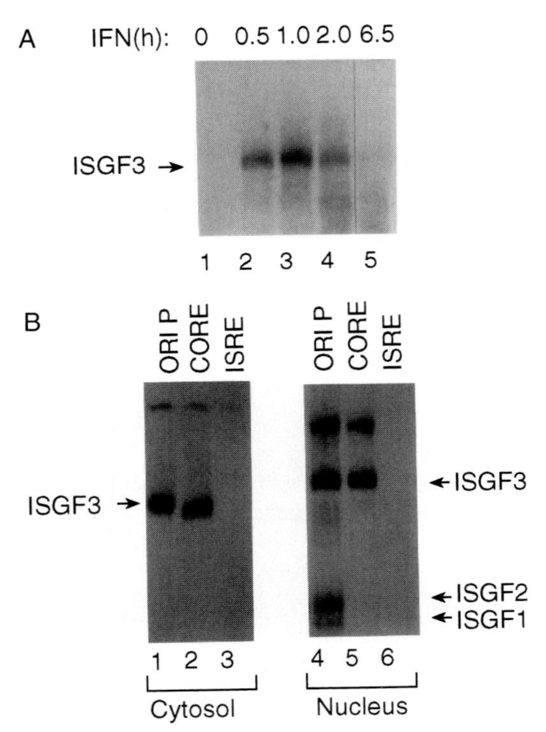

Fig. 21 ISGF3 is induced by interferon α in both cytoplasm and nucleus. (A) Nuclear extracts of human fibroblasts treated with interferon α for the indicated times were analyzed by gel retardation with a labeled ISRE as a probe. (B) Nuclear and cytoplasmic extracts from HeLa cells primed with interferon γ for 16 hr and treated with interferon α for 15 min were subjected to competition gel retardation analysis. The slower mobility complex present in lanes 4 and 5 is observed with extracts that contain high amounts of ISGF3 (either nuclear or cytoplasmic) and may result from dimer formation. The sequence of the double-stranded oligonucleotide that contained an ISRE core equivalent was 5′-AGGAATTTCCCACTTTCACTTCTC-3′ and its complement; the ISRE oligonucleotide sequence was 5′-GGCTTCAGTTTCGGTTTCCCTTTCCCGG-3′ and its complement. Reproduced with permission from Levy, D. E., D. S. Kessler, R. Pine, and J. E. Darnell, Jr. (1989). Cytoplasmic activation of ISGF3, the positive regulator of interferon-alpha-stimulated transcription, reconstituted *in vitro*. *Genes Dev.* **3**:1362–1371.

chromatography. The column fraction that was most enriched for the factor was subjected to preparative EMSA. Proteins in the retarded band were cut out, further separated by SDS–PAGE, and detected by silver staining. Fu *et al.* (21) found that the factor consisted of four polypeptides of apparent molecular mass of 113 kDa, 91 kDa, 84 kDa, and 48 kDa, respectively. Preparative experiments at a larger scale led to partial sequencing of each of the polypeptides and to molecular cloning of the corresponding genes (see Sections III,B and IV,B). To purify sufficient protein for partial sequencing, the relevant oligonucleotide was attached to macroporous beads to perform DNA affinity chromatography essentially according to (153). For preparative purposes, column chromatography is usually preferred over preparative EMSA because the sample can easily be precleared of unspecific DNA binding proteins by chromatography over a DNA column with an irrelevant oligonucleotide, and larger volumes of nuclear lysates can be processed.

The EMSA is a very sensitive, relatively simple, and rapid method for the detection of DNA binding activities in a sample. The specificity of the interaction can be probed with appropriate controls such as displacements of the labeled oligonucleotide from the complex with an excess of cold oligonucleotide of the same sequence. The assay does not provide, however, any information concerning the nature of the interacting compounds, the stoichiometry of the complex, the nature, the site, or the binding constant of the interaction. Because the assay requires nondenturing gel conditions, even the size of the complex can only be estimated very approximately. To obtain more structural information about the complex, Fu and co-workers further fractionated by SDS–PAGE the proteins comigrating with the mobility-shifted oligonucleotide (21). The four polypeptides that were detected were likely to be a part of the DNA binding complex. Comigration with the labeled oligonucleotides does not, however, conclusively prove that all the proteins present are part of the complex, nor did the results suggest that the four polypeptides constituted the whole complex. It is plausible that some additional components dissociated under the EMSA conditions. Furthermore, the approach used did not indicate the stoichiometry of the complex, did not reveal which one(s) of the polypeptides directly interacted with DNA, did not explain how the complex was stabilized, and could not identify potential low molecular weight cofactors.

This case illustrates some of the difficulties encountered in investigating protein : ligand complexes. Currently, there is no single technique that can answer all or even a majority of questions that are essential for understanding the structure and function of protein : ligand complexes. The following describes selected techniques that have been used successfully. Collectively, they provide a powerful toolbox to study protein : ligand complexes. The fact that some of the techniques have only been developed over the past few years is an encouraging indication that technology in this field is progressing rapidly, and that comprehensive analysis of protein : ligand complexes at high sensitivity therefore might be within reach in the near future.

C. Analysis of Protein : Ligand Complexes by Coprecipitation

One of the simplest and most powerful methods for the analysis of protein : ligand complexes is the coprecipitation assay. This technique is adapted from immune precipitations, which involve the use of antibodies to bind in solution to proteins containing their cognate binding sequence. The formed complexes are then precipitated by adding a macromolecular, insoluble component that can bind to the antibody. Precipitating agents include fixed, protein A bearing *Staphylococcus aureus* cells (Cowan I strain), purified protein A covalently attached to macromolecular beads, or a secondary antibody (which is directed against the primary antibody) attached to macromolecular beads. If the initial binding is conducted under stringent conditions established by high detergent or salt concentrations in the sample solution, only the proteins that are specifically bound by the primary antibody will be precipitated. If the initial binding is conducted under less stringent conditions, other proteins complexed with the ligand of the primary antibody will be coprecipitated (see Fig. 20). The components of these precipitates are usually separated by 1-D or 2-D gel electrophoresis, and often the samples from which the immunoprecipitation is performed are radiolabeled. This approach need not be limited to antibody binding assays. Any ligand to which a specific binding molecule is known can be used. The method also does not have to be conducted entirely in solution. The ligands of interest can be attached to a column. As exemplified in the ISGF3 case study, this format is usually chosen if the method is scaled up for preparative purposes. There are too many detailed protocols for immuno- and coprecipitation to cite; however, a good place to start is the book by Harlow and Lane (508) and the Current Protocols in Protein Science series (509), which provides detailed accounts of all aspects of the use of antibodies in biological research.

The power of the technique comes from its ability to selectively precipitate associated proteins from extremely complex samples, including whole-cell lysates. Proteins, which are components of a biological system, can therefore be detected and subsequently analyzed without the need for a specific biochemical assay, because specific physical association can in many cases be interpreted as being involved in the same process. However, it is important to confirm that any protein interactions that are detected by this method are in fact specific. There are at least the following three sources of nonspecific interactions, which can lead to misleading or ambiguous results: (i) nonspecific binding of proteins to the precipitating agent, (ii) nonspecific binding of proteins to a portion of the primary antibody separate to the binding site, or a weak cross-reaction with the binding site, and (iii) nonspecific interactions of proteins with the specifically precipitated protein. Although it is relatively straightforward to diagnose and eliminate the first two potential problems, the third one is more difficult to recognize. Problems with proteins nonspecifically binding to the precipitating reagents are typically dealt with by preclearing of the sample or by quenching of the specific interaction with an excess of ligand. Essentially all precipitation protocols call for a "preclearing" step. The sample to be analyzed is first incubated with an irrelevant antiserum (often a preimmune serum) and then subjected to a precipitation using the same reagents as for the subsequent precipitation with the specific antiserum. The beads are then removed prior to adding the specific antibody. Therefore, any protein that may nonspecifically adhere either to immunoglobulin or to the precipitating beads is removed. In some cases it is of advantage to have a slightly "dirty" immunoprecipitation, provided that the specifically precipitated proteins can be unambiguously identified. Franza *et al.* showed that in 2-DE separations of precipitates, the presence of very low levels of some nonspecfically bound proteins facilitated correlation of patterns from different experiments (147,510). Preincubation of the sample with excess quantities of the target against which the antibody was raised provides an independent way to verify whether proteins are precipitated as the result of a specific interaction with the antibody. Since many antisera are raised against synthetic peptides or bacterially expressed fusion proteins, availability of quenching molecules is frequently not a problem.

The ideal result from the preceding two steps would be that all nonspecific binding is eliminated by preclearing and therefore, in the presence of a competing ligand, no precipitated proteins would be detected. In most precipitation experiments this is not the case, and the results have to be interpreted with caution or additional control experiments need to be performed.

Verification of the specificity of the interaction between coprecipitated proteins and the specifically precipitated protein is more difficult. Neither the preclearing nor the competition strategy is helpful. Washing of the precipitates under conditions of increasing stringency can qualitatively assess the stability of the precipitated complex. Frequently it is assumed or implied that specificity of binding correlates with affinity, i.e., the more stable the complex, the more specific the interaction. This assumption should be applied cautiously, since numerous weak but specific interactions have been found to be functionally important.

Lectins, oligonucleotides, and other molecules capable of undergoing specific interactions, such as SH2 domains, have been used in precipitation experiments similar to the ones with antibodies. Lectins are used for the determination of carbohydrate moieties and in some cases their linkages [e.g., (424)]; oligonucleotides are used to examine proteins that bind to specific or random DNA sequences [e.g., (147)]; and other binding proteins have been used to detect specifically interacting procedures [e.g., (511)].

As an alternative to separating precipitated complexes by denaturing gel electrophoresis "native" or nondenaturing gel electrophoresis has been used successfully, primarily for the detection of protein : DNA complexes. The assay, initially described by Fried and Crothers (512), is commonly referred to as the EMSA or "gel-shift assay." ^{32}P-labeled oligonucleotides are incubated with nuclear lysates and separated by nondenaturing gel electrophoresis, and the gels are autoradiographed. The presence of complexes binding to the oligonucleotides is detected by slower electrophoretic mobility of a (small) fraction of the oligonucleotides compared to the mobility of free probes. A number of parameters, including specificity of the interaction and nature of the interacting proteins, can be examined by variations of this assay. The specificity of the DNA : protein interaction can be verified by competition with various sequence-related or nonspecific oligonucleotides. Provided that antisera specific for the potential DNA-binding proteins are available, the protein composition of the DNA : protein complex can be probed by a "Super-shift" assay. This involves further reduction of the electrophoretic mobility of the DNA : protein complex by binding of an antibody to one of the components [e.g., (513,514)]. Alternatively, the DNA : protein complex can be transferred from the gel to a membrane and probed with antibodies (515,516). As illustrated in the ISGF3 case study, to date, applications of these techniques have been strictly descriptive. The amounts of proteins recovered from a band in a EMSA experiment were too low for further analysis. The rapid progress in the sensitivity of protein analytical techniques (Section IV) suggests that the amounts of protein in such bands will soon be within the sensitivity range of advanced protein identification methods, and that therefore the massive expensive and difficult scaling-up steps can be avoided.

D. Analysis of Protein: Ligand Complexes by Probing Blots with Specific Ligands

"Western blotting" or "immunoblotting" (517) is one of the most widely used techniques in protein biochemistry for the detection of a specific protein in a complex biological sample. Following gel electrophoresis, all proteins present are blotted onto a membrane, and the specific protein is detected by binding of suitably tagged antibodies raised against that antigen (see Fig. 20). Immunoblotting and the use of lectins to probe blotted glycoproteins [e.g., (518)] are the two most widely used techniques that are based on protein: ligand interactions on membranes. A vast literature is readily available for suitable protocols [e.g., (470,509,519,520)], and these techniques are not further covered in this chapter.

In addition to antibodies and lectins, a wide range of ligands have been successfully used to detect ligand-binding proteins on membranes. They include metal ions and small organic molecules such as GTP and heparin, proteins, DNA, RNA, and intact cells. Initially, most of the protocols attempted to probe proteins after separation *in situ* in the gel slice. It was found that this required extensive washing to remove the SDS and buffer salts prior to incubation with the ligand, and that detection of low energy radioisotopes was quenched by the gel slice, thus limiting detection sensitivity. Hence, probing of electroblotted proteins was found more suitable and provided increased sensitivity of detection.

Depending on structural requirements of the binding site, specific protein: ligand interactions may require refolding of the polypeptide after denaturing gel electrophoresis. The fact that proteins electroblotted out of SDS–polyacrylamide gels can be renatured at all is remarkable and was convincingly demonstrated by Ferrell and Martin (521), who showed that serine and threonine kinases could be refolded to resume kinase activity on a nitrocellulose membrane after separation by SDS–PAGE. Renaturation of kinases with specificity for tyrosines has been accomplished in only a few cases (522,523), illustrating the unpredictability of protein renaturation. Most renaturation protocols are deviations of the basic method by Ferrell and Martin (521,524). They often involve sequential washes of the blot in decreasing concentrations of urea or guanidinium chloride or incubation in a Tween 20-containing buffer also containing an excess of a nonspecific protein such as bovine serum albumin (BSA) to quench potential ligand binding sites. If proteins cannot be renatured, nondenaturing (native) gel electrophoresis provides an alternative separation method, which is compatible with protein transfer to a membrane for subsequent ligand blotting. Native gel electrophoresis is sometimes of lower resolution, less general, and less reproducible than SDS–PAGE.

As with any other assay to detect protein: ligand interactions, probing of electroblotted proteins with soluble ligands requires appropriate control experiments to determine the specificity of the observed interaction. Competition between a (radio)labeled ligand and an excess of unlabeled ligand or a related molecule and destruction of a tertiary binding sites by heat denaturation of the ligand prior to probing (525) have been used to establish the specificity of interactions. The latter strategy obviously fails in cases in which the protein : ligand interaction is dependent on a linear epitope.

1. Probing Blotted Proteins with Metal Ions

The simplest ligand used to probe gel-separated proteins on membranes is $^{45}Ca^{2+}$ (526). If membranes were incubated in $^{45}CaCl_2$ in an imidazole/KCl/$MgCl_2$ buffer (527), specific interactions were reported without protein renaturation after SDS–PAGE or 2-D PAGE separation, and no difference in efficiency of detection was found whether or not the nitrocellulose blot had been dried prior to probing (527). A variety of calcium-binding proteins, including those containing EF hands as well as heavy metal binding proteins such as metallothionen and other cadmium-binding proteins, were detected using this assay. Using $^{109}CdCl_2$ as a probe for proteins electroblotted to nitrocellulose, Aoki *et al.* (528) demonstrated a detection sensitivity for metallothioneins that was higher than the sensitivity by Amido Black staining of a replicate nitrocellulose membrane. For both Ca- and Cd-binding proteins, specificity of the interaction was verified either by quenching nonspecific binding with the related divalent cations Mg^{2+} and Zn^{2+}, respectively, or by competition with excess of homologous "cold" ligand.

2. Probing Blotted Proteins with Small Organic Molecules

Guanosine nucleotide triphosphate (GTP) and heparin are examples of small molecules that have been used effectively in ligand blotting. Radiolabeled [α-^{32}P]GTP has been used to detect the *ras*-related low-molecular-weight GTP binding proteins on membranes following 1-D and 2-D electrophoresis (529). Other GTP binding proteins such as the heterotrimeric G proteins were not detected by this method. The basic protocol for probing blotted proteins with [α-^{32}P]GTP uses a reducing Tris buffer containing Tween 20, $MgCl_2$, and nonspecific "blocking" proteins such as BSA or skin milk (529). Detailed evaluation of relevant parameters indicated that for this assay nitrocellulose was a superior membrane to PVDF (530), that the assay was best conducted on a nitrocellulose blot that had not been dried between transfer and assay (531), and that the addition of unlabeled ATP in the blocking and incubation buffers enhanced the specificity of the signal (531). The quality of the ATP used was important because some lots contained small quantities of GTP that competed with the radiolabeled GTP. The specificity of the interaction was confirmed by preincubation of the membrane in "cold" GTP, GDP, cyclic GMP, and ATP and comparison of the labeling patterns obtained by subsequent probing with [α-^{32}P]GTP (532).

A more general assay, which detects most if not all GTP binding proteins, has been described by Peter (533–535). Eukaryotic cells were permeabilized and incubated with [α-^{32}P]GTP, which, following hydrolysis to GDP, displaced the endogenous nucleotide of GTP-binding proteins. The labeled GDP was then covalently attached to the protein by means of oxidation of the GDP ribose moiety, resulting in Schiff base formation with the ε-amino group of an adjacent lysine residue. The bond was then stabilized by cyanoborohydride-catalyzed reduction. Although specific and general for all GTP-binding proteins, this method does alter the electrophoretic mobility of proteins in IEF and 2-D gel electrophoresis.

^{125}I-labeled heparin has been used to probe for heparin binding proteins separated by electrophoresis. Interestingly, it was noted that heparin-binding proteins are difficult to renature on membranes following electroblotting and that it was easier to establish the protein : heparin interaction in the gel (536,537). The successful gel overlay protocol calls for fixation of the gel in 2-propanol/acetic acid and incubation in a Tris/NaCl buffer, which may contain BSA, to renature the proteins prior to incubation with ^{125}I-labeled heparin (536,537). The method, used in conjunction with synthetic peptide competitors and specific protein mutants, was effective in characterizing the heparin-binding properties of acidic (537) and basic fibroblast growth factor (536).

3. *Probing Blotted Proteins with Protein Ligands*

In contrast to the ligands described earlier, radiolabeled proteins are frequently difficult to prepare in quantities sufficient for membrane probing experiments. Therefore, several strategies for the detection of protein:protein interactions on membranes have been developed. They either explore different possibilities for incorporating radiolabels directly into the protein probe or employ conditions that permit detection of specifically bound, unlabeled probes on the membrane. The sensitivity and dynamic range of detection, ease of use, and general applicability of each of these methods has to be considered prior to its application. Specifically, the following approaches have been described: (i) direct radiolabeling of the protein probe by covalent modification, most frequently by ^{125}I, (ii) noncovalent association of a radiolabeled substrate with the protein probe, (iii) metabolic radiolabeling of the protein probe by *in vitro* translation, (iv) expression of the protein probe as an affinity tagged fusion protein, which in turn is detected by interaction with an affinity tag-specific reagent or reaction, and (v) use of an antibody raised directly against the protein probe.

Radioiodinated proteins have been used to probe for interacting proteins in gels as well as after electroblotting on nitrocellulose membranes. Otto *et al.* (538) followed a gel overlay protocol similar to the one described earlier for the detection of heparin-binding proteins. Using the radioiodinated M_r 130,000 actin binding protein vinculin, they detected proteins of M_r 220,000, 130,000 (vinculin), and 42,000 (actin) in extracts from HeLa cells and a variety of other proteins in extracts from chicken gizzard. The polyacrylamide gel was fixed in methanol/acetic acid, washed with 10% (v/v) ethanol, and preincubated in a renaturation buffer consisting of Hepes/NaCl/EGTA/NP-40/gelatin/BSA prior to addition of the ^{125}I labeled vinculin (538). The gel was then washed for 3 days before it was dried and autoradiographed. This protocol was updated by Crawford *et al.* (539) to confirm the interaction between zyxin and α-actinin. To reduce the time required by this length procedure, Crawford *et al.* (539) used ^{125}I-labeled zyxin to probe interacting proteins blotted onto nitrocellulose. They discovered that α-actinin was the most prominent zyxin-binding protein. Elimination of the signal by heat denaturation of the ^{125}I-labeled zyxin probe prior to incubation established the specificity of the interaction.

The search for proteins specifically interacting with GTP-bound small GTP binding proteins represents an interesting case because the assay not only detected such proteins but also, in some cases, revealed their function as GTPase activating proteins (GAPs). Glutathionine *S*-transferase (GST) fusion proteins with the small GTP binding proteins Rac1, Cdc42, or RhoA were preincubated with [γ-^{32}P]GTP and used as probes on a nitrocellulose blot. Proteins that simply bound the p21 proteins showed as dark bands, whereas interacting GAP proteins promoted hydrolysis of the bound GTP and release of the radiolabel from the complex and therefore showed as bands lighter than the background (540). By this method six GAPs with differing molecular weight, specificity, and tissue distribution were detected that interacted with members of the *rho* p21 family (540).

In vitro translation of cDNAs using a labeled amino acid is one of the simplest methods for generating a labeled protein probe for use in ligand blotting experiments. Most frequently, some of the incorporated amino acids are radiolabeled, although biotinylated amino acids have also been used. There are now several excellent coupled *in vitro* transcription/translation kits available. Provided that the cDNA of interest is in a plasmid with an appropriate promoter, the experiment can be conducted in a single reaction tube. Using *in vitro* translated spliceosome-associated protein (SAP) 62 as a probe and 2-DE separated prespliceosome proteins on nitrocellulose as the sample, Bennett and Reed demonstrated a direct interaction between the probe and SAP 114 (541). If proteins in the reticulocyte lysate used for *in vitro* translation are interfering with the assay, the protein probe can be translated as a fusion protein and easily purified from the reaction mixture. As demonstrated by Zhang *et al.* in their study of U2 small nuclear ribonucleoprotein auxiliary factors (542), generation of labeled probes by *in vitro* translation also provides the ability to rapidly examine the interactions of truncated forms of the probe with the target protein.

Overexpression of genetically engineered protein probes containing a specific affinity tag provides for rapid and simple purification of the probe and for detection of the tagged protein at high sensitivity. Numerous affinity tags, including GST, a *myc* or ribonuclease S peptide, maltose binding protein, constitutive biotinylation of a fusion protein, or oligohistidine (His_6) have been used successfully [e.g., (543–545)]. This procedure obviates the need to raise antibodies to individual proteins for each assay, provided that the tag does not interfere with binding to the target protein or correct folding of the protein. The array of tags available and the choice to place them either at the N- or C-terminus of the protein suggest that a suitable tag will be found for most situations. The use of affinity-tagged proteins as probes for protein interactions is popular in signal transduction research, where interactions between SH2 and SH3 domain-containing proteins with protein ligands are extremely important. Taylor and Shalloway (511) probed plots of gel-separated whole-cell lysates, or protein fractions prepared by selective precipitation with GST-SH3 and GST-SH2 fusion proteins or with GST alone. After electroblotting, the nitrocellulose blot was blocked in Tris/NaCl/Tween 20/BSA and probed with GST fusion proteins at 2 μg/ml in a similar buffer, to reveal proteins that specifically interacted with either or both SH2 and SH3 domains (511).

Fusion proteins involved in protein:protein interactions on membranes are typically detected with antibodies specific for the affinity tag. For most of the popular tags, including GST, *myc* or ribonuclease S peptide, and maltose binding protein, specific antibodies are readily commercially available. The difficulty in raising an antibody specific for the His_6 limited the use of oligohistidine extended proteins as probes on membranes until nitriloacetic acid (NTA) conjugated alkaline phosphatase (546) or biotinyl-NTA (547) were developed as specific reagents. The significance of the His_6 tag system lies in the fact that the protein can be purified from lysates of the overexpressing bacteria on Ni-chelate chromatography, which affords very mild, nondenaturing elution conditions. However, denaturing conditions can be employed, if necessary, which is not the case for many of the other tags.

4. *Probing Blotted Proteins with DNA and RNA*

Probing blotted proteins with DNA or RNA is frequently also called "Southwestern" or "Northwestern" blotting. In this technique radiolabeled or biotinylated oligonucleotides are used to probe the membrane-immobilized proteins. Dejgaard and Celis (548) have demonstrated that the technique is also compatible with 2-DE. They separated hnRNPs by 2-DE and detected hnRNP proteins K and L by their interaction with poly(rC), and hnRNPs C and M by their interaction with poly(rU). Blotted proteins were renatured in a Tris/NaCl/Ficoll 400/PVP-40 buffer and incubated with ^{32}P-labeled oligonucleotides. The technique complements the EMSA described previously. In an EMSA, a number of protein components can interact in the form of a multicomponent complex, and not all of the proteins in the complex necessarily bind the oligonucleotide probe directly. Furthermore, the EMSA does not provide any information on the size or the state of phosphorylation of a protein. In contrast, the size of the interacting protein is revealed by Southwestern blotting and the interacting protein can be further identified by Western blotting. If the method is performed on proteins separated by 2-DE, potentially even the state of phosphorylation of the oligonucleotide-binding species can be deduced. The method fails, however, if the nucleic acid binding proteins cannot be sufficiently renatured or if the interaction requires the presence of (protein) cofactors.

5. *Probing Blotted Proteins with Cells*

The most complex class of ligands used to probe blots of gel-separated proteins are intact cells. This technique, termed "cell blotting," has been used to detect lectins (549), extracellular matrix proteins, and other integral membrane proteins such as cell-adhesion molecules (550,551). As with any of the methods described in this section, success relies upon the ability to sufficiently renature the target proteins. To minimize problems with protein renaturation, Rao *et al.* (549) used native PAGE at pH 4.5. Similarly, Seshi (550,551) used a lithium dodecyl sulfate (LDS)–PAGE system run at low temperature (0–4°C) in an attempt to preserve both the tertiary and quarternary structure. Rao *et al.* used erythrocytes to probe nitrocellulose blots in search of lectins. The inherent red color of erythrocytes obviated the need for staining of the cells (549). In cases in which colorless cells were used, adsorbed cells need to be stained for detection. Seshi *et al.* showed that lymphocytes absorbed to the membrane could be either stained with hematoxylin and detected microscopically on the blot, or stained with propidium iodide and detected with a long-wavelength UV light

(550,551). Frequently, cell–cell contacts are established by multiple weak interactions involving different receptor : ligand pairs. Although cell blotting only examines relatively strong interactions of a cell with a single protein ligand, it might nevertheless prove extremely useful for the interactions of cell surface structures with specific protein ligands.

E. Analysis of Protein : Ligand Complexes with Biosensors

Analysis of protein interactions by means of optical biosensors, which employ surface plasmon resonance (SPR) detection or integrated optics incorporating waveguiding films, is a relatively new and rapidly growing field. The reader is referred to a number of excellent reviews (552–556), which cover this topic in far more detail than is possible in this chapter.

Optical biosensors measure the change in refractive index of the medium very close to the surface of the probe. It has been found that the refractive index is directly proportional to both the amount and molecular weight of macromolecules that are bound to the surface of the probe. Therefore, immobilization of a biomolecule to the sensor surface, typically gold-plated glass, will establish a baseline signal that increases upon binding of a ligand to the immobilized molecule. The principle of operation is illustrated in Fig. 22 (552,556). One of the interacting components is coupled to the sensor surface either directly or indirectly through an adaptor molecule, and a baseline signal is established. The second, interacting component is then injected over the surface at a constant flow rate, and binding is recorded in real time by an increase in the SPR signal. An example of the type of raw data produced by an SPR system in a hypothetical experiment is shown in Fig. 23. The first segment of the resonance signal is the baseline. This is followed by a rise in signal as a consequence of injecting a sample containing component(s) able to bind to the immobilized probe. The rate of increase in signal is an indicator of the binding rate. A signal increase rate of zero during the injection indicates that a steady state has been reached and equilibrium data can be calculated. The decreasing signal after the infusion is stopped is an indicator for the dissociation of the complex. The slope of the curve allows the calculation of the dissociation rate. Multimolecular assemblies formed on the sensor by sequential injection of components, which will add on to the complex already formed on the sensor surface, are recorded as a cumulative signal increase. If all the sequential binding reactions proceed to equilibrium, each binding event can be quantified separately. Finally, the bound ligand is eluted and the signal returns to the baseline. The turnaround time for such an experiment is on the order of 10 min, and if the analyte is replenished at a constant flow rate, multiple parameters, including association rates, dissociation rates, equilibrium kinetics, and stoichiometry, can be calculated (552,556).

The search for ligands of orphan receptors and quantitation of interactions between purified components are examples of areas in which biosensors have been used. Using an SPR system, Bartley *et al.* (557) identified a molecule termed B61 as the ligand for the ECK receptor protein-tyrosine kinase by screening concentrated conditioned media from 25 cell lines for compounds that bound to the immobilized extracellular domain of the ECK receptor. Media that read positive in the assay were used as the source for affinity chromato-

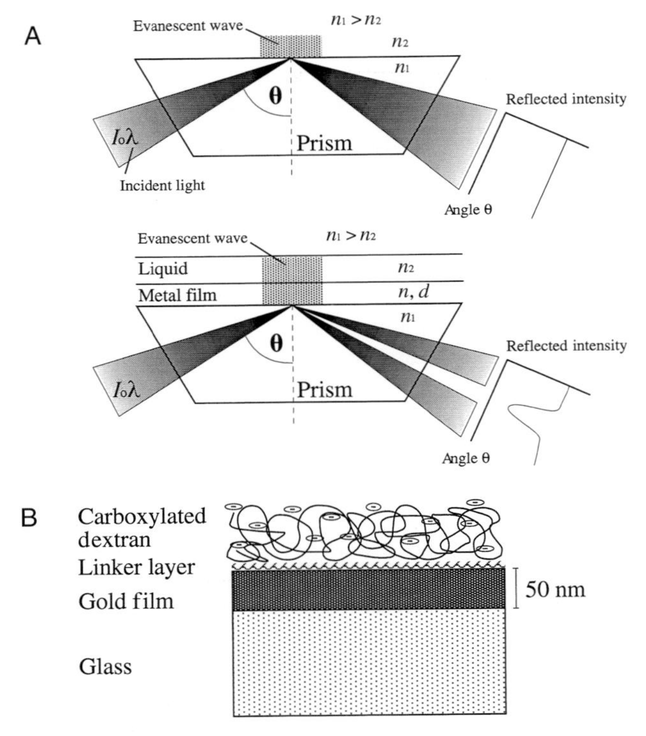

Fig. 22 Surface plasmon resonance theory and schematic of a sensor chip. (A) illustrates the theory behind SPR, which is associated with the phenomenon of total internal reflection, which occurs at the boundary between substances of different refractive index (n) e.g., glass and water. As shown in the upper portion of the panel, light incident on the internal surface of the material of higher refractive index (n_1) is reflected internally; however, the electromagnetic field of the incident light does cross a short way beyond the surface as an *evanescent wave* (extending into the medium a few hundred nanometers). Inserting a thin metal film (e.g., gold) at the interface allows surface plasmon resonance to occur. This is because gold and silver have free electron clouds—plasmons—that, in effect, absorb energy from the evanescent wave, which results in a drop in intensity of the reflected light at a particular angle of incidence (lower portion). The evanescent wave profile depends upon the refractive index of the medium it probes (n_2). When the refractive index changes, so does the evanescent wave, which in turn changes the plasmon vector length—effecting a new incident angle for resonance. In summary, light incident at a specific angle to an interface of two refractive indices, between which a thin film of gold is placed, will be absorbed. The angle at which that absorption occurs is very sensitive to refractive index changes in the external medium, but only in the region close to the metal surface (within 300 nm or so). All proteins, independent of their amino acid composition, alter the refractive index of water by a similar amount per unit mass—thus, as the protein content of the 300-nm layer changes, so does the refractive index. There is a linear correlation between the surface concentration of protein and the resonance angle shift: one resonance unit is equivalent to 1 pg/mm^2 mass (556). (B) shows a schematic of the sensor chip as used in the BIAcore. In this case a carboxylated dextran matrix, ~100 nm of which extends into the liquid, providing a hydrophilic environment in which the interactions can occur, is attached to the gold film via an intermediate linker layer. The thin gold film is interfaced with a glass prism (556). Redrawn from R. Granzow (Pharmacia Biosensor, Piscataway) with permission. Copyright Pharmacia Biosensor AB.

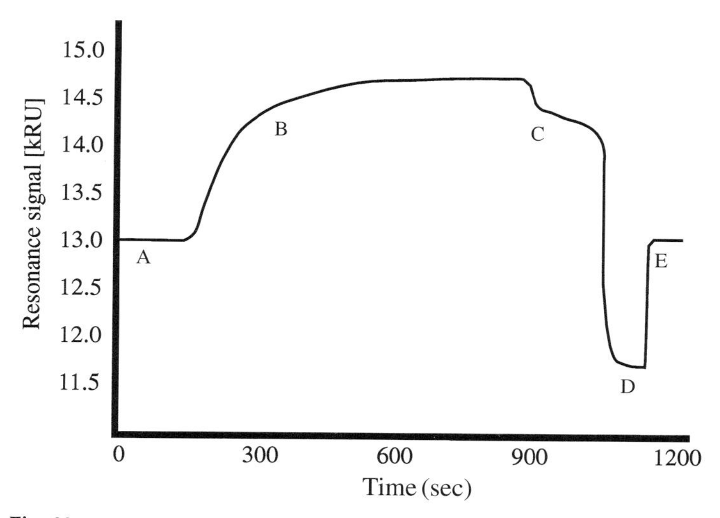

Fig. 23 Sensorgram from a surface plasmon resonance based instrument with a flow-through cell. The first component of the sensorgram curve (A) represents the baseline signal. Following injection (at constant flow rate) of a ligand, which binds to the surface, an increase in the SPR signal is recorded. The steepness of the slope (B) allows calculation of the binding rate. When the injection is completed, the rate of decrease of the signal (C) can be used to calculate the dissociation rate. If the binding rate becomes zero during the injection, equilibrium data can be extracted from the steady-state situation. The mass increase required to saturate the surface allows calculation of the stoichiometry of binding. The ligand is then eluted (D) and the baseline is resumed (E). Redrawn from R. Granzow (Pharmacia Biosensor, Piscataway) with permission. Copyright Pharmacia Biosensor AB.

graphic purification of the ligand. The affinity column was prepared by immobilizing the same extracellular domain of the ECK receptor that was used in the SPR assay (557). Fisher *et al.* used SPR to study DNA binding proteins with respect to their DNA sequence specificity, association rates, complex stoichiometry and stability, and effect of phosphorylation (558). Double-stranded oligonucleotides were immobilized on the sensor surface and probed with recombinant ETS1 oncoproteins p42 and p51 (558).

Advantages of biosensor technology for the analysis of interacting biomolecules explicitly outlined by Chaiken *et al.* (552) include the following: (i) The interaction is monitored directly as it occurs on the prepared surface of the sensor; (ii) no labeling of any of the components is required; (iii) sequential interactions can be monitored and quantitated, and the surface can be regenerated between injections; (iv) sample requirements are small, typically 1 μg/injection or less and, at approximately 10 pmol of sample detected, the sensitivity is high; and (v) interactions where the soluble component is in a complex mixture can be analyzed. The method by itself does not, however, identify the interacting compound if inhomogeneous samples are probed.

As with any other technique that probes physiologically relevant interactions, prevention of denaturation is a concern. There is a range of chemistries available to attach purified probe molecules, including proteins, to the sensor. Suitable attachment conditions for most situations should, therefore, be available. Furthermore, the deposition can be monitored in real time by the SPR system so that the optimal extent of sensor loading can be determined and future sensors for the same experiment can be produced at a high reproducibility. The demonstration that binding events involving low-mass (200 Da) compounds can be observed by biosensors (556) make this approach amenable to new areas of research such as small-molecule screening and screening of degenerate libraries.

F. Genetic Approach of Yeast Two-Hybrid Analysis

The method of yeast two-hybrid analysis described by Fields and Song (148) is a very powerful approach for the identification of proteins that potentially interact physiologically with a particular protein or protein domain. Yeast two-hybrid analysis yields large numbers of candidate proteins. Some of the techniques described earlier can then be used to verify whether interactions observed in the genetic screen are reciprocated *in vitro* and *in vivo*. The two-hybrid system has three major applications: (i) testing known proteins for their ability to interact, (ii) defining domains, orientation of domains, or amino acids critical for an interaction, and (iii) screening libraries for proteins that bind a specific target protein (559).

The method exploits the ability of a pair of interacting proteins to bring a transcription activation domain into close proximity with the DNA-binding site that regulates the expression of an adjacent reporter gene. This is illustrated in Fig. 24. In the yeast *Saccharomyces cerevisiae,* the GAL4 protein is a transcrip-

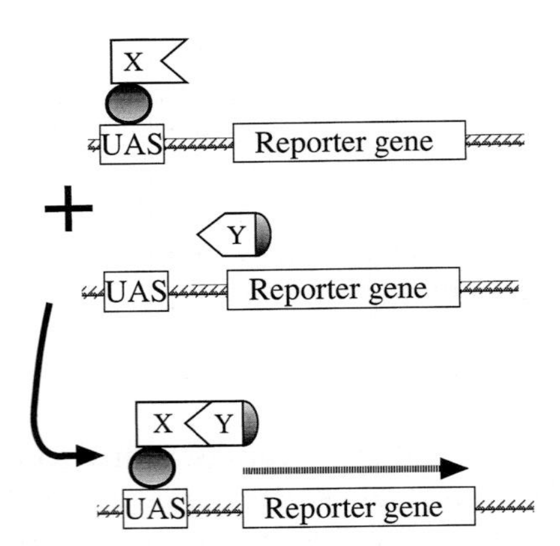

Fig. 24 Schematic of the two-hybrid system. A hybrid protein is generated that includes the DNA-binding domain of the GAL4 protein (filled circle) and the bait protein (which does not contain an activation domain) (labeled X). This construct can bind the upstream activation sequence (UAS) for the yeast *GAL* genes, but cannot transactivate the reporter gene unless the activation domain is present. An activation domain hybrid is constructed with protein Y (which could be encoded by a library of different genes) and the activation domain of GAL4 (filled semicircle). This hybrid will not activate transcription because it cannot bind the UAS. By producing both hybrids in the same transformant, if proteins X and Y bind noncovalently, transcription from the UAS will be activated, which can be detected by measuring the activity produced from the product of the reporter gene. Adapted from Chien, C. T., P. L. Bartel, R. Sternglanz, and S. Fields. (1991). The two-hybrid system: A method to identify and clone genes for proteins that interact with a protein of interest. *Proc. Natl. Acad. Sci. USA* **88:**9578–9582; and Fields, S., and R. Sternglanz. (1994). The two-hybrid systems: An assay for protein–protein interactions. *Trends Genet.* **10:**286–292.

tional activator required for the expression of genes that encode enzymes for galactose utilization. The GAL4 proteins consists of two separable and functionally essential domains in a single polypeptide, the N-terminal sequence-specific DNA-binding domain (residues 1–147 which binds to UAS_G (upstream activation sequence) and the C-terminal domain (residues 768–881) containing acidic regions, which are necessary to activate transcription (148). Hybrid proteins containing these essential parts of GAL4 in different polypeptides can then be constructed so that either fusion protein individually expressed in a cell will not activate transcription of a reporter gene under the control of UAS_G. In the example illustrated in Fig. 24, protein X is fused to the GAL4 DNA-binding domain, and protein Y is fused to the GAL4 activating region. If the fusion proteins are coexpressed so that protein X and Y can form a protein : protein complex, then the proximity of the two domains of the GAL4 protein is reconstituted and transcription of the reporter gene under the control of UAS_G occurs (148). Most commonly, the reporter gene is *lacZ*, which leads to production of β-galactosidase activity; this activity can be monitored colorimetrically by blue coloration of colonies if a suitable, chromogenic galactosidase substrate is present. Such colonies can then be isolated for plasmid purification and sequencing of the insert, thereby allowing identification of the interacting proteins by their sequence.

Large-scale screening for proteins that can bind to a specific target protein is currently the most powerful application of the two-hybrid system. The target protein is usually fused to the DNA-binding domain and a cDNA library is ligated to the activation domain. Even though only approximately one-sixth of the inserts are likely to be in the correct orientation and reading frame, with the typical complexity of $>10^6$ inserts used, a relatively large fraction of potential protein : protein interactions in a cell can be screened rapidly. cDNA libraries that have been successfully used for two-hybrid screening include both oligo(dT)- and random-primed, as well as size-selected libraries likely to encode only small protein domains (559).

With the yeast two-hybrid system, as with any method, there are caveats that have to be understood to avoid erroneous conclusions to be drawn. Fields and Sternglanz (559) discussed the following situations:

1. The fusion protein needs to be correctly folded in yeast and, since the fusion proteins are synthesized in the cytoplasm, imported into the nucleus.
2. Protein : protein interactions that depend on posttranslational modifications such glycosylation, disulfide bond formation, or protein phosphorylation might not be reproduced in yeast. We have shown that coexpression of a tyrosine kinase in yeast was sufficient to establish SH2 domain or phosphotyrosine binding domain-dependent interactions, suggesting that in special cases this limitation can be overcome (572).
3. Because of the orientation of the standard hybrid constructs, the transcription factor domain may block accessibility if the N-terminal domain of either hybrid construct is required for the protein : protein interaction. This can potentially be overcome by reversing the orientation of the hybrid construct.
4. Proteins that activate transcription when fused to the DNA-binding domain can cause positive signals without involvement of a protein : protein interaction. If two defined proteins are being assayed, this is easily solved by reversing the fusion partners. The problem is more significant in random

library screens where random transcriptional activation is not easily distinguished from hybrid-induced positives.

5. Transcription is optimal when the activation domain hybrid is in excess over the transcriptional domain hybrid. Different expression levels or instability of one of the hybrids can therefore negatively affect the results.
6. Protein : protein complexes frequently consist of more than two subunits. The two-hybrid system might therefore only detect a part of a functional protein complex, or an interaction might be missed in cases in which the simultaneous presence of several proteins is required.
7. Finally, protein : protein interactions observed in the two-hybrid system do not necessarily occur *in vivo*. Coexpression of two proteins in the same cell at the same time and transport to the same subcellular compartment are just a few of the preconditions that have to be fulfilled for a physiological protein : protein interactions to occur. In addition, when screening entire libraries, there are often false positives due to homodimerization of the bait with clones of itself.

Despite this list of caveats, the two-hybrid method is extremely powerful and has the ability to screen entire libraries very rapidly for proteins interacting with a specific target protein. The plasmids that produce the positive signal can be sequenced, providing the ability to rapidly clone the corresponding gene, which can then be expressed to determine whether the genetic interaction can be replicated in various *in vitro* or *in vivo* assays, including the ones described earlier in this section.

G. Summary

In this section we have described a number of approaches to the analysis of protein : ligand complexes that are currently widely used. In previous sections, which describe methods for protein identification and for determining post-translational modifications, the objective of the experiments was straightforward and easily defined. For the analysis of protein : ligand complexes, the objectives are more difficult to define. Depending on the project or the situation, different parameters, including composition of the complex, stoichiometry of subunits, binding constants, the nature of the bonds and topography of contact areas, the state of modification, and processing of the proteins in the complex are of particular interest.

None of the methods currently available can answer these questions by themselves; applied in combination they are, however, capable of answering most questions concerning protein : ligand complexes. Therefore, it is essential that the experimentalist know the strengths and limitations of each method and be capable of choosing and applying the optimal technique for the situation.

The most significant limitations common to all the techniques described here is the difficulty in distinguishing between true and artifactual interactions, the difficulty in detecting the whole complexity of the protein : ligand structure, and difficulties in detecting protein : ligand interactions that are stabilized by multiple, weak interactions. Suitable control experiments that we described are quite effective in distinguishing true and false positives in *in vitro* experiments. Verification of the specificity of protein/ligand interactions *in vivo* is more difficult. Frequently, as discussed for the ISGF3 case study, this requires an independent track of experiments.

The same study also impressively illustrated that a combination of techniques can lead to the discovery of a functionally important protein : ligand complex and to a basic understanding of the mechanism of action of a process. The biochemical methods for detecting and characterizing protein complexes are dramatically enhanced by the rapidly progressing protein analytical techniques for protein identification and determination of posttranslational modifications, which were described in the previous sections.

Each one of the methods described here is undergoing rapid improvement; principally, new approaches such as the use of ESI-MS (e.g., (560–564)) or MALDI-MS (e.g., (565,566)) for the analysis of protein complexes are being explored, and novel combinations of different established techniques create new insights into the structure and function of protein : ligand structures.

It can, therefore, be anticipated that the analytical capabilities for investigating protein : ligand complexes will improve rapidly in the near future. There remain, however, very formidable analytical challenges, such as the activation of regulatory proteins by the noncovalent association with small ligands (e.g., ceramide activated protein kinase (567)), which requires the simultaneous measurement of the structure and the activity of the protein : ligand complex, and the analysis of systems that are stabilized by the cooperative action of multiple weak interactions, as exemplified by cell–cell contacts.

VII. Conclusion

With this chapter we have attempted to discuss contemporary technology for the analysis of regulated biological systems on the protein level. Using the example of the interferon-induced signal transduction pathway, we have described several typical stages in such projects, including the detection and identification of the elements in a biological sample that participate in a particular process, the determination of posttranslational processing and modifications that are induced in a process-dependent manner, and the induction of process-specific protein : ligand complexes.

In each one of these areas rapid and impressive technological achievements have been reported over the past few years. They are mainly based on the development of mass spectrometric methods, on the possibility of interfacing mass spectrometers with high-performance separation techniques for biomolecules, and on the complementary use of diverse types of data to solve a specific problem.

It has to be stressed, however, that the analytical approaches described in this chapter provide but an average of the events observed in the biological source, typically millions of unsynchronized cells. To understand the molecular mechanisms of complex regulated systems, analysis of regulatory events in a single cell in a time-dependent fashion is required, a task that is well beyond the capabilities of current analytical techniques. The average of events observed by the approaches described in this chapter can, however, be considered as a library of possible events related to a biological system. Higher sensitivity techniques capable of analyzing single cells will have to be devised to determine which of the possible events are occurring under specific regulatory conditions in a single cell.

Acknowledgments

We would like to thank our colleagues who made data available prior to publication, and Dr. Tony Polverino for critical review of the manuscript. We would like to thank Danette Barron (Amgen Inc.) for preparing Figs. 1, 2, 6, 15, and 21, and Drs. Mann (EMBL) and Yates (U. Wash.) for the original artwork for Figs. 11 and 12 S. D. P. would like to thank Drs. Hsieng Lu and Mike Rohde for encouragement. R. A. would like to thank the National Science and Technology Center for Molecular Biotechnology and the Canadian Genetics Disease Network of Centers of Excellence for financial support.

References

1. Huxley, A. F., and A. M. Gordon. (1962). Striation patterns in active and passive shortening of muscle. *Nature.* **193:**280.
2. Szent-Gyœrgyi, A. (1941). *Studies Inst. Med. Chem., Univ. Szeged.* **1:**17–26.
3. Szent-Gyœrgyi, A. (1947). Chemistry of Muscular Contraction. Academic Press, New York.
4. Allen, B. G., and M. P. Walsh (1994). The biochemical basis of the regulation of smooth-muscle contraction. *Trends Biochem-Sci.* **19:**362–368.
5. Fleischmann, R. D., M. D. Adams, O. White, R. A. Clayton, E. F. Kirkness, A. R. Kerlavage, C. J. Bult, J. F. Tomb, A. A. Dougherty, J. M. Merrick, *et al.* (1995). Whole-genome random sequencing and assembly of *Haemophilus influenzae* Rd. *Science* **269:**496–512.
6. Williams, N. (1995). Closing in on the complete yeast genome sequence. *Science* **268:**1560–1561.
7. Fraser, C. M., J. D. Gocayne, O. White, M. D. Adams, R. A. Clayton, R. D. Fleischmann, C. J. Bult, A. R. Kerlavage, G. Sutton, J. M. Kelley, *et al.* (1995). The minimal gene complement of *Mycoplasma genitalium. Science* **270:**397–403.
8. Collins, F., and D. Galas. (1993). A new five-year plan for the U.S. Human Genome Project. *Science* **262:**43–46.
9. Adams, M. D., A. R. Kerlavage, R. D. Fleischmann, R. A. Fuldner, C. J. Bult, N. HY. Lee, E. F. Kirkness, K. G. Weinstock, J. D. Gocayne, O. White, *et al.* (1995). Initial assessment of human gene diversity and expression patterns based upon 83 million nucleotides of cDNA sequence. *Nature* **377**(Suppl.):3–174.
10. Sen, G. C., and P. Lengyel. (1992). The interferon system. A bird's eye view of its biochemistry. *J. Biol. Chem.* **267:**5017–5020.
11. Marx, J. (1992). Taking a direct path to the genes. *Science* **257:**744–745.
12. Larner, A. C., G. Jonak, Y. S. Cheng, B. Korant, E. Knight, and J. E. Darnell, Jr. (1984). Transcriptional induction of two genes in human cells by beta interferon. *Proc. Natl. Acad. Sci. USA* **81:**6733–6737.
13. Larner, A. C., A. Chaudhuri, and J. E. Darnell, Jr. (1986). Transcriptional induction by interferon. New protein(s) determine the extent and length of the induction. *J. Biol. Chem.* **261:**453–459.
14. Levy, D., A. Larner, A. Chaudhuri, L. E. Babiss, and J. E. Darnell, Jr. (1986). Interferon-stimulated transcription: Isolation of an inducible gene and identification of its regulatory region. *Proc. Natl. Acad. Sci. USA* **83:**8929–8933.
15. Reich, N., B. Evans, D. Levy, D. Fahey, E. Night, Jr., and J. E. Darnell, Jr. (1987). Interferon-induced transcription of a gene encoding a 15-kDa protein depends on an upstream enhancer element. *Proc. Natl. Acad. Sci. USA* **84:**6394–6398.
16. Levy, D. E., D. S. Kessler, R. Pine, and J. E. Darnell, Jr. (1989). Cytoplasmic activation of ISGF3, the positive regulator of interferon-alpha-stimulated transcription, reconstituted *in vitro. Genes Dev.* **3:**1362–1371.
17. Levy, D. E., D. S. Kessler, R. Pine, N. Reich, and J. E. Darnell, Jr. (1988). Interferon-induced nuclear factors that bind a shared promoter element correlate with positive and negative transcriptional control. *Genes Dev.* **2:**383–393.
18. Kessler, D. S., D. E. Levy, and J. E. Darnell, Jr. (1988). Two interferon-induced nuclear factors

bind a single promoter element in interferon-stimulated genes. *Proc. Natl. Acad. Sci. USA* **85:**8521–8525.
19. Lew, D. J., T. Decker, and J. E. Darnell, Jr. (1989). Alpha interferon and gamma interferon stimulate transcription of a single gene through different signal transduction pathways. *Mol. Cell Biol.* **9:**5404–5411.
20. Reich, N. C., and J. E. Darnell, Jr. (1989). Differential binding of interferon-induced factors to an oligonucleotide that mediates transcriptional activation. *Nucleic Acids Res.* **17:**3415–3424.
21. Fu, X. Y., D. S. Kessler, S. A. Veals, D. E. Levy, and J. E. Darnell, Jr. (1990). ISGF3, the transcriptional activator induced by interferon alpha, consists of multiple interacting polypeptide chains. *Proc. Natl. Acad. Sci. USA* **87:**8555–8559.
22. Schindler, C., X. Y. Fu, T. Improta, R. Aebersold, and J. E. Darnell, Jr. (1992). Proteins of transcription factor ISGF-3: One gene encodes the 91- and 84-kDa ISGF-3 proteins that are activated by interferon alpha. *Proc. Natl. Acad. Sci. USA* **89:**7836–7839.
23. Fu, X. Y., C. Schindler, T. Improta, R. Aebersold, and J. E. Darnell, Jr. (1992). The proteins of ISGF-3, the interferon alpha-induced transcriptional activator, define a gene family involved in signal transduction. *Proc. Natl. Acad. Sci. USA* **89:**7840–7843.
24. Veals, S. A., C. Schindler, D. Leonard, X. Y. Fu, R. Aebersold, J. E. Darnell, Jr., and D. E. Levy. (1992). Subunit of an alpha-interferon-responsive transcription factor is related to interferon regulatory factor and Myb families of DNA-binding proteins. *Mol. Cell Biol.* **12:**3315–3324.
25. Shuai, K., G. R. Stark, I. M. Kerr, and J. E. Darnell, Jr. (1993). A single phosphotyrosine residue of Stat91 required for gene activation by interferon-gamma. *Science* **261:**1744–1746.
26. Fu, X. Y. (1992). A transcription factor with SH2 and SH3 domains is directly activated by an interferon alpha-induced cytoplasmic protein tyrosine kinase(s). *Cell* **70:**323–335.
27. Darnell, J. E., Jr., I. M. Kerr, and G. R. Stark. (1994). Jak-STAT pathways and transcriptional activation in response to IFNs and other extracellular signaling proteins. *Science* **264:**1415–1421.
28. Pellegrini, S., and C. Schindler. (1993). Early events in signalling by interferons. *Trends Biochem. Sci.* **18:**338–342.
29. Schindler, C., and J. E. Darnell, Jr. (1995). Transcriptional responses to polypeptide ligands—The Jak-Stat pathway. *Annu. Rev. Biochem.* **64:**621–651.
30. Velculescu, V. E., L. Zhang, B. Vogelstein, and K. W. Kinzler. (1995). Serial analysis of gene expression [see comments]. *Science* **270:**484–487.
31. Schena, M., D. Shalon, R. W. Davis, and P. O. Brown. (1995). Quantitative monitoring of gene expression patterns with a complementary DNA microarray [see comments]. *Science* **270:**467–470.
32. Qureshi, S. A., M. Saldittgeorgieff, and J. E. Darnell. (1995). Tyrosine-phosphorylated Stat1 and Stat2 plus a 48-kDa protein all contact DNA in forming interferon-stimulated-gene factor-3. *Proc. Natl. Acad. Sci. USA* **92:**3829–3833.
33. O'Farrell, P. H. (1975). High resolution two-dimensional gel electrophoresis of proteins. *J. Biol. Chem.* **250:**4007–4021.
34. Klose, J. (1975). *Humangenetik* **26:**231–243.
35. Klose, J., and E. Zeindl. (1984). An attempt to resolve all the various proteins in a single human cell type by two-dimensional electrophoresis: I. Extraction of all cell proteins. *Clin. Chem.* **30:**2014–2020.
36. Klose, J. (1989). Systematic analysis of the total proteins of a mammalian organism: Principles, problems and implications for sequencing the human genome, *Electrophoresis* **10:**140–152.
37. Klose, J., and U. Kobalz. (1995). 2-Dimensional electrophoresis of proteins—an updated protocol and implications for a functional analysis of the genome. *Electrophoresis* **16:**1034–1059.
38. Garrels, J. I. (1979). Two-dimensional gel electrophoresis and computer analysis of proteins synthesized by clonal cell lines. *J. Biol. Chem.* **254:**7961–7977.
39. Miller, M. J., N.-H. Xuong, and E. P. Geiduschek. (1979). *Proc. Natl. Acad. Sci. USA* **76:**5222–5225.
40. Patton, W. F. (1995). Biologist's perspective on analytical imaging systems as applied to protein gel electrophoresis. *J. Chromatogr. A* **698:**55–87.
41. Myrick, J. E., P. F. Lemkin, M. K. Robinson, and K. M. Upton. (1993). Comparison of the Bio Image Visage 2000 and the GELLAB-II two-dimensional electrophoresis image analysis systems. *Appl. Theor. Electrophor.* **3:**335–346.
42. Anderson, N. L., J. Taylor, A. E. Scandora, B. P. Coulter, and N. G. Anderson. (1981). The TYCHO system for computer analysis of two-dimensional gel electrophoresis patterns. *Clin. Chem.* **27:**1807–1820.

43. Miller, M. J., A. D. Olson, and S. S. Thorgeirsson. (1984). Computer analysis of two-dimensional gels: Automatic matching. *Electrophoresis* **5:**297–303.
44. Appel, R. D., D. F. Hochstrasser, M. Funk, J. R. Vargas, C. Pellegrini, A. F. Muller, and J. R. Scherrer. (1991). The MELANIE project: From a biopsy to automatic protein map interpretation by computer. *Electrophoresis* **12:**722–735.
45. Lemkin, P. F., Y. Wu, and K. Upton. (1993). An efficient disk based data structure for rapid searching of quantitative two-dimensional gel databases. *Electrophoresis* **14:**1341–1350.
46. Vincens, P., N. Paris, J.-L. Pujol, C. Gaboriaud, T. Rabilloud, J.-L. Pennetier, P. Matherat, and P. Tarroux. (1986). HERMeS: A second generation approach to the automatic analysis of two-dimensional electrophoresis gels. Part 1: Data acquisition. *Electrophoresis* **7:**347–356.
47. Solomon, J. E., and M. G. Harrington. (1993). A robust, high-sensitivity algorithm for automated detection of proteins in two-dimensional electrophoresis gels. *Comput. Appl. Biosci.* **9:**133–139.
48. Garrels, J. I. (1989). The QUEST system for quantitative analysis of two-dimensional gels. *J. Biol. Chem.* **264:**5269–5282.
49. Monardo, P. J., T. Boutell, J. I. Garrels, and G. I. Latter. (1994). A distributed system for two-dimensional gel analysis. *Comput. Appl. Biosci.* **10:**137–143.
50. Patterson, S. D., and G. I. Latter. (1993). Evaluation of storage phosphor imaaging for quantitative analysis of 2-D gels using the Quest II system. *BioTechniques* **15:**1076–1083.
51. Rodriquez, L. V., D. M. Gersten, L. S. Ramagli, and D. A. Johnston. (1993). Towards stoichiometric silver staining of proteins resolved in complex two-dimensional electrophoresis gels: Real-time analysis of pattern development. *Electrophoresis* **14:**628–637.
52. Neumann, U., H. Khalaf, and M. Rimpler. (1994). Quantitation of electrophoretically separated proteins in the submicrogram range by dye elution. *Electrophoresis* **15:**916–921.
53. Harrington, M. G., K. H. Lee, M. Yun, T. Zewert, J. E. Bailey, and L. Hood. (1993). Mechanical precision in two-dimensional electrophoresis can improve protein spot positional reproducibility. *Appl. Theor. Electrophor.* **3:**347–353.
54. Nokihara, K., N. Morita, and T. Kuriki. (1992). Applications of an automated apparatus for two-dimensional electrophoresis, Model TEP-1, for microsequence analyses of proteins. *Electrophoresis* **13:**701–707.
55. Patton, W. F., M. G. Pluskal, W. M. Skea, J. L. Buecker, M. F. Lopez, R. Zimmerman, L. M. Belanger, and P. D. Hatch. (1990). Development of a dedicated two-dimensional gel electrophoresis system that provides optimal pattern reproducibility and polypeptide resolution. *BioTechniques* **8:**518–527.
56. Bjellqvist, B., K. Ek, P. G. Righetti, E. Gianazza, A. Gorg, R. Westermeier, and W. Postel. (1982). Isolectric focusing in immobilized pH gradients: Principles, methodology and some applications. *J. Biochem. Biophys. Methods* **6:**317–339.
57. Gorg, A. (1993). Two dimensional electrophoresis with immobilized pH gradients: Current state. *Biochem. Soc. Trans.* **21:**130–132.
58. Corbett, J. M., M. J. Dunn, A. Posch, and A. Georg. (1994). Positional reproducibility of protein spots in two-dimensional polyacrylamide gel electrophoresis using immobilised pH gradient isoelectric focusing in the first dimension: An interlaboratory comparison. *Electrophoresis* **15:**1205–1211.
59. Blomberg, A., L. Blomberg, J. Norbeck, S. J. Fey, P. M. Larsen, M. Larsen, P. Roepstorff, H. Degand, M. Boutry, A. Posch, and A. Gorg. (1995). Interlaboratory reproducibility of yeast protein-patterns analyzed by immobilized pH gradient 2-dimensional gel-electrophoresis. *Electrophoresis* **16:**1935–1945.
60. Gersten, D. M., L. S. Ramagli, D. A. Johnston, and L. V. Rodriguez. (1992). The use of radioactive bacteriophage proteins as X–Y markers for silver stained two-dimensional electrophoresis gels and quantification of the patterns. *Electrophoresis* **13:**87–92.
61. Bjellqvist, B., B. Basse, E. Olsen, and J. E. Celis. (1994). Reference points for comparisons of two-dimensional maps of proteins from different human cell types defined in a pH scale where isoelectric points correlate with polypeptide compositions. *Electrophoresis* **15:**529–539.
62. Dean, D. P., M. T. Cronan, J. M. Merenda, J. P. Gardner, P. A. Connelly, and J. E. Celis. (1994). "Spot transfer", elution and comigration with known proteins allows accurate tranferral of protein identifications between distinct two-dimensional electrophoretic systems. *Electrophoresis* **15:**540–543.
63. Chiari, M., C. Micheletti, M. Nesi, M. Fazio, and P. G. Righetti. (1994). Towards new formulations for polyacrylamide matrices: *N*-Acryloylaminoethoxyethanol, a novel monomer combining high hydrophilicity with extreme hydrolytic stability. *Electrophoresis* **15:**177–86.

64. Harrington, M. G., and T. E. Zewert. (1994). Poly(ethylene glycol)methacrylate–acrylamide copolymer media for hydrophobic protein and low temperature electrophoresis. *Electrophoresis* **15:**195–199.
65. Harrington, M. G., K. H. Lee, J. E. Bailey, and L. E. Hood. (1994). Sponge-like electrophoresis media: Mechanically strong materials compatible with organic solvents, polymer solutions and two-dimensional electrophoresis. *Electrophoresis* **15:**187–194.
66. Chiari, M., and P. G. Righetti. (1995). New types of separation matrices for electrophoresis. *Electrophoresis* **16:**1815–1829.
67. Sinha, P., E. Koettigen, R. Westermeier, and P. G. Righetti. (1992). Immobilized pH 2.5-11 gradients for two-dimensional electrophoresis. *Electrophoresis* **13:**210–214.
68. Bjellqvist, B., C. Pasquali, F. Ravier, J. C. Sanchez, and D. Hochstrasser. (1993). A nonlinear wide-range immobilized pH gradient for two-dimensional electrophoresis and its definition in a relevant pH scale. *Electrophoresis* **14:**1357–1365.
69. Bossi, A., P. G. Righetti, G. Vecchio, and S. Severinsen. (1994). Focusing of alkaline proteases (subtilisins) in pH 10-12 immobilized gradients. *Electrophoresis* **15:**1535–1540.
70. Bjellgvist, B., J.-C. Sanchez, C. Pasquali, F. Ravier, N. Paquet, S. Frutiger, G. J. Hughes, and D. Hochstrasser. (1993). Micropreparative two-dimensional electrophoresis allowing the separation of samples containing milligram amounts of proteins. *Electrophoresis* **14:**1375–1378.
71. Latham, K. E., J. I. Garrels, C. Chang, and D. Solter. (1991). Quantitative analysis of protein synthesis in mouse embryos. I. Extensive reprogramming at the one- and two-cell stages. *Development* **112:**921–932.
72. Latham, K. E., R. S. Beddington, D. Solter, and J. I. Garrels. (1993). Quantitative analysis of protein synthesis in mouse embryos. II: Differentiation of endoderm, mesoderm, and ectoderm. *Mol. Reprod. Dev.* **35:**140–150.
73. Latham, K. E., J. I. Garrels, C. Chang, and D. Solter. (1992). Analysis of embryonic mouse development: Construction of a high-resolution, two-dimensional gel protein database. *Appl. Theor. Electrophor.* **2:**163–170.
74. Latham, K. E., J. I. Garrels, and D. Solter. (1994). Alterations in protein synthesis following transplantation of mouse 8-cell stage nuclei to enucleated 1-cell embryos. *Dev. Biol.* **163:**341–350.
75. Andersen, S. O., K. Rafn, T. N. Krogh, P. Hojrup, and P. Roepstorff. (1995). Comparison of larval and pupal cuticular proteins in *Tenebrio molitor*. *Insect Biochem. Mol. Biol.* **25:**177–187.
76. Neumann, N. F., W. S. Pon, A. Nowicki, W. M. Samuel, and M. Belosevic. (1994). Antigens of adults and third-stage larvae of the meningeal worm, *Parelaphostrongylus tenuis* (Nematoda, Metastrongyloidea). *J. Vet. Diagn. Invest.* **6:**222–229.
77. Harrington, M. G., J. A. Coffman, F. J. Calzone, L. E. Hood, R. J. Britten, and E. H. Davidson. (1992). Complexity of sea urchin embryo nuclear proteins that contain basic domains. *Proc. Natl. Acad. Sci. USA* **89:**6252–6256.
78. Mooradian, A. D., and K. E. Meredith. (1992). The effect of age on protein composition of rat cerebral microvessels. *Neurochem. Res.* **17:**665–670.
79. Bentlage, H., R. de Coo, H. ter Laak, R. Sengers, F. Trijbels, W. Ruitenbeek, W. Schlote, K. Pfeiffer, S. Gencic, G. von Jagow, and H. Schägger. (1995). Human diseases with defects in oxidative phosphorylation. 1. Decreased amounts of assembled oxidative phosphorylation complexes in mitochondrial encephalomyopathies. *Eur. J. Biochem.* **227:**909–915.
80. Bini, L., B. Magi, C. Cellesi, A. Rossolini, and V. Pallini. (1992). Two-dimensional electrophoresis analysis of human serum proteins during the acute-phase response. *Electrophoresis* **13:**743–746.
81. Knecht, M., V. Regitz Zagrosek, K. P. Pleissner, P. Jungblut, C. Steffen, A. Hildebrandt, and E. Fleck. (1994). Characterization of myocardial protein composition in dilated cardiomyopathy by two-dimensional gel electrophoresis. *Eur. Heart J.* **15**(Suppl. D):37–44.
82. Myrick, J. E., S. P. Caudill, M. K. Robinson, and I. L. Hubert. (1993). Quantitative two-dimensional electrophoretic detection of possible urinary protein biomarkers of occupational exposure to cadmium. *Appl. Theor. Electrophor.* **3:**137–146.
83. Harrington, M. G., R. Aebersold, B. M. Martin, C. R. Merril, and L. Hood. (1993). Identification of a brain-specific human cerebrospinal fluid glycoprotein, beta-trace protein. *Appl. Theor. Electrophor.* **3:**229–234.
84. Ji, H., R. H. Whitehead, G. E. Reid, R. L. Moritz, L. D. Ward, and R. J. Simpson. (1994). Two-dimensional electrophoretic analysis of proteins expressed by normal and cancerous human

crypts: Application of mass spectrometry to peptide-mass fingerprinting. *Electrophoresis* **15:**391–405.

85. Easty, D., I. R. Hart, K. Patel, C. Seymour, M. Yacoub, A. Domscheit, S. Gunther, W. Postel, A. Gorg, and M. J. Dunn. (1991). Changes in protein expression during melanoma differentiation determined by computer analysis of 2-D gels. *Clin. Exp. Metastasis* **9:**221–230.
86. Saunders, F. K., D. A. Winfield, J. R. Goepel, B. W. Hancock, R. M. Sharrard, and M. H. Goyns. (1994). 2D-gel analysis of protein synthesis profiles of different stages of chronic lymphocytic leukaemia. *Leuk. Lymphhoma* **14:**319–322.
87. Wirth, P. J., L. D. Luo, Y. Fujimoto, and H. C. Bisgaard. (1992). Two-dimensional electrophoretic analysis of transformation-sensitive polypeptides during chemically, spontaneously, and oncogene-induced transformation of rat liver epithelial cells. *Electrophoresis* **13:**305–320.
88. Vandekerckhove, J., G. Bauw, K. Vancompernolle, B. Honore, and J. Celis (1990). Comparative two-dimensional gel analysis and microsequencing identifies gelsolin as one of the most prominent downregulated markers of transformed human fibroblast and epithelial cells. *J. Cell Biol.* **111:**95–102.
89. Honore, B., H. H. Rasmussen, A. Celis, H. Leffers, P. Madsen, and J. E. Celis. (1994). The molecular chaperones HSP28, GRP78, endoplasmin, and calnexin exhibit strikingly different levels in quiescent keratinocytes as compared to their proliferating normal and transformed counterparts: cDNA cloning and expression of calnexin. *Electrophoresis* **15:**482–490.
90. Gromov, P. S., and J. E. Celis. (1994). Several small GTP-binding proteins are strongly downregulated in simian virus 40 (SV40) transformed human keratinocytes and may be required for the maintenance of the normal phenotype. *Electrophoresis* **15:**474–481.
91. Wirth, P. J., L.-D. Luo, T. Benjamin, T. N. Hoang, A. D. Olson, and D. C. Parmlee. (1993). The rat liver epithelial (RLE) cell nuclear protein database. *Electrophoresis* **14:**1199–1215.
92. Wirth, P. J. (1994). Two-dimensional polyacrylamide gel electrophoresis in experimental hepatocarcinogenesis studies. *Electrophoresis* **15:**358–371.
93. Lin, C. S., R. H. Aebersold, S. B. Kent, M. Varma, and J. Leavitt. (1988). Molecular cloning and characterization of plastin, a human leukocyte protein expressed in transformed human fibroblasts. *Mol. Cell. Biol.* **8:**4659–4668.
94. Giometti, C. S., S. L. Tollaksen, C. Chubb, C. Williams, and E. Huberman. (1995). Analysis of proteins from human breast epithelial-cells using 2-dimensional gel-electrophoresis. *Electrophoresis* **16:**1215–1224.
95. Schmid, H. R., D. Schmitter, P. Blum, M. Miller, and D. Vonderschmitt. (1995). Lung-tumor cells—a multivariate approach to cell classification using 2-dimensional protein pattern. *Electrophoresis* **16:**1961–1968.
96. Zeindl Eberhart, E., P. Jungblut, and H. M. Rabes. (1994). Expression of tumor-associated protein variants in chemically induced rat hepatomas and transformed rat liver cell lines determined by two-dimensional electrophoresis. *Electrophoresis* **15:**372–381.
97. Giometti, C. S., M. A. Gemmell, J. Taylor, S. L. Tollaksen, R. Angeletti, and S. Gluecksohn Waelsch. (1992). Evidence for regulatory genes on mouse chromosome 7 that affect the quantitative expression of proteins in the fetal and newborn liver. *Proc. Natl. Acad. Sci. USA* **89:**2448–2452.
98. Giometti, C. S., S. L. Tollaksen, and D. Grahn. (1994). Altered protein expression detected in the F1 offspring of male mice exposed to fission neutron. *Mutat. Res.* **320:**75–85.
99. Rabilloud, T., P. Vincens, D. Asselineau, J. L. Pennetier, M. Darmon, and P. Tarroux. (1994). Computer analysis of two-dimensional electrophoresis gels as a tool in cell biology: Study of the protein expression of human keratinocytes from normal to tumor cells. *Cell. Mol. Biol.* **40:**17–27.
100. Strahler, J. R., X. X. Zhu, N. Hora, Y. K. Wang, P. C. Andrews, N. A. Roseman, J. V. Neel, L. Turka, and S. M. Hanash. (1993). Maturation stage and proliferation-dependent expression of dUTPase in human T cells. *Proc. Natl. Acad. Sci. USA* **90:**4991–4995.
101. Hendil, K. B., P. Kristensen, and W. Uerkvitz. (1995). Human proteasomes analysed with monoclonal antibodies. *Biochem. J.* **305:**245–252.
102. Nystrom, T. (1994). The glucose-starvation stimulon of *Escherichia coli:* Induced and repressed synthesis of enzymes of central metabolic pathways and role of acetyl phosphate in gene expression and starvation survival. *Mol. Microbiol.* **12:**833–843.
103. Coles, A. M., H. A. Crosby, and J. H. Pearce. (1991). Analysis of the human serological response to *Chlamydia trachomatis* 60-kDa proteins by two-dimensional electrophoresis and immunoblotting. *FEMS Microbiol. Lett.* **65:**299–303.

104. Voris, B. P., and D. A. Young. (1981). Glucocorticoid-induced proteins in rat thymus cells. *J. Biol. Chem.* **256:**11319–11329.
105. Villa, P., M. Miehe, M. Sensenbrenner, and B. Pettmann. (1994). Synthesis of specific proteins in trophic factor-deprived neurons undergoing apoptosis. *J. Neurochem.* **62:**1468–1475.
106. Robaye, B., A. P. Doskeland, N. Suarez-Huerta, S. O. Doskeland, and J. E. Dumont. (1994). Apoptotic cell death analyzed at the molecular level by two-dimensional gel electrophoresis. *Electrophoresis* **15:**503–510.
107. Baxter, G. D., and M. F. Lavin. (1992). Specific protein dephosphorylation in apoptosis induced by ionizing radiation and heat shock in human lymphoid tumor lines. *J. Immunol.* **148:**1949–1954.
108. Patterson, S. D., J. S. Grossman, P. D'Andrea, and G. I. Latter. (1995). Reduced Numatrin/B23/Nucleophosmin labeling in apoptotic Jurkat T-lymphoblasts. *J. Biol. Chem.* **270:**9429–9436.
109. Amess, B., and A. M. Tolkovsky. (1995). Programmed cell-death in sympathetic neurons—a study by 2-dimensional polyacrylamide-gel electrophoresis using computer image-analysis. *Electrophoresis* **16:**1255–1267.
110. Lee, C., and J. A. Sensibar. (1987). Proteins of the rat prostate. II. Synthesis of new proteins in the ventral lobe during castration-induced regression. *J. Urol.* **138:**903–980.
111. Leger, J. G., M. L. Montpetit, and M. P. Tenniswood. (1987). Characterization and cloning of androgen-repressed mRNAs from rat ventral prostate. *Biochem. Biophys. Res. Commun.* **147:**196–203.
112. Wadewitz, A. G., and R. A. Lockshin (1988). Programmed cell death: Dying cells synthesize a co-ordinated, unique set of proteins in two different episodes of cell death. *FEBS Lett.* **241:**19–23.
113. Sensibar, J. A., B. Alger, A. Tseng, L. Berg, and C. Lee. (1990). Proteins of the rat prostate. III. Effect of testosterone on protein synthesis by the ventral prostate of castrated rats. *J. Urol.* **143:**161–166.
114. Caudill, S. P., J. E. Myrick, and M. K. Robinson. (1993). Exploratory data analysis in two-dimensional electrophoresis. *Appl. Theor. Electrophor.* **3:**133–136.
115. Rovensky, P., and I. Lefkovits. (1994). Data management in a two-dimensional gel electrophoresis laboratory: "Gelscript 1" softward. *Electrophoresis* **15:**977–983.
116. Nugues, P. M. (1993). Two-dimensional electrophoresis image interpretation. *IEE Trans. Biomed. Eng.* **40:**760–770.
117. Hughes, G. J., S. Frutiger, N. Paquet, F. Ravier, C. Pasquali, J. C. Sanchez, R. James, J. D. Tissot, B. Bjellqvist, and D. F. Hochstrasser. (1992). Plasma protein map: An update by microsequencing. *Electrophoresis* **13:**707–714.
118. Yun, M., W. Wu, L. Hood, and M. Harrington. (1992). Human cerebrospinal fluid protein database: Edition 1992. *Electrophoresis* **13:**1002–1013.
119. Jungblut, P., A. Otto, E. Zeindl Eberhart, K. P. Plessner, M. Knecht, V. Regitz Zagrosek, E. Fleck, and B. Wittmann Liebold. (1994). Protein composition of the human heart: The construction of a myocardial two-dimensional electrophoresis database. *Electrophoresis* **15:**685–707.
120. Corbett, J. M., C. HY. Wheeler, C. S. Baker, M. HY. Yacoub, and M. J. Dunn. (1994). The human myocardial two-dimensional gel protein database: Update 1994. *Electrophoresis* **15:**1459–1465.
121. Hochstrasser, D. F., S. Frutiger, N. Paquet, A. Bairoch, F. Ravier, C. Pasquali, J. C. Sanchez, J. D. Tissot, B. Bjellqvist, R. Vargas, R. D. Appel, and G. J. Hughes. (1992). Human liver protein map: A reference database established by microsequencing and gel comparison. *Electrophoresis* **13:**992–1001.
122. Golaz, O., G. J. Hughes, S. Frutiger, N. Paquet, A. Bairoch, C. Pasquali, J.-C. Sanchez, J.-D. Tissot, R. D. Appel, C. Walzer, L. Balant, and D. F. Hochstrasser. (1993). Plasma and red blood cell protein maps: Update 1993. *Electrophoresis* **14:**1223–1231.
123. Kettman, J. R., C. Coleclough, and I. Lefkovits. (1994). Lymphocyte Proteinpaedia stage two: T-cell polypeptides from a partitioned cDNA library revealed by the dual decay method. *Int. Arch. Allergy Immunol.* **103:**131–142.
124. Hanash, S. M., J. R. Strahler, Y. Chan, R. Kuick, D. Teichroew, J. V. Neel, N. Hailat, D. R. Keim, J. Gratiot Deans, D. Ungar, *et al.* (1993). Data base analysis of protein expression patterns during T-cell ontogeny and activation. *Proc. Natl. Acad. Sci. USA* **90:**3314–3318.
125. Celis, J. E., B. Gesser, H. H. Rasmussen, P. Madsen, H. Leffers, K. Dejgaard, B. Honore, E. Olsen, G. Ratz, J. B. Lauridsen, *et al.* (1990). Comprehensive two-dimensional gel protein databases offer a global approach to the analysis of human cells: The transformed amnion cells (AMA) master database and its link to genome DNA sequence data. *Electrophoresis* **11:**989–1071.

126. Celis, J. E., and E. Olsen. (1994). A qualitative and quantitative protein database approach identifies individual and groups of functionally related proteins that are differentially regulated in simian virus 40 (SV40) transformed human keratinocytes: An overview of the functional changes associated with the transformed phenotype. *Electrophoresis* **15:**309–344.
127. Celis, J. E., H. H. Rasmussen, E. Olsen, P. Madsen, H. Leffers, B. Honore, K. Dejgaard, P. Gromov, H. Vorum, A. Vassilev, *et al.* (1994). The human keratinocyte two-dimensional protein database (update 1994): Towards an integrated approach to the study of cell proliferation, differentiation and skin diseases. *Electrophoresis* **15:**1349–1458.
128. Burggraf, D., K. Andersson, C. Eckerskorn, and F. Lottspeich. (1992). Towards a two-dimensional database of common human proteins. *Electrophoresis* **13:**729–732.
129. Anderson, N. L., R. Esquer Blasco, J. P. Hofmann, and N. G. Anderson. (1991). A two-dimensional gel database of rat liver proteins useful in gene regulation and drug effects studies. *Electrophoresis* **12:**907–930.
130. Anderson, N. L., R. Esquerblasco, J. P. Hofmann, L. Meheus, J. Raymackers, S. Steiner, F. Witzmann, and N. G. Anderson. (1995). Updated 2-dimensional gel database of rat-liver proteins useful in gene-regulation and drug effect studies. *Electrophoresis* **16:**1977–1981.
131. Giometti, C. S., J. Taylor, and S. L. Tollaksen. (1992). Mouse liver protein database: A catalog of proteins detected by two-dimensional gel electrophoresis. *Electrophoresis* **13:**970–991.
132. Garrels, J. I., and B. R. Franza, Jr. (1989). The REF52 protein database. Methods of database construction and analysis using the QUEST system and characterizations of protein patterns from proliferating and quiescent REF52 cells. *J. Biol. Chem.* **264:**5283–5298.
133. Garrels, J. I., and B. R. Franza, Jr. (1989). Transformation-sensitive and growth-related changes of protein synthesis in REF52 cells. A two-dimensional gel analysis of SV40-, adenovirus-, and Kirsten murine sarcoma virus-transformed rat cells using the REF52 protein database. *J. Biol. Chem.* **264:**5299–5312.
134. Garrels, J. I., B. Futcher, R. Kobayashi, G. I. Latter, B. Schwender, T. Volpe, J. R. Warner, and C. S. McLaughlin. (1994). Protein identifications for a *Saccharomyces cerevisiae* protein database. *Electrophoresis* **15:**1466–1486.
135. Boucherie, H., G. Dujardin, M. Kermorgant, C. Monribot, P. Slonimski, and M. Perrot. 1995. Two-dimensional protein map of *Saccharomyces cerevisiae*—construction of a gene-protein index. *Yeast* **11:**601–613.
136. VanBogelen, R. A., P. Sankar, R. L. Clark, J. A. Bogan, and F. C. Neidhardt. (1992). The gene-protein database of *Escherichia coli:* Edition 5. *Electrophoresis* **13:**1014–1054.
137. Santaren, J. F., J. Van Damme, M. Puype, J. Vandekerckhove, and A. Garcia-Bellido. (1993). Identification of *Drosophila* wing imaginal disc proteins by two-dimensional gel analysis and microsequencing. *Exp. Cell Res.* **206:**220–226.
138. Giometti, C. S., S. L. Tollaksen, S. Mukund, Z. H. Zhou, K. R. Ma, X. H. Mai, and M. W. W. Adams. (1995). 2-Dimensional gel-electrophoresis mapping of proteins isolated from the hyperthermophile *Pyrococcus furiosus. J. Chromatogr. A* **698:**341–349.
139. Niederreiter, M., M. Gimona, F. Streichsbier, J. E. Celis, and J. V. Small (1994). Complex protein composition of isolated focal adhesions: A two-dimensional gel and database analysis. *Electrophoresis* **15:**511–519.
140. Xu, C., D. R. Rigney, and D. J. Anderson. (1994). Two-dimensional electrophoretic profile of human sperm membrane proteins. *J. Androl.* **15:**595–602.
141. Boutell, T., J. I. Garrels, B. R. Franza, P. J. Monardo, and G. I. Latter. (1994). REF52 on Global Gel Navigator: An Internet-accessible two-dimensional gel electrophoresis data base. *Electrophoresis* **15:**1487–1490.
142. Appel, R. D., J. C. Sanchez, A. Bairoch, O. Golaz, F. Ravier, C. Pasquali, G. J. Hughes, and D. F. Hochstrasser. (1994). The SWISS-2DPAGE database of two-dimensional polyacrylamide gel electrophoresis. *Nucleic Acids Res.* **22:**3581–3582.
143. Simpson, R. J., A. Tsugita, J. E. Celis, J. I. Garrels, and H. W. Mewes. (1992). Workshop on two-dimensional gel protein databases. *Electrophoresis* **13:**1055–1061.
144. Jungblut, P., H. Baumeister, and J. Klose. (1993). Classification of mouse liver proteins by immobilized metal affinity chromatography and two-dimensional electrophoresis. *Electrophoresis* **14:**638–643.
145. Ramsby, M. I., G. S. Makowski, and E. A. Khairallah. (1994). Differential detergent fractionation of isolated hepatocytes: Biochemical, immunochemical and two-dimensional gel electrophoresis characterization of cytoskeletal and noncytoskeletal compartments. *Electrophoresis* **15:**265–277.

146. Celis, J. E., H. H. Rasmussen, H. Leffers, P. Madsen, B. Honore, K. Dejgaard, P. Gromov, E. Olsen, H. J. Hoffmann, M. Nielsen, B. Gesser, M.Puype, J. Van Damme, and J. Vanderkerckhove. (1993). Human cellular protein patterns and their link to genome DNA mapping and sequencing data: Towards an integrated approach to the study of gene expression. *In* "Genetic Engineering," pp. 21–40. Plenum Press, New York.

147. Franza, B. R., Jr., S. F. Josephs, M. Z. Gilman, W. Ryan, and B. Clarkson. (1987). Characterization of cellular proteins recognizing the HIV enhancer using a microscale DNA-affinity precipitation assay. *Nature* **330**:391–395.

148. Fields, S., and O. Song. (1989). A novel genetic system to detect protein–protein interactions. *Nature* **340**:245–246.

149. Cordwell, S. J., M. R. Wilkins, A. Cerpa-Poljak, A. A. Gooley, M. Duncan, K. L. Williams, and I. Humphery-Smith. (1995). Cross-species identifiction of proteins separated by two-dimensional gel electrophoresis using matrix-assisted laser desorption ionization/time-of-flight mass spectrometry and amino acid composition. *Electrophoresis* **16**:438–443.

150. Wasinger, V. C., S. J. Cordwell, A. Cerpa-Poljak, J. X. Yan, A. A. Gooley, M. R. Wilkins, M. W. Duncan, R. Harris, K. L. Williams, and I. Humphery-Smith. (1995). Progress with gene-product mapping of the mollicutes: *Mycoplasma genitalium*. *Electrophoresis* **16**:1090–1094.

151. Wilkins, M. R., J.-C. Sanchez, A. A. Gooley, R. D. Appel, I. Humphery-Smith, D. F. Hochstrasser, and K. L. Williams. (1995). Progress with proteome projects: Why all proteins expressed by a genome should be identified and how to do it. *Biotech. Gen. Eng. Reviews* **13**:19–50.

152. Patterson, S. D., and R. Aebersold. (1995). Mass spectrometric approaches for the identification of gel-separated proteins. *Electrophoresis* **16**:1791–1814.

153. Kadonaga, J. T., and R. Tjian. (1986). Affinity purification of sequence-specific DNA binding proteins. *Proc. Natl. Acad. Sci. USA* **83**:5889–5893.

154. Aebersold, R. H., J. Leavitt, R. A. Saavedra, L. E. Hood, and S. B. Kent. (1987). Internal amino acid sequence analysis of proteins separated by one- or two-dimensional gel electrophoresis after *in situ* protease digestion on nitrocellulose. *Proc. Natl. Acad. Sci. USA* **84**:6970–6974.

155. Pawson, T., and G. D. Gish. (1992). SH2 and SH3 domains: From structure to function. *Cell* **71**:359–362.

156. Pluskal, M. G., M. B. Przekop, M. R. Kavonian, C. Vecoli, and D. A. Hicks. (1986). Immobilon™ PVDF transfer membrane: A new membrane substrate for Western blotting of proteins. *BioTechniques* **4**:272–283.

157. Matsudaira, P. (1987). Sequence analysis from picomole quantities of proteins electroblotted onto polyvinylidene difluoride membranes. *J. Biol. Chem.* **262**:10035–10038.

158. Warlow, R. S., A. Gooley, R. Rajasekariah, N. Pszarac, and R. S. Walls. (1995). A preparative method for sequencing proteins and peptides: *In situ* gel staining with subsequent passive elution onto polyvinylidene difluoride membranes. *Electrophoresis* **16**:84–91.

159. Walsh, B. J., A. A. Gooley, K. L. Williams, and S. N. Breit. (1995). Identification of macrophage activation associated proteins by two-dimensional gel electrophoresis and microsequencing. *J. Leukoc. Biol.* **57**:507–512.

160. Vandekerckhove, J., G. Bauw, M. Puype, J. Van Damme, and M. Van Montagu. (1985). Protein-blotting on polybrene-coated glass-fiber sheets. *Eur. J. Biochem.* **152**:9–19.

161. Aebersold, R. H., D. B. Teplow, L. E. Hood, and S. B. H. Kent. (1986). Electroblotting onto activated glass. High efficiency preparation of proteins from analytical sodium dodecyl sulfate–polyacrylamide gels for direct sequence analysis. *J. Biol. Chem.* **261**:4229–4238.

162. Burkhart, W. A., M. B. Moyer, W. M. Bodnar, A. M. Everson, V. G. Valladares, and J. M. Bailey. (1995). Direct collection onto Zitex and PVDF for Edman sequencing: Elimination of polybrene. *In* "Techniques in Protein Chemistry VI," pp. 169–176. Academic Press, San Diego.

163. Baker, C. S., M. J. Dunn, and M. H. Yacoub. (1991). Evaluation of membranes used for electroblotting of proteins for direct automated microsequencing. *Electrophoresis* **12**:342–348.

164. Eckerskorn, C., and F. Lottspeich. (1993). Structural characterization of blotting membranes and the influence of membrane parameters for electroblotting and subsequent amino acid sequence analysis of proteins. *Electrophoresis* **14**:831–838.

165. Brown, J. L., and W. K. Roberts. (1976). Evidence that approximately eighty percent of the soluble proteins from Ehrlich ascites cells are amino-terminally acetylated. *J. Biol. Chem.* **251**:1009–1014.

166. Brown, J. L. (1979). A comparison of the turnover of amino-terminally acetylated and non-acetylated mouse L-cell proteins. *J. Biol. Chem.* **254**:1447–1449.

167. Bailey, J. M., and J. E. Shively. (1994). A chemical method for the C-terminal sequence analysis of proteins. *Methods: Comp. Methods Enzymol.* **6:**334–350.
168. Bailey, J. M., O. Tu, G. Issai, and J. E. Shively. (1995). C-terminal sequence analysis of polypeptides containing C-terminal proline. *In* "Techniques in Protein Chemistry VI," pp. 239–247. Academic Press, San Diego.
169. Bozzini, M., J. Zhao, P.-M. Yuan, D. Ciolek, Y.-C. Pan, J. Horton, D. R. Marshak, and V. L. Boyd. (1995). Applications using an alkylation method for carboxy-terminal protein sequencing. *In* "Techniques in Protein Chemistry VI," pp. 229–237. Academic Press, San Diego.
170. Miller, C. G., D. H. Hawke, J. Tso, and S. Early. 1995. Automated C-terminal protein sequence analysis using the Hewlett-Packard G1009A C-terminal protein sequencing system. *In* "Techniques in Protein Chemistry VI," pp. 219–227. Academic Press, San Diego.
171. Tempst, P., and L. Riviere. (1989). Examination of automated polypeptide sequencing using standard phenyl isothiocyanate reagent and subpicomole high-performance liquid chromatographic analysis. *Anal. Biochem.* **183:**290–300.
172. Tempst, P., S. Geromanos, C. Elicone, and H. Erdjument-Bromage. (1994). Improvements in microsequencer performance for low picomole sequence analysis. *Methods: Comp. Methods Enzymol.* **6:**248–261.
173. Elicone, C., M. Lui, S. Geromanos, H. Erdjument-Bromage, and P. Tempst. (1994). Microbore reversed-phase high-performance liquid-chromatographic purification of peptides for combined chemical sequencing-laser-desorption mass-spectrometric analysis. *J. Chromatogr. A* **676:**121–137.
174. Erdjument-Bromage, H., M. Lui, D. M. Sabatini, S. H. Snyder, and P. Tempst. (1994). High-sensitivity sequencing of large proteins—partial structure of the rapamycin-Fkbp12 target. *Protein Sci.* **3:**2435–2446.
175. Totty, N. F., M. D. Waterfield, and J. J. Hsuan. (1992). Accelerated high-sensitivity microsequencing of proteins and peptides using a miniature reaction cartridge. *Protein Science* **1:**1215–1224.
176. Moritz, R. L., and R. J. Simpson. (1992). Application of capillary reversed-phase high-performance liquid chromatography to high-sensitivity protein sequence analysis. *J. Chromatogr.* **599:**119–130.
177. Blacher, R. W., and J. H. Wieser. (1993). On-line microbore HPLC detection of femtomole quantities of PTH-amino acids. *In* "Techniques in Protein Chemistry IV," pp. 427–433. Academic Press, San Diego.
178. Rohde, M. F., C. Clogston, L. A. Merewether, P. Derby, and K. D. Nugent. (1995). Protein sequence analysis using microbore PTH separations. *In* "Techniques in Protein Chemistry VI," pp. 201–208. Academic Press, San Diego.
179. Bailey, J. M. (1995). Chemical methods of protein sequence analysis. *J. Chromatogr. A* **705:**47–65.
180. Aebersold, R., J. D. Watts, H. D. Morrison, and E. J. Bures. (1991). Determination of the site of tyrosine phosphorylation at the low picomole level by automated solid-phase sequence analysis. *Anal. Biochem.* **199:**51–60.
181. Crankshaw, M. W., and G. A. Grant. (1993). Identification of modified PTH-amino acids in protein sequence analysis. The Association of Biomolecular Resource Facilities, St. Louis. 38 pp.
182. L'Italien, J. J., and S. B. H. Kent. (1984). *J. Chromatogr.* **283:**149–156.
182. Palacz, Z., J. Salnikow, S.-W. Jin, and B. Wittmann-Liebold. (1984). 4-(*N-tert*-Butyloxycarbonylaminomethyl) phenylisothiocyanate: Its synthesis and use in microsequencing. *FEBS Lett.* **176:**365–370.
184 Jin, S. W., G. X. Chen, Z. Palacz, and B. Wittmann Liebold. (1986). A new sensitive Edman-type reagent: 4-(*N*-1-Dimethylaminonaphthalene-5-sulfonylamino)phenyl isothiocyanate. Its synthesis and application for micro-sequencing of polypeptides. *FEBS Lett.* **198:**150–154.
185. Hirano, H., and B. Wittmann Liebold. (1986). Protein pico-sequencing with 4-([5-(dimethylamino)-1-naphthylsulfonyl]amino) phenyl isothiocyanate. *Biol. Chem. Hoppe-Seyler* **367:**1259–1265.
186. Miyano, H., T. Nakajima, and K. Imai. (1987). Micro-scale sequence analysis from the N-terminus of peptides using the fluorogenic Edman reagent 4-*N,N*-dimethylamino-1-naphthyl isothiocyanate. *Biomed. Chromatogr.* **2:**139–144.
187. Tsugita, A., M. Kamo, C. S. Jone, and N. Shikama. (1989). Sensitization of Edman amino acid derivatives using the fluorescent reagent, 4-aminofluorescein. *J. Biochem.* **106:**60–65.
188. Farnsworth, V., and K. Steinberg. (1993). Automated subpicomole protein sequencing using an alternative postcleavage conversion chemistry. *Anal. Biochem.* **215:**190–199.

189. Imakyure, O., M. Kai, T. Mitsui, H. Nohta, and Y. Ohkura. (1993). Fluorogenic reagents for amino acids in high-performance liquid chromatography, phenanthraoxazolylphenylisothiocyanates. *Anal. Sci.* **9:**647–652.
190. Imakyure, O., M. Kai, and Y. Ohkura. (1994). A fluorogenic reagent for amino acids in liquid chromatography, (4-(2-cyanoisoindoyly)phenylisothiocyanate). *Anal. Chim. Acta.* **291:**197–204.
191. Muramoto, K., K. Nokihara, A. Ueda, and H. Kaqmiya. (1994). Gas-phase microsequencing of peptides and proteins with a fluorescent Edman-type reagent, fluorescein isothiocyanate. *Biosci. Biotech. Biochem.* **58:**300–304.
192. Chang, J. Y. (1977). High-sensitivity sequence analysis of peptides and proteins of 4-NN-dimethylaminoazodenzene 4′-isothiocyanate. *Biochem. J.* **163:**517–520.
193. Silver, J., and L. Hood. (1975). Automated microsequence analysis by use of radioactive phenylisothiocyanate. *Anal. Biochem.* **67:**392–396.
194. Ender, B. E., and H. G. Gasen. (1988). The synthesis of a phosphorus-32-labeled Edman reagent for the sensitive identification of amino-acid derivatives. *Hoppe-Seyler's Z. Physiol. Chem.* **365:**839–845.
195. Aebersold, R., E. J. Bures, M. Namchuk, M. H. Goghari, B. Shushan, and T. C. Covey. (1992). Design, synthesis, and characterization of a protein sequencing reagent yielding amino acid derivatives with enhanced detectability by mass spectrometry. *Protein Sci.* **1:**494–503.
196. Bures, E. J., H. Nika, D. T. Chow, H. D. Morrison, D. Hess, and R. Aebersold. (1995). Synthesis of the protein-sequencing reagent 4-(3-pyridinylmethylaminocarboxypropyl) phenyl isothiocyanate and characterization of 4-(3-Pyridinylmethylaminocarboxypropyl) phenylthiohydantoins. *Anal. Biochem.* **224:**364–372.
197. Nika, H., D. T. Chow, E. J. Bures, D. Hess, H. D. Morrison, and R. Aebersold. (1994). Automated subpicomole level protein and peptide sequencing. *J. Prot. Chem.* **13:**439–441.
198. Hess, D., H. Nika, D. T. Chow, E. J. Bures, H. D. Morrison, and R. Aebersold. (1995). Liquid-chromatography electrospray-ionization mass-spectrometry of 4-(3-pyridinylmethylaminocarboxypropyl) phenylthiohydantoins. *Anal. Biochem.* **224:**373–381.
199. Tull, D., D. L. Burgoyne, D. T. Chow, S. G. Withers, and R. A. Aebersold. (1996). A mass spectrometry based approach for probing enzyme active sites: Identification of Glu 127 in *Cellulomanas fimi* exoglycanase as the residue modified by *N*-bromoacetyl cellobiosylamine. *Anal. Biochem.* **234:**119–125.
200. Bailey, J. M., O. Tu, C. Basic, G. Issai, and J. E. Shively. (1994). Strategies for increasing the sensitivity of N-terminal sequence analysis. *In* "Techniques in Protein Chemistry VI," pp. 169–178. Academic Press, San Diego.
201. Basic, C., J. M. Bailey, and T. D. Lee. (1995). An electrospray ionization study of some novel alkylamine thiohydantoin amino acid derivatives. *J. Am. Soc. Mass Spectrom.* **6:**1211–1220.
202. Stolowitz, M. L., C.-S. Kim, S. R. Marsh, and L. Hood. (1993). Thiobenzoylation method of protein sequencing: Gas chromatography/mass spectrometric detection of 5-acetoxy-2-phenylthiazoles. *In* "Methods in Protein Sequence Analysis," pp. 37–44. Plenum Press, New York.
203. Chait, B. T., R. Wang, R. C. Beavis, and S. B. H. Kent. (1993). Protein ladder sequencing. *Science* **262:**89–92.
204. Wang, R., B. T. Chait, and S. B. H. Kent. (1994). Protein ladder sequencing: Towards automation. *In* "Techniques in Protein Chemistry V," pp. 19–26. Academic Press, San Diego.
205. Bartlet-Jones, M., W. A. Jeffery, H. F. Hansen, and D. J. C. Pappin. (1994). Peptide ladder sequencing by mass-spectrometry using a novel, volatile degradation reagent. *Rapid Commun. Mass Spectrom.* **8:**737–742.
206. Inglis, A. S. (1991). Chemical procedures for C-terminal sequencing of peptides and proteins. *Anal. Biochem.* **195:**183–196.
207. Casagranda, F., and J. F. Wilshire. (1994). C-Terminal sequencing of peptides. The thiocyanate degradation method. *Methods Mol. Biol.* **32:**335–349.
208. Tsugita, A., K. Takamoto, M. Kamo, and H. Iwadate. (1992). C-Terminal sequencing of proteins. A novel partial acid hydrolysis and analysis by mass spectrometry. *Eur. J. Biochem.* **206:**691–696.
209. Klarskov, K., K. Breddam, and P. Roepstorff. (1989). C-Terminal sequence determination of peptides degraded with carboxypeptidases of different specificities and analyzed by 252-Cf plasma desorption mass spectrometry. *Anal. Biochem.* **180:**28–37.
210. Schar, M., K. O. Bornsen, and E. Gassmann. (1991). Fast protein sequence determination with matrix-assisted laser desorption and ionization mass spectrometry. *Rapid Commun. Mass Spectrom.* **5:**319–326.

211. Thiede, B., B. WittmannLiebold, M. Bienert, and E. Krause. (1995). MALDI-MS for C-terminal sequence determination of peptides and proteins degraded by carboxypeptidase-Y and carboxypeptidase-P. *FEBS Lett.* **357:**65–69.
212. Woods, A. S., A. Y. C. Huang, R. J. Cotter, G. R. Pasternack, D. M. Pardoll, and E. M. Jaffee. (1995). Simplified high-sensitivity sequencing of a major histocompatibility complex class I-associated immunoreactive peptide using matrix-assisted laser-desorption ionization mass-spectrometry. *Anal. Biochem.* **226:**15–25.
213. Aldrich, C. J., A. Decloux, A. S. Woods, R. J. Cotter, M. J. Soloski, and J. Forman. (1994). Identification of a tap-dependent leader peptide recognized by alloreactive T-cells specific for a class Ib antigen. *Cell* **79:**649–658.
214. Patterson, D. H., G. E. Tarr, F. E. Regnier, and S. A. Martin. (1995). C-Terminal ladder sequencing via matrix-assisted laser-desorption mass-spectrometry coupled with carboxypeptidase-Y time-dependent and concentration-dependent digestions. *Anal. Chem.* **67:**3971–3978.
215. Lottspeich, F., and A. Henschen. (1982). Proteins, peptides and amino acids. *In* "HPLC in Biochemistry." VCH-Verlag, Weinheim.
216. Hunkapiller, M. W., J. E. Strickler, and K. J. Wilson. (1984). Contemporary methodology for protein structure determination. *Science* **226:**304–311.
217. Irvine, G. B. (1994). Amino acid analysis. *Methods Mol. Biol.* **32:**257–265.
218. Vorm, O., and P. Roepstorff. (1994). Peptide sequence information derived by partial acid-hydrolysis and matrix-assisted laser desorption/ionization mass-spectrometry. *Biol. Mass Spectrom.* **23:**734–740.
219. Knierman, M. D., J. E. Coligan, and K. C. Parker. (1994). Peptide fingerprints after partial acid-hydrolysis—analysis by matrix-assisted laser-desorption ionization mass-spectrometry. *Rapid Commun. Mass Spectrom.* **8:**1007–1010.
220. Biemann, K. (1990) Sequencing of peptides by tandem mass spectrometry and high-energy collision-induced dissociation. *Methods Enzymol.* **193:**455–479.
221. Hunt, D. F., J. R. Yates, III, J. Shabanowitz, S. Winston, and C. R. Hauer. (1986). Protein sequencing by tandem mass spectrometry. *Proc. Natl. Acad. Sci. USA* **83:**6233–6237.
222. Lewis, K. C., D. M. Dohmeier, J. W. Jorgenson, S. L. Kaufman, F. Zarrin, and F. D. Dorman. (1994). Electrospray-condensation particle counter: A molecule-counting LC detector for macromolecules. *Anal. Chem.* **66:**2285–2292.
223. Wilm, M. S., and M. Mann. (1994). Electrospray and Taylor–Cone theory, Dole's beam of macromolecules at last? *Int. J. Mass Spectrom. Ion Processes* **136:**167–180.
224. Andren, P. E., and R. M. Caprioli. (1995). *In vivo* metabolism of substance P in rat striatum utilizing microdialysis/liquid chromatography/micro-electrospray mass spectrometry. *J. Mass Spectrom.* **30:**817–824.
225. Gale, D. C., and R. D. Smith. (1993). Small volume and low flow-rate electrospray ionization mass spectrometry of aqueous samples. *Rapid Commun. Mass Spectrom.* **7:**1017–1021.
226. Andren, P. E., M. R. Emmett, and R. M. Caprioli. (1994). Micro-electrospray: Zeptomole/attomole per microliter sensitivity for peptides. *Anal. Chem.* **5:**867–869.
227. Valaskovic, G. A., N. L. Kelleher, D. P. Little, D. J. Aaserud, and F. W. McLafferty. (1995). Attomole-sensitivity electrospray source for large-molecule mass spectrometry. *Anal. Chem.* **67:**3802–3805.
228. Wahl, J. H., D. R. Goodlett, H. R. Udseth, and R. D. Smith. (1993). Use of small-diameter capillaries for increasing peptide and protein detection sensitivity in capillary electrophoresis-mass spectrometry. *Electrophoresis* **14:**448–457.
229. Davis, M. T., D. C. Stahl, S. A. Hefta, and T. D. Lee. (1995). A microscale electrospray interface for online, capillary liquid-chromatography tandem mass-spectrometry of complex peptide mixtures. *Anal. Chem.* **67:**4549–4556.
230. Mann, M., and M. Wilm. (1995). Electrospray mass spectrometry for protein characterization. *Trends Biochem. Sci.* **20:**219–224.
231. Calvio, C., G. Neubauer, M. Mann, and A. I. Lamond. (1995). Identification of hnRNP P2 as Tls/Fus using electrospray mass-spectrometry. *RNA* **1:**724–733.
232. Spengler, B., D. Kirsch, R. Kaufmann, and E. Jaeger. (1992). Peptide sequencing by matrix-assisted laser-desorption mass spectrometry. *Rapid Commun. Mass Spectrum.* **6:**105–108.
233. Spengler, B., D. Kirsch, and R. Kaufmann. (1992). Fundamental aspects of postsource decay in matrix-assisted laser desorption mass spectrometry. 1. Residual gas effects. *J. Phys. Chem.* **96:**9678–9684.

234. Kaufmann, R., D. Kirsch, and B. Spengler. (1994). Sequencing of peptides in a time-of-flight mass spectrometer: Evaluation of postsource decay following matrix-assisted laser desorption ionization (MALDI). *Int. J. Mass Spectrom. Ion Processes* **131**:355–385.
235. Kellner, R., G. Talbo, T. Houthaeve, and M. Mann. (1995). Edman degradation and MALDI sequencing enables N- and C-terminal sequence analysis of peptides. *In* "Techniques in Protein Chemistry VI," pp. 47–54. Academic Press, San Diego.
236. Huberty, M. C., J. E. Vath, W. Yu, and S. A. Martin. (1993). Site-specific carbohydrate identification in recombinant proteins using MALD-TOF MS. *Anal. Chem.* **65**:2791–2800.
237. Zhou, J., W. Ens, N. Poppe-Schriemer, K. G. Standing, and J. B. Westmore. (1993). Cleavage of interchain disulfide bonds following matrix-assisted laser desorption. *Int. J. Mass Spectrom. Ion Processes* **126**:115–122.
238. Yu, W. J. E. Vath, M. C. Huberty, and S. A. Martin. (1993). Identification of the facile gas-phase cleavage of the Asp–Pro and Asp–Xxx peptide bonds in matrix-assisted laser desorption time-of-flight mass spectrometry. *Anal. Chem.* **65**:3015–3023.
239. Talbo, G., and M. Mann. (1994). Distinction between phosphorylated and sulfated peptides by matrix assisted laser desorption ionization reflector mass spectrometry at the sub picomole level. *In* "Techniques in Protein Chemistry V," pp. 105–113. Academic Press, San Diego.
240. Spengler, B., F. Lutzenkirchen, and R. Kaufmann. (1993). On-target deuteration for peptide sequencing by laser mass spectrometry. *Org. Mass Spectrom.* **28**:1482–1490.
241. James, P., M. Quadroni, E. Carafoli, and G. Gonnet. (1994). Protein identification in DNA databases by peptide mass fingerprinting. *Protein Sci.* **3**:1347–1350.
242. Kaufmann, R., D. Kirsch, J. L. Tourmann, J. Machold, F. Hucho, Y. Utkin, and V. Tsetlin. (1995). Matrix-assisted laser-desorption ionization (MALDI) and post-source decay (PSD) product ion mass analysis localize a photolabel cross-linked to the delta-subunit of NACHR protein by neurotoxin-II. *Eur. Mass Spectrom.* **1**:313–325.
243. Yates, J. R., III, A. L. McCormack, J. B. Hayden, and M. P. Davey. (1994). Sequencing peptides derived from the Class II major histocompatibility complex by tandem mass spectrometry. *In* "Cell Biology: A Laboratory Handbook," Vol. 3, pp. 380–388. Academic Press, San Diego.
244. Kaufmann, R., B. Spengler, and F. Lutzenkirchen. (1993). Mass spectrometric sequencing of linear peptides by product-ion analysis in a reflection time-of-flight mass spectrometer using matrix-assisted laser desorption ionization. *Rapid Commun. Mass Spectrom.* **7**:902–910.
245. Cornish, T. J., and R. J. Cotter. (1994). A curved field reflectron time-of-flight mass-spectrometer for the simultaneous focusing of metastable product ions. *Rapid Commun. Mass Spectrom.* **8**:781–785.
246. Fabris, D., M. M. Vestling, M. M. Cordero, V. M. Doroshenko, R. J. Cotter, and C. Fenselau. (1995). Sequencing electroblotted proteins by tandem mass spectrometry. *Rapid Commun. Mass Spectrom.* **9**:1051–1055.
247. Vestling, M. M., and C. Fenselau. (1994). Poly(vinylidene difluoride) membranes as the interface between laser desorption mass spectrometry, gel electrophoresis, and *in situ* proteolysis. *Anal. Chem.* **66**:471–477.
248. Vestling, M. M., and C. C. Fenselau. (1994). Protease digestions of PVDF membranes for matrix-assisted laser desorption mass spectrometry. *In* "Techniques in Protein Chemistry V," pp. 59–67. Academic Press, San Diego.
249. Loo, R. R. O., N. Dales, and P. C. Andrews. (1994). Surfactant effects on protein structure examined by electrospray ionization mass spectrometry. *Protein Sci.* **3**:1975–1983.
250. Bhown, A. S., J. E. Mole, F. Hunter, and J. C. Bennett. (1980). High-sensitivity sequence determination of proteins quantitatively recovered from sodium dodecyl sulfate gels using an improved electrodialysis procedure. *Anal. Biochem.* **103**:184–190.
251. Hunkapiller, M. W., E. Lujan, F. Ostrander, and L. E. Hood. (1983). Isolation of microgram quantities of proteins from polyacrylamide gels for amino acid sequence analysis. *Methods Enzymol.* **91**:227–236.
252. Hager, D. A., and R. R. Burgess. (1980). Elution of proteins from sodium dodecyl sulfate–polyacrylamide gels, removal of sodium dodecyl sulfate, and renaturation of enzymatic activity: Results with sigma subunit of *Escherichia coli* RNA polymerase, wheat germ DNA topoisomerase, and other enzymes. *Anal. Biochem.* **109**:76–86.
253. Clauser, K. R., S. C. Hall, D. M. Smith, J. W. Webb, L. E. Andrews, H. M. Tran, L. B. Epstein, and A. L. Burlingame. (1995). Rapid mass spectrometric peptide sequencing and mass matching for characterization of human melanoma proteins isolated by two-dimensional PAGE. *Proc. Natl. Acad. Sci. USA* **92**:5072–5076.

254. Hall, S. C., D. M. Smith, F. R. Masiarz, V. W. Soo, H. M. Tran, L. B. Epstein, and A. L. Burlingame. (1993). Mass spectrometric and Edman sequencing of lipocortin I isolated by two-dimensional SDS/PAGE of human melanoma lysates. *Proc. Natl. Acad. Sci. USA* **90:**1927–1931.
255. Haebel, S., C. Jensen, S. O. Andersen, and P. Roepstorff. (1995). Isoforms of a cuticular protein from larvae of the meal beetle. *Tenebrio molitor,* studied by mass spectrometry in combination with Edman degradation and two-dimensional polyacrylamide gel electrophoresis. *Protein Sci.* **4:**394–404.
256. Konigsberg, W. H., and L. Henderson. (1983). Removal of sodium dodecyl sulfate from proteins by ion-pair extraction. *Methods Enzymol.* **91:**254–259.
257. Breme, U., J. Breton, C. Visco, F. Orsini, and P. G. Righetti. (1995). Characterization of proteins by sequential isoelectric-focusing on immobilized pH gradients and electrospray mass-spectrometry. *Electrophoresis* **16:**1381–1384.
258. Kawasaki, H., Y. Emori, and K. Suzuki. (1990). Production and separation of peptides from proteins stained with Coomassie brilliant blue R-250 after separation by sodium dodecyl sulfate–polyacrylamide gel electrophoresis. *Anal. Biochem.* **191:**332–336.
259. Kawasaki, H., and K. Suzuki. (1990). Separation of peptides dissolved in a sodium dodecyl sulfate solution by reversed-phase liquid chromatography: Removal of sodium dodecyl sulfate from peptides using an ion-exchange precolumn. *Anal. Biochem.* **186:**264–268.
260. Rosenfeld, J., J. Capdevielle, J. C. Guillemot, and P. Ferrara. (1992). In-gel digestion of proteins for internal sequence analysis after one- or two-dimensional gel electrophoresis. *Anal. Biochem.* **203:**173–179.
261. Hellman, U., C. Wernsted, J. Gonez, and C. H. Heldin. (1995). Improvement of an in-gel digestion procedure for the micropreparation of internal protein-fragments for amino acid sequencing. *Anal. Biochem.* **224:**451–455.
262. Xiang, F., and R. C. Beavis. (1994). A method to increase contaminant tolerance in protein matrix-assisted laser desorption/ionization by the fabrication of thin protein-doped polycrystalline films. *Rapid Commun. Mass Spectrom.* **8:**199–204.
263. Merewether, L. A., C. L. Clogston, S. D. Patterson, and H. S. Lu. (1995). Peptide mapping at the 1μg level: In-gel vs PVDF digestion techniques. *In* "Techniques in Protein Chemistry V," pp. 153–160. Academic Press, San Diego.
264. Kirchner, M., J. Fernandez, Q. A. Shakey, F. Gharahdaghi, and S. Mische. (1996). Enzymatic digestion of PVDF-bound proteins: A survey of sixteen non-ionic detergents. *In* "Techniques in Protein Chemistry VII," pp. 287–298. Academic Press, San Diego.
265. Cano, L., K. M. Swiderick, and J. E. Shively. (1995). Comparison of ESI-MS, LSIMS, and MALDI-TOF-MS for the primary structure analysis of a monoclonal antibody. *In* "Techniques in Protein Chemistry VI," pp. 21–30. Academic Press, San Diego.
266. Moritz, E., O. Vorm, M. Mann, and P. Roepstorff. (1994). Identification of proteins in polyacrylamide gels by mass spectrometric peptide mapping combined with database search. *Biol. Mass Spectrom.* **23:**249–261.
267. Moritz, R., J. Eddes, H. Ji, G. E. Reid, and R. J. Simpson. (1995). Rapid separation of proteins and peptides using conventional silica-based supports: Identification of 2-D gel proteins following in-gel proteolysis. *In* "Techniques in Protein Chemistry VI," pp. 311–319. Academic Press, San Diego.
268. Swiderick, K. M., M. L. Klein, S. A. Hefta, and J. E. Shively. (1995). Strategies for the removal of ionic and nonionic detergents from protein and peptide mixtures for on- and off-line liquid chromatography mass spectrometry (LC-MS). *In* "Techniques in Protein Chemistry VI," pp. 267–275. Academic Press, San Diego.
269. Stoney, K., and K. Nugent. (1995). Online sample preparation of complex biological samples prior to analysis by HPLC, LC/MS and/or protein sequencing. *In* "Techniques in Protein Chemistry VI," pp. 277–284. Academic Press, San Diego.
270. Jeno, P., T. Mini, S. Moes, E. Hintermann, and M. Horst. (1995). Internal sequences from proteins digested in polyacrylamide gels. *Anal. Biochem.* **224:**75–82.
271. Brune, D. C. (1992). Alkylation of cysteine with acrylamide for protein sequence analysis. *Anal. Biochem.* **207:**285–290.
272. Bauw, G., J. Van Damme, M. Puype, J. Vandekerckhove, B. Gesser, G. P. Ratz, J. B. Lauridsen, and J. E. Celis. (1989). Protein-electroblotting and microsequencing strategies in generating protein data bases from two-dimensional gels. *Proc. Natl. Acad. Sci. USA* **86:**7701–7705.
273. Bauw, G., H. H. Rasmussen, M. van den Bulcke, J. van Damme, M. Puype, B. Gesser, J. E.

Celis, and J. Vandekerckhove. (1990). Two-dimensional gel electrophoresis, protein electroblotting and microsequencing: A direct link between proteins and genes. *Electrophoresis* **11:**528–536.
274. Lombard-Platet, G., and P. Jalinot. (1993). Funnel-well SDS-PAGE: A rapid technique for obtaining sufficient quantities of low-abundance proteins for internal sequence analysis. *BioTechniques* **15:**668–672.
275. Rasmussen, H. H., J. Van Damme, M. Bauw, M. Puype, B. Gesser, J. E. Celis, and J. Vandekerkhove. (1991). Protein-electroblotting and microsequencing in establishing integrated human protein databases. *In* "Methods in Protein Sequence Analysis," pp. 103–114. Birkhauser Verlag, Denmark.
276. Vandekerkhove, J., M. Rider, H.-H. Rasmussen, S. De Boeck, M. Puype, J. Van Damme, B. Gesser, and J. Celis. (1993). Routine amino acid sequencing on 2D-gel separated proteins: A protein elution and concentration gel system. *In* "Methods in Protein Sequence Analysis," pp. 11–19. Plenum Press, New York.
277. Rider, M. H., M. Puype, J. Van Damme, K. Gevaert, S. De Boeck, J. D'Alayer, H. H. Rasmussen, J. E. Celis, and J. Vanderkerchove. (1995). An agarose-based gel-concentration system for microsequence and mass spectrometric characterization of proteins previously purified in polyacrylamide gels starting at low picomole levels. *Eur. J. Biochem.* **230:**258–265.
278. Gold, M. R., T. Yungwirth, C. L. Sutherland, R. J. Ingham, D. Vianzon, R. Chiu, I. van Oostveen, H. D. Morrison, and R. Aebersold. (1994).Purfication and identification of tyrosine-phosphorylated proteins from B lymphocytes stimulated through the antigen receptor. *Electrophoresis* **15:**441–453.
279. Vestling, M. M., and C. Fenselau. (1995). Surfaces for interfacing protein gel electrophoresis directly with mass spectrometry. *Mass Spectrom. Rev* **14:**169–178.
280. Henzel, W. J., T. M. Billeci, J. T. Stults, S. C. Wong, C. Grimley, and C. Watanabe. (1993). Identifying proteins from two-dimensional gels by molecular mass searching of peptide fragments in protein sequence databases. *Proc. Natl. Acad. Sci. USA* **90:**5011–5015.
281. Pappin, D. J. C., P. Hojrup, and A. J. Bleasby. (1993). Rapid identification of proteins by peptide-mass fingerprinting. *Curr. Biol.* **3:**327–332.
282. Zhang, W., A. J. Czernik, T. Yungwirth, R. Aebersold, and B. T. Chait. (1994). Matrix-assisted laser desorption mass spectrometric peptide mapping of proteins separated by two-dimensional gel electrophoresis: Determination of phosphorylation in synapsin I. *Protein Sci.* **3:**677–686.
283. Hess, D., T. C. Covey, R. Winz, R. W. Brownsey, and R. Aebersold. (1993). Analytical and micropreparative peptide mapping by high performance liquid chromatography/electrospray mass spectrometry of proteins purified by gel electrophoresis. *Protein Sci.* **2:**1342–1351.
284. Tempst, P., A. J. Link, L. R. Riviere, M. Fleming, and C. Elicone. (1990). Internal sequence analysis of proteins separated on polyacrylamide gels at the submicrogram level: improved methods, applications and gene cloning strategies. *Electrophoresis* **11:**537–553.
285. Fernandez, J., M. DeMott, D. Atherton, and S. M. Mische (1992). Internal sequence analysis: Enzymatic digestion for less than 10 μg of protein bound to polyvinylidene difluoride or nitrocellulose membranes. *Anal. Biochem.* **201:**255–264.
286. Fernandez, J., L. Andrews, and S. M. Mische. (1994). A one-step enzymatic digestion procedure for PVDF-bound proteins that does not require PVP-40. *In* "Techniques in Protein Chemistry V," pp. 215–222. Academic Press, San Diego.
287. Fernandez, J., F. Gharahdaghi, and S. M. Mische. (1995). Enzymatic digestion of PVDF-bound proteins in the presence of glucopyranoside detergents: Applicability to mass spectrometry. *In* "Techniques in Protein Chemistry VI," pp. 135–142. Academic Press, San Diego.
288. Sutton, C. W., K. S. Pemberton, J. S. Cottrell, J.M. Corbett, C. H. Wheeler, M. J. Dunn, and D. J. Pappin. (1995). Identification of myocardial proteins from two-dimensional gels by peptide mass fingerprinting. *Electrophoresis* **16:**308–316.
289. Patterson, S. D., D. Hess, T. Yungwirth, and R. Aebersold. (1992). High-yield recovery of electroblotted proteins and cleavage fragments from a cationic polyvinylidene fluoride-based membrane. *Anal. Biochem.* **202:**193–203.
290. Aebersold, R., S. D. Patterson, and D. Hess. (1992). Strategies for the isolation of peptides from low-abundance proteins for internal sequence analysis. *In* "Techniques in Protein Chemistry III," pp. 87–96. Academic Press, San Diego.
291. Patterson, S. D. (1995). Matrix-assisted laser-desorption/ionization mass spectrometric approaches for the identification of gel-separated proteins in the 5–50 pmol range. *Electrophoresis* **16:**1104–1114.

292. Schagger, H., and G. von Jagow. (1987). Tricine–sodium dodecyl sulfate–polyacrylamide gel electrophoresis for the separation of proteins in the range from 1 to 100 kDa. *Anal. Biochem.* **166:**368–379.
293. Reid, G. E., H. Ji, J. S. Eddes, R. L. Mortiz, and R. J. Simpson. (1995). Nonreducing two-dimensional polyacrylamide gel electrophoretic analysis of human colonic proteins. *Electrophoresis* **16:**1120–1130.
294. Chait, B. T., and S. B. H. Kent, (1992). Weighing naked proteins: Practical high-accuracy mass measurement of peptides and proteins. *Science* **257:**1885–1894.
295. Henzel, W. J., J. T. Stults, and C. Watanabe. (1989). *In* "Abstracts of the 3rd Protein Soc. Symp., Seattle, July 29–Aug. 2."
296. Yates, J. R., III, P. R. Griffin, T. Hunkapiller, S. Speicher, and L. E. Hood. (1991). *In* "Abstracts of the 5th Protein Soc. Symp., Baltimore, June 22–26."
297. James, P., M. Quadroni, E. Carafoli, and G. Gonnett. (1993). Protein identification by mass profile fingerprinting. *Biochem. Biophys. Res. Commun.* **195:**58–64.
298. Mann, M., P. Hojrup, and P. Roepstorff. (1993). Use of mass spectrometric molecular weight information to identify proteins in sequence databases. *Biol. Mass Spectrom.* **22:**338–345.
299. Yates, J. R., III, S. Speicher, P. R. Griffin, and T. Hunkapiller. (1993). Peptide mass maps: A highly informative approach to protein identification. *Anal. Biochem.* **214:**397–408.
300. Patterson, S. D. (1994). From electrophoretically separated protein to identification: Strategies for sequence and mass analysis. *Anal. Biochem.* **221:**1–15.
301. Cottrell, J. S. (1994). Protein identification by peptide mass fingerprinting. *Pept. Res.* **7:**115–124.
302. Rasmussen, H. H., E. Mortz, M. Mann, P. Roepstorff, and J. E. Celis. (1994). Identification of transformation sensitive proteins recorded in human two-dimensional gel protein databases by mass spectrometric peptide mapping alone and in combination with microsequencing. *Electrophoresis* **15:**406–416.
303. Winz, R., D. Hess, R. Aebersold, and R. W. Brownsey. (1994). Unique structural features and differential phosphorylation of the 280-kDa component (isozyme) of rat liver acetyl-CoA carboxylase. *J. Biol. Chem.* **269:**14438–14445.
304. Daga, A., V. Micol, D. Hess, R. Aebersold, and G. Attardi. (1993). Molecular characterization of the transcription termination factor from human mitochondria. *J. Biol. Chem.* **268:**8123–8130.
305. Pappin, D. J. C., and A. J. Bleasby. (1994). Further development of a peptide-mass database for the rapid identification of proteins by mass spectrometric techniques. *Protein Sci.* **3** (Suppl. 1):81.
306. Pappin, D. J. C., D. Rahman, H. F. Hansen, M. Bartlet-Jones, W. Jeffery, and A. J. Bleasby. (1995). Chemistry, mass spectrometry and peptide-mass databases: Evolution of methods for the rapid identification and mapping of cellular proteins. *In* "Mass Spectrometry in the Biological Sciences," pp. 135–150. Humana Press, Totowa, NJ.
307. McCloskey, J. A. (1990). Introduction of deuterium by exchange for measurement by mass spectrometry. *Methods Enzymol.* **193:**329–338.
308. Sepetov, N. F., O. L. Issakova, M. Lebl, K. Swiderek, D. C. Stahl, and T. D. Lee. (1993). The use of hydrogen-deuterium exchange to facilitate peptide sequencing by electrospray tandem mass spectrometry. *Rapid Commun. Mass Spectrom.* **7:**58–62.
309. Knapp, D. R. (1990). Chemical derivatization for mass spectrometry. *Methods Enzymol.* **193:**314–329.
310. Craig, A. G., W. H. Fischer, J. E. Rivier, J. M. McIntosh, and W. R. Gray. (1995). MS based scanning methodologies applied to *Conus* venom. *In* "Techniques in Protein Chemistry VI," pp. 31–38. Academic Press, San Diego.
311. Mann, M., and M. Wilm. (1994). Error-tolerant identification of peptides in sequence databases by peptide sequence tags. *Anal. Chem.* **66:**4390–4399.
312. Eng, J. K., A. L. McCormack, and J. R. I. Yates. (1994). An approach to correlate tandem mass spectral data pf peptides with amino acid sequences in a protein database. *J. Am. Soc. Mass Spectrom.* **5:**976–989.
313. Yates, J. R., III, J. K. Eng, A. L. McCormack, and D. Schieltz. (1995). Method to correlate tandem mass-spectra of modified peptides to amino-acid-sequences in the protein database. *Anal. Chem.* **67:**1426–1436.
314. Dirksen, M. L., and A. Chrambach. (1972). *Separ. Sci.* **7:**744–772.
315. Chiari, M., P. G. Righetti, A. Negri, F. Ceciliani, and S. Ronchi. (1992). Preincubation with cysteine prevents modification of sulfhydryl groups in proteins by unreacted acrylamide in a gel. *Electrophoresis* **13:**882–884.

316. Chiari, M., A. Manocchi, and P. G. Righetti. (1990). Formation of a cysteine–acrylamide adduct in isoelectric focusing gels. *J. Chromatogr.* **500:**697–704.
317. Ploug, M., A. L. Jensen, and V. Barkholt. (1989). Determination of amino acid compositions and NH_2-terminal sequences of peptides electroblotted onto PVDF membranes from tricine–sodium dodecyl sulfate–polyacrylamide gel electrophoresis: Application to peptide mapping of human complement component C3. *Anal. Biochem.* **181:**33–39.
318. le Maire, M., S. Deschamps, J. V. Moller, J. P. Le Caer, and J. Rossier. (1993). Electrospray ionization mass spectrometry on hydrophobic peptides electroeluted from sodium dodecyl sulfate–polyacrylamide gel electrophoresis application to the topology of the sarcoplasmic reticulum Ca^{2+} ATPase. *Anal. Biochem.* **214:**50–57.
319. Bonaventura, C., J. Bonaventura, R. Stevens, and D. Millington. (1994). Acrylamide in polyacrylamide gels can modify proteins during electrophoresis. *Anal. Biochem.* **222:**44–48.
320. Klarskov, K., D. Roecklin, B. Bouchon, J. Sabatie, A. Van Dorssalaer, and R. Bischoff. (1994). Analysis of recombinant *Schistosoma mansoni* antigen rSmp28 by on-line liquid chromatography-mass spectrometry combined with sodium dodecyl sulfate polyacrylamide gel electrophoresis. *Anal. Biochem.* **216:**127–134.
321. Ploug, M., B. Stoffer, and A. L. Jensen. (1992). *In situ* alkylation of cysteine residues in a hydrophobic membrane protein immobilized on polyvinylidene difluoride membranes by electroblotting prior to microsequence and amino acid analysis. *Electrophoresis* **13:**148–153.
322. Moos, M., Jr., N. Y. Nguyen, and T. Y. Liu. (1988). Reproducible high yield sequencing of proteins electrophoretically separated and transferred to an inert support. *J. Biol. Chem.* **263:**6005–6008.
323. Dunbar, B., and S. B. Wilson. (1994). A buffer exchange procedure giving enhanced resolution to polyacrylamide gels prerun for protein sequencing. *Anal. Biochem.* **216:**227–228.
324. Fantes, K. H., and I. G. S. Furminger. (1967). Proteins, persulphate and disc electrophoresis. *Nature* **215:**750–751.
325. Van Dorsselaer, A., F. Bitsch, B. Green, S. Jarvis, P. Lepage, R. Bischoff, H. V. J. Kolbe, and C. Roitsch. (1990). Application of electrospray mass spectrometry to the characterization of recombinant proteins up to 44 kDa. *Biomed. Environ. Mass Spectrom.* **19:**692–704.
326. Krishna, R. G., and F. Wold. (1993). Post-translational modification of proteins. *Adv. Enzymol. Relat. Areas Mol. Biol.* **67:**265–298.
327. Krishna, R. G., and F. Wold. (1993). Post-translational modifications of proteins. *In* "Methods in Protein Sequence Analysis," pp. 167–172. Plenum Press, New York.
328. Cornish-Bowden, A. (1980). Critical values for testing the significance of amino acid composition indexes. *Anal. Biochem.* **105:**233–238.
329. Shaw, G. (1993). Rapid identification of proteins. *Proc. Natl. Acad. Sci. USA.* **90:**5138–5142.
330. Church, F. C., H. E. Swaisgood, and G. L. Catignani. (1984). Compositional analysis of proteins following hydrolysis by immobilized proteases. *J. Appl. Biochem.* **6:**205–211.
331. Eckerskorn, C., and F. Lottspeich. (1990). Combination of two-dimensional gel electrophoresis with microsequencing and amino acid composition analysis: Improvement of speed and sensitivity in protein characterization. *Electrophoresis* **11:**554–561.
332. Nakagawa, S., and T. Fukuda. (1989). Direct amino acid analysis of proteins electroblotted onto polyvinylidene difluoride membrane from sodium dodecyl sulfate–polyacrylamide gel. *Anal. Biochem.* **181:**75–78.
333. Tous, G. I., J. L. Fausnaugh, O. Akinyosoye, H. Lackland, P. Winter-Cash, F. J. Victoria, and S. Stein. (1989). Amino acid analysis on polyvinylidene difluoride membranes. *Anal. Biochem.* **179:**50–55.
334. Lottspeich, F., C. Eckerskorn, and R. Grimm. (1994). Amino acid analysis on microscale from electroblotted proteins. *In* "Cell Biology: A Laboratory Handbook," Vol. 3, pp. 417–421. Academic Press, New York.
335. Gharahdaghi, F., D. Atherton, M. DeMott, and S. M. Mische. (1992). Amino acid analysis of PVDF bound proteins. *In* "Techniques in Protein Chemistry III," pp. 249–260. Academic Press, San Diego.
336. Latter, G. I., E. Metz, S. Burbeck, and J. Leavitt. (1983). Measurement of amino acid composition of proteins by computerized microdensitometry of two-dimensional electrophoresis gels. *Electrophoresis* **4:**122–126.
337. Latter, G. I., S. Burbeck, J. Fleming, and J. Leavitt. (1984). Identification of polypeptides on two-dimensional gels by amino acid composition. *Clin. Chem.* **30:**1925–1932.

338. Amemiya, Y., and J. Miyahara. (1988). Imaging plate illuminates many fields. *Nature* **336:**89–90.
339. Johnston, R. F., S. C. Pickett, and D. L. Barker. (1990). Autoradiography using storage phosphor technology. *Electrophoresis* **11:**355–360.
340. Garrels, J. I., B. R. J. Franza, S. D. Patterson, K. Latham, D. Solter, C. Chang, and G. Latter. (1993). Protein databases constructed by quantitative two-dimensional gel electrophoresis. *In* "Methods in Protein Sequence Analysis, pp. 247–253. Plenum, New York.
341. Hobohm, U., T. Houthaeve, and C. Sander. (1994). Amino acid analysis and protein database compositional search as a rapid and inexpensive method to identify proteins. *Anal. Biochem.* **222:**202–209.
342. Jungblut, P., M. Dzionara, J. Klose, and B. Wittmann Leibold. (1992). Identification of tissue proteins by amino acid analysis after purification by two-dimensional electrophoresis. *J. Protein Chem.* **11:**603–612.
343. Sibbald, P. R., H. Sommerfeldt, and P. Argos. (1991). Identification of proteins in sequence databases from amino acid composition data. *Anal. Biochem.* **198:**330–333.
344. Eckerskorn, C., P. Jungblut, W. Mewes, J. Klose, and F. Lottspeich. (1988). Identification of mouse brain proteins after two-dimensional electrophoresis and electroblotting by microsequence analysis and amino acid composition analysis. *Electrophoresis* **9:**830–838.
345. Tempst, P., H. Erdjument-Bromage, P. Casteels, S. Geromanos, M. Lui, M. Powell, and R. W. Nelson. (1995). MALDI-TOF mass spectrometry in the protein biochemistry lab: From characterization of cell cycle regulators to the quest for novel antibiotics. *In* "Mass Spectrometry in the Biological Sciences," pp. 105–133. Humana Press, Totowa, NJ.
346. Hess, D., and R. Aebersold. (1994). Internal sequence analysis of proteins separated by polyacrylamide gel electrophoresis. *Methods: Comp. Methods Enzymol.* **6:**227–238.
347. Wong, S. C., C. Grimley, A. Padua, J. H. Bourell, and W. J. Henzel. (1993). Peptide mapping of 2-D gel proteins by capillary HPLC. *In* "Techniques in Protein Chemistry IV," pp. 371–378. Academic Press, San Diego.
348. Moritz, R. L., G. E. Reid, L. D. Ward, and R. J. Simpson. (1994). Capillary HPLC: A method for protein isolation and peptide mapping. *Methods: Comp. Methods Enzymol.* **6:**213–226.
349. Johnson, R. S., and K. A. Walsh. (1992). Sequence analysis of peptide mixtures by automated integration of Edman and mass spectrometric data. *Protein Sci.* **1:**1083–1091.
350. Thomas, D., S. D. Patterson, and R. A. Bradshaw. (1995). Shc binds to F-actin and translocates to the cytoskeleton upon nerve growth factor stimulation in PC12 cells. *J. Biol. Chem.* **270:**28924–28931.
351. Patterson, S. D., D. Thomas, and R. A. Bradshaw. (1996). Application of combined mass spectrometry and partial amino acid sequence to the identification of gel-separated proteins. *Electrophoresis* **17:**877–891.
352. Hunter, T. 1987. 1001 Protein kinases. *Cell* **50:**823–829.
353. Cobb, M. H., and E. J. Goldsmith. (1995). How MAP kinases are regulated. *J. Biol. Chem.* **270:**14843–14846.
354. Vojtek, A. B., and J. A. Cooper. (1995). Rho family members: Activators of MAP kinase cascades. *Cell* **82:**527–529.
355. Charbonneau, H., and N. K. Tonks. (1992). 1002 Protein phosphatases? *Annu. Rev. Cell Biol.* **8:**463–493.
356. Kolesnick, R., and D. W. Golde. (1994). The sphingomyelin pathway in tumor necrosis factor and interleukin-1 signaling. *Cell* **77:**325–328.
357. Schmidt, H. H., and U. Walter. (1994). NO at work. *Cell* **78:**919–925.
358. Menniti, F. S., K. G. Oliver, J. W. Putney, Jr., and S. B. Shears. (1993). Inositol phosphates and cell signaling: New views of InsP5 and InsP6. *Trends Biochem. Sci.* **18:**53–56.
359. Moolenaar, W. H. (1995). Lysophosphatidic acid, a multifunctional phospholipid messenger. *J. Biol. Chem.* **270:**12949–12952.
360. Majerus, P. W. (1992). Inositol phosphate biochemistry. *Annu. Rev. Biochem.* **61:**225–250.
361. Lalli, E., and P. Sassone Corsi. (1994). Signal transduction and gene regulation: The nuclear response to cAMP. *J. Biol. Chem.* **269:**17359–17362.
362. Hannun, Y. A. (1994). The sphingomyelin cycle and the second messenger function of ceramide. *J. Biol. Chem.* **269:**3125–3128.
363. Divecha, N., and R. F. Irvine. (1995). Phospholipid signaling. *Cell* **80:**269–278.
364. Clapham, D. E. (1995). Calcium signaling. *Cell* **80:**259–268.
365. Patterson, S. D., and J. I. Garrels. (1994). Two-dimensional gel electrophoresis of post-transla-

tional modifications. *In* "Cell Biology: A Laboratory Handbook," Vol. 3, pp. 249–257. Academic Press, New York.

366. Torres, C. R., and G. W. Hart. (1984). Topography and polypeptide distribution of terminal *N*-acetylglucosamine residues on the surfaces of intact lymphocytes. *J. Biol. Chem.* **259:**3308–3317.
367. Hart, G. W., R. S. Haltiwanger, G. D. Holt, and W. G. Kelly. (1989). Glycosylation in the nucleus and cytoplasm. *Annu. Rev. Biochem.* **58:**841–874.
368. Wells, K., and J. S. Cordingley. (1991). Detecting proteins containing 3,4-dihydroxyphenylalanine by silver staining of polacrylamide gels. *Anal. Biochem.* **194:**237–242.
369. Wold, F. (1981). *In vivo* chemical modification of proteins. *Annu. Rev. Biochem.* **50:**783–814.
370. Bradshaw, R. A., and A. E. Stewart. (1994). Analysis of protein modifications: Recent advances in detection, characterization and mapping. *Curr. Opin. Biotech.* **5:**85–93.
371. Schindler, C., K. Shuai, V. R. Prezioso, and J. E. Darnell, Jr. (1992). Interferon-dependent tyrosine phosphorylation of a latent cytoplasmic transcription factor. *Science* **257:**809–813.
372. Boyle, W. J., P. van der Geer, and T. Hunter. (1991). Phosphopeptide mapping and phosphoamino acid analysis by two-dimensional separation on thin-layer cellulose plates. *Methods Enzymol.* **201:**110–149.
373. van der Geer, P., and T. Hunter. (1994). Phosphopeptide mapping and phosphoamino acid analysis by electrophoresis and chromatography on thin-layer cellulose plates. *Electrophoresis* **15:**544–554.
374. Wettenhall, R. E., R. H. Aebersold, and L. E. Hood. (1991). Solid-phase sequencing of ^{32}P-labeled phosphopeptides at picomole and subpicomole levels. *Methods Enzymol.* **201:**186–199.
375. Fischer, E. H., and E. G. Krebs. (1989). Commentary on "the phosphorylase b to a converting enzyme of rabbit skeletal muscle." *Biochem. Biophys. Acta* **1000:**297–301.
376. Huang, J. M., Y. F. Wei, Y. H. Kim, L. Osterberg, and H. R. Matthews. (1991). Purification of a protein histidine kinase from the yeast *Saccharomyces cerevisiae.* The first member of this class of protein kinases. *J. Biol. Chem.* **266:**9023–9031.
377. Duclos, B., S. Marcandier, and A. J. Cozzone. (1991). Chemical properties and separation of phosphoamino acids by thin-layer chromatography and/or electophoresis. *Methods Enzymol.* **201:**10–21.
378. Johnson, L. N., and D. Barford. (1990). Glycogen phosphorylase. The structural basis of the allosteric response and comparison with other allosteric proteins. *J. Biol. Chem.* **265:**2409–2412.
379. Watts, J. D., J. S. Sanghera, S. L. Pelech, and R. Aebersold. (1993). Phosphorylation of serine 59 of p56lck in activated T cells. *J. Biol. Chem.* **268:**23275–23282.
380. Joung, I., T. Kim, L. A. Stolz, G. Payne, D. G. Winkler, C. T. Walsh, J. L. Strominger, and J. Shin. (1995). Modification of Ser59 in the unique N-terminal region of tyrosine kinase p56lck regulates specificity of its Src homology 2 domain. *Proc. Natl. Acad. Sci USA* **92:**5778–5782.
381. Wange, R. L., N. Isakov, T. R. Burke, Jr., A. Otaka, P. P. Roller, J. D. Watts, R. Aebersold, and L. E. Samelson. (1995). F2(Pmp)2-TAM zeta 3, a novel competitive inhibitor of the binding of ZAP-70 to the T cell antigen receptor, blocks early T cell signaling. *J. Biol. Chem.* **270:**944–948.
382. Watts, J. D., M. Affolter, D. L. Krebs, R. L. Wange, L. E. Samelson, and R. Aebersold. (1994). Identification by electrospray ionization mass spectrometry of the sites of tyrosine phosphorylation induced in activated Jurkat T cells on the protein tyrosine kinase ZAP-70. *J. Biol. Chem.* **269:**29520–29529.
383. Lis, H., and N. Sharon. (1993). Protein glycosylation. Structural and functional aspects. *Eur. J. Biochem.* **218:**1–27.
384. Endo, T., D. Groth, S. B. Prusiner, and A. Kobata. (1989). Diversity of oligosaccharide structures linked to asparagines of the scrapie prion protein. *Biochemistry* **28:**8380–8388.
385. Rush, R. S., P. L. Derby, D. M. Smith, C. Merry, G. Rogers, M. F. Rohde, and V. Katta. (1995). Microheterogeneity of erythropoietin carbohydrate structure. *Anal. Chem.* **67:**1442–1452.
386. Hatton, M. W. C., L. Marz, and E. Regoeczi. (1983). On the significance of heterogeneity of plasma glycoprotein possessing *N*-glycans of the complex type. *Trends Biochem. Sci.* **8:**287–291.
387. True, D. D., M. S. Singer, L. A. Lasky, and S. D. Rosen. (1990). Requirement for sialic acid on the endothelial ligand of a lymphocyte homing receptor. *J. Cell Biol.* **111:**2757–2764.
388. Rosen, S. D. (1990). The LEC-CAMs: An emerging family of cell–cell adhesion receptors based upon carbohydrate recognition. *Am. J. Respir. Cell Mol. Biol.* **3:**397–402.
389. Berg, E. L., L. M. McEvoy, C. Berlin, R. F. Bargatze, and E. C. Butcher. (1993). L-selectin-mediated lymphocyte rolling on MAdCAM-1. *Nature* **366:**695–698.

390. Corral, L., M. S. Singer, B. A. Macher, and S. D. Rosen. (1990). Requirement for sialic acid on neutrophils in a GMP-140 (PADGEM) mediated adhesive interaction with activated platelets. *Biochem. Biophys. Res. Commun.* **172:**1349–1356.
391. Kearse, K. P., and G. W. Hart. (1991). Lymphocyte activation induces rapid changes in nuclear and cytoplasmic glycoproteins. *Proc. Natl. Acad. Sci. USA* **88:**1701–1705.
392. Haltiwanger, R. S., W. G. Kelly, E. P. Roquemore, M. A. Blomberg, L.-Y. D. Dong, L. Kreppel, T.-Y. Chou, and G. W. Hart. (1992). Glycosylation of nuclear and cytoplasmic proteins is ubiquitous and dynamic. *Biochem. Soc. Trans.* **20:**264–269.
393. Schultz, A. M., L. E. Henderson, and S. Oroszlan. (1988). Fatty acylation of proteins. *Annu. Rev. Cell Biol.* **4:**611–647.
394. Paige, L. A., M. J. S. Nadler, M. L. Harrison, J. M. Cassady, and R. L. Geahlen. (1993). Reversible palmitoylation of the protein-tyrosine kinase p56lck. *J. Biol. Chem.* **268:**8669–8674.
395. Dizhoor, A. M., L. H. Ericsson, R. S. Johnson, S. Kumar, E. Olshevskaya, S. Zozulya, T. A. Neubert, L. Stryer, J. B. Hurley, and K. A. Walsh. (1992). The NH2 terminus of retinal recoverin is acylated by a small family of fatty acids. *J. Biol. Chem.* **267:**16033–16036.
396. Neubert, T. A., R. S. Johnson, J. B. Hurley, and K. A. Walsh. (1992). The rod transducin alpha subunit amino terminus is heterogeneously fatty acylated. *J. Biol. Chem.* **267:**18274–18277.
397. Johnson, R. S., H. Ohguro, K. Palczewski, J. B. Hurley, K. A. Walsh, and T. A. Neubert. (1994). Heterogeneous *N*-acylation is a tissue- and species-specific posttranslational modification. *J. Biol. Chem.* **269:**21067–21071.
398. Linder, M. E., P. Middleton, J. R. Hepler, R. Taussig, A. G. Gilman, and S. M. Mumby. (1993). Lipid modifications of G proteins: α subunits are palmitoylated. *Proc. Natl. Acad. Sci. USA.* **90:**3675–3679.
399. James, G., and E. N. Olson. (1990). Fatty acylated proteins as components of intracellular signaling pathways. *Biochemistry* **29:**2623–2634.
400. Muszbek, L., and M. Laposata. (1993). Myristoylation of proteins in platelets occurs predominantly through thioester linkages. *J. Biol. Chem.* **268:**8251–8255.
401. James, G., annd E. Olson. (1990). Experimental approaches to the study of reversible protein acylation in mammalian cells. *Methods: Companion Methods Enzymol.* **1:**270–275.
402. Cox, A. D., and C. J. Der. (1992). Protein prenylation: More than just glue? *Curr. Opin. Cell. Biol.* **4:**1008–1016.
403. Franco, M., P. Chardin, M. Chabre, and S. Paris. (1993). Myristoylation is not required for GTP-dependent binding of ADP-ribosylation factor ARF1 to phospholipids. *J. Biol. Chem.* **268:**24531–24534.
404. Ames, J. B., T. Tanaka, L. Stryer, and M. Ikura. (1994). Secondary structure of myristoylated recoverin determined by three-dimensional heteronuclear NMR: Implications for the calcium-myristoyl switch. *Biochemistry* **33:**10743–10753.
405. McLaughlin, S., and A. Aderem. (1995). The myristoyl-electrostatic switch: A modulator of reversible protein-membrane interactions. *Trends Biochem. Sci.* **20:**272–276.
406. Zheng, J., D. R. Knighton, N. H. Xuong, S. S. Taylor, J. M. Sowadski, and L. F. Ten Eyck. (1993). Crystal structures of the myristylated catalytic subunit of cAMP-dependent protein kinase reveal open and closed conformations. *Protein Sci.* **2:**1559–1573.
407. Berthiaume, L., I. Deichaite, S. Peseckis, and M. D. Resh. (1994). Regulation of enzymatic activity by active site fatty acylation. A new role for long chain fatty acid acylation of proteins. *J. Biol. Chem.* **269:**6498–6505.
407a. Hunter T. and B. M. Sefton, eds. (1991). *Methods Enzymol.* **200**.
407b. Hunter, T. and B. M. Sefton, eds. (1991). *Methods Enzymol.* **201**.
408. LeGendre, N., and P. Matsudaira. (1989). Purification of proteins and peptides by SDS-PAGE. *In* "A Practical Guide to Protein and Peptide Purification for Microsequencing," pp. 49–69. Academic Press, San Diego.
409. Hildebrandt, E., and V. A. Fried. (1989). Phosphoamino acid analysis of protein immobilized on polyvinylidene difluoride membrane. *Anal. Biochem.* **177:**407–412.
410. Kamps, M. P., and B. M. Sefton. (1989). Acid and base hydrolysis of phosphoproteins bound to Immobilon facilitates analysis of phosphoamino acids in gel-fractionated proteins. *Anal. Biochem.* **176:**22–27.
411. Murthy, L. R., and K. Iqbal. (1991). Measurement of picomoles of phosphoamino acids by high-performance liquid chromatography. *Anal. Biochem.* **193:**299–305.
412. Meyer, H. E., B. Eisermann, M. Heber, E. Hoffmann-Posorske, H. Korte, C. Weigt, A. Wegner,

T. Hutton, A. Donella-Deana, and J. W. Perich. (1993). Strategies for nonradioactive methods in the localization of phosphorylated amino acids in proteins. *FASEB J.* **7:**776–782.

413. Kozma, L. M., A. J. Rossomando, and M. J. Weber. (1991). Comparison of three methods for detecting tyrosine phosphorlated proteins. *Methods Enzymol.* **201:**28–43.
414. Contor, L., F. Lamy, and R. E. Lecocq. (1987). Use of electroblotting to detect and analyze phosphotyrosine-containing peptides separated by two-dimensional gel electrophoresis. *Anal. Biochem.* **160:**414–420.
415. Heffetz, D. M. Fridkin, and Y. Zick. (1989). Antibodies directed against phosphothreonine residues as potent tools for studying protein phosphorylation. *Eur. J. Biochem.* **182:**343–348.
416. Elbein, A. D. (1987). Inhibitors of the biosynthesis and processing of *N*-linked oligosaccharide chains. *Annu. Rev. Biochem.* **56:**497–534.
417. Parchment, R. E., C. M. Ewing, and J. H. Shaper. (1986). The use of galactosyltransferase to probe nitrocellulose-immobilized glycoproteins for nonreducing terminal *N*-acetylglucosamine residues. *Anal. Biochem.* **154:**460–469.
418. O'Shannessy, D. J., P. J. Voorstad, and R. H. Quarles. (1987). Quantitation of glycoproteins on electroblots using the biotin–streptavidin complex. *Anal. Biochem.* **163:**204–209.
419. Heimgartner, U. B. Kozulic, and K. Mosbach. (1989). Polyacrylic polyhydrazides as reagents for detection of glycoproteins. *Anal. Biochem.* **181:**182–189.
420. Heimgartner, U., B. Kozulic, and K. Mosbach. (1990). Reversible and irreversible cross-linking of immunoglobulin heavy chains through their carbohydrate residues. *Biochem. J.* **267:**585–591.
421. Weitzhandler, M., and M. Hardy. (1990). Sensitive blotting assay for the detection of glycopeptides in peptide maps. *J. Chromatogr.* **510:**225–232.
422. Doerner, K. C., and B. A. White. (1990). Detection of glycoproteins separated by nondenaturing polyacrylamide gel electrophoresis using the periodic acid–Schiff stain. *Anal. Biochem.* **187:**147–150.
423. Devine, P. L., and J. A. Warren. (1990). Glycoprotein detection on immobilon PVDF transfer membrane using the periodic acid/Schiff reagent. *BioTechniques* **8:**492–495.
424. Lis, H., and N. Sharon. (1986). Lectins as molecules and as tools. *Annu. Rev. Biochem.* **55:**35–67.
425. Valenzano, K. J., L. M. Kallay, and P. Lobel. (1993). An assay to detect glycoproteins that contain mannose 6-phosphate. *Anal. Biochem.* **209:**156–162.
426. Bartles, J. R., and A. L. Hubbard. (1984). ^{125}I-wheat germ agglutinin blotting: Increased sensitivity with polyvinylpyrrolidone quenching and periodate oxidation/reduction. *Anal. Biochem.* **140:**284–292.
427. Patterson, S. D., and K. Bell. (1990). The carbohydrate side chains of the major plasma serpins of horse and wallaby: Analyses of enzymatic and chemically treated (including "Smith degradation") protein blots by lectin binding. *Biochem Int.* **20:**429–436.
428. Hansler, M., T. Arendt, and K. Lange. (1992). Effects of bile salt exposure on pancreatic duct barrier function and protein release determined by two-dimensional electrophoresis and lectin-affinoblotting. *Electrophoresis* **13:**747–748.
429. Krauss, M. R., and P. J. Collins. (1989). Protein database development: Identification of glycosylated and phosphorylated proteins in unfractionated rat fibroblast lysates. *Electrophoresis* **10:**158–163.
430. Jadach, J., and G. A. Turner. (1993). An ultrasensitive technique for the analysis of glycoproteins using lectin blotting with enhanced chemiluminescence. *Anal. Biochem.* **212:**293–295.
431. Gravel, P., O. Golaz, C. Walzer, D. F. Hochstrasser, H. Turler, and L. P. Balant. (1994). Analysis of glycoproteins separated by two-dimensional gel electrophoresis using lectin blotting revealed by chemiluminescence. *Anal. Biochem.* **221:**66–71.
432. Mumby, S. M., and J. E. Buss. (1990). Metabolic radiolabeling techniques for identification of prenylated and fatty acylated proteins. *Methods: Companion Methods Enzymol.* **1:**216–220.
433. Vidal, M., B. Murgue, F. Basse, and A. Bienvenue. (1991). Fatty acylation of human platelet proteins: Evidence for myristoylation of a 50 kDa peptide. *Biochem. Int.* **23:**1175–1184.
434. Paige, L. A., G.-Q. Zheng, S. A. DeFrees, J. M. Cassady, and R. L. Geahlen. (1990). Metabolic activation of 2-substituted derivatives of myristic acid to form potent inhibitors of myristoyl CoA : protein *N*-myristoyltransferase. *Biochemistry* **29:**10566–10573.
435. Peseckis, S. M., I. Deichaite, and M. D. Resh. (1993). Iodinated fatty acids as probes for myristate processing and function. Incorporation into pp60src. *J. Biol. Chem.* **268:**5107–5114.
436. Clarke, S. (1992). Protein isoprenylation and methylation at carboxyl-terminal cysteine residues. *Annu. Rev. Biochem.* **61:**355–386.

437. Sinensky, M., and R. J. Lutz. (1992). The prenylation of proteins. *BioEssays.* **14:**25–31.
438. Kim, C. M., J. L. Goldstein, and M. S. Brown. (1992). cDNA cloning of MEV, a mutant protein that facilitates cellular uptake of mevalonate, and identification of the point mutation responsible for its gain of function. *J. Biol. Chem.* **267:**23113–23121.
439. Sanford, J., J. Codina, and L. Birnbaumer. (1991). γ-Subunits of G proteins, but not their α- or β-subunits, are polyisoprenylated. *J. Biol. Chem.* **266:**9570–9579.
440. Pollard, K. M., E. K. L. Chan, B. J. Grant, K. F. Sullivan, E. M. Tn, and C. A. Glass. (1990). *In vitro* postranslational modification of lamin B cloned from a human T-cell line. *Mol. Cell Biol.* **10:**2164–2175.
441. Mercier, J.-C., F. Grosclaude, and B. Ribadeau-Duman. (1971). Primary structure of bovine s1 casein. Complete sequence. *Eur. J. Biochem.* **23:**41–47.
442. Shannon, J. D., and J. W. Fox. (1995). Identification of phosphorylation sites by Edman degradation. *In* "Techniques in Protein Chemistry VI," pp. 117–123. Academic Press, San Diego.
443. Coull, J. M., D. J. Pappin, J. Mark, R. Aebersold, and H. Koster. (1991). Functionalized membrane supports for covalent protein microsequence analysis. *Anal. Biochem.* **194:**110–120.
444. Aebersold, R., G. D. Pipes, R. E. Wettenhall, H. Nika, and L. E. Hood. (1990). Covalent attachment of peptides for high sensitivity solid-phase sequence analysis. *Anal. Biochem.* **187:**56–65.
445. Dadd, C. A., R. G. Cook, and C. D. Allis. (1993). Fractionation of small tryptic phosphopeptides by alkaline PAGE followed by amino acid sequencing. *BioTechniques* **14:**266–273.
446. Huddleston, M. J., R. S. Annan, M. F. Bean, and S. A. Carr. (1993). Selective detection of phosphopeptides in complex mixtures by electrospray liquid chromatography/mass spectrometry. *J. Am. Soc. Mass Spectrom.* **4:**710–717.
447. Huddleston, M. J., R. S. Annan, M. F. Bean, and S. A. Carr. (1994). Selective detection of Thr-, Ser-, and Tyr-phosphopeptides in complex digests by electrospray LC-MS. *In* "Techniques in Protein Chemistry V," pp. 123–130. Academic Press, San Diego.
448. Covey, T., B. Shushan, R. Bonner, W. Schroder, and F. Hucho. (1991). *In* "Methods in Protein Sequence Analysis, pp. 249–256. Birkhauser Press, Basel.
449. Ding, J., W. Burkhart, and D. B. Kassel. (1994). Identification of phosphorylated peptides from complex mixtures using negative-ion orifice-potential stepping and capillary liquid chromatography/electrospray ionization mass spectrometry. *Rapid Commun. Mass Spectrom.* **8:**94–98.
450. Yip, T.-T., and T. W. Hutchens. (1992). Mapping and sequence-specific identification of phosphopeptides in unfractionated protein digest mixtures by matrix-assisted laser desorption/ionization time-of-flight mass spectrometry. *FEBS Lett.* **308:**149–153.
451. Yip, T.-T., and T. W. Hutchens. (1993). Protein phosphorylation: Sequence-specific identification of *in vivo* phosphorylation sites by MALDI-TOF mass spectrometry. *In* "Techniques in Protein Chemistry IV," pp. 201–210. Academic Press, San Diego.
452. Wang, Y. K., P.-C. Liao, J. Allison, D. A. Gage, P. C. Andrews, D. M. Lubman, S. M. Hanash, and J. R. Strahler. (1993). Phorbol 12-myristate 13-acetate-induced phosphorylation of Op18 in Jurkat T cells. *J. Biol. Chem.* **268:**14269–14277.
453. Craig, A. G., C. A. Hoeger, C. L. Miller, T. Goedken, J. E. Rivier, and W. H. Fischer. (1994). Monitoring protein kinase and phosphatase reactions with matrix-assisted laser desorption/ionization mass spectrometry and capillary zone electrophoresis: Comparison of the detection efficiency of peptide–phosphopeptide mixtures. *Biol. Mass Spectrom.* **23:**519–528.
454. Jonscher, K., G. Currie, A. L. McCormack, and J. R. D. Yates. (1993). Matrix-assisted laser desorption of peptides and proteins on a quadrupole ion trap mass spectrometer. *Rapid Commun. Mass Spectrom.* **7:**20–26.
455. Doroshenko, V. M., and R. J. Cotter. (1995). High-performance collision-induced dissociation of peptide ions formed by matrix-assisted laser desorption/ionization in a quadrupole ion trap mass spectrometer. *Anal. Chem.* **67:**2180–2187.
456. Qin, J., and B. T. Chait. (1995). Preferential fragmentation of protonated gas-phase peptide ions adjacent to acidic amino-acid-residues. *J. Am. Chem. Soc.* **117:**5411–5412.
457. Jonscher, K. R., and J. R. Yates, III. (1993). *In* "Proceedings of the 41st ASMS Conference on Mass Spectrometry and Allied Topics, Sante Fe," pp. 695a–695b.
458. Qin, J., and B. T. Chait. (1995). Practical matrix-assisted laser desorption/ionization ion trap mass spectrometry. *In* "Proceedings of the 43rd ASMS Conference on Mass Spectrometry and Allied Topics, Atlanta."

459. Amankwa, L. N., K. Harder, F. Jirik, and R. Aebersold. (1995). High-sensitivity determination of tyrosine-phosphorylated peptides by online enzyme reactor and electrospray-ionization mass-spectrometry. *Protein Sci.* **4:**113–125.
460. Amankwa, L. N., and W. G. Kuhr. (1992). Trypsin-modified fused-silica capillary microreactor for peptide mapping by capillary zone electrophoresis. *Anal. Chem.* **64:**1620–1613.
461. Consler, T. G., B. L. Persson, H. Jung, K. H. Zen, K. Jung, G. G. Prive, G. E. Verner, and H. R. Kaback. (1993). Properties and purification of an active biotinylated lactose permease from *Escherichia coli*. *Proc. Natl. Acad. Sci. USA* **90:**6934–6938.
462. Resing, K. A., R. S. Johnson, and K. A. Walsh. (1995). Mass spectrometric analysis of 21 phosphorylation sites in the internal repeat of rat profilaggrin, precursor of an intermediate filament associated protein. *Biochemistry* **34:**9477–9487.
463. Liao, P. C., J. Leykam, P. C. Andrews, D. A. Gage, and J. Allison. (1994). An approach to locate phosphorylation sites in a phosphoprotein: Mass mapping by combining specific enzymatic degradation with matrix-assisted laser desorption/ionization mass spectrometry. *Anal. Biochem.* **219:**9–20.
464. Nuwaysir, L. M., and J. T. Stults. (1993). Electrospray ionization mass spectrometry of phosphopeptides isolated by on-line immobilized metal-ion affinity chromatography. *J. Am. Soc. Mass Spectrom.* **4:**662–669.
465. Andersson, L., and J. Porath. (1986). Isolation of phosphoproteins by immobilized metal (Fe^{3+}) affinity chromatography. *Anal. Biochem.* **154:**250–254.
466. Kassel, D. B., T. G. Consler, M. Shalaby, P. Sekhri, N. Gordon, and T. Nadler. (1995). Direct coupling of an automated 2-dimensional microcolumn affinity chromatography-capillary HPLC system with mass spectrometry for biomolecule analysis. *In* "Techniques in Protein Chemistry VI," pp. 39–46. Academic Press, San Diego.
467. Affolter, M., J. D. Watts, D. L. Krebs, and R. Aebersold. (1994). Evaluation of 2-dimensional phosphopeptide maps by electrospray-ionization mass-spectrometry of recovered peptides. *Anal. Biochem.* **223:**74–81.
468. Watts, J. D., M. Affolter, D. L. Krebs, R. L. Wange, L. E. Samelson, and R. Aebersold. (1995). Electrospray ionization mass spectrometric investigation of signal transduction pathways: Determination of sites of inducible protein phosphorylation in activated T-cells. *In* "Biochemical and Biotechnological Applications of Electrospray Ionization Mass Spectrometry," Vol. 619, pp. 381–407. ACS Symposium Series, Washington, D.C.
469. Faye, L., and M. J. Crispeels. (1985). Characterization of *N*-linked oligosaccharides by affinoblotting with Concanavalin A-peroxidase and treatment of the blots with glycosidases. *Anal. Biochem.* **149:**218–224.
470. Beisiegel, U. (1986). Protein blotting. *Electrophoresis* **7:**1–18.
471. Thompson, S., J. A. E. Latham, and G. A. Turner. (1987). A simple, reproducible and cheap batch method for the analysis of serum glycoproteins using Sepharose-coupled lectins and silver staining. *Clin. Chim. Acta* **167:**217–223.
472. Hayes, B. K., K. D. Greis, and G. W. Hart. (1995). Specific isolation of *O*-linked *N*-acetylglucosamine glycopeptides from complex mixtures. *Anal. Biochem.* **228:**115–122.
473. Shao, M.-C., and C. C. Q. Chin. (1992). Method for the detection of glycopeptides at the picomole level in HPLC peptide maps. *Anal. Biochem.* **207:**100–105.
474. Weitzhandler, M., D. Kadlecek, N. Avdalovic, J. G. Forte, D. Chow, and R. R. Townsend. (1993). Monosaccharide and oligosaccharide analysis of proteins transferred to polyvinylidene fluoride membranes after sodium dodecyl sulfate–polyacrylamide gel electrophoresis. *J. Biol. Chem.* **268:**5121–5130.
475. Weitzhandler, M., D. Kadlecek, N. Avdalovic, and R. R. Townsend. (1993). Monosaccharide and oligosaccharide analysis of recombinant erythropoietin electro-transferred onto polyvinylidene fluoride membrane. *In* "Techniques in Protein Chemistry IV," pp. 135–142. Academic Press, San Diego.
476. Ogawa, H., M. Ueno, H. Uchibori, I. Matsumoto, and N. Seno. (1990). Direct carbohydrate analysis of glycoproteins electroblotted onto polyvinylidene difluoride membrane from sodium dodecyl sulfate–polyacrylamide gel. *Anal. Biochem.* **190:**165–169.
477. Jackson, P. (1994). The analysis of fluorophore-labeled glycans by high resolution polyacrylamide gel electrophoresis. *Anal. Biochem.* **216:**243–252.
478. Kawashima, H., T. Murata, K. Yamamoto, A. Tateishi, T. Irimura, and T. Osawa. (1992). A simple method for the release of asparagine-linked oligosaccharides from a glycoprotein purified by SDS-polyacrylamide gel electrophoresis. *J. Biochem.* **111:**620–622.

479. Gooley, A. A., and K. L. Williams. (1994). Towards characterizing *O*-glycans: The relative merits of *in vivo* and *in vitro* approaches in seeking peptide motifs specifying *O*-glycosylation sites. *Glycobiology* **4:**413–417.
480. Wilson, I. B., Y. Gavel, and G. von Heijne. (1991). Amino acid distributions around *O*-linked glycosylation sites. *Biochem. J.* **275:**529–534.
481. Reinhold, V. N., B. B. Reinhold, and C. E. Costello. (1995). Carbohydrate molecular weight profiling, sequence, linkage, and branching data: ES-MS and CID. *Anal. Chem.* **67:**1772–1784.
482. Derby, P. L., T. W. Strickland, and M. F. Rohde. (1993). Site specific heterogeneity of *N*-linked oligosaccharides on recombinant human erythropoietin. *In* "Techniques in Protein Chemistry IV," pp. 161–168. Academic Press, San Diego.
483. Reason, A. J., H. R. Morris, M. Panico, R. Marais, R. H. Treisman, R. S. Haltiwanger, G. W. Hart, W. G. Kelly, and A. Dell. (1992). Localization of *O*-GlyNAc modification on the serum response transcription factor. *J. Biol. Chem.* **267:**16911–16921.
484. Gooley, A. A., B. J. Classon, R. Marschalek, and K. L. Williams. (1991). Glycosylation sites identified by detection of glycosylated amino acids released from Edman degradation: The identification of Xaa-Pro-Xaa-Xaa as a motif for Thr-*O*-glycosylation. *Biochem. Biophys. Res. Commun.* **178:**1194–1201.
485. Pisano, A., J. W. Redmond, K. L. Williams, and A. A. Gooley. (1993). Glycosylation sites identified by solid-phase Edman degradation: *O*-Linked glycosylation motifs on human glycophorin A. *Glycobiology* **3:**429–435.
486. Pisano, A., N. H. Packer, J. W. Redmond, K. L. Williams, and A. A. Gooley. (1994). Characterization of *O*-linked glycosylation motifs in the glycopeptide domain of bovine kappa-casein. *Glycobiology* **4:**837–844.
487. Gooley, A. A., N. H. Packer, A. Pisano, J. W. Redmond, K. L. Williams, A. Jones, M. Loughnan, and P. F. Alewood. (1995). Characterization of individual *N*- and *O*-linked glycosylation sites using Edman degradation. *In* "Techniques in Protein Chemistry VI," pp. 83–90. Academic Press, San Diego.
488. Carr, S. A., M. J. Huddleston, and M. F. Bean. (1993). Selective identification and differentiation of *N*- and *O*-linked oligosaccharides in glycoproteins by liquid chromatography-mass spectrometry. *Protein Sci.* **2:**183–196.
489. Huddleston, M. J., M. F. Bean, and S. A. Carr. (1993). Collisional fragmentation of glycopeptides by electrospray ionization LC/MS and LC/MS/MS: Methods for selective detection of glycopeptides in protein digests. *Anal. Chem.* **65:**877–884.
490. Harvey, D. J. (1993). Quantitative aspects of the matrix-assisted laser desorption mass spectrometry of complex oligosaccharides. *Rapid Commun. Mass Spectrom.* **7:**614–619.
491. Sutton, C. W., J. A. O'Neill, and J. S. Cottrell. (1994). Site-specific characterization of glycoprotein carbohydrates by exoglycosidase digestion and laser desorption mass spectrometry. *Anal. Biochem.* **218:**34–46.
492. Edge, C. J., T. W. Rademacher, M. R. Wormald, R. B. Parekh, T. D. Butters, D. R. Wing, and R. A. Dwek. (1992). Fast sequencing of oligosaccharides: the reagent-array analysis method. *Proc. Natl. Acad. Sci. USA* **89:**6338–6342.
492a. "Methods: A companion to Methods in Enzymology," Vol. 1, issue 3 (1990).
493. Kamiya, Y., A. Sakurai, S. Tamura, and N. Takahashi. (1979). *Agric. Biol. Chem.* **43:**1049–1053.
494. Farnsworth, C. C., P. J. Casey, W. N. Howald, J. A. Glomset, and M. H. Gelb. (1990). Structural characterization of prenyl groups attached to proteins. *Methods: Comparison Methods Enzymol.* **1:**231–240.
495. Farnsworth, C. C., S. L. Wolda, M. H. Gelb, and J. A. Glomset. (1989). Human lamin B contains a farnesylated cysteine residue. *J. Biol. Chem.* **264:**20422–20429.
496. Xie, H., H. K. Yamane, R. C. Stephenson, O. C. Ong, B. K.-K. Fung, and S. Clarke. (1990). Analysis of prenylated carboxyl-terminal cysteine methyl esters in proteins. *Methods: Companion Methods Enzymol.* **1:**276–282.
497. Hrycyna, C. A., S. K. Sapperstein, S. Clarke, and S. Michaelis. (1991). The *Saccharomyces cerevisiae STE14* gene encodes a methyltransferase that mediates C-terminal methylation of a-factor and RAS proteins. *EMBO J.* **10:**1699–1709.
498. Callahan, F. E., H. A. Norman, T. Srinath, J. B. St. John, R. Dhar, and A. K. Mattoo. (1989). Identification of covalently bound fatty acids on acylated proteins immobilized on nitrocellulose paper. *Anal. Biochem.* **183:**220–224.
499. Crise, B., and J. K. Rose. (1992). Identification of palmitoylation sites on CD4, the human immunodeficiency virus receptor. *J. Biol. Chem.* **267:**13593–13597.

500. Kunz, B. C., K. A. Muczynski, C. F. Welsh, S. J. Stanley, S.-C. Tsai, R. Adamik, P. P. Chang, J. Moss, and M. Vaughan. (1993). Characterization of recombinant and endogenous ADP-ribosylation factors synthesized in Sf9 insect cells. *Biochemistry* **32:**6643–6648.
501. Muszbek, L., and M. Laposata. (1989). Covalent modification of platelet proteins by palmitate. *Blood* **74:**1339–1347.
502. Kokame, K. Y. Fukuda, T. Yoshizawa, T. Takao, and Y. Shimonishi. (1992). Lipid modification at the N terminus of photoreceptor G-protein α-subunit. *Nature.* **359:**749–752.
503. Okubo, K., N. Hamasaki, K. Hara, and M. Kageura. (1991). Palmitoylation of cysteine 69 from the COOH-terminal of Band 3 protein in the human eryhthrocyte membrane. *J. Biol. Chem.* **266:**16420–16424.
504. Stults, J. T., P. R. Griffin, D. D. Lesikar, A. Naidu, B. Moffat, and B. J. Benson. (1991). Lung surfactant protein SP-C from human, bovine, and canine sources contains palmityl cysteine thioester linkages. *Am. J. Physiol.* **261:**L118–L125.
505. Moscarello, M. A., H. Pang, C. R. Pace Asciak, and D. D. Wood. (1992). The N terminus of human myelin basic protein consists of C2, C4, C6, and C8 alkyl carboxylic acids. *J. Biol. Chem.* **267:**9779–9782.
506. Bean, M. F., R. S. Annan, M. E. Hemling, M. Mentzer, M. J. Huddleston, and S. A. Carr. (1995). LC-MS methods for selective detection of posttranslational modifications in proteins: Glycosylation, phosphorylation, sulfation, and acylation. *In* "Techniques in Protien Chemistry VI," pp. 107–116. Academic Press, San Diego.
507. Phizicky, E. M., and S. Fields. (1995). Protein–protein interactions—methods for detection and analysis. *Microbiological Reviews* **59:**94–123.
508. Harlow, E., and D. Lane. (1988). "Antibodies: A Laboratory Manual." Cold Spring Harbor Laboratory Press, Cold Spring Harbor. 726 pp.
509. Colligan, J., B. Dunn, H. Ploegh, D. Speicher, and P. Wingfield. (1995). "Current protocols in protein science." *In* "Current Protocols." Wiley, New York.
510. Franza, B. R., Jr., F. J. Rauscher, III, S. F. Josephs, and T. Curran. (1988). The Fos complex and Fos-related antigens recognize sequence elements that contain AP-1 binding sites. *Science* **239:**1150–1153.
511. Taylor, S. J., and D. Shalloway. (1994). An RNA-binding protein associated with Src through its SH2 and SH3 domains in mitosis. *Nature* **368:**867–871.
512. Fried, M., and D. M. Crothers. (1981). Equilibria and kinetics of lac repressor–operator interactions by polyacrylamide gel electrophoresis. *Nucleic Acids Res.* **9:**6505–6525.
513. Hermann, T., X. K. Zhang, M. Tzukerman, K. N. Wills, G. Graupner, and M. Pfahl (1991). Regulatory functions of a non-ligand-binding thyroid hormone receptor isoform. *Cell Regul.* **2:**565–574.
514. Ozes, O. N., and M. W. Taylor. (1994). Reversal of interferon-gamma-resistant phenotype by poly(I:C): possible involvement of ISGF2 (IRF1) in interferon-gamma-mediated induction of the IDO gene. *J. Interferon Res.* **14:**25–32.
515. Demczuk, S., M. Harbers, and B. Vennstrom. (1993). Identification and analysis of all components of a gel retardation assay by combination with immunoblotting. *Proc. Natl. Acad. Sci. USA* **90:**2574–2578.
516. Novak, U., and L. Paradiso. (1995). Identification of proteins in DNA–protein complexes after blotting of EMSA gels. *BioTechniques* **19:**54–55.
517. Towbin, H., T. Staehelin, and J. Gordon. (1979). Electrophoretic transfer of proteins from polyacrylamide gels in nitrocellulose sheets: Procedure and some applications. *Proc. Natl. Acad. Sci. USA* **76:**4350–4354.
518. Bar-Nun, S., and J. M. Gershoni. (1994). Protein-blot analysis of glycoproteins and lectin overlays. *In* "Cell Biology: A Laboratory Handbook," Vol. 3, pp. 323–331. Academic Press, New York.
519. Gershoni, J. M., and G. E. Palade. (1983). Protein blotting: Principles and applications. *Anal. Biochem.* **131:**1–15.
520. Gershoni, J. M. (1988). Protein blotting: A tool for the analytical biochemist. *In* "Advances in Electrophoresis, Vol. 1, pp. 141–175. VCH-Verlag, Weinheim, Germany.
521. Ferrell, J. E., Jr., and G. S. Martin. (1989). Thrombin stimulates the activities of multiple previously unidentified protein kinases in platelets. *J. Biol. Chem.* **264:**20723–20729.
522. Geahlen, R. L., and M. L. Harrison. (1986). Detection of a novel lymphocyte protein-tyrosine kinase by renaturation in polyacrylmide gels. *Biochem. Biophys. Res. Commun.* **134:**963–969.

523. Sarmay, G., I. Pecht, and J. Gergely. (1994). Protein-tyrosine kinase activity tightly associated with human type II Fc gamma receptors. *Proc. Natl. Acad. Sci. USA* **91:**4140–4144.
524. Ferrell, J. E., Jr., and G. S. Martin. (1991). Assessing activities of blotted protein kinases. *Methods Enzymol.* **200:**430–435.
525. Crawford, A. W., and M. C. Beckerle. (1994). Blot overlay assay: A method to detect protein–protein interactions. *In* "Cell Biology: A Laboratory Handbook," Vol. 3, pp. 301–308. Academic Press, New York.
526. Maruyama, K., T. Mikawa, and S. Ebashi. (1984). Detection of calcium binding proteins by ^{45}Ca autoradiography on nitrocellulose membrane after sodium dodecyl sulfate gel electrophoresis. *J. Biochem.* **95:**511–519.
527. Hoffman, H. J., and J. E. Celis. (1994). Calcium overlay assay. *In* "Cell Biology: A Laboratory Handbook," Vol. 3, pp. 309–312. Academic Press, New York.
528. Aoki, Y., M. Kunimoto, Y. Shibata, and K. T. Suzuki. (1986). Detection of metallothionein on nitrocellulose membrane using Western blotting technique and its application to identification of cadmium binding proteins. *Anal. Biochem.* **157:**117–122.
529. Maltese, W. A., K. M. Sheridan, E. M. Repko, and R. A. Erdman (1990). Post-translational modification of low molecular mass GTP-binding proteins by isoprenoid. *J. Biol. Chem.* **265:**2148–2155.
530. Chen, L.-M., Y. Liang, J.-H. Tai, and Y. Chern. (1994). Comparison of nitrocellulose and PVDF membranes in GTP-overlay assay and Western blot analysis. *BioTechniques* **16:**600–601.
531. Gromov, P. S., and J. E. Celis. (1994). Blot overlay assay for the identification of GTP-binding protiens. *In* "Cell Biology: A Laboratory Handbook," Vol. 3, pp. 313–316. Academic Press, New York.
532. Basson, M. D., J. R. Goldenring, L. H. Tang, J. J. Lewis, P. Padfield, J. D. Jamieson, and I. M. Modlin. (1991). Redistribution of 23 kDa tubulovesicle-associated GTP-binding proteins during parietal cell stimulation. *Biochem. J.* **279:**43–48.
533. Peter, M. E., B. Wittmann Liebold, and M. Sprinzl. (1988). Affinity labeling of the GDP/GTP binding site in *Thermus thermophilus* elongation factor Tu. *Biochemistry* **27:**9132–9139.
534. Peter, M. E., C. Hall, A. Ruhlmann, J. Sancho, and C. Terhorst. (1992). The T-cell receptor zeta chain contains a GTP/GDP binding site. *EMBO J.* **11:**933–941.
535. Peter, M. E., and L. A. Huber. (1994). Two-dimensional gel-based mapping of *in situ* crosslinked GTP-binding proteins. *In* "Cell Biology: A Laboratory Handbook, Vol. 3, pp. 317–322. Academic Press, New York.
536. Murphy, P. R., N. Katsumata, Y. Sato, C. K. Too, and H. G. Friesen. (1990). In-gel ligand blotting with ^{125}I-heparin for detection of heparin-binding growth factors. *Anal. Biochem.* **187:**197–201.
537. Mehlman, T., and W. H. Burgess. (1990). Detection and characterization of heparin-binding proteins with a gel overlay procedure. *Anal. Biochem.* **188:**159–163.
538. Otto, J. J. (1983). Detection of vinculin-binding proteins with an ^{125}I-vinculin gel overlay technique. *J. Cell. Biol.* **97:**1283–1287.
539. Crawford, A. W., J. W. Michelsen, and M. C. Beckerle. (1992). An interaction between zyxin and alpha-actinin. *J. Cell. Biol.* **116:**1381–1393.
540. Manser, E., T. Leung, C. Monfries, M. Teo, C. Hall, and L. Lim. (1992). Diversity and versatility of GTPase activating proteins for the p21rho subfamily of ras G proteins detected by a novel overlay assay. *J. Biol. Chem.* **267:**16025–16028.
541. Bennett, M., and R. Reed. (1993). Correspondence between a mammalian spliceosome component and an essential yeast splicing factor. *Science* **262:**105–108.
542. Zhang, M., P. D. Zamore, M. Carmo Fonseca, A. I. Lamond, and M. R. Green. (1992). Cloning and intracellular localization of the U2 small nuclear ribonucleoprotein auxiliary factor small subunit. *Proc. Natl. Acad. Sci. USA* **89:**8769–8773.
543. Niman, H. L., R. A. Houghten, L. E. Walker, R. A. Reisfeld, I. A. Wilson, J. M. Hogle, and R. A. Lerner. (1983), Generation of protein-reactive antibodies by short peptides is an event of high frequency: Implications for the structural basis of immune recognition. *Proc. Natl. Acad. Sci. USA* **80:**4949–4953.
544. Witzgall, R., E. O'Leary, and J. V. Bonventre. (1994). A mammalian expression vector for the expression of GAL4 fusion proteins with an epitope tag and histidine tail. *Anal. Biochem.* **223:**291–298.
545. Winder, S. J., and J. Kendrick-Jones. (1995). Protein production in three different expression vectors from a single polymerase chain reaction product. *Anal. Biochem.* **231:**271–273.

546. Botting, C. H., and R. E. Randall. (1995). Reporter enzyme-nitrilo-triacetic acid-nickel conjugates—reagents for detecting histidine-tagged proteins. *Biotechniques* **19:**362–363.
547. O'Shannessy, D. J., K. C. O'Donnell, J. Martin, and M. Brighamburke. (1995). Detection and quantitation of hexa-histidine-tagged recombinant proteins on Western blots and by a surface-plasmon resonance biosensor technique. *Anal. Biochem.* **229:**119–124.
548. Dejgaard, K., and J. E. Celis. (1994). Two-dimensional Northwestern blotting. *In* "Cell Biology: A Laboratory Handbook," Vol. 3, pp. 339–344. Academic Press, New York.
549. Rao, U. J., P. R. Ramasarma, D. R. Rao, and K. V. Prasad. (1994). Detection of lectin activity on Western blots using erythrocytes. *Electrophoresis* **15:**907–910.
550. Seshi, B. (1994). Cell adhesion to proteins separated by lithium dodecyl sulfate–polyacrylamide gel electrophoresis and blotted onto a polyvinylidene difluoride membrane: A new cell-blotting technique. *J. Immunol. Methods* **176:**185–201.
551. Seshi, B. (1994). Discovery of novel hematopoietic cell adhesion molecules from human bone marrow stromal cell membrane protein extracts by a new cell-blotting technique. *Blood* **83:**2399–2409.
552. Chaiken, I., S. Rose, and R. Karlsson. (1992). Analysis of macromolecular interactions using immobilized ligands. *Anal. Biochem.* **201:**197–210.
553. Vadgama, P., and P. W. Crump. (1992). Biosensors: Recent trends. *Analyst* **117:**1657–1670.
554. Griffiths, D., and G. Hall. (1993). Biosensors—what real progress is being made? *Trends Biotechnol.* **11:**122–130.
555. Panayotou, G., M. D. Waterfield, and P. End. (1993). Riding the evanescent wave. *Curr. Biol.* **3:**913–915.
556. Szabo, A., L. Stolz, and R. Granzow. (1995). Surface Plasmon Resonance and its use in biomolecular interaction analysis (BIA). *Curr. Opin. Struc. Biol.* **5:**699–705.
557. Bartley, T. D., R. W. Hunt, A. A. Welcher, W. J. Boyle, V. P. Parker, R. A. Lindberg, H. S. Lu, A. M. Colombero, R. L. Elliott, B. A. Guthrie, *et al.* (1994). B61 is a ligand for the ECK receptor protein-tyrosine kinase. *Nature* **368:**558–560.
558. Fisher, R. J., M. Fivash, J. Casas Finet, J. W. Erickson, A. Kondoh, S. V. Bladen, C. Fisher, D. K. Watson, and T. Papas. (1994). Real-time DNA binding measurements of the ETS1 recombinant oncoproteins reveal significant kinetic differences between the p42 and p51 isoforms. *Protein Sci.* **3:**257–266.
559. Fields, S., and R. Sternglanz. (1994). The two-hybrid systems: An assay for protein–protein interactions. *Trends Genet.* **10:**286–292.
560. Ganem, B., Y. T. Li, and J. D. Henion. (1991). Detection of noncovalent receptor-ligand complexes by mass spectrometry. *J. Am. Chem. Soc.* **113:**6294–6296.
561. Katta, V., and B. T. Chait. (1991). Observation of the heme-globin complex in native myoglobin by electrospray-ionization mass spectrometry. *J. Am. Chem. Soc.* **113:**8534–8535.
562. Bruce, J. E., G. A. Anderson, R. D. Chen, X. H. Cheng, D. C. Gale, S. A. Hofstadler, B. L. Schwartz, and R. D. Smith. (1995). Bio-affinity characterization mass-spectrometry. *Rapid Commun. Mass Spectrom.* **9:**644–650.
563. Schwartz, B. L., D. C. Gale, R. D. Smith, A. Chilkoti, and P. S. Stayton. (1995). Investigation of noncovalent ligand-binding to the intact streptavidin tetramer by electrospray-ionization mass-spectrometry. *J. Mass Spectrom.* **30:**1095–1102.
564. Schwartz, B. L., J. E. Bruce, G. A. Anderson, S. A. Hofstadler, A. L. Rockwood, R. D. Smith, A. Chilkoti, and P. S. Stayton. (1995). Dissociation of tetrameric ions of noncovalent streptavidin complexes formed by electrospray-ionization. *J. Am. Soc. Mass Spectrom.* **6:**459–465.
565. Woods, A. S., J. C. Buchsbaum, T. A. Worrall, J. M. Berg, and R. J. Cotter. (1995). Matrix-assisted laser desorption/ionization of noncovalently bound compounds. *Anal. Chem.* **67:**4462–4465.
566. Tang, X. D., J. H. Callahan, P. Zhou, and A. Vertes. (1995). Noncovalent protein–oligonucleotide interactions monitored by matrix-assisted laser desorption/ionization mass-spectrometry. *Anal. Chem.* **67:**4542–4548.
567. Liu, J., S. Mathias, Z. Yang, and R. N. Kolesnick. (1994). Renaturation and tumor necrosis factor-alpha stimulation of a 97-kDa ceramide-activated protein kinase. *J. Biol. Chem.* **269:**3047–3052.
568. Chien, C. T., P. L. Bartel, R. Sternglanz, and S. Fields. (1991). The two-hybrid system: A method to identify and clone genes for proteins that interact with a protein of interest. *Proc. Natl. Acad. Sci. USA* **88:**9578–9582.
569. Yuen, S., D. Sheer, K. Hsi, and R. Mattaliano. (1990). *Appl. Biosystems Res. News:*1–11.
570. Eckerskorn, C., W. Mewes, H. Goretzki, and F. Lottspeich. (1988). A new siliconized-glass fiber as support for protein-chemical analysis of electroblotted proteins. *Eur. J. Biochem.* **176:**509–519.

571. Aebersold, R. (1997). High sensitivity sequence analysis of proteins separated by polyacrylamide gel electrophoresis. *In* ''Advances in Electrophoresis, Vol. 4, pp. 81–168. VCH-Verlag, Weinheim, Germany.
572. Lioubin, M. N., P. A. Algate, S. Tsai, K. Carlberg, R. Aebersold, and L. R. Rohrschneider (1996). p150^{ship}, a signal transduction molecule with inositol polyphosphate-5-phosphatase activity. *Genes Dev.* **10:**1084–1095.

Chapter 2

Posttranslational Modifications

Radha Gudepu Krishna and Finn Wold*

I. Introduction

In characterizing a protein we need to establish both the covalent structure and the folded, three-dimensional organization of this covalent structure into the active protein. Since the protein may also be associated with other structures, such as membranes, nucleic acids, and proteins, its *in vivo* compartment and associations are also a significant part of defining the end product of protein biosynthesis. With the recent progress in molecular biology and statistical prediction of structure, the most common current approach to the examination of protein structure is to isolate the mRNA for the protein in question and determine its base sequence, and then to deduce the amino acid sequence corresponding to the determined base sequence. This information is then sub-

* Deceased.

jected to any one or several of the available programs that predict folded structures and membrane associations, and as a result a reasonably good estimate of the complete structure can be arrived at in amazingly short time.

One obvious problem with the preceding approach is that it omits certain possible steps in protein biosynthesis, namely the co- and posttranslational reactions that modify the polypeptide chain and the amino acid side chains, so that the final covalent structure is not a simple, accurate translation of the mRNA. Thus, only for a few proteins does the observed polypeptide chain correspond precisely to the deduced structure starting, as specified by the virtually universal initiation codon AUG, with Met as the N-terminal amino acid, and ending with a C-terminal residue corresponding to the codon preceding one of the termination codons, UAA, UAG, or UGA. For most proteins the observed structure will differ from the predicted one; as a minimum the initiation Met (or formyl-Met in bacteria) is likely to have been removed, or, if the protein contains more than two Cys residues, disulfide bonds between unknown pairs of Cys may have to be accounted for in the covalent structure. For a substantial number of proteins the differences from the predicted structure are considerably more complicated than those indicated by these minimum modifications. Thus, peptide bond cleavage may go well beyond just the removal of the initiator amino acid; whole peptide segments may be removed from different parts of the encoded sequence, and even more perplexing, the peptides resulting from proteolytic cleavages may be ligated into new and scrambled sequences quite different from the starting structure. In addition to peptide cleavage and transpeptidations, amino acids may also be added to the encoded sequence. Similarly, for amino acid side-chain modifications such as disulfide bond formation, most of the amino acids can be derivatized to the extent that although the genetic code specifies 20 (or if we include selenocysteine, 21) amino acids, and the protein sequence deduced from mRNA can contain only the linear array of these 20–21 primary amino acids, the actual number of amino acids found in proteins is much closer to 200 than to 20. If the specificity of all these reactions, the bases by which a single residue out of a large number of like residues is selected for modification, were known, the prediction of covalent structure could include the co- and posttranslational modifications, but, as will be discussed later, such knowledge is not available at this stage.

We think about the area of protein processing in terms of a "black box," emphasizing that we know the start and the finish of the process, but that everything between takes place in the unknown of the closed, black box. We can describe the basic blueprint (the mRNA-specified sequence) for a given protein with certainty; for isolated proteins we can often also describe the actual structure of the final product with the high degree of precision that would be the result of, for example, high-resolution X-ray crystallographic analysis. How the complete linear sequence between the initiation codon and the termination codon assembled at a given site became the characterized three-dimensional structure isolated at a different site is a consequence of a series of unknown reactions taking place in an unknown sequence in parallel with the transport of the processing and/or processed protein along an often unknown pathway. The nature and the dynamics of these processing steps are what we refer to as the black box of protein synthesis. Experimentally, the classical approaches to the questions hiding in the black box would involve purifying the various processing enzymes and characterizing them with suitable substrates to establish their specificity and site of action. The main problem with this approach is to find suitable substrates. The finished protein has already been modified in what are generally irreversible reactions and is not of any use. Trying to generate a reasonable substrate by other means is also complicated in that we do not know what the actual structure was of the intermediate that was acted on *in vivo;* the true substrate may have to be folded in a certain way, or be temporarily associated with other proteins (chaperones). If chaperones are involved, they could specify that the modification must take place in a certain compartment under the proper conditions of reduction / oxidation (red / ox), pH, cofactors, etc., that exist in that compartment. In the absence of a clear definition of what these variables may be, it would be an extremely unlikely event to put together the proper set of *in vitro* conditions needed to study the modification reaction.

It is probably reasonable to make certain general predictions for the events in the black box. The most obvious one is that the three separate processing steps, transport, noncovalent folding, and covalent modification of peptide bonds and amino acid side chains, must go hand in hand in constituting the complete protein maturation. The step-by-step manner in which they are integrated will undoubtedly vary from protein to protein, but the interrelatedness of the steps must be universal. A series of intriguing questions arise regarding the effect of covalent modification reactions on the noncovalent interactions involved in transport and folding, the effect of compartment location on folding and covalent modifications, and the effect of folding on transport and covalent modification; some of these questions may be subjected to experimental exploration. It is interesting to note that there are currently very few concrete data related to the effect of covalent amino acid modifications on protein folding and transport. It is common practice to apply the very impressive predictive programs to the linear structures deduced from cDNA sequences, and arrive at amazingly good three-dimensional (3D) structures and postulates about compartment (membrane) interactions for these structures. However, the statistical and theoretical data that are the bases for the predictions are derived for the original 20 primary amino acids, and the predictions for any protein that has been covalently modified during biosynthesis may consequently not be reliable. Another generality pertains to the biological function of the modifications. It is the inferential view that Nature would not take the trouble to evolve the apparatus required for a given processing reaction, unless there was a very good reason for it. It is thus taken for granted that the many protein modification reactions have a significant biological function; the actual definition of what that function is has proven to be quite elusive for many of the reactions.

In this brief introduction several concepts, questions, and key words have been brought up as natural points of concentration in a chapter on posttranslational modifications of proteins. The main goal of this chapter is to elaborate on these points along the following lines. We intend to establish *what they are* by briefly reviewing the different posttranslational modifications and describing special aspects of the individual derivatives or families (glycosylations, Lys cross-links) of derivatives. To this end, we have made an attempt to compile a fairly complete listing of known covalent derivatives of proteins; this has been done primarily by expanding previous reviews (Uy and Wold, 1977; Wold, 1981, 1988; Alix, 1988; Krishna and Wold, 1993 a, b; Graves *et al.*, 1994) and updating them. In the next couple of sections, we focus on *how they are made.* First we consider general aspects such as the reversibility and the enzymology and the compartmentalization of the different reactions, along with the possible cofactor (chaperone) involvement, attempting to establish whether certain features may have general applicability. Next we consider the specificity of the individual modification reactions, attempting to find by which rules specific residues or peptide bonds are selected for modification. This very important aspect of posttranslational modifications is obviously the key to the model building from mRNA-deduced sequences using empirically derived parameters for secondary / tertiary structure preferences and spatial restraints. If we know the specificity rules, we also know with certainty that a given primary amino acid in a special protein environment will be converted to a secondary amino acid derivative; when we now introduce this secondary amino acid into the sequence, a more accurate model structure can obviously be derived from the mRNA data. The final part of the chapter will examine the question of *why they are made,* the biological functions of posttranslational reactions. This topic includes the most diverse and complex aspects of posttranslational modification of proteins, but it is also perhaps the most important one, in that it has to tackle the most fundamental aspects of protein synthesis, structure, and function. Throughout the chapter we have tried to use recent references; as a result some of the original references, including perhaps the report of the discovery of a given modification, have been eliminated. We assume that all the newer references will have proper reference to the background material.

The final introductory point is a semantic one. A certain amount of ambiguity has evolved in the usage of the terms "posttranslational" and "processing." "Posttranslational," when used along with "cotranslational," has come to mean specifically that the reactions take place after the polypeptide chain has been released from the polysome; it is distinct from "cotranslational," which refers to the reactions that occur while the nascent chain is still attached to the polysome. This nomenclature has obviously been useful and popular; perhaps it should be completed by creating a separate category to accommodate the "pretranslational" reactions that take place at the level of amino acyl-tRNA. All of these terms are in fact quite ambiguous in that they are based on the implied definition of "translation" as being synonymous to polymerization. In a strict sense, the actual translation from the language of nucleotides to the language of amino acids takes place when the appropriate pair of anticodon and amino acid are linked together in the amino acyl-tRNA, and all the reactions beyond this point are in principle posttranslational. In this chapter we will use "posttranslational" in this generic sense and try to avoid to use the other more specific terms. To many groups "processing" has taken on a quite specific meaning as the general term for proteolytic cleavages. Although proteolytic cleavages certainly represent important steps in protein processing, they are not the only ones. Again we will use "processing" in the generic sense to describe all the activities of covalent modification, folding, and transport that we assigned previously to the black box of protein synthesis.

II. Types of Posttranslational Derivatives

In deciding which amino acid derivatives to include as constituents of proteins, we may encounter gray areas such as that for short peptides of uncertain biosynthetic origin. On the one hand, it is clear that a number of short peptides, such as hormones and neuropeptides, are synthesized as multifunctional large polypeptide precursors whose sequences are encoded by mRNA and are assembled by the regular ribosomal synthetic apparatus. On the other hand, there are short polypeptide antibiotics and cell wall constituents that are assembled in step-by-step amino acid activation and condensation catalyzed by specific enzymes in the absence of genetic information and ribosomes, and these should not be included in consideration of *in vivo* modifications of proteins. These latter types of peptides frequently contain rather exotic amino acid derivatives, including D-amino acids. A cyclic heptapeptide toxin from the mushroom *Amanita virosa* contains L-Ala, L-Val, D-Ser, D-Thr, 2,3-*trans*-3,4-*trans*-3,4-dihydroxy-L-Pro, and 2′-(methylsulfonyl)-L-Trp, and either γ,δ-dihydroxy-L-Leu or γ,δ,δ'-trihydroxy-L-Leu (Faulstich *et al.*, 1980). The presence of D-amino acids has in the past been used as a criterion in deciding that a given peptide with its unique amino acid derivatives is the product of soluble enzymes rather than being derived from a protein synthesized on the ribosome–mRNA complex. The commitment step in proteins synthesis, the formation of the amino acyl-tRNA, has been well established to have an absolute specificity for L-α-amino (imino) acids, and since we did not know of any case of enzyme-catalyzed inversion of enantiomers in proteins (as is discussed later, slow, nonenzymatic racemization does occur in proteins to a certain extent), it was reasonable to generalize that the presence of D-amino acids in stoichiometric amounts in a peptide (protein) precluded the biosynthesis of that peptide by the normal protein synthetic apparatus. As so often is the case in science, exceptions to the rule are found, and it is now clear that a number of relatively short peptides derived from higher molecular weight precursors contain D-amino acids as true posttranslational products of the corresponding primary L-amino acids (Kreil, 1994). These are included in the lists that follow. The nonribosomally synthesized peptides with modified amino acids still exist and must be considered as possibilities when short peptides are encountered, but the presence of D-amino acids can no longer be used as an indicator for the biosynthetic source of the peptides.

A major part of this chapter could have been devoted to the ingenuity, serendipity, alert observation, and elegant experimentation that led to the discovery of the various posttranslational reactions. Common methodology involving amino acid analysis, nucleic acid and amino acid sequencing, mass spectrometry, and NMR has been discussed in other publications (Graves *et al.*, 1994; Krishna and Wold, 1993b, 1997), and these topics will not be included here. To summarize all the current information on methodology, we will simply state the following basic premise: *It is necessary to know both the encoded mRNA sequence and the complete covalent protein structure to establish with certainty what the nature and extent of the posttranslational modifications are.*

A. Modifications Involving Peptide Bonds

1. *Peptide Bond Cleavage: Limited Proteolysis*

Proteolytic cleavage is undoubtedly the most extensive and common of all covalent modifications of proteins, and it seems safe to predict that probably all protein products isolated from living cells have been modified by proteolysis from the precursor sequence encoded in the mRNA. The most obvious basis for this prediction is the fact that the universal initiation codon for protein synthesis, AUG, specifies methionine as the initiator amino acid in all eukaryotic proteins; in prokaryotes, the Met-$tRNA_f$ is formylated to give *N*-formyl-Met-$tRNA_f$, and *N*-formyl-Met consequently is the starting amino acid. Because only very few proteins are isolated with the starting amino acid still attached, it is clear that some trimming at the amino-terminal sequence has taken place during and/or after the assembly of the polymer. The initial trimming of either the amino-terminal Met in eukaryotes or the formyl or formyl-Met group in prokaryotes does not appear to have any dramatic effect on biological activity; for iso-1-cytochrome *c* a large number (about 100) of N-terminal variations, including additions, deletions, and single amino acid replacements observed in natural (92 different species) and mutationally altered structures, are fully active (Hampsey *et al.*, 1986). The reactions appear to be quite general, but follow certain rules of specificity (Arfin and Bradshaw, 1988). Thus, in yeast it has been shown that the Met is removed fast if the second residue in the chain is small (a radius of gyration of 1.29 Å) and slower if Pro is in the third position. With larger side chains in the second position the Met remains (Moerschell *et al.*, 1990). The removal of the formyl group, Met, or formyl-Met in prokaryotes probably also follow specificity rules, but they are not well defined at this stage. It is very unusual to observe unprocessed proteins with the initiation formyl-Met still attached; in one such case, pertaining to the *Escherichia coli* Trp operon cloned in a multicopy plasmid, a mixture of formylated and normal, deformylated α subunit was observed. This is consistent with a model of a fixed level of soluble, deformylating enzyme unable to keep up with the overproduction of new proteins in the modified system (Sugino *et al.*, 1980). The trimming and acetylation of the N terminus of eukaryotic proteins will be further discussed later under N-terminal modifications.

In addition to the N-terminal trimming, extensive internal peptide bond cleavage takes place in the maturation of many proteins. Many of these cleavages are quite well understood in terms of their biological functions, their enzymology, and their specificity. It may be useful to think of them as belonging to two major classes of proteolytic processing, that associated with transport (topogenesis) and that associated with activation. In general, the former one involves the nascent chains or the completed chains immediately after release from the polysomes and before the protein is fully folded; the latter one involves the folded precursor proprotein. In the former class are the various signal peptidases that remove signal peptides from the N-terminal end of the preprotein chain specified by the mRNA; in the second class are the cases of zymogen → enzyme transformations in digestive enzymes (Neurath and Walsh, 1976) in the blood clotting cascade (Hedner and Davie, 1989), and in complement activation (Müller-Eberhard, 1988), prohormone → hormone activation (Docherty and Steiner, 1982), polyprotein → neuropeptides/peptide hormones conversion (Lynch and Snyder, 1986), and the many cases of macromolecular assembly activated by proteolysis (Kräusslich and Wimmer, 1988). The proteases involved in many of these reactions show exquisite specificity; in some of the virus assemblies it appears that a single peptide bond in a single protein is cleaved in the activation process. All the posttranslational proteolytic reactions are in fact quite specific and well-regulated processes.

Several of these proteolytic processing steps can be illustrated by a brief review of the biosynthesis of insulin (Docherty and Steiner, 1982). This disulfide-bonded two-chain structure is encoded as the precursor preproinsulin. In the early stages of polymerization while the nascent chain is still attached to the membrane-associated polysomes, two cleavages take place: the removal of the N-terminal Met, to expose (in the rat) a new N-terminal Ala, followed by the removal of the 23-amino acids-long signal sequence exposing Phe as the new N-terminal residue. These steps convert the encoded preproinsulin to the 84-amino acids chain proinsulin (Phe^{1}-Asn^{84}). The removal of the signal peptide is associated with the translocation of proinsulin into the lumen of the endoplasmic reticulum (ER) (Sanders and Schekman, 1992). The next step involves an energy-requiring transport of the completed polypeptide chain to the Golgi; it is likely that the coordinated folding and disulfide bond formation (6 -SH to 3 -S—S- bonds: Cys^{3}-Cys^{70}, Cys^{19}-Cys^{83}, Cys^{69}-Cys^{74}) take place during the ER–Golgi transport. The proinsulin molecule is next packaged in secretory granules and transported to the plasma membrane for release. During this last step, two additional peptide bond cleavages occur at positions 32–33, and 63–64, to release the inactive 31-amino acids-long "propeptide" precursor (Glu^{33}-Arg^{63}) and a disulfide-linked pair of polypeptides, the B-chain precursor (Phe^{1}-Arg^{32}) and the completed A-chain (Gly^{64}-Asn^{84}). While still in the storage granule, the basic amino acids, Lys and Arg, are removed from the C-terminal end of the peptides by a carboxypeptidase-like activity that also appears to be involved in the production of peptide hormones and neuropeptides from polyprotein precursors, and the final processing products consequently are the propeptide Glu^{33}-Gln^{61}, after removal of Lys-62 and Arg-61 and insulin B-chain (Phe^{1}-Ala^{30}), after removal of Arg-31 and Arg-32, cross-linked by disulfide bonds Cys^{3}-Cys^{70} and Cys^{19}-Cys^{83} to the A-chain (Gly^{64}-Asn^{84}) with an internal disulfide bond, Cys^{69}-Cys^{74}. Both active hormone and propeptide are finally release to the extracellular compartment. In this case the rat mRNA specified a polypeptide chain of 108 amino acids, from which five single amino acids, a 23-amino acid N-terminal peptide, and a 29-amino acid interior peptide are removed before the biologically active product is produced. The utility of this particular processing reaction is proposed to be associated with the economical production of a two-chain protein. The monomolecular folding and disulfide formation in proinsulin is much more efficient than the bimolecular association, disulfide formation, and folding involving a mixture of reduced chains A and B.

2. *Peptide Bond (Intramolecular) Isomerization: Isoaspartate Formation, Prolyl Peptide cis–trans Isomerization*

The side chains of some of the amino acids in proteins can under certain circumstances interact with the peptide bond to cause the formation of, for example, esters from Ser residues, thioesters from Cys (Wallace, 1993), and isoaspartate from Asp or Asn (Geiger and Clarke, 1987). As will be seen later, these nonenzymatic conversions may in turn lead to other posttranslational events. In the case of isoaspartate, the isopeptide bond involving the β-carboxyl group of Asp linked to the α-amino group of the next amino acid in the chain with the α-carboxyl of Asp free is a stable derivative that appears in several proteins that have long half-lives. The *cis–trans* isomerization of peptide bonds does not involve covalent bond cleavage, but is a conformational adjustment of importance to the folding process. The planar peptide bond generally favors the *trans* conformation, but for Xxx-Pro peptide bonds, the *cis* and *trans* forms are about equal, and *cis* conformation is often observed for Pro peptides. The *cis–trans* conversion for these peptides is also quite slow, and in the folding of a protein into its native structure, the *cis–trans* isomerization of Xxx-Pro bonds may well be one of the rate-limiting steps; an enzyme, prolyl isomerase, is known to catalyze the isomerization (see Schmid, 1992).

3. *Peptide Bond Formation; Transpeptidation: Peptide Bond Splicing with Peptide Deletion and/or Permutation*

Among the most intriguing posttranslational modification reactions are those in which the mRNA-encoded peptide sequence is proteolytically cleaved and then reassembled and ligated in an order that is different than the original one, or in the same order but after eliminating some internal segment(s). The phenomenon, as a naturally occurring processing event, was first observed for the plant lectin concanavalin A (Bowles and Pappin, 1988), and in subsequent years some other cases have been observed in yeast, in an archaebacterium and a eubacterium. (For a review, see Wallace, 1993). The process may turn out to be a rare one, but since knowledge of both cDNA and protein sequence is needed to recognize posttranslational splicing, other cases may well be found in the future.

For concanavalin A the process from the proprotein to the mature lectin can be described in a series of steps. The protein is actually synthesized as a preproprotein, inserted into the ER with removal of the signal peptide, and transported through the Golgi and secreted as a major storage protein in jack beans. As illustrated in Fig. 1A, the conversion of the proprotein, a 261-amino acid residues long, presumably completely folded, polypeptide (Ser^{1}-----Val^{261}), to the mature protein involves proteolytic cleavages in 4 different locations and 1 peptide bond formation. The final products are concanavalin A with 237 amino acids (Ala^{135}---Asn^{254}-Ser^{1}---Asn^{119}) and 3 peptides, containing 4, 9, and 11 amino acids respectively.

Both the kinetics and the energetics need to be considered in arriving at a mechanism for these reactions. It was noted early that all the cleavages take place at the C-terminal side of Asn residues, and since jack beans contain an Asn-specific protease, it was a reasonable conjecture that the cleavage reactions could be catalyzed by this enzyme. In this case, since the condensation of a free amino and carboxyl group to form a peptide bond in dilute aqueous solution obviously is a very unfavorable reaction, it was proposed that the second cleavage is a transpeptidation that releases the C-terminal peptide and retains the N-terminal one as an activated enzyme-bound acyl derivative that can be transferred to the α-amino group of Ser-1 in an energetically favored reaction. It should be recognized that the 3D folding of the precursor may by itself overcome the energy barrier against peptide bond formation by placing the reacting amino and carboxyl groups in the proper juxtaposition to favor condensation. As will be seen for other cases of protein splicing, a strong argument can be made that the splicing reactions are autocatalytic processes, and that no enzyme in fact may be required. The fact that the concanavalin A gene can be expressed in *Escherichia coli* to yield the normal mature lectin in the absence of an Asn-protease (Yamauchi and Minamikawa, 1990) is consistent with a possible autocatalytic process. In this case, an alternative active intermediate must obviously be considered; as shown in Fig. 1B, a mechanism involving a C-terminal aminosuccinimide intermediate has been proposed, and aminosuccinide has indeed been shown to be present in the excised sequence from model peptides (Shao *et al.*, 1995).

In the case of other spliced proteins discussed by Wallace (1993), a yeast ATPase, an eubacterial RecA protein, and an archaebacterial DNA polymerase, no circular permutation is observed; in fact, the process is completely analogous to RNA splicing, except that the intervening sequences are eliminated at the protein level. For the ATPase and the RecA protein the situation is that a proprotein A-B-C is cleaved and spliced to yield the mature ATPase or RecA of the structure A-C and the intervening sequence B. In the case of the 69-kDa ATPase (A-C), the 50-kDa B was found to be a DNA endonuclease; for the 38-kDa RecA, B was 47 kDa and also showed some homology to known endonuclease structures. The archaebacterial DNA polymerase proprotein is more complex, with an A-B-C-D-E structure that is cleaved and spliced to give the polymerase A-C-E and two intervening sequences, B and D. D codes for an endonuclease; no function has yet been found for B. The strongest argument that these are indeed processed at the protein level is the isolation and characterization of the intact "intervening (protein) sequences" and, in certain mutants, of the unprocessed proprotein. The strongest arguments that they may be products of autocatalytic processing are that they are generally processed well in heterologous organisms, and that the chemistry of the cleavage sites is consistent with a common intramolecular transpeptidation reaction using the OH of Ser and Thr and the SH of Cys as the nucleophiles to cleave the peptide bond and to form internal esters or thioesters as the activated acyl groups forming the new peptide in the spliced product. In the discussion by Wallace (1993), he points out that the N-terminal cleavage sites (A-B, and in the case of the DNA polymerase also C-D) have either Xaa*-Cys-Xaa or Asn(Asp)*-Ser-Xaa to be matched with His-Asn*-Cys-Yaa and His-Asn*-Ser(Tre)-Yaa, respectively, at the C-terminal cleavage sites (B-C and D-E). He proposes that the cleavages occur at the bonds indicated by the asterisk to yield spliced sequences Xaa-Cys-Yaa or Asn(Asp)-Ser(Thr)-Yaa in the mature proteins. An aminosuccinimide intermediate is not possible for most of these splicing reactions, and an alternative autocatalytic mechanism, involving intramolecular ester/thioester formation involving Ser, Thr, and Cys, has been proposed and is illustrated in

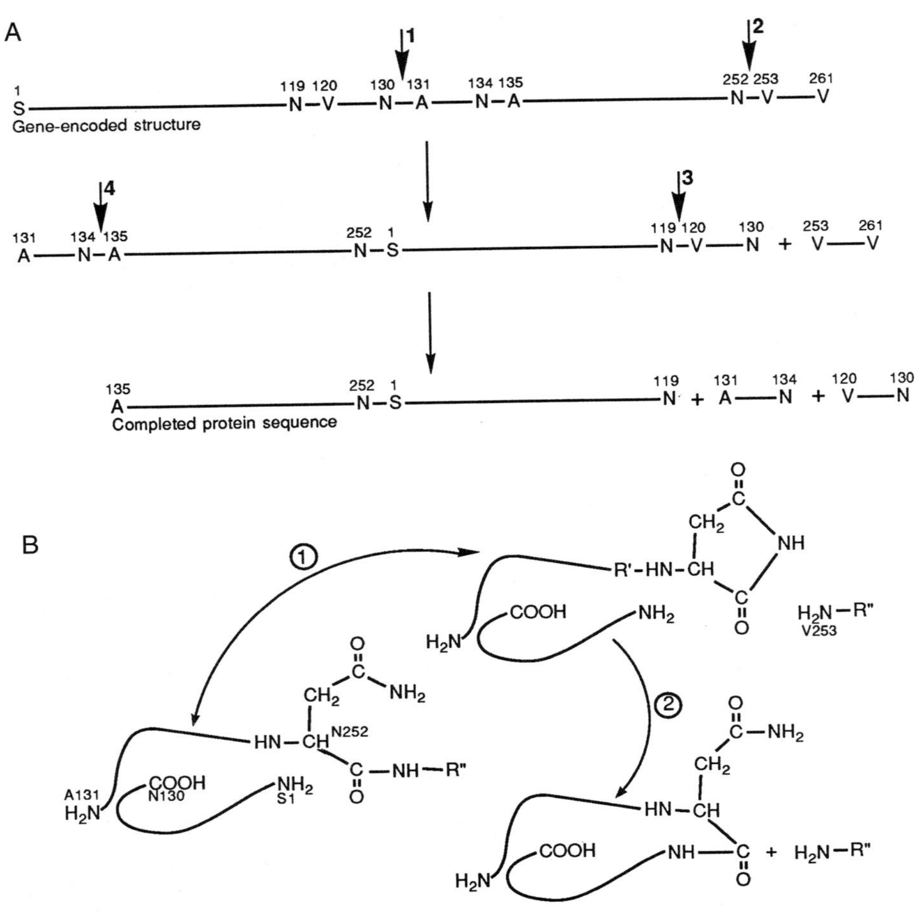

Fig. 1 Protein splicing: reactions and mechanisms. (A) The conversion of the gene-encoded 261 amino acid-containing proconcanavalin A (Ser[1]----Val[261]) to the 237 amino acid-containing concanavalin A (Ala[135]---Asn[252]-Ser[1]---Asn[130]) through peptide cleavage and transpeptidation. The first cleavage is at Asn[130]-Ala[131] to yield two peptides, Ser[1]----Asn[130] and Ala[131]-----Val[261]. The next cleavage is at Asn[252]-Val[253] to eliminate the C-terminal peptide Val[253]-----Val[261]. One attractive way of visualizing the rest of this step is that the Ala[131]-----Asn[252] is retained in a form conserving an activated acyl intermediate of Asn[252], which can then be transferred directly to the N terminus of Ser-1 to yield the peptide Ala[131]-----Asn[252]-Ser[1]-----Asn[130]. In the third step the bond at Asn[119]-Val[120] is cleaved to remove the C-terminal peptide, Val[120]-----Asn[130], and in the fourth step a cleavage at

Fig. 1C. It is clear that peptide-bound residues such as Asn and Asp, Ser, Thr, and Cys have unique reactivities that may result in many of the posttranslational derivatives that will be discussed elsewhere in this chapter, such as D-amino acids and various derivatives of dehydroalanine, and that must be seriously considered both for autocatalytic peptide hydrolysis and peptide splicing.

It is impossible to leave this topic without mentioning the fact that circularly permuted proteins created by genetic manipulations are beginning to yield exciting new information relevant to the laws that govern folding as well as to detailed structure/function relationships. As an example, Yang and Schachman

C

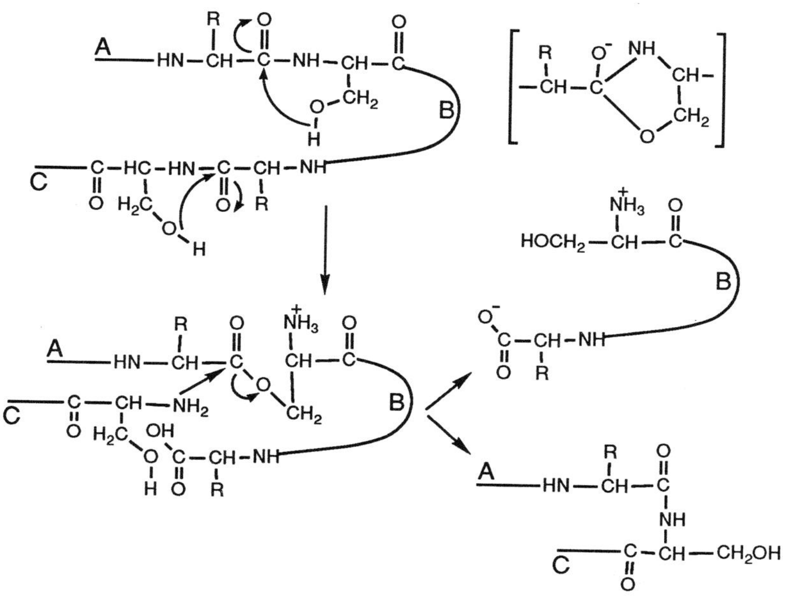

Asn^{134}-Ala^{135} removes the N-terminal peptide Ala^{131}--Asn^{134}. (B) Proposed mechanisms for nonenzymatic protein splicing through an aminosuccinimide intermediate. [Illustrated with data taken from Wallace (1993) and Shao *et al.* (1995).] All the cleavages in the processing of concanavalin A (Con A) take place at the C-terminal end of Asn, and the proposed mechanism is thus quite plausible, especially since the occurrence of C-terminal succinimide has been demonstrated in cleaving models peptides (Shao *et al.*, 1995). With reference to the steps shown in (A), the second step after the peptide bond 130-131 is shown to have been cleaved. In reaction 1, Asn-252 forms a succinimide intermediate with the release of the C-terminal peptide H_2N-R″ (residues 253–261); the active succinimide is now poised to react either with the released amino group of H_2N—R″ (reaction 1 in reverse) or with the original N-terminal α-amino group of Ser1, which through the folding of the precursor protein presumably is in the appropriate position to react (reaction 2). In the former case the product is simply the starting peptides; in the latter case the new peptide bond between the original residues Asn-252 and Ser-1 is formed. (C) Another proposed mechanism (Wallace, 1993) involving an intermediate ester (Ser and Thr) or thioester (Cys) is illustrated using Ser as an example; it can apply to any of the other splicing examples discussed in the text, where many of the cleavages take place in the absence of a suitably located Asn residue. As illustrated, esters are formed (through the putative intermediate shown in brackets) cleaving the peptide bond at the N-terminal side of Ser. The ester bond between peptides B and C is shown as having reacted with water to liberate the N-terminal NH_2 of peptide C, which then becomes an appropriately poised nucleophile to attack the ester between peptides A and B. The products of the reaction are the spliced protein A-C and the liberated protein B.

(1993) and Zhang and Schachman (1996) have mutated aspartate transcarbamoylase cDNA to yield protein products in which the normal N and C termini are linked in peptide bonds (they are in close proximity in the native structure), and in which new N and C termini are introduced at different locations; these authors have characterized the catalytic and regulatory properties and structural details of these mutant proteins. It appears that the general rule for these types of mutations are that if the new termini are in exterior loop regions with little secondary structure, the properties of the mutant proteins are close to those of the wild-type one; if the peptide chain is opened in a helix or sheet, however, the mutant proteins are insoluble and inactive.

B. Modifications Involving Amino and Carboxyl Termini

In Sections II, B and C the different covalent derivatives of amino acids known to us at this time will be listed with very little discussion of the individual reactions. The two major reaction types, spontaneous and enzyme-catalyzed, will be identified whenever possible, and special features of certain reactions will also be emphasized as needed.

The N Terminus: $H_3\overset{+}{N}$—

N-Formyl- (C1)	Adams and Capecchi (1966); Kreil and Kreil-Kiss (1967); Sugino *et al.* (1980)
N-Acetyl- (C2)	Tsunasawa and Sakiyama (1984); Augen and Wold (1986); Arfin and Bradshaw (1988); Krishna *et al.* (1991)
N-Acyl- (C2, C4, C6, C8, C10)	Moscarello *et al.* (1992)
N-Lauroyl- (C12)	Neubert *et al.* (1992)
N-Myristoyl- (C14)	Carr *et al.* (1982); Rudnick *et al.* (1993); Johnson *et al.* (1994)
N-Tetradeca (mono and di)enoyl- (C14 : 1; C14 : 2)	Neubert *et al.* (1992)
N-Aminoacyl-	Kaji *et al.* (1965); Kaji (1976)
N-α-Ketoacyl-	van Poelje and Snell (1990)
N-Methyl-	Chen *et al.* (1977b); Lederer *et al.* (1977); Pettigrew and Smith (1977); Siegel (1988)
N-Pyrrolidone carboxyl-	Orlowski and Meister (1971); Awade *et al.* (1994)
N-Glucuronyl-	Lin and Kolattukudy (1980); Kolattukudy (1984)
N-Glycosyl-	Kennedy and Baynes (1984); Monnier (1989)

The most common group of derivatives of the N terminus is the result of N^{α}-acylation, and several different acyl groups have been observed. The majority of these are the result of enzymatically catalyzed reactions, and many of the acyltransferases involved have been characterized, along with their various acyl donors, such as acyl-CoA. The formyl group is found almost exclusively in prokaryotes; one eukaryotic example has been observed as N^{α}-formyl-Gly in a fraction of honeybee melittin (Kreil and Kreil-Kiss, 1967). In prokaryotes the formyl group is introduced as a posttranslational modification at the level of Met-tRNA$_{fMet}$ and the specificity is thus determined entirely by the unique structure of tRNA$_{fMet}$; Met attached to another tRNA, Met-tRNA$_{Met}$, is not a substrate for the transferase that transfers the formyl group from formyltetrahydrofolate to the amino group of Met-tRNA$_{fMet}$. The acetyl group is found in eukaryotes; in this case the transfer of the acetyl group from acetyl-CoA to the N-terminal α-amino group in most cases takes place on the nascent polypeptide chain. The specificity of the reaction follows some complicated rules closely interlinked with action of other enzymes such as amino peptidases and perhaps acetylaminoacyl-peptide hydrolase, acting on the N-terminus of the growing or sometimes the recently completed peptide chains. As pointed out in Section I, A, it is clear that these specificity determinants involve the nature of the N-terminal amino acid and its immediate neighbor in the sequence (Arfin and Bradshaw, 1988); 95% of the acetylated N-terminals have one of 5 amino acids, Ala, Ser, Gly, Thr, Met, as their N-terminal acetylamino acid. However, it also appears that there are signals encoded in the extended chain that may determine the ultimate outcome of the N-terminal acetylation (Augen and Wold, 1986; Krishna *et al.*, 1991; Sokolik *et al.*, 1994).

A couple of acylation reactions are not catalyzed by acyl transferases. Thus, the appearance of an α-keto acid at the N terminus of a protein is not the result of the introduction of the acid by a transferase, but rather of the unusual cleavage of an interior peptide bond with the loss of ammonia to leave the new N terminus as an α-keto acid. The best studied cases involve the conversion of an internal Ser to an N-terminal pyruvate, and the reaction mechanism is well documented to be as illustrated here (van Poelje and Snell, 1990): The Ser OH replaces the NH of the peptide bond to yield an ester, which in turn, by β-elimination, causes cleavage of the bond to yield an N-terminal dehydroalanine. The dehydroalanine finally rearranges to the imine, which is hydrolyzed to give pyruvate and ammonia. The reaction appears to be autocatalytic and is obviously related to the one discussed in Fig. 1C.

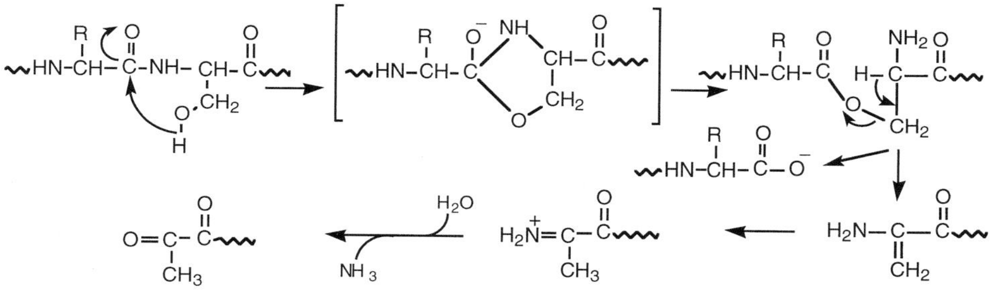

In another acylation reaction the α-amino group of Glu is acylated by the γ-carboxyl group of Glu itself to yield the cyclic pyrrolidone carboxylate as the N-terminal amino acid. This reaction can take place with N-terminal Glu as well as Gln, and it is a spontaneous reaction as well as an enzyme-catalyzed one. As expected, the spontaneous reaction favors Gln as the starting material, and the spontaneous cyclization of the N terminus during routine sequencing is a major cause of abrupt termination of sequencing. The fact that a given peptide may give quite reasonable yields of PTH-Gln and subsequent PTH-amino acids, whereas another, under the same conditions, may terminate after a very low yield of PTH-Gln, suggests that the amino acid sequence in each peptide determines the chemical stability of the N-terminal Gln. Enzymatically, the cyclic derivative can be made from Glu with, for example, glutamine synthase; activation of the γ-carboxyl group is required in the form of ATP. With Gln as substrate the reaction becomes a simple transfer catalyzed by γ-glutamyl-cyclotransferase. In this case the activated γ-glutamyl group can be transferred to other acceptors; these reactions will be discussed later under Gln modifications.

Another nonenzymatic reaction is the glycosylation of the N terminus in the presence of high concentrations of reducing sugars that may occur naturally under special conditions. The Amadori rearrangement of the resulting Schiff base between the reducing sugar and the amino group yields stable covalent derivatives. The ε-amino group of Lys is a common site for these reactions and the chemical mechanism will be considered in more detail under Lys.

The C Terminus: $-\overset{O}{\overset{\|}{C}}-O^-$

-Amide	Kreil (1984); Bradbury and Smyth (1991)
O-(ADP-ribosyl)-	Hayaishi and Ueda (1984); Adamietz and Hilz (1984)
O-Methyl-	Inglese *et al.* (1992); Clarke (1992)
-(*N*-Ethanolamine-glycan-phosphoinositides)	Ferguson and Williams (1988); Englund (1993); Udenfriend and Kodukula (1995)
-(N^α-Tyr)	Flavin and Murofushi (1984); Webster *et al.* (1987)

There are only a few well-characterized derivatives of the carboxyl-terminal end, and at the present time all of these derivatives appear to be quite specialized and restricted to only a few proteins. The unsubstituted amides (α-$CONH_2$) are found almost exclusively in short, physiologically active peptides such as insect toxins (Suchanek *et al.*, 1980; Kreil, 1984) and hormones (Lowry and Chadwick, 1970). The amide group is associated with several different amino acids [Asp, Glu, Gly, His, Met, Phe, Pro, Tyr, Val (Uy and Wold, 1977)] but the peptides are derived from large polyprotein precursors, and the amino acid at the C-terminal side of the residue which is to become the C-terminal amide appears to be Gly in all cases. After the proteolytic cleavage to yield the Gly-terminated peptide, this peptide is acted on by a peptidylglycine hydroxylase (the ascorbic acid, copper, and oxygen required for the amidation are presumably used in this reaction), and by a second enzyme that catalyzes the cleavage of the product into peptide amide and glyoxylate:

$$\text{Peptide-}\overset{O}{\overset{\|}{C}}-\underset{H}{\underset{|}{N}}-\underset{H}{\underset{|}{\overset{H}{\overset{|}{C}}}}-\overset{O}{\overset{\|}{C}}-O^- \longrightarrow \text{Peptide-}\overset{O}{\overset{\|}{C}}-\underset{H}{\underset{|}{N}}-\underset{H}{\underset{|}{\overset{OH}{\overset{|}{C}}}}-\overset{O}{\overset{\|}{C}}-O^- \longrightarrow \text{Peptide-}\overset{O}{\overset{\|}{C}}-\underset{H}{\underset{|}{\overset{H}{\overset{|}{N}}}} + \underset{H}{\underset{|}{\overset{O}{\overset{\|}{C}}}}-\overset{O}{\overset{\|}{C}}-O^-$$

The ADP-ribosylation of carboxyl-terminal Lys of histone H1 (Riquelme *et al.*, 1979; Ogata *et al.*, 1980 a, b; Hayaishi and Ueda, 1982, 1984; Adamietz and Hilz, 1984) will be considered along with other ADP-ribosylation reactions, and the *N*-ethanolamine–glycan–phosphoinositides derivatives and the α-carboxymethylation reactions, which appear to be associated with S-isoprenylation (Clarke, 1992), will be discussed along with other membrane anchors in the discussion of functions later. The derivative in which a Tyr residue is added in peptide linkage to the α-COOH group of the α chain of tubulin in the absence of ribosomes and mRNA will also be discussed under functions; at this point it will only be noted that the reaction is quite analogous to the biosynthesis of glutathione and other short peptides in the use of ATP as activator:

$$\alpha\text{-Tubulin-Gly-Glu-Glu} + \text{Tyr} + \text{ATP} \rightarrow \alpha\text{-Tubulin-Gly-Glu-Glu-Tyr} + \text{ADP} + \text{P}_i$$

The findings that some α-tubulin mRNA contain Tyr as the C-terminal residue, and that an enzyme capable of removing the Tyr exists suggest that a cyclic removal and addition of the Tyr may take place, and that the reaction thus may have a regulatory function. The C-terminal ends of tubulins is the site for several other posttranslational modifications (e.g., γ-glutamyl derivatives; see under Glu), and appear to be involved in the regulation of microtubule assembly and deassembly.

C. Modifications Involving Individual Amino Acid Side Chains

Arginine: $H_2N{-}\underset{\omega}{\overset{\overset{+}{N}H_2}{\overset{\|}{C}}}{-}NH{-}CH_2{-}CH_2{-}CH_2{-}$

N^{ω}-(ADP-ribosyl)-	Oppenheimer (1984); Ueda and Hayaishi (1985)
N^{ω}-Methyl-	Paik and Kim (1990)
N^{ω}-Dimethyl-	Paik and Kim (1990)
N^{ω},$N^{\omega'}$-Dimethyl-	Paik and Kim (1990)
Ornithine	Sletten and Aakesson (1971)
Citrulline	Rothnagel and Rogers (1984)
N^{ω}-Phosphoryl	Fujitaki and Smith (1984)

The cross-linked Arg derivative, pentosidine (Grandhee and Monnier, 1991) is listed and discussed under Lys.

Derivatives of the guanidinium group of arginine have been found in a variety of proteins in both eukaryotes and prokaryotes. Only the cross-linked derivative, pentosidine, is produced in a nonenzymatic reaction; the various enzymes involved in the other reactions are well characterized. As for most of the posttranslational reactions, the majority of the Arg modifications are irreversible. The ADP-ribosyl derivatives may have some exceptions to this generality (Pope *et al.*, 1986); the reversibility of ADP-ribosylation reactions will be discussed under function later. An interesting feature of the ADP-ribosylation catalyzed by bacterial toxins is that the product formed in the reaction is the α-ribosyl anomer that rapidly equilibrates to a mixture of α- and β-anomers. This epimerization may be relevant to the specificity of the putative *N*-glycosidase involved in the reverse reaction. The *N*-phosphoryl derivative has been identified in myelin basic protein (Fujitaki and Smith, 1984). Apparently, *N*-phosphoryl derivatives may be much more common than expected, and those of Arg, His, and Lys are quite acid labile and are destroyed by the conditions commonly employed in the study of other phosphate derivatives. Citrulline is a special derivative found primarily in structural proteins of hair and skin.

Asparagine: $H_2N—\overset{O}{\overset{\|}{C}}—\underset{\beta}{CH_2}—$

N-Glycosyl-	Kornfeld and Kornfeld (1985); Tanner and Lehle (1987); Paulson (1989)
Aspartate	Robinson *et al.* (1973); Van Kleef *et al.* (1975); De Jong *et al.* (1988)
N-Methyl-	Klotz and Glazer (1985)
N^{ε}-(β-Aspartyl)lysine	Klostermeyer (1984)
erythro-β-Hydroxy-	Stenflo *et al.* (1987)
N-(ADP-ribosyl)-	Sekine *et al.* (1989)

There do not appear to be many derivatives of Asn, but the actual number would in fact be quite large if we could list all the structures of asparagine-linked oligosaccharides that exist in nature. The details of the biosynthesis of these glycoproteins are beyond the scope of this chapter, but a few points can be made here and again later in connection with specificity and function of these fascinating structures. In all eukaryotes a common precursor, Glc_3-Man_9-$GlcNAc_2$-, is assembled on a lipid carrier and transferred by an oligosaccharyltransferase (Silberstein and Gilmore, 1996) to Asn residues in the nascent chain as it enters the ER (Fig. 2). It is required but not sufficient that the Asn is in a -Asn-Xxx-Ser(Thr)- sequence for the transfer to take place (Struck and Lennarz, 1980). Once the precursor has been attached to the appropriate Asn residues, the further processing of the oligosaccharide may follow widely different pathways depending on the organism, cell, and organelle in which the processing proceeds. It is clear that one main determinant of the processing path is the complement of enzymes available in the processing compartment(s). Glucosidases to remove the three Glc residues in the precursor appear to be present in all cells, but as indicated in Fig. 2, the subsequent steps may be unique for individual cells and organisms. Another general processing decision that may apply across the board is whether the glycan is to be modified at all. In glycoproteins from virtually all sources to date, some glycans are found whose structures, $Man_{5\text{-}9}$-$GlcNAc_2$-, are quite close to the starting ones and can be considered as essentially "unprocessed." If the glycan in a given location in a given glycoprotein is selected for processing, it will be converted to the characteristic end product of the compartment of action, large $Man_{\sim 300}$-$GlcNAc_2$- in yeast extracellular glycoproteins (Cohen *et al.*, 1980), complex oligosaccharides containing other sugars, such as Fuc, Gal, sialic acid, and additional GlcNAc in higher animals, and Xyl and Fuc in plants and some invertebrates (Fig. 2). Some of the derivatives may even be phosphorylated (Cohen *et al.*, 1980; Tabas and Kornfeld, 1980). Some cases of expressing glycoprotein mRNAs in heterologous eukaryotic organisms suggest that the signal "processing" vs "no processing" may be at least partly understood by all organisms. The number of different glycosylated Asn derivatives is indeed impressively large. In order to fully appreciate the impact of protein glycosylation, it will be necessary to establish how the glycan and the protein interact: to what extent the glycan modifies protein structure and to what extent the protein modifies the glycan structure. Some very significant recent results from the exploration of glycoproteins by X-ray crystallography (Shanaan *et al.*, 1991) and NMR (Lommerse *et al.*, 1995) herald new insights in this area.

It has been assumed all along that the N-glycosylation represents a permanent modification. However, an enzyme has recently been isolated from fibroblasts that can remove the glycan from the intact glycoprotein (Suzuki *et al.*, 1994). The existence of such an enzyme could obviously have important *in vivo* regulatory implications, but it must be kept in mind that the glycosylation/deglycosylation would be a single-cycle event. The enzyme is a glycoamidase that hydrolyzes the C–N bond of the glycosylated Asn, thus leaving Asp as the product of the reaction. It is interesting to note that this represents one way of converting Asn to Asp in proteins; the spontaneous, nonenzymatic hydrolysis reaction is the other one. The rate at which the spontaneous deamidation of a given asparagine residue takes place under physiological conditions is determined by the neighboring amino acids in the polypeptide sequence (Robinson *et al.*, 1973). The reaction is slow enough to affect mostly long-lived proteins, but it still seems reasonable to consider the asparagine residues in proteins as built-in clocks determining the lifetime of the protein according to the rate of the spontaneous conversion of the neutral amide to the charged acid. It is clear that this conversion indeed does take place *in vivo* (Midelfort and Mehler, 1972) and in fact represents a significant change in long-lived proteins such as α- and β-crystallin from lens (Van Kleef *et al.*, 1975).

The remaining derivatives are rather rare: *N*-methyl-Asn has been observed in phycobiliproteins; the Asp-Lys cross-link, N^{ε}-(β-aspartyl)lysine, in colostrum; β-hydroxy-Asn in vitamin K-dependent protein S; and the ADP-ribosyl derivative only in a single protein (the *rho* gene product) as a product of the action of *Clostridium botulinum* ADP-ribosyltransferase.

Aspartate: 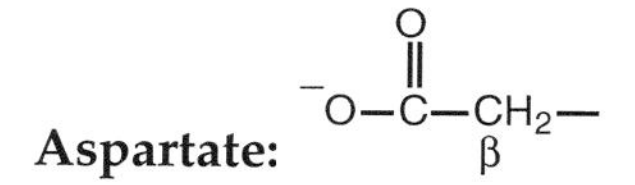

D-Asp (racemization)	Helfman and Bada (1976)
β-Carboxy-	Koch *et al.* (1984)
erythro-β-Hydroxy-	Drakenberg *et al.* (1983); Stenflo *et al.* (1989)
β-Methylthio-	Kowalak and Walsh (1996)
O-Phosphoryl-	Degani and Boyer (1973)
O-Methyl-	Clarke (1988)

Of the reactions of Asp, the slow formation of D-Asp, like the deamidation of Asn and the isomerization of the Asp peptide bond, is a spontaneous reaction, and the end product of the reaction presumably should be something approaching a racemic mixture. This slow reaction should be significant only in proteins with long half-lives. Analysis of hard-tissue proteins show that about 0.1% of L-Asp is converted to D-Asp per year (Helfman and Bada, 1976); in the human lens protein (α-crystallin) the rate is about 0.14% per year in normal individuals, but significantly higher in individuals with cataracts (Masters *et al.*, 1977). It should be noted that there are also enzymatically catalyzed inversions of several different amino acids in proteins; in these cases the inversion is complete to yield exclusively the D-enantiomer in the product (Kreil, 1994) (see later discussion under "D-amino acids").

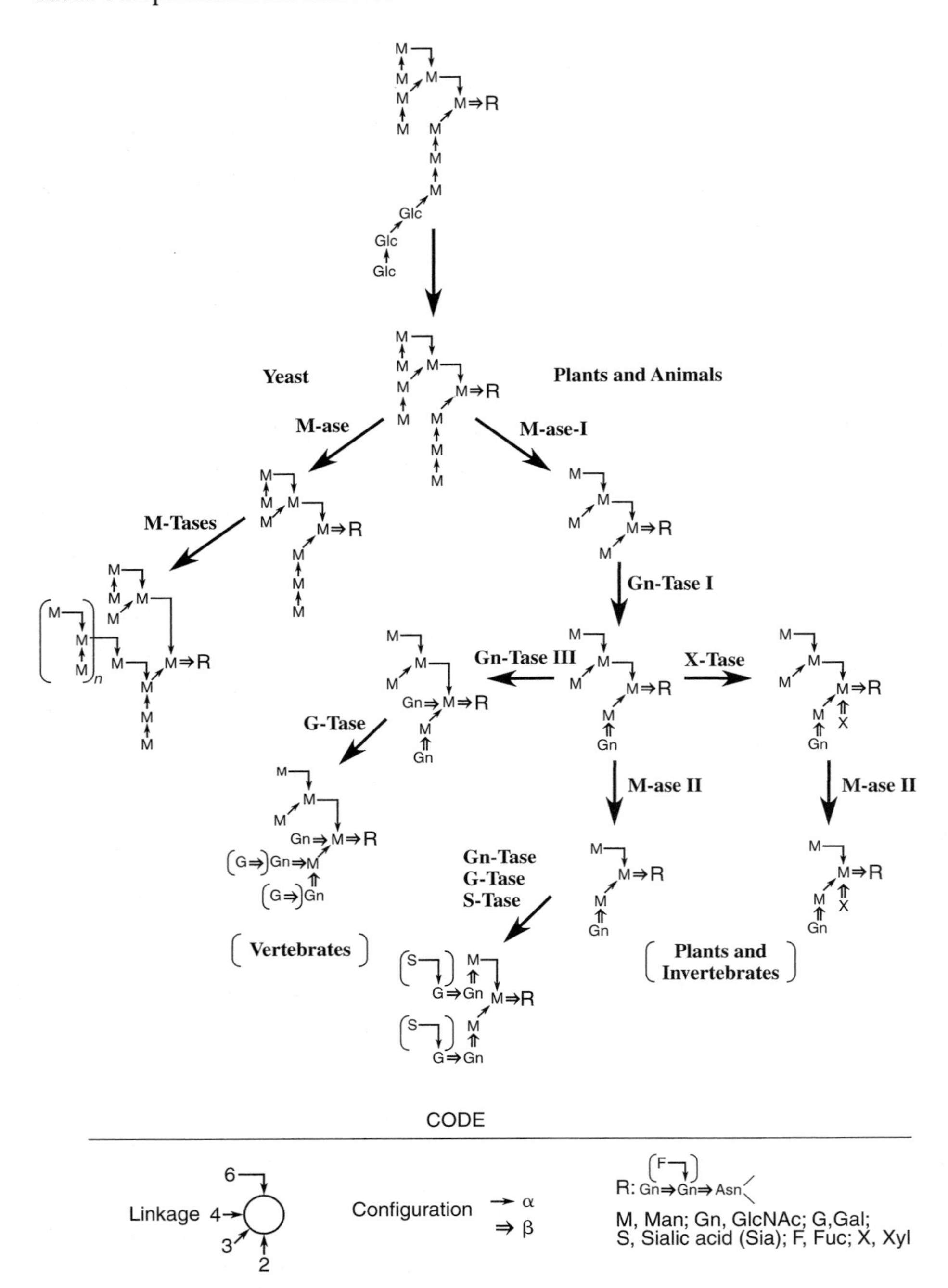

Fig. 2 Typical Asn-linked glycan structures. [Modified from the scheme by Shao *et al.* (1994).] The universal precursor contains 9 Man and 3 Glc residues linked to the 2 GlcNAc residues that, together with protein-bound Asn, make up the R group in the reported structures. The lipid-bound precursor is transferred intact to an appropriate Asn acceptor to yield the universal precursor structure at the top of the figure. In the processing of this precursor, the 3 Glc residues are first removed from the precursor in all cells. The resulting Man_9 intermediate is next handled in a variety of different ways, depending on the organism or cell in which the processing takes place

Enzyme-catalyzed reactions include β-carboxy-Asp (Koch *et al.*, 1984) and β-hydroxy-Asp. The former has been found in prokaryotic and eukaryotic ribosomal proteins and has been proposed to be synthesized by a vitamin K-dependent carboxylase like the one catalyzing the formation of γ-carboxy-Glu in Ca-binding, blood clotting, and calcification proteins. β-Hydroxy-Asp is often associated with vitamin K-dependent proteins, and is, for example, found together with γ-carboxy-Glu in protein C (Drakenberg *et al.*, 1983). The enzyme catalyzing the hydroxylation is a dioxoglutarate-dependent dioxygenase (Stenflo *et al.*, 1989). The methylthio derivative of Asp has been found in an *E. coli* ribosomal protein. The derivative, β-aspartylphosphate (O^4-phosphoaspartate) has been identified as the transient phosphorylated derivative in the Na^+,K^+-ATPase of plasma membranes and the Ca^{2+}-ATPase of sarcoplasmic reticulum (Bastide *et al.*, 1973; Degani and Boyer, 1973). It should be noted that the transmethylase catalyzing the transfer of the methyl group for *S*-adenosyl-Met to protein-bound Asp does not use L-Asp as a substrate at all. The methylation takes place either at the β-carboxyl group of D-Asp or at the α-carboxyl group of L-iso-Asp. It has consequently been proposed that the Asp methylase has a role in marking or perhaps salvaging abnormal Asp residues arising from spontaneously modified Asp or Asn (Aswad, 1984; Clarke, 1988).

Cysteine: $HS\text{-}CH_2\text{-}$

Cystine	Creighton (1984); Freedman (1992)
S-γ-Glutamyl-	Tack *et al.* (1980); Sottrup-Jensen (1989)
S-(2-Histidyl)-	Lerch (1984)
S-(3-Tyr)	Ito *et al.* (1991)
S-(*sn*-1-Glyceryl)-	Lampen and Nielsen (1984)
S-(*sn*-1-Diacylglyceryl)-	Hayashi and Wu (1990)
S-(*sn*-1-{2,3-Di-*O*-[3′,7′,11′,15′-tetramethylhexadecyl]}glyceryl)-	Sagami *et al.* (1995)
S-Palmitoyl-	Towler *et al.* (1988); Hancock *et al.* (1989); McIlhinney (1990)

and depending on the structure of the protein matrix in which it is located. In *Saccharomyces cerevisiae*, the best-studied yeast species, an $\alpha(1 \rightarrow 2)$-linked Man is removed and then several Man residues (up to several hundreds) are added to give oligo- and polymannose glycans. In plants and animals several Man residues are removed, and GlcNac, Gal, Sia, and sometimes fucose (Fuc) and/or xylose (Xyl) are added in well-defined sequences of steps to yield different groups of structures, such as the short-chain glycans containing Xyl in plants and some invertebrates; the hybrid structures containing 5 Man and additional GlcNAc, Gal, and perhaps Sia in, for example, oviduct cells; and multiantennary complex glycans in other vertebrate cells. The ones shown here are biantennary, containing 2 chains of Sia-Gal-GlcNAc, but they can also be tri- and tetra-antennary, and, as indicated by the parenthetic F in the R group, may also contain Fuc. The structures shown are only the most common ones; the total number of different structures is much larger, and at any given glycosylation site, the glycans are generally heterogeneous, suggesting a mixture of incompletely processed derivatives. The variety of structures obviously reflects the presence of different enzyme activities in different cells, but in addition it is also clear that the substrate quality of the glycan is a major determinant for the processing specificity. Thus, two identical glycan precursors located at two different glycosylation sites in the same protein may be processed in completely different manners; one may be converted to a complex glycan while the other may remain essentially unprocessed as a oligomannose structure. The protein environment of each glycan is clearly important in determining the structure of the finished glycoprotein product.

S-Farnesyl-	Glomset *et al.* (1990); Seabra *et al.* (1991); Inglese *et al.* (1992); Furfine *et al.* (1995)
S-Geranylgeranyl-	Glomset *et al.* (1990); Seabra *et al.* (1991); Yamane *et al.* (1990)
S-Heme	Margoliash and Schejter (1966)
S-Phycocyanobilin	Glazer (1984)
S-*p*-Coumaroyl	Hoff *et al.* (1994)
S-(6-Flavin [FMN])	Singer and McIntire (1984)
S-(8α-Flavin [FAD])	Singer and McIntire (1984)
S-Coenzyme A	Thorneley (1992)
S-(ADP-ribosyl)-	West *et al.* (1985)
S-Glycosyl-	Weiss *et al.* (1971)
Dehydroalanine	Gavaret *et al.* (1979); Scaloni *et al.* (1994)
Lysinoalanine	Steinert and Idler (1979)
Lanthionine	Steinert and Idler (1979)
Selenocysteine	Listed and discussed under its parent primary amino acid, Ser

It should be noted that although dehydroalanine is properly listed as a derivative of Cys, it can also be derived from Ser and from Tyr [in which case dehydroalanine is produced during thyroxine formation in thyroglobulin (Steinert and Idler, 1979; Gavaret *et al.*, 1979)]; thus, several of the products from reactions of dehydroalanine (lysinoalanine, lanthionine, alaninohistidine) can in principle and do in fact have their origin in amino acids other than Cys.

The most obvious and common derivative of Cys is the oxidized, disulfide-containing cystine. This reaction is important in stabilizing the three-dimensional folding of individual domains by intrachain cross-links (Baldwin, 1975; Lukens, 1976; Creighton, 1984; Freedman, 1992), as well as in cross-linking between protomers in the formation and maturation of multichain structures (Lukens, 1976; Steinert and Idler, 1979; Olson and Lane, 1989). It is likely that the catalytic agents responsible for the reaction must be active in many compartments and cells, but the main site for disulfide bond formation is probably the ER, where it proceeds in parallel with the folding process itself. Although the disulfide formation can proceed spontaneously, the biological process is likely to be enzyme-catalyzed. Thus, the enzyme protein disulfide isomerase is very abundant in the ER, and is a nonspecific catalyst for the formation and interchange of disulfide bonds in proteins. The genetics, the kinetic properties, and the distribution of the enzyme have been well studied in eukaryotes, and its direct involvement in folding dynamics has been documented (Freedman, 1992).

Other cross-links formed involving the thiol group of Cys include a thioester, *S*-γ-glutamyl-Cys, and two thioethers, *S*-(2-histidyl)-Cys and *S*-(3-Tyr)-Cys. In addition, there are some presumably spontaneous cross-links; lysinoalanine and lanthionine, that are products of dehydroalanine derived from Cys.

A large number of prosthetic groups are attached to proteins through Cys. These include the *S*-cysteinylheme of cytochrome *c* (Margoliash and Schejter, 1966), the *S*-cysteinyl phycocyanobilins in various phycocyanin derivatives

(Williams and Glazer, 1978; Lagarias *et al.*, 1979; Glazer, 1984), the thioester, *S*-*p*-coumaroyl-Cys (Hoff *et al.*, 1994), 8α-(*S*-cysteinyl)flavin thiohemiacetal (Singer and McIntire, 1984), 6-(*S*-cysteinyl)riboflavin 5′-phosphate (Ghisla *et al.*, 1980), and the mixed disulfide of coenzyme A and Cys in a bacterial flavodoxin (Thorneley *et al.*, 1992). Cys is also glycosylated: *S*-Galactosyl-Cys (Lote and Weiss, 1971), *S*-glucosyl-Cys (Weiss *et al.*, 1971), and *S*-(ADP-ribosyl)-Cys (West *et al.*, 1985) have been reported.

A fascinating group of derivatives belong to the general class of "membrane anchors." The phosphoinositide derivative of the C-terminal carboxyl group has already been mentioned as one member of this class. In addition there are glycerol derivatives in which the Cys S replaces one of the OH groups of the glycerol with no (most likely a precursor) or two fatty acids esterified to the other OH groups, *S*-(*sn*-1-glyceryl)-Cys (Lampen and Nielsen, 1984) and *S*-(*sn*-1-diacylglyceryl)-Cys (Hayashi and Wu, 1990), or with polyisoprene units ether-linked to the two OH groups as in *S*-(*sn*-1-{2,3-di-*O*-[3′,7′,11′,15′-tetramethylhexadecyl]}glyceryl)-Cys (Sagami *et al.*, 1995). A single fatty acid may also form a thioester with Cys, *S*-palmitoyl-Cys. Finally in this group are the thioether-linked polyisoprene derivatives, *S*-farnesyl-Cys and *S*-geranylgeranyl-Cys (Glomset *et al.*, 1990). A good deal of information is becoming available on these fascinating reactions. Thus, the enzyme farnesyltransferase has been isolated and its properties studied (Furfine *et al.*, 1995). The function of these derivatives in the association of the modified proteins with membranes will be considered later (Section IV, D).

Since the thiol group is involved in the catalytic apparatus of several enzymes, a number of transient covalent Cys derivatives such as thiohemiacetals and thioesters have been demonstrated or deduced as transient catalytic intermediates; these derivatives are not included here. It is quite possible that derivatives such as the (peptide)-cysteine-*S*-*S*-cysteine (free) in α_1-antitrypsin (Glaser and Karic, 1976), the cysteine *S*-phosphate (Pigiet and Conley, 1978), cysteine *S*-sulfonate, and the cysteine *S*-sulfide (persulfide) (Tsang and Schiff, 1976) observed in bacterial thioredoxin and in sulfotransferase may also fit in the category of catalytic intermediates.

Glutamate: $^{-}O-\overset{\overset{\large O}{\|}}{C}-\underset{\gamma}{CH_2}-CH_2-$

O-(ADP-ribosyl)-	Ogata *et al.* (1980 a, b); Ueda and Hayaishi (1985)
γ-Carboxy-	Stenflo and Suttie (1977); Carlisle and Suttie (1980); Suttie (1985)
O-Methyl-	Koshland (1981); Clarke (1985); Clarke (1988)
N^{α}-(γ-Glutamyl)-$Glu_{1\text{-}5}$	Eddé *et al.* (1990)
N^{α}-(γ-Glutamyl)-$Gly_{3\text{-}34}$	Redeker *et al.* (1994)
N-(γ-Glutamyl)ethanolaminephosphate	Whiteheart *et al.* (1989)
S-γ-Glutamyl-Cys is listed under Cys	

The derivative O^{γ}-ADP-ribosylglutamate has been identified in histones H1 and H2B (Ogata *et al.*, 1980 a, b) and in several animal cells (Hayaishi and

Ueda, 1984), as well as in plants (Whitby *et al.*, 1979). In the nucleus the reaction generally involves poly(ADP)-ribosylation, and long chains of ADP-ribose are attached to carboxyl groups in reversible fashion. These reactions will be discussed further later.

γ-Carboxyglutamate was originally found in five plasma glycoproteins, including prothrombin, and provided the missing link between vitamin K deficiency and faulty blood clotting, when vitamin K was established to be the required cofactor for the carboxylase that introduces the extra carboxyl group (Carlisle and Suttie, 1980). The fact that both the unique Ca^{2+} and Ba^{2+} binding properties of prothrombin and its activity in clotting are lost upon removal of the γ-carboxyl group [by heating *in vacuo* (Tuhy *et al.*, 1979)] represented strong evidence that the malonate-like side chain is involved in Ca^{2+} binding as an essential component of biological function. γ-Carboxyglutamate has also been found in other tissues, primarily those associated with calcification (Hauschka *et al.*, 1975; Stenflo and Suttie, 1977) and has been reported to be present in ribosomal proteins of both prokaryotes and eukaryotes as well (Van Buskirk and Kirsch, 1978). The functions of these derivatives will be discussed later.

O^{γ}-Methylglutamate has received a good deal of attention because of its role in the dynamics of bacterial chemotaxis (Koshland, 1981; Stock and Simms, 1988) and ameba movement (Mato and Marin-Cao, 1979). Also, the presence of alkali-labile methyl groups in proteins of animal cells (Paik and Kim, 1990; Clarke, 1985, 1988) suggests that the methyl ester of Glu is involved in functions other than chemotaxis.

Some derivatives in which several Glu residues (1 to 5) or Gly residues (3–34) are added to the γ-carboxyl group of Glu appear to be unique to the C-terminal part of tubulin. The logical precursor for these derivatives is a protein-bound Gln, but apparently Glu itself is the precursor *in vivo*. The fact that the α-carboxyl group is activated to participate in the formation of Glu-Glu bonds in the outer Glu residues of the modified derivatives is consistent with a mechanism in which each carboxyl group must be activated for the reaction to take place. The attachment of ethanolamine-phosphoglycerol to specific Glu residues has been observed in eukaryotic elongation factor 1α.

A final *in vivo* modification, in which Glu is converted to Gln, is an example of a "silent" modification reaction with a primary amino acid as product. This reaction is unique in that the modification takes place at the level of amino acyl-tRNA and in that the encoded gene actually specifies Gln as the amino acid to be incorporated. The reason for the transformation is that there is no activating machinery for Gln in yeast mitochondria (Martin *et al.*, 1977) (nor apparently in some prokaryotes), and glutaminyl-tRNA can thus be produced only from the amidation of preformed glutamyl-tRNA.

Glutamine: $H_2N{-}\overset{O}{\overset{\|}{C}}{-}\underset{\gamma}{CH_2}{-}CH_2{-}$

Glutamate	Robinson *et al.* (1973); Wold (1985)
N^{ε}-(γ-Glutamyl)lysine	Loewy (1984); Greenberg *et al.* (1991)
N-(γ-Glutamyl)-L-ornithine	Lou (1975)
N-(γ-Glutamyl)polyamine	Beninati *et al.* (1985); Piacentini *et al.* (1988)

N,N-(Bis-γ-glutamyl)polyamine	Beninati *et al.* (1985); Piacentini *et al.* (1988)
N^5-Methyl-	Lhoest and Colson (1977)

The deamidation of Gln to Glu is observed in proteins. Like the Asn to Asp conversion, it appears to be a spontaneous reaction, the rate of which depends on the sequence environment of the amide residue that is hydrolyzed (Robinson *et al.*, 1973). It is possible to rationalize the relative reactivity of Asn and Gln residues by assuming that the inductive effect of the electron-withdrawing peptide bond is expressed more strongly over the single methylene of Asn and Asp than over the two methylenes of Gln and Glu. This should make Asp a stronger acid than Glu, and by analogy Asn a stronger electrophile than Gln. Thus, Asn should be more susceptible to nucleophilic substitutions and more likely to lose an amide proton in a base-catalyzed reaction, whereas Gln should be easier to protonate at low pH and more reactive in reactions requiring protonation (Wold, 1985).

Most of the glutamine derivatives are products of transamidation reactions in which the amide -NH_2 is replaced by an amine -NHR. The enzymes responsible for the various reactions all belong to a family of transglutaminases; their occurrence and specificity and the biological significance of their products have been extensively reviewed (Folk, 1980, 1983; Loewy, 1984; De Jong *et al.*, 1988; Lorand, 1988; Greenberg *et al.*, 1991). It is interesting to note that if no amine acceptor is available, the enzyme will donate the γ-glutamyl moiety to water; the hydrolysis of Gln to Glu, which can take place spontaneously, can consequently also be enzyme catalyzed. The common product of the action of transglutaminases is a substituted amide derivative; a simple one if the amino group comes from a soluble amine, or a cross-linked one if the amino group comes from a protein-bound donor. Lys is probably the most common donor to yield the "isopeptide bond," N^ε-(γ-glutamyl)lysine cross-link. In the case of polyamine donors, spermine, spermidine, and putrescine have been observed both monovalently attached and bivalently attached to yield another type of cross-link.

N^5-Methylglutamine has been reported in ribosomal protein L3 in *E. coli* (Lhoest and Colson, 1977). It is assumed that this is the product of a direct methylation of Gln, but other reactions could also be involved. One of these is transglutaminase-catalyzed incorporation of methylamine; another is the spontaneous addition of methylamine to an activated Glu residue. Such a reaction has been demonstrated in, for example, α_2-macroglobulin (Swenson and Howard, 1979); it was subsequently suggested that the uniquely reactive residue is a thiolester, *S*-(γ-glutamyl)cysteine, involving a neighboring Cys residue (Tack *et al.*, 1980).

Histidine:

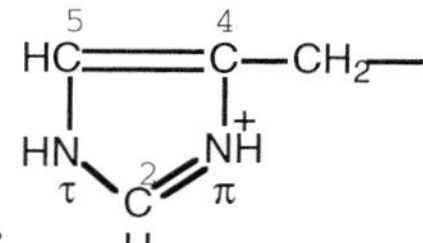

Diphthamide	Van Ness *et al.* (1980 a, b); Moehring and Moehring (1984)
N^τ-(ADP-ribosyl)diphthamide	Van Ness *et al.* (1980 a, b); Bodley *et al.* (1984); van Heynigen (1992)

N-Phosphoryl-	Smith *et al.* (1976); Chen *et al.* (1977a)
N^{π}-Methyl-	Young and Munro (1978); Huszar (1984)
4-Iodo- and diiodo-	Wolff and Covelli (1969)
N^{τ}- And N^{π}-(8α-flavin [FAD])	Singer and McIntire (1984)
N^{π}-(8α-Flavin [FMN])	Willie *et al.* (1996)

Several other derivatives of His reacting with other amino acids to form cross-links are listed elsewhere: *S*-(2-histidyl)-Cys under Cys; alanino(τ and π)histidine under Ser; the His-Lys cross-links aldol histidine and histidinohydroxymerodesmosine under Lys.

A most fascinating His derivative is diphthamide with the structure 2-[3-carboxamido-3-(trimethylammonio)propyl]histidine (see Fig. 7; Section III, B); it is found in elongation factor 2 in all eukaryotes and in archaebacteria. It is one of the few examples of a posttranslational modification that involves a single site in a single protein. Diphthamide is the acceptor molecule for ADP-ribosylation catalyzed by diphtheria toxin, and apparently also by cellular ADP-ribosyltransferases. The ribosyl moiety in the ADP-ribosylated derivative is linked to the N^{τ}-(*N*-1) of the 2-substituted imidazole ring (Van Ness *et al.*, 1978, 1980 a, b; Bodley *et al.*, 1984; Althaus *et al.*, 1985; Chen *et al.*, 1985; Iglewski *et al.*, 1985; Ueda and Hayaishi, 1985; Shall, 1988) to give N^{τ}-ADP-ribosyl-2-[3-carboxamido-3-(trimethylammonio)propyl]histidine. The ADP-ribosylated derivative does not function in protein synthesis. Mutants in which His is not converted to diphthamide are resistant to the bacterial toxins, but the mutants are active in protein synthesis (Chen *et al.*, 1985).

$N^{\tau(\pi)}$-Phosphorylhistidine has been demonstrated in myelin basic protein (Smith *et al.*, 1976) and in nuclear proteins, notably histone H4 (Chen *et al.*, 1977a), N^{τ}-(methyl)histidine in muscle proteins (Young and Munro, 1978; Huszar, 1984), histone fractions (Gershey *et al.*, 1969), and 4-(iodo)histidine in thyroglobulin (Wolff and Covelli, 1969). His is also a site for prosthetic group attachment, and both the N^{τ}- and N^{π} derivatives of 8α-(histidyl)flavin are known (Singer and McIntire, 1984). Cell-free synthesis of one of these derivatives has been reported (Hamm and Decker, 1980).

Lysine: $H_3\overset{+}{N}-\underset{\varepsilon}{CH_2}-\underset{\delta}{CH_2}-CH_2-CH_2-$

N^{ε}-Acetyl-	Allfrey *et al.* (1984)
N^{ε}-(N^{α}-Monomethylalanyl)-	Chen and Chen-Schmeisser (1977)
N^{ε}-Murein (peptidoglycan)	Hantke and Braun (1973)
N^{ε}-Lipoyl-	Hale and Perham (1980)
N^{ε}-Biotinyl-	Goss and Wood (1984)
N^{ε}-Ubiquitinyl-	Hershko (1988, 1991)
N^{ε}-Phosphoryl-	Chen *et al.* (1977a)
N^{ε}-Phosphopyridoxyl-	Tanase *et al.* (1979)
N^{ε}-Retinyl-	Khorana (1988)
N^{ε}-Glycosyl-	Baynes *et al.* (1984)
N^{ε}-Mono-, di-, trimethyl-	Paik and Dimaria (1984); Huszar (1984)
Hypusine: N^{ε}-(4-amino-2-hydroxybutyl)-	Park *et al.* (1993 a, b)
Allysine	Mirelman and Siegel (1979); Guay and Lamy (1979)

δ-Hydroxy-	Eyre (1987); Kivirikko *et al.* (1992)
δ-Hydroxyallysine	Eyre (1987)
Cross-links	Eyre *et al.* (1984); Guay and Lamy (1979); Narayana and Page (1976)
(desmosines, syndesines, pyridinolines)	Fujimoto *et al.* (1978); Fukae and Mechanic (1980)
δ-Glycosyloxy-	Levine and Spiro (1979)

The primary ε-amino function of Lys is an excellent site for chemical modifications of a protein, and it is not at all surprising that a large number of the reagents that have been developed by protein chemists over the years are directed toward the Lys amino group (see Fig. 3). Nature has also done well in utilizing this functional group to yield a large number of naturally occurring Lys derivatives in various proteins. As expected, acylation of the Lys side chains is quite common giving rise to both common and unusual acyl derivatives such as N^{ε}-phosphoryllysine (Chen *et al.*, 1977a), N^{ε}-acetyllysine (Vidali *et al.*, 1988), N^{ε}-(N^{α}-monomethylalanyl)lysine, in which the modified Lys is the amino-terminal residue in *E. coli* ribosomal protein Sll (Chen and Chen-Schmeisser, 1977), and the N^{ε}-(diaminopimelyl)lysine, which represents a linkage between murein and lipoprotein in gram-negative bacteria (Hantke and Braun, 1973). In addition, two coenzymes are covalently attached to enzymes through N^{ε}-acylation in N^{ε}-biotinyllysine and N^{ε}-lipoyllysine. An important acyl derivative is the ubiquitinyllysine, which, as will be discussed later under function, marks proteins for removal and destruction.

Cofactors are also linked to Lys via their carbonyl functions as Schiff bases such as N^{ε}-retinallysine aldimine or N^{ε}-(phosphopyridoxal)lysine aldimine. These reactions can take place spontaneously in the absence of specific enzymes; other reactions in the same category are the spontaneous glycation reactions through which high concentration of reducing sugars can yield Schiff bases either with the N-terminal amino group or with the N^{ε}-amino group. The reaction between amino groups and the reducing group of sugars, the browning or Maillard reaction, is well known, but is quite complex (Monnier, 1989). In general the assumed initial Schiff base undergoes an acid-catalyzed Amadori rearrangement to yield relatively stable ketoamine products (see Fig. 4, also Section III, A). The further degradation of these Amadori products may yield compounds with dicarbonyl groups that are more reactive than the original reducing sugar, and can stimulate further reaction in an autocatalytic fashion. The final products can consequently be quite complex and may give rise to cross-links (see pentosidine in the lower part of Fig. 3). The fact that these reactions clearly can take place *in vivo* and can be linked to diabetes and aging makes them highly relevant.

The primary amino group of Lys can be also be alkylated, and all three methylated derivatives, N^{ε}-methyllysine, N^{ε}-dimethyllysine, and N^{ε}-trimethyllysine, are known in a variety of proteins in both prokaryotes and eukaryotes. Enzymes responsible for these methylation reactions (protein-lysine methylases) have been isolated and characterized from a number of sources (Paik and Dimaria, 1984); some appear capable of using a variety of protein acceptors, whereas others are as specific as the *Neurospora* methylase that apparently methylates only Lys-72 in cytochrome *c* (Durban *et al.*, 1978). *Neurospora* cyto-

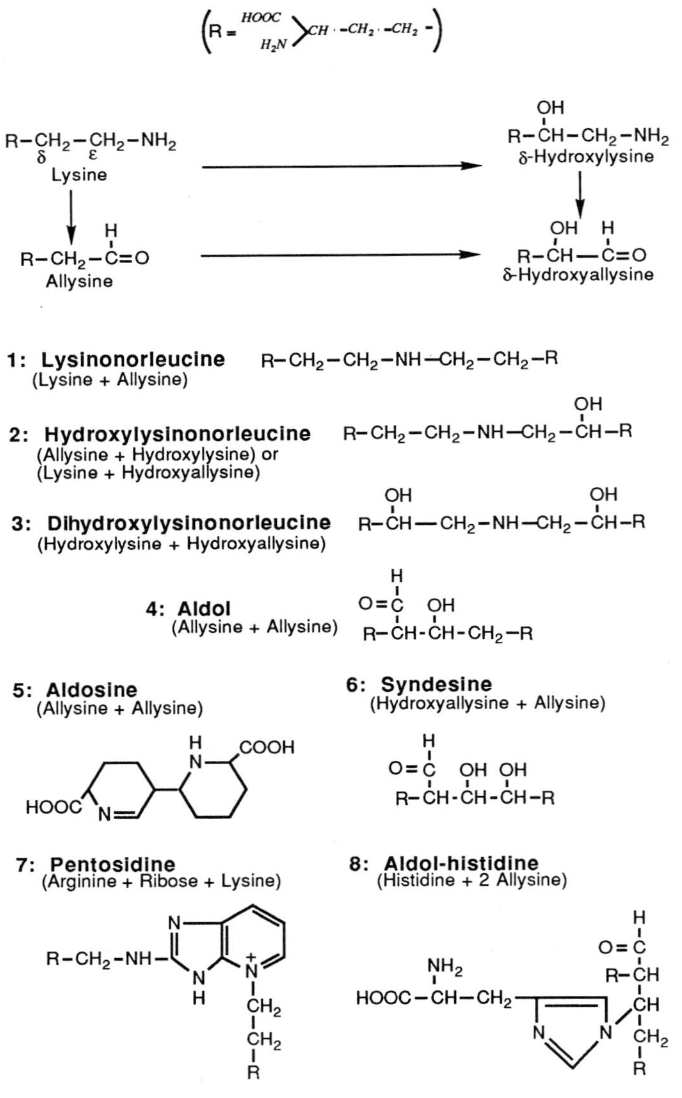

(R = HOOC, H_2N >CH-CH_2-CH_2-)

R-CH_2-CH_2-NH_2 (δ, ε)
Lysine

R-CH(OH)-CH_2-NH_2
δ-Hydroxylysine

R-CH_2-C(H)=O
Allysine

R-CH(OH)-C(H)=O
δ-Hydroxyallysine

1: Lysinonorleucine (Lysine + Allysine) R-CH_2-CH_2-NH-CH_2-CH_2-R

2: Hydroxylysinonorleucine (Allysine + Hydroxylysine) or (Lysine + Hydroxyallysine) R-CH_2-CH_2-NH-CH_2-CH(OH)-R

3: Dihydroxylysinonorleucine (Hydroxylysine + Hydroxyallysine) R-CH(OH)-CH_2-NH-CH_2-CH(OH)-R

4: Aldol (Allysine + Allysine) R-CH(CHO)-CH(OH)-CH_2-R

5: Aldosine (Allysine + Allysine)

6: Syndesine (Hydroxyallysine + Allysine) R-CH(CHO)-CH(OH)-CH(OH)-R

7: Pentosidine (Arginine + Ribose + Lysine) R-CH_2-NH-

8: Aldol-histidine (Histidine + 2 Allysine) HOOC-CH(NH_2)-CH_2-

Fig. 3 Various protein cross-links derived from lysine, δ-hydroxylysine, allysine, and δ-hydroxyallysine. Many of the known derivatives have been isolated after reduction with $NaBH_4$. Compounds **1–6** contain two precursor amino acids and either are the condensation products of the primary amino group of lysine or hydroxylysine and the carbonyl of allysine or hydroxyallysine, or are the aldol condensation products of allysine and hydroxyallysine. Compound **7**, pentosidine, is also a two-amino acid cross-link, in which Lys is linked to Arg through a molecule of ribose (Sell and Monnier, 1989). Other sugars such as glucose, fructose, or ascorbate can replace ribose in the initial nonenzymatic glycation of the lysine ε-amino group (Grandhee and Monnier, 1991). Compounds **8–14** are cross-links containing three amino acids. One, **8**, contains His as the third building block; the others are condensation products of 2 allysines with either lysine or hydroxylysine, either in open-chain merodesmosines or in the ring structures of pyridinolines. Compounds **15–17** are cross-links containing 4 amino acid precursors, either 2 allysines, 1 hydroxylysine and 1 histidine, or 3 allysines and 1 lysine in the 2 desmosines. These derivatives have been studied primarily in collagens and elastins. Since hydroxylysine is not present in elastin, the hydroxylysine-containing derivatives are not found in elastin. Several recent reviews of these compounds and their biosynthesis are available (Pokharna *et al.*, 1995; Mecham and Davis, 1994; Nakamura and Suyama, 1994; Eyre, 1987).

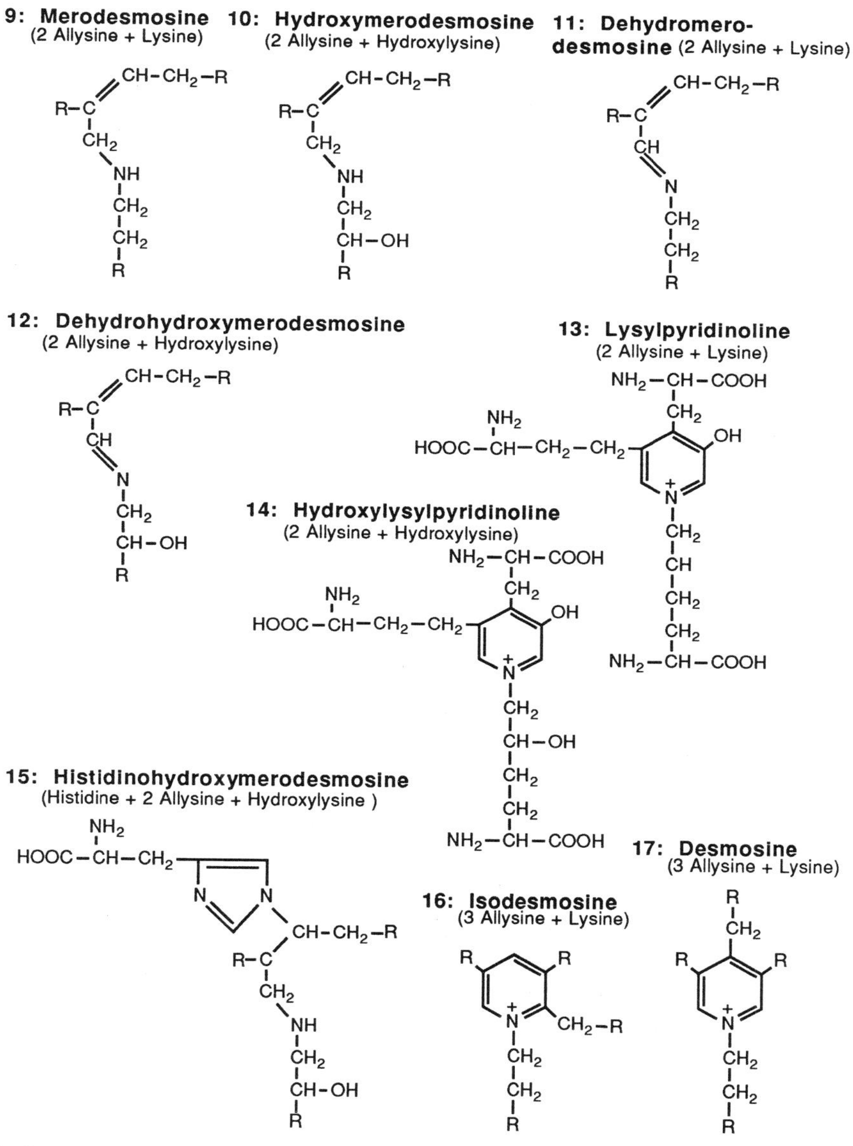

Fig. 3 *(continued)*

chrome *c* contains trimethyllysine in this position. Hypusine is a remarkable example of a unique alkyl derivative. Like diphthamide, which is found only in elongation factor 2, hypusine is also found at a single site in a single protein, initiation factor 5A, in all eukaryotes. The basis for the specificity of its synthesis will be discussed later (Fig. 7, Section III, B).

A large number of known Lys derivatives are found as cross-links in various structural proteins such as collagen and elastin. In collagen, but not in elastin, δ-hydroxylysine is a key starting material along with Lys itself. It is formed

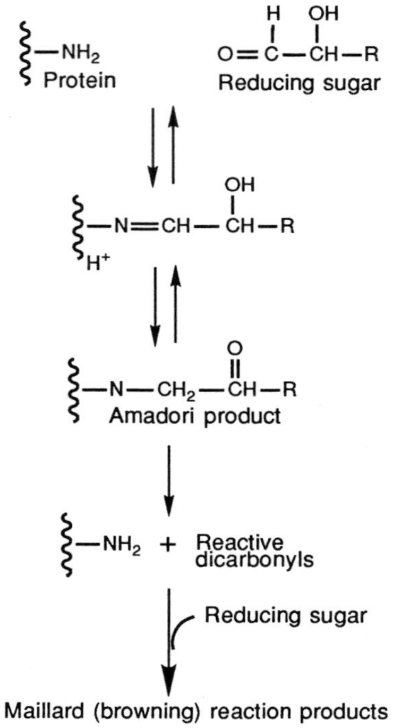

Protein —NH_2 + $O{=}C(H){-}CH(OH){-}R$ Reducing sugar

⇅

—$N{=}CH{-}CH(OH){-}R$ (H^+)

⇅

—$N{-}CH_2{-}C({=}O){-}R$ Amadori product

↓

—NH_2 + Reactive dicarbonyls

↓ Reducing sugar

Maillard (browning) reaction products

Fig. 4 Some steps involved in the Maillard reaction. [Redrawn from the representation by Monnier (1989).] The figure shows the initial two reversible reactions leading to a relatively stable ketoamine, the Amadori product of the glycation reaction. This product represents a steady-state intermediate; it can, through the reverse reactions, return to the original primary amine and reducing sugar, or it can react to yield dicarbonyls and free amine. The reactive dicarbonyls propagate autocatalytic reactions, and along with the free amines and the reducing sugars, they produce a very complex mixture of end products as the Maillard or browning reaction products.

from Lys by lysylhydroxylase, which requires α-ketoglutarate, O_2, and peptide-bound Lys as substrates, and Fe^{2+} and a reducing agent such as ascorbate as cofactors (Bornstein, 1974; Gallop and Paz, 1975). It appears that the hydroxylation reaction takes place in the early stages of collagen biosynthesis, perhaps even at the stage of nascent chains. A methylated derivative of hydroxylysine, N^{ε}-trimethylhydroxylysine, is known (Nakajima and Volcani, 1970). The cross-linked derivatives are derived from protein allysine (α-aminoadipic acid semialdehyde) and its relative, δ-hydroxyallysine, the products of oxidative deamination of Lys and δ-hydroxylysine side chains, respectively. The oxidase has been well characterized (Siegel and Fu, 1976) and is found quite widely in nature (Mirelman and Siegel, 1979); allysine and some of its derivatives have been demonstrated in bacteria (Diedrich and Schnaitman, 1978; Mirelman and Siegel, 1979) and in animal cells (Guay and Lamy, 1979). The oxidase products, with their reactive carbonyl function, now become the initiators of a series of complex chemical reactions leading to a group of cross-links, each involving either two, three, or four amino acid residues. Several of the reactions involved could proceed spontaneously, but it is likely that specific enzymes (e.g., oxidoreductases) are involved as well. The key intermediates and products in these reactions are presented in Fig. 3.

A final set of Lys derivatives are obtained from glycosylation of the δ-hydroxyl group of δ-hydroxylysine to yield glycosides such as O^{δ}-(glucosylgalactosyl)hydroxylysine and O^{δ}-(galactosyl)hydroxylysine.

Methionine: $H_3C{-}S{-}CH_2{-}CH_2{-}$

Sulfoxide Brot *et al.* (1984); Kikuchi and Tamiya (1992)

A large family of structural proteins, the hinge-ligament proteins of molluscan bivalves, is characterized by high levels of Met sulfoxide, and it has been proposed that the derivative is required to keep the protein hydrophilic and thus to promote the proper swelling and elasticity of the ligament (Kikuchi and Tamiya, 1992). Aside from this possible *in vivo* function, it has been assumed that Met sulfoxide in proteins is the result of an undesirable, spontaneous oxidation, leading to inactivation of the protein. The isolation of an enzyme, Met(O)-peptide reductase, from both prokaryotic and eukaryotic sources (Brot *et al.*, 1984) is consistent with the idea of a repair function for the enzyme, as well as with the notion that the sulfoxide may be quite a common *in vivo* occurrence.

Phenylalanine: (benzene ring)—CH_2— (β)

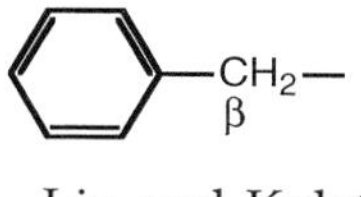

β-Glycosyloxy- Lin and Kolattukudy (1980); Kolattukudy (1984)

O^{β}-Glucosyl-β-hydroxyphenylalanine has been identified as a naturally occurring amino acid derivative in the fungal enzyme cutinase (Lin and Kolattukudy, 1979; Kolattukudy, 1984). It is assumed that the precursor of this derivative is free β-hydroxyphenylalanine. This nonglycosylated precursor has not been observed, however.

Proline: H_2C(4)—CH_2(3); H_2C(5)—$\overset{+}{N}_1H_2$—CH(2)—

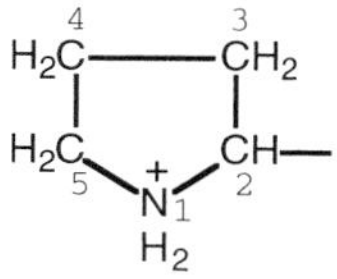

3-Hydroxy-	Bornstein (1974)
4-Hydroxy-	Kivirikko *et al.* (1989, 1992)
3,4-Dihydroxy-	Nordwig and Pfab (1969)
4-Glycosyloxy-	
O^4-Arabinosylhydroxy-	Lamport (1984)
O^4-Galactosylhydroxy-	Lamport (1984)
O^4-Glucosylhydroxy-	Lamport (1984)

The main derivatives of Pro are the different hydroxyprolines: the major derivative in proteins, 4-hydroxyproline (Bornstein, 1974; Kivirikko *et al.*, 1989), as well as the minor derivative, 3-hydroxyproline (Bornstein, 1974) and the rather rare 3,4-dihydroxyproline (Nordwig and Pfab, 1969). Monohydroxyprolines, like the δ-hydroxylysine, are essential components of collagen, and the hydroxylation is catalyzed by enzymes very similar to the Lys hydroxylase, which require Fe^{2+}, reducing agents, α-ketoglutarate, and O_2 for their action. The major proline hydroxylase activity has been shown to exist almost exclusively in the lumen of the ER. It is a heterotetramer containing two hydroxylase subunits and two protein disulfide-isomerase subunits; one possible function of the latter with its C-terminal Lys-Asp-Glu-Leu (KDEL) "address label" is to keep the complex in the ER (Kivirikko *et al.*, 1989). It has been proposed that this enzyme is specific for 4-hydroxylation, and that a separate enzyme is required for the 3-hydroxylation (Bornstein, 1974). A hydroxylase preparation from earthworm cuticle (which contains relatively high levels of 3-hydroxyproline) may hydroxylate both positions (Nordwig and Pfab, 1969). As in the case of hydroxylysine, hydroxyproline represents a new potential site for further modification, and glycosylated hydroxyprolines have been reported in plant cell-wall material. Whereas animal collagens contain relatively little carbohydrate (always on the hydroxylysines), the analogous plant extracellular matrix proteins, extensins, contain as much as 50% carbohydrate, and this is almost exclusively short chains of arabinose glycosidically attached to 4-O of hydroxyproline (see Fig. 5). More complex glycoproteins containing other hydroxyproline-bound sugars are also known (Lamport, 1984).

Serine: HO—CH_2—

Selenocysteine	Stadtman (1996)
O-Phosphoryl-	Taylor *et al.* (1990); Cohen (1989); Stewart and Sharp (1984)
O-Pantetheinephosphoryl-	Vagelos (1973); Pfeifer *et al.* (1995)
O-(GlcNAc-1-phosphoryl)-	Gustafson and Gander (1984)
O-(Glycerol-1-phosphoryl)-	Stimson *et al.* (1996)
Alanino(τ- or π-histidine)	Sass and Marsh (1983, 1984)
Lanthionine	Kaletta *et al.* (1991)
O-Acetyl-	Rudman *et al.* (1979)
O-Fatty acyl-	Towler *et al.* (1988)
O-Methyl-	Sheid and Pedrinan (1975); Swanson and Applebury (1983)
O-Glycosyl-	Jentoft (1990); Hardingham and Fosang (1992)

Selenocysteine (SeCys) has been observed in several prokaryotic reductases and in at least one eukaryotic enzyme, gutathione peroxidase. Because of its lability, it was originally difficult to demonstrate unequivocally that this is a true protein amino acid, but the biosynthesis and function of this amino acid are now well documented. It is discussed here because it is derived from Ser. Free SeCys itself apparently does not exist, but a $tRNA^{Sec}$ for the incorporation of SeCys does exist and can be charged with Ser. The resulting Ser-$tRNA^{Sec}$ is not incorporated; the termination codon TGA (UGA in the message) terminates the protein synthesis. In the presence of selenium, however, the Ser-$tRNA^{Sec}$ is converted to SeCys-$tRNA^{Sec}$, and this permits the read-through of the termination codon with the incorporation of SeCys. If the termination codon is replaced with a Cys codon, the enzymes can be synthesized with Cys in place of SeCys; these products show much decreased activity. It is probably appropriate to ask whether SeCys in fact should be considered as the 21st primary amino acid. It has a codon, a specific tRNA; the only thing missing is the amino acid itself, which can only be produced from precursor Ser attached to the tRNA.

One of the most widely distributed posttranslational modifications in Nature is O^{β}-phosphoserine, the main phosphorylated derivative in phosphoproteins. As will be discussed later, the majority of these derivatives are of the reversible kind, with the presence or absence of the phosphate group directly involved in the regulation of the biological activity of the phosphoprotein involved. The specificity and regulation of a large number of protein kinases responsible for the transfer of phosphate from ATP to the acceptor protein have been studied extensively (Taylor *et al.*, 1990), as have the catalysts of the

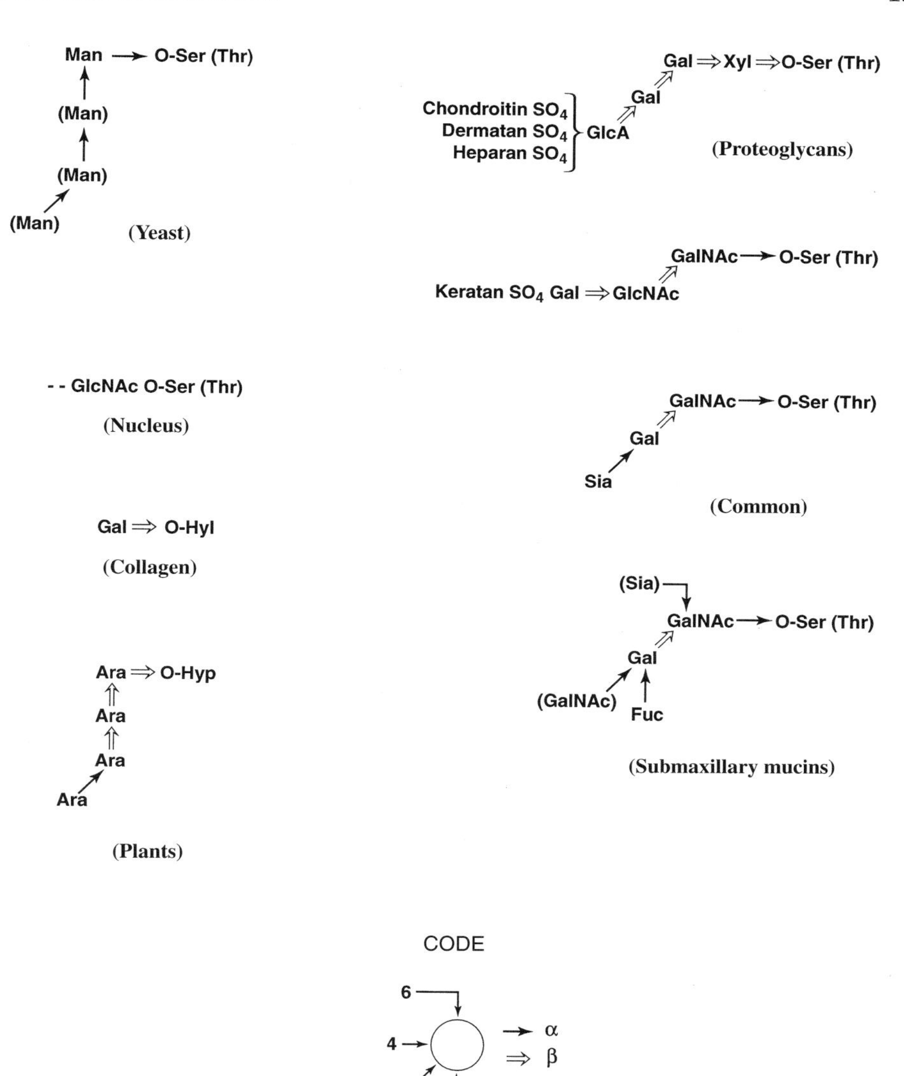

Fig. 5 Some typical *O*-linked glycans. Ser (Thr) are the common attachment points, but in some proteins secondary amino acids such as hydroxylysine (Hyl) and hydroxyproline (Hyp) can also be glycosylated, as shown in this figure. The xylosyl Ser derivatives are the anchoring sites for several proteoglycans. The trisaccharide Sia-Gal-GalNAc appears to be the most common glycan associated with glycoproteins; in *Saccharomyces* only Man derivatives are observed.

reverse reactions, protein phosphatases (Cohen, 1989). Serine phosphate is also formed as a *O*-phosphoseryl-tRNA, which may be incorporated directly into phosphoproteins such as casein and phosvitin as stable components (Steward and Sharp, 1984). It is likely that phosphate esters also could be stable components of certain phosphoproteins (e.g., the milk protein casein). The suggestion that a phosphodiester link exists between a Ser and a Thr residue in an *Azotobac-*

ter flavodoxin could be consistent with a permanent structural role for the phosphate group (Live and Edmondson, 1988). In another phosphate diester derivative of Ser, O^{β}-(4′-phosphopantetheine)serine, the phosphate diester provides the covalent linkage between the prosthetic group and the protein in the acyl carrier proteins of bacteria, yeast, and higher plants (Vagelos, 1973). In other derivatives, the serine phosphate is linked to the reducing end of *N*-acetylglucosamine through the acid-labile glycosylphosphate bond in O^{β}-(*N*-acetylglucosaminylphospho)serine (Gustafson and Gander, 1984) in a slime mold protease, or to glycerol in pilin from *Neisseria meningitides* (Stimson *et al.*, 1996).

An interesting derivative derived nonenzymatically from phosphoserine is the crosslink τ- or π-histidinoalanine. β-Elimination of Ser or, more readily, PSer yields dehydroalanine, which can react with an available protein nucleophile to give a cross-link (see, for example, lysinoalanine and lanthionine under Cys). Histidinoalanine has been identified as a very minor component of bone and dentin collagen, especially in collagen from older individuals. A phosphoprotein in the extrapallial fluid of certain bivalve molluscs contains substantial amounts of both isomers ($N^{\tau}:N^{\pi} = 3:1$). In this case the cross-link between phosphoprotein monomers leads to the formation of large phosphoprotein particles (40 nm in diameter and up to 30×10^6 Da mass). The fact that these particles sequester calcium and phosphate suggest that they may be physiologically significant. It is not known whether the β-elimination and subsequent alkylation are enzymatically catalyzed reactions (Sass and Marsh, 1983, 1984). Lanthionine, another product of Ser-derived dehydroalanine, has been found in bacterial antibiotics as a true posttranslational derivative.

Other ester derivatives of Ser are the *N,O*-diacetyl derivatives reported in a melanotropic peptide (Rudman *et al.*, 1979) and the fatty acyl esters (mostly palmitate) found along with thioesters in many eukaryotic proteins. The proposed presence of methyl ethers in proteins is based on the observation of alkali- and acid-stable *O*-methyl derivatives in seminal plasma proteins (Sheid and Pedrinan, 1975) and in proteins in photoreceptor rod outer segments (Swanson and Applebury, 1983).

Serine is also a major glycosylation site in proteins (Paulson, 1989; Jentoft, 1990). In terms of the sugar linked directly to Ser by the glycoside bond, there are at least four distinct families of glycosylated Ser derivatives, (see Fig. 5). One is derived from O^{β}-(*N*-acetylgalactosaminyl)serine and is probably the most common one in animal proteins and also in proteoglycans (Hardingham and Fosang, 1992); another, derived from O^{β}-mannosylserine (Nakajima and Ballou, 1974), is the common family in *Saccharomyces cerevisiae;* and those derived from O^{β}-galactosylserine and from O^{β}-xylosylserine are common derivatives in proteoglycans (Hardingham and Fosang, 1992). Based on lectin binding experiments, it has been proposed that a derivative containing O^{β}-(*N*-acetylglucosaminyl) as the Ser-bound sugar exists in both nuclear and cytoplasmic proteins (Hart *et al.*, 1989) and may constitute a fifth family of glycosylated Ser residues. Several enzymes have uniquely reactive Ser residues that participate directly in the catalytic function of the enzyme, frequently by forming transient covalent intermediates of the reaction. These derivatives are not included in this review.

Threonine: H_3C—CH— / HO

O-Phosphoryl-	Taylor *et al.* (1990); Cohen (1989)
O-Glycosyl-	Jentoft (1990); Hardingham and Fosang (1992)
O-Fatty acyl-	Towler *et al.* (1988)
O-Methyl-	Sheid and Pedrinan (1975); Swanson and Applebury (1983)
β-Methyllanthionine	Kaletta *et al.* (1991)

Many of the enzymes responsible for derivatization of Ser residues in proteins apparently can use Thr equally well as the site of modification, and many of the derivatives discussed earlier for Ser also exist for Thr: To emphasize this point, the same references are listed for the two amino acids. This does not mean that they have been unequivocally shown to be identical in all their reactions; in fact, they are not. But it may well be that the similarities are much more impressive than the differences. In presenting the common *O*-glycosylation products in Fig. 5 we have followed the common procedure of presenting Ser and Thr as equivalent. The proposed *O*-methyl ethers (Sheid and Pedrinan, 1975; Swanson and Applebury, 1983) listed previously for Ser could also be Thr derivatives. As in the case of bacterial lanthionine, the product of the addition of a Cys thiol to dehydrated Thr, β-methyllanthionine, has also been observed in bacterial peptides (Kaletta *et al.*, 1991).

Tryptophan:

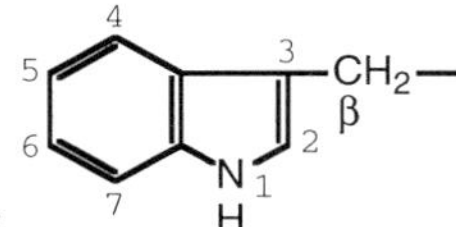

2,4′-Bis-Trp-6′,7′-dione	McIntire *et al.* (1991); Klinman and Mu (1994)
C2-Aldohexopyranosyl-	Hofsteenge *et al.* (1994); de Beer *et al.* (1995)
6-Bromo-	Jimenez *et al.* (1997)

Three derivatives of Trp have been well documented during the last few years. The first is the cross-link between two Trp residues in which one of the Trp residues also contains a 6,7-dione. This tryptophantryptoquinone is the cofactor in methylamine dehydrogenases found in the periplasm of methylotrophic bacteria. As expected, the gene specifies Trp for each of the two positions involved in the derivative. The second derivative is observed in a ribonuclease isolated from human urine. The encoded residue is Trp, but the isolated peptides were all found to have a mass 162 higher than that calculated for the Trp-containing peptides. Fragmentation analysis and NMR established the structure as C2-aldohexopyranosyltryptophan; more recent work has established that the deriviative is an α-mannopyranose (de Beer *et al.*, 1995). Other human ribonucleases have been found to have modified Trp residues in the same position, but the nature of the modification has not been established. The third derivative is a brominated Try in a biologically active octapeptide from the venom of the deep-sea species *Conus radiatus*. The derivative has been shown to be a true posttranslational product in which the 6-bromotryptophan residue is encoded as Try (UGG) in the prepropeptide.

HO–(ring: 3 2 / 4 1 / 5 6)–CH_2— β

Tyrosine:

3-Iodo-(bromo-, chloro-)	Hunt (1984)
3,5-Diiodo-(bromo-, chloro-)	Hunt (1984)
Thyronine	Hunt (1984); Nunez (1984)
3,4-Dihydroxy-Phe(DOPA)	Waite and Benedict (1984); Waite and Rice-Ficht (1987)
3,4,6-Trihydroxy-Phe(TOPA)	Janes *et al.* (1990); Klinman and Mu (1994)
O-Phosphoryl-	Martensen (1984); Hunter and Cooper (1985)
O-Sulfuryl-	Niehrs *et al.* (1994); Rosenquist and Nicholas (1993)
O-Adenylyl-	Shapiro and Stadtman (1968); Rhee (1984)
O-Uridylyl-	Adler *et al.* (1975); Rhee (1984)
O-(8α-Flavin [FAD])	Singer and McIntire (1984)
3,3′-Bityr	Amado *et al.* (1984); Guptasharma and Balasubramanyam (1992)
3,3′; 5,5′-TriTyr	Garcia-Castineiras *et al.* (1978)
Isodi-Tyr	Fry (1984)
β-Glycosyloxy-	Lin and Kolattukudy (1980)
6-(N^{ε}-Lys)-3,4-dihydroxy-Phe (lysine-tyrosylquinone)	Wang *et al.* (1996)

A variety of halogenated derivatives of Tyr have been established in invertebrate scleroproteins and in vertebrate thyroglobulins; they have been listed separately by Hunt (1984). The mono- and diiodo-Tyr derivatives in thyroglobulin are intermediates in the biosynthesis of the thyroxin family of hormones. This reaction is now believed to involve a β-elimination of mono- and diiodityrosines in thyroglobulin, which leaves a protein-bound dehydroalanine (Gavaret *et al.,* 1979) and provides the reactive mono- or diiodophenol radical that can react with the phenolic hydroxyl of another residue of diiodotyrosine to give the two most common forms of the hormone, 3,5,3′-triiodothyronine and 3,5,3′,5′-tetraiodothyronine (thyroxine) as residues in the thyroglobulin polypeptide chain (Nunez, 1984).

Ring hydroxylation is also a fairly common Tyr reaction; dihydroxyphenylalanine (DOPA) is present in quite large quantities in adhesive proteins of several marine invertebrates, and recently the trihydroxyphenylalanine derivative, TOPA, has also been shown to be present in eukaryotic oxidases and dehydrogenases. The cDNA show that the encoded precursor for TOPA is Tyr. A TOPA derivative in which the 6-OH is replaced by the N^{ε}-amino group of a protein Lys has been reported (Wang *et al.,* 1996); the resulting cross-linked derivative is the redox cofactor in lysyl oxidase.

Several derivatives of the phenolic hydroxyl group are known; the monoester tyrosine O^4-phosphate has emerged as an extremely important component of the regulatory machinery of all living cells. The Tyr-specific protein kinases, in contrast to the Ser/Thr kinases, are membrane bound and are key actors in a variety of signal transductions. Tyrosine O^4-sulfate (Niehrs *et al.,* 1994; Rosenquist and Nicholas, 1993) is another family of Tyr esters. In contrast to the phosphate esters, the sulfate esters appear to be quite stable in that there do not seem to be sulfatases ready to remove them in a regulatory cycle. It should be noted that the acidity of the Tyr phenolic group gives the esters some character of anhydrides, and the inherent chemical properties reflect this fact. A unique group of derivatives involve phosphate diesters. Thus, the covalently bound regulatory effectors that appear to be unique to glutamine synthetase of gram-negative bacteria, O^4-adenylyltyrosine (Shapiro and Stadtman, 1968) and O^4-uridylyltyrosine (Adler *et al.*, 1975), are diesters, and so are the derivatives through which protein-Tyr is linked to the 5′-phosphate of RNA [5′-(O^4-tyrosylphospho)-RNA] in polio virus (Ambros and Baltimore, 1978) or to the 5′ phosphate of DNA [5′-(O^4-tyrosylphospho)-DNA] in *E. coli* DNA topoisomerase I and in *Micrococcus luteus* DNA gyrase (Tse *et al.*, 1980). The latter derivative, as a putative intermediate in the reaction catalyzed by the gyrase, should perhaps not be included in a list of true *in vivo* modifications.

The derivative 8α-(O^4-tyrosyl)flavin (Singer and McIntire, 1984) represents another example of a covalently linked cofactor. Oxidative phenolic coupling of Tyr rings to yield cross-links such as 3,3′-bityrosine and 3,3′;5′3″-trityrosine (Anderson, 1963) appears to be a natural occurrence in several systems, such as the fertilization membrane of sea urchin eggs (Foerder and Shapiro, 1977), adhesive substances of sea mussels (DeVore and Gruebel, 1978), and human lens protein (Garcia-Castineiras *et al.*, 1977). Although it is clear that subjecting a variety of proteins to the action of peroxidase at alkaline pH *in vitro* will lead to the artificial production of these cross-links (Aeschbach *et al.*, 1976), the available evidence favors this reaction as a natural protein processing step. The final derivative of Tyr is analogous to that discussed earlier for Phe, O^{β}-glycosyl-β-hydroxytyrosine (Lin and Kolattukudy, 1980). Again it is assumed that this derivative requires the formation of β-hydroxytyrosine as a precursor, although that compound has not been observed in proteins.

D-*Amino Acids.* It appears, not surprisingly, that the side chains of the aliphatic amino acids, alanine, (glycine), isoleucine, leucine, and valine, have not been subject to chemical modification. However, they cannot all be excluded from the list of modified amino acids, since it is becoming clear that at least Ala, Ile, and Leu, along with Met, Phe, and Try, are among the amino acids that are converted from the L to the D enantiomer in proteins (Kreil, 1994; Jimenez *et al.*, 1997). These D-amino acids have been found in physiologically active peptides from amphibian skin and from molluscs, always in the second position of the peptide sequence. The peptides have been shown to be derived from high molecular weight precursors, synthesized on polysomes, and the cDNA of the precursor show that the D-amino acids are encoded as the normal L enantiomers. In some cases only the D-amino acid is observed in a given peptide; in other cases the D and the L enantiomers are observed together. This is pertinent to the question of whether the reaction constitutes a quantitative conversion or a racemization followed perhaps by faster clearance of the L-amino acid-containing isomer. It seems likely that the inversion must be catalyzed, but nothing is known about the putative racemases. The analytical procedures for the determination of D-amino acids in a protein are quite unique in that the all the chemical and all the physical properties, except for optical rotation, are identical for the D and L enantiomers. The solution to this problem is to produce diastereoisomers by reacting with a chiral reagents (Bada, 1984; Scaloni *et al.*, 1991) and to separate and analyze the two diastereoisomers formed from a mixture of D- and L-amino acids (L-leucyl-L-serine and L-leucyl-

D-serine are readily separated by HPLC). It is also possible to separate racemic mixtures of amino acid by so-called chiral chromatography on columns with chiral packing (Davankov *et al.*, 1980); the constant problem will continue to be the need to separate and identify a small amount of the D enantiomer in the presence of large amounts of the L enantiomer.

D. Are Posttranslational Reactions Faithfully Executed in Recombinant Proteins?

This is an important issue that has received a good deal of attention over the past several years. Part of the basic problem is quite obvious when we consider that the machinery required for most eukaryotic modification simply does not exist in prokaryotes, and vice versa, and also take into account that even within eukaryotes the reactions may not be compatible. Thus, to achieve the faithful expression of a liver glycoprotein, for example, yeast would not have the enzymes needed to make the typical complex glycans of liver, but might well produce a glycoprotein containing all polymannose glycans instead. It is possible that this incompatibility can be at least in part alleviated by the trick of coexpressing processing enzymes along with the desired protein product. Thus, when the clone for a murine protein kinase was expressed in *E. coli*, the native N-terminal myristoyl group was missing in the product, but when plasmid was constructed to contain both the gene for the kinase and the gene for yeast myristoyltransferase, the myristoylated kinase could be produced in good yield (Duronio *et al.*, 1990). The problems associated with the search for the right expression vectors for posttranslationally modified proteins is well illustrated by the work to produce recombinant protein C (Yan *et al.*, 1990). Without going into the final detail of the work, the important message is that it was possible to establish that the human kidney cell line 293 satisfies all the requirements for the production in high yields of a processed and active protein C. The cells carry out all the required posttranslational modifications: (1) the formation of γ-carboxyglutamate, Gla, at the first nine Glu residues in the sequence, (2) the β-hydroxylation of Asp-71, (3) the N-linked glycosylation at 4 glycosylation sites, (4) the disulfide bond formation, and (5) several peptide bond cleavages, to remove an 18-residue signal peptide, a 24-residue propeptide, and also the internal dipeptide 156–157 to yield the final two-chain zymogen. A final proteolytic cleavage at 169–170 removes the peptide 158–169 and yields the active serine protease, which acts as a potent anticoagulant and antithrombotic agent. A large number of other systems were explored before the 293 cells were found, but none of them had all the desired properties; even if all the posttranslational reactions took place, the rate of synthesis and the yield of product might be too low.

One fascinating aspect of this work on protein C, with bearing on the question asked previously, began with the observation that the protein produced in the 239 cells actually had higher anticoagulant activity than the plasma protein C, and continued with the discovery of some unique structural features in the glycans from the recombinant protein. The content of sialic acid was about two times lower and the content of fucose five times higher in the recombinant protein as compared to the plasma protein C, and on further analysis a unique nonreducing terminal trisaccharide, GalNAcβ(1-4)

[Fucα(1-3)]GlcNAcβ(1-•), was found in the kidney cell product (Yan *et al.*, 1993). It is not known whether the same structure is present at a lower level in the plasma protein. Finally, it has been found that protein C inhibits selectin-mediated cell adhesion, and the inhibition has nothing to do with the protease activity, but rather is an expression of the glycans. The mixture of the free glycans from the 239 cell-produced protein strongly inhibits the adhesion, suggesting that the unique fucosylated structures may play a special role in the process (Grinnell *et al.*, 1994).

Considering all the different aspects of this one case history, it appears that the answer to the question about faithful execution of posttranslational reactions in recombinant proteins will turn out to be a complicated one. There is a simple yes or no answer based on the use of expression vectors that either have or do not have the necessary machinery to carry out the reactions, but most of the answers are much more ambiguous and tentative. As illustrated previously, the first step is to define the basic requirements in terms of the structural modifications that are to be monitored, and also the associated biological activity requirements that must be satisfied. The success of the experiment is evaluated in terms of these initial definitions. Then a broader spectrum of biological activities may be evaluated, and more detail may be introduced in the assessment of the modification reactions themselves (Are all the modification sites reacted? Are the actual products identical in every detail?); then, suddenly, subtle differences are discovered that may not seriously affect the general behavior and use of the recombinant product, but nevertheless demonstrate less than completely faithful execution of the modification reactions. The glycosylation reactions will undoubtedly always be the trickiest ones in this regard, and considering the fact that even in homologous production of glycoproteins the glycosylation process yields heterogeneous glycans reflecting the relative reactivity of each glycosylation site, and the enzyme makeup of the individual cells and tissues that make the glycoproteins, variations in heterologous systems should perhaps be expected. A recent comparative study of the properties of recombinant human γ-interferon produced in three different expression systems, Chinese hamster ovary (CHO) cells, the mammary gland of transgenic mice, and baculovirus-infected Sf9 (*Spodoptera frugiperda*, fall armyworm ovary) insect cells, confirmed the very significant influence of the expression system on posttranslational processing. Extensive differences were observed both in the C-terminal trimming and in the structure of the incorporated glycans in the three systems (James *et al.*, 1996).

There is a second aspect to the preceding question that also must be considered: is it really the same question stated in a different way, namely, is it possible that the artificial conditions of heterologous expressions may lead to unique modifications not observed in the wild-type native expression? There is probably not sufficient data available at this stage to answer this question, but there are several observations (summarized briefly by Tu *et al.*, 1995) that may be consistent with the possibility that artificial posttranslational reactions can occur in the course of overproducing recombinant proteins. The modifications involve peptide chain extensions and truncations as well as faulty amino acid incorporations and unique reactions. Briefly, cases of both N- and C-terminally truncated murine interleukin-6 (Tu *et al.*, 1995; Zhang *et al.*, 1992), murine interleukin-1α (Daumy *et al.*, 1989), and human tissue factor (Paborsky *et al.*, 1989) have been observed when these proteins are expressed in *E. coli;* similarly, extensions of the polypeptide chain have also been observed for interleukin-6 (Danley *et al.*, 1991; Tu *et al.*, 1995). In one case the C-terminal extension is a short tripeptide, Gln-Lys-Leu, and an examination of the plasmid containing the interleukin gene suggests that the extension is due to an unexpected suppressor mutation that lead to a read-through of the original termination codon (Danley *et al.*, 1991). In the other case the C-terminal "tag peptide" found in 5–10% of a recombinant murine interleukin-6 contains 11 amino acid residues and has been identified as one that is encoded by a small stable RNA of *E. coli* (10Sa RNA); when the interleukin gene is expressed in an *E. coli* mutant missing the 10Sa RNA gene, no extension is observed (Tu *et al.*, 1995). Another study has provided the mechanistic basis for this reaction (Keiler *et al.*, 1996). The 10Sa RNA encodes a 10-amino acid sequence (ANDENYALAA), and in addition it has tRNA properties and can be charged with Ala. In cases of damaged (truncated) mRNA lacking a termination codon, the protein synthesis stops at the last codon and no product is released. The Ala-charged 10Sa RNA now enters, placing the Ala–RNA complex in the amino acyl site of the ribosome. Ala is incorporated into the polypeptide chain, and the polymerization continues with the encoded 10 amino acid sequence in 10Sa RNA; the product is released at the UAA termination codon. The product, the original mRNA-encoded protein + A (from the tRNA component of 10Sa RNA) + the ANDENYALAA sequence, turns out to be marked for destruction by an *E. coli* protease that recognizes C-terminal sequences like the one acquired by the product. Presumably, when the modified protein is overproduced, small amounts of the product can be observed. This mechanism explains the low yield of the *E. coli*-produced interleukin-6 product, and also emphasizes the importance of maintaining an intact termination sequence in the recombinant messages.

Other unusual modifications that may be associated with heterologous overproduction have been reported. Thus, if a low-frequency Arg codon, AGA, was used for all six Arg residues in synthetic insulin-like growth factor, Lys was incorporated instead of Arg in the protein produced in *E. coli.* If the preferred *E. coli* Arg codon, CGT, was used, only Arg was incorporated (Seetharam *et al.*, 1988). Codon usage thus may have to be considered carefully in choosing the proper expression vectors. In another system involving expression of a bovine somatotropin gene by *E. coli*, norleucine was found to be incorporated instead of Met (Bogosian *et al.*, 1989). In this case the growth conditions were such that the organism made fairly large quantities of norleucine, large enough that this amino acid could compete successfully with Met for incorporation. By eliminating intracellular norleucine synthesis through elimination of the Leu operon or by increasing the Met concentration by addition of large quantities of Met, the misincorporation could be prevented. In yet other cases, N-terminal Met removal from recombinant proteins may become incomplete and take place at varied rates in *E. coli.* The Met aminopeptidase of *E. coli* has specificity requirements similar to those of the yeast enzyme (Moerschell *et al.*, 1990), and consequently the second residue in a given sequence will affect the rate at which N-terminal Met is removed. In the cases of recombinant interleukin-2 and ricin, aminopeptidase hyperproducing strains of *E. coli* were required to achieve complete Met removal (Ben-Bassat *et al.*, 1987). Finally, in the case of human growth hormone produced in *E. coli*, a completely new posttranslational derivative was observed. The derivative, a trisulfide with an extra sulfur atom between the Cys^{182}-Cys^{189} bridge, was characterized by mass spectrometry and by the release of H_2S upon treatment with excess Cys, and appears to be unique to the recombinant form; growth hormone isolated from the pituitary never shows the presence of this isoform (Jespersen *et al.*, 1994).

As stated at the outset, it is too early to draw any definite conclusions as to the reality of artificial modifications in recombinant proteins. However, it seems prudent to consider this issue in designing the proper experiments for the production of biologically active proteins by recombinant techniques.

III. Types of Posttranslational Reactions: How Modifications Are Made

It would be a major undertaking to try to describe and categorize every one of the reactions involved in posttranslational modification. The product of the effort would cover a very broad area of reaction mechanisms, and the overall impression might be rather like admiring the individual color splashes of a Monet painting at very close range. To really appreciate the impressionistic display, one needs to step back and see how the patterns of individual spots fit together to produce the complete picture. One of the first patterns that emerges from a broad overview of the reactions is that there are spontaneous ones and enzyme-catalyzed ones. These two broad categories need to be considered from somewhat different angles when we look at them as end products of evolutionary selection, and also consider the specificity determinants involved in their production.

The issue of whether a given reaction is or is not catalyzed by specific enzymes is a fairly tricky one, and in fact, for many of the reactions to be discussed here the reaction mechanism is not clearly established. One of the best criteria for a nonenzymatic reaction is the demonstration that it can be carried out *in vitro* in the absence of any added enzyme; this would conclusively show that the reaction can take place in the absence of enzymes, but it does not really show that the *in vivo* reaction takes place in the absence of enzymes. Another good criterion is the ability to produce fully matured proteins in heterologous systems that do not have the enzyme(s) that might be needed. As mentioned previously for concanavalin A, the reactions involved in the circular permutation of the protein (Bowles and Pappin, 1988) require peptide bond cleavages at the C-terminal end of Asn residues, and since an Asn-specific protease is indeed present in jack beans, it was natural to conclude that the circular permutation in this case is an enzymatic process. However, the fact that the protein is produced in *E. coli*, which does not have such a protease, strongly suggests that the enzyme is not required for the processing. In spite of lingering uncertainties, these simple kinds of criteria will be used in defining the nonenzymatic reactions in the following.

A. Nonenzymatic Reactions

The nonenzymatic reactions can perhaps be considered to have evolved by selecting for certain structures in the substrates themselves. Considering that many of the nonenzymatic reactions may be undesirable, the structures may represent compromises of placing the reactive moiety in an environment where an inevitable reaction will proceed as slowly as possible. The deamidation of Asn and Gln, the racemization of, for example, Asp and Ser, the formation of dehydroalanine from, for example, Cys and phospho-Ser leading to a variety of condensation products, the slow oxidation of several amino acids such as Cys, His, and Met, and the slow cleavage and permutation of peptide bonds may all be in this category. It would be ideal if they did not take place at all, but in the proper structures the spontaneous reactions can at least be slow enough that the products can be observed only in long-lived proteins. On the other hand, some of these same reactions may also be desirable, and the selection then presumably has been for structures to give optimal rate of the nonenzymatic modifications in the maturation of the proteins involved. The *in vitro* data of Robinson *et al.* (1973), in which they showed that the half-life of Asn in pentapeptides under physiological conditions (pH 7.4, ionic strength 0.2, and at 37°C) varied from 18 days for the peptide Gly-Arg-Asn-Ala-Gly to 507 days for the peptide Gly-Ile-Asn-Ala-Gly, demonstrates that the half-lives are significantly affected by relatively minor changes in the primary structure of the amide's environment; this in turn is a required condition if we are to consider the possibility that the half-life of Asn (and Gln) may represent built-in clocks for the destruction of proteins. Under these circumstances the selection for short half-lives in the evolution of a given protein could be as normal and "desirable" as the one proposed previously for the selection of long half-lives. The definition of undesirable and desirable here is clearly not derived from strictly defined scientific facts; it rather represents an assumption based on the observations that sometimes the derivatives are formed slowly under circumstances where the organisms may have developed special tools to repair the presumed undesirable reactions (see, for example, the methylation of Asp), and sometimes they are formed fast, in high yield, and can be rationalized in terms of a biological function to be desirable ones.

The most obvious examples of "desirable" nonenzymatic reactions are the various cross-links observed in structural proteins (see Fig. 3 for Lys derivatives and under Ser and Cys for other cross-links derived from dehydroalanine), and perhaps also the permuted products of transpeptidation observed for a small number of proteins (Wallace, 1993). As a general rule, these reactions proceed quite smoothly with simple acid or base catalysis; however, many of them, as condensation reactions, whether they are spontaneous or enzyme catalyzed, are highly unfavored thermodynamically in dilute aqueous solutions. One way Nature can counter this thermodynamic obstacle is to arrange for the reaction to take place in a compartment in which the concentration of the two reactants exceeds that of the water to extent that the equilibrium of the reaction favors condensation, and that does indeed appear to be the way these reactions take place in proteins. In the case of intramolecular reactions the individual protein molecules are folded up in such a way that the two reactive moieties are brought into close proximity; in intermolecular reactions multiple chains must be associated in such a way that they favor specific reactions in the same mode of high local effective reactant concentrations. All the models that are considered for protein cross-linking reactions are based on this general concept, and in the case of some of them, notably the circularly permuted proteins such as concanavalin A (Bowles and Pappin, 1988) and also the mutant Asp transcarbamoylase protein (Yang and Schachman, 1993), the crystallography data in comparison to that for other related proteins provide direct evidence for the general model. If a fully folded molecule is the precursor of the modification reactions, it is probably reasonable to assume that the reactions are quite specific at the sites where the reactants are brought into proper juxtaposition. In addition, it appears that some of the reactions may be rather nonspecific and simply occur because the reactive residues are present in very high concentrations. The histidinoalanine derivative found at the level of 6% of the total amino acids in the phosphoprotein in extrapallial fluid of some molluscs (Sass and Marsh, 1983, 1984) may be an example of nonspecific cross-link formation between monomers containing as much as 29% phospho-Ser and 35% His.

The family of protein derivatives involving the reaction of the reducing group of sugars with protein amino groups (and other nucleophiles) (see the Maillard or "browning" reaction under Lys, Fig. 4) is interesting in that the derivatives are the products of true bimolecular reactions, in which the soluble sugars have to find the reactive protein groups before the reaction can take place. It is not surprising that these products are found in significant amounts *in vivo* only if high concentration of sugar is encountered and mostly as accumulated derivatives in long-lived proteins. The chemistry of the most common steps in the reaction is shown in Fig. 4; the main point to note is that the Amadori products allow stable derivatives to form from the initial, readily reversible Schiff base and amino deoxyketose derivatives, and that similar rearrangements can take place in the condensation products involving hydroxylysine and hydroxyallysine in connective-tissue cross-links (Fig. 3). The stable Maillard products have received a good deal of attention as possible indicators of diabetes and aging. Thus, the content of glucose derivatives of the N^{ε} group of Lys or of the N^{α} group in abundant proteins such as serum albumin and hemoglobin has been shown to increase after incidents of hyperglycemia. Nonrenewable proteins such as eye lens proteins and collagens show significant content of stable Maillard products with age; it is not clear whether the content keeps increasing or whether a steady state is established; it does seem clear that the aging process by this criterion starts considerably earlier in diabetics (Monnier, 1989).

B. Enzymatic Reactions

The enzymatic reactions have presumably evolved by selecting for enzymes capable of affecting specific, desirable reactions either reversibly or irreversibly on a number of protein substrates to yield products that have unique features of both structure and function. By making the enzymes involved in these reactions, Nature has avoided expanding the genetic dictionary to include the extra amino acids, and one can argue that even fairly sophisticated, multienzyme pathways to make a given secondary amino acid may be more economical than establishing the machineries required for the synthesizing that amino acid and for incorporating it into proteins by mRNA-directed biosynthesis. This point becomes particularly significant when we note that several of the posttranslational reactions are limited to specific tissues and cells, and that the majority of living systems would not need the expanded genetic dictionary. If all the known amino acid derivatives were to be encoded, the available three-letter code would be exhausted, and all the subtle regulatory aspects of selective codon usage might be lost.

Another way of classifying posttranslational reactions is in terms of their reversibility. Recognizing that all reactions in principle are reversible, the term is used here to signify readily observable reversible reactions; it does not seem unreasonable for this discussion to consider a reaction with a log K_{eq} between -1 and 1 to be a reversible one, and one with a log K_{eq} of >3 as an irreversible one. With these simple generalizations the reactions can readily be placed in three separate groups in terms of the reversibility and the reaction paths (Fig. 6). In the following discussion we are including only the enzymatic reactions; most of the spontaneous reactions either are completely reversible, and consequently of little significance here, or are unidirectional to yield stable derivatives, many of which are excreted as distinct compounds after degradation of the proteins.

1. *Irreversible, Unidirectional Reactions*

The first of the three groups is the one in which the encoded structure (the polypeptide chain, the N or C terminus, or the primary amino acid side chains) has been modified in enzyme-catalyzed reactions involving one or several steps to yield a permanently modified structure. As in the case of the spontaneous

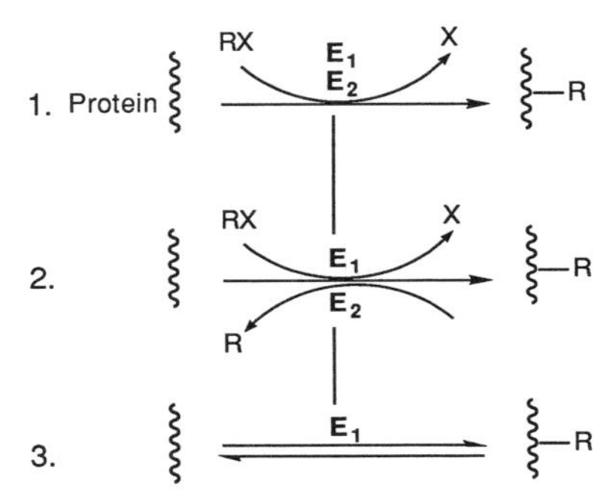

Fig. 6 Major types of posttranslational reactions. The known enzyme-catalyzed reactions can be categorized in terms of the reversibility of the reactions. One set of reactions (1) are strictly irreversible and unidirectional, and represent true, permanent conversions of one primary amino acid residue to a secondary amino acid derivative, or cleavages of peptide bonds. The reactions in 1 may involve a single enzyme (e.g., preoteolytic cleavage) or may require several enzymes working in sequence (e.g., the biosyntheses of diphthamide and hypusine illustrated in Fig. 7). Many of the stable products of reaction types 1 are excreted intact after the *in vivo* degradation of the parent proteins; the enzymes needed for their degradation are not normal constituents of the catabolic machinery. The reactions of type 2 are also irreversible, but the presence of a second enzyme (or multienzyme complex) that can catalyze the reverse reaction, also irreversibly, restoring the original unmodified protein substrate, makes the overall reaction bidirectional. Since the biological activities of the modified and the unmodified forms of the protein substrates are different, these reactions represent the activation/deactivation switches of biological activity in living cells. Phosphorylation/dephosphorylation (catalyzed by complementary kinases/phosphatases) are the most obvious and best-studied examples of the type 2 reactions. Other examples are briefly considered in the text. The reactions of type 3 are rare. Only one reaction, the reversible isomerization of disulfide and sulfhydryl groups catalyzed by protein disulfide isomerase, can be listed.

reactions, many, but not all, of the products are stable compounds that can be isolated after degradation of the protein. It must be recognized here that the harsh treatments of acid and base hydrolysis at high temperature that may be used in determining the amino acid composition of a protein in the laboratory may destroy some of the derivatives and thus preclude their identification. The majority of the known amino acid derivatives fit in this category, supporting the notion that they represent Nature's way to increase the number of protein amino acids from the 21 encoded ones (including selenocysteine as a primary amino acid) to a much larger number by producing unique stable amino acids posttranslationally. All the enzyme-catalyzed peptide bond cleavages belong in this category of unidirectional reactions. There may well be reversible ones as well, like the ones involved in the transpeptidation reactions discussed previously, but if instead of new permuted structures they are simply reversible reactions, they will clearly not be observed at all. As has already been pointed out, the amino acid modifications in the first group of reactions may involve a single enzyme such as the transferases involved in incorporating various groups onto the protein. The simple acylations and alkylations by which the protein may acquire its fatty acid and methyl groups and the more complex glycosylations by which the glycoproteins obtain their glycans are examples of such single enzyme transfer reactions. It should be noted that the total process may be quite a bit more complex than the single-enzyme designation implies. The donor molecules have to be constructed and activated, and as in the case of N-linked glycosylation, the process of assembling the complete glycan containing 3 Glc, 9 Man, and 2 GlcNAc attached to a lipid carrier (see Fig. 2) is not an insignificant one. Nevertheless, the final step, the transfer of the new group into the protein, is catalyzed by a single enzyme, and this enzyme is the one that selects each specific residue that is to be modified.

In other cases fairly complex sets of enzymes are required for a given modification to take place; the proposed reactions for the biosynthesis of diphthamide and hypusine are given in Fig. 7 to illustrate a couple of the more complex, multienzyme reactions. For these reactions it is also possible to identify a single enzyme as the one that determines the specificity of the modification reaction (E_1 for diphthamide and E_2 for hypusine). In the case of the two examples given in Fig. 7, that specificity is exquisite; only a single residue in a single protein is modified. Before leaving this group of reactions, it should be noted that a given posttranslational reaction may cause a second apparently quite unrelated reaction to take place in an unexpected precursor–product relationship. Perhaps the proteolytic cleavage, the carboxylmethylation, and the *S*-isoprenylation of C-terminal Cys in some proteins illustrate this type of coupled reactions best.

2. *Irreversible, Bidirectional Reactions*

The reactions in the second group might be considered to be reversible ones, in that the modified residue can readily be converted back to the starting unmodified state. However, the pathways for the forward and the reverse reactions are different, each one is an irreversible, unidirectional reaction catalyzed by a unique enzyme, and the overall effect becomes the indicated irreversible, bidirectional reactions (Fig. 9). A characteristic feature of all these reactions is that the properties (e.g., biological activity) of the modified and the unmodi-

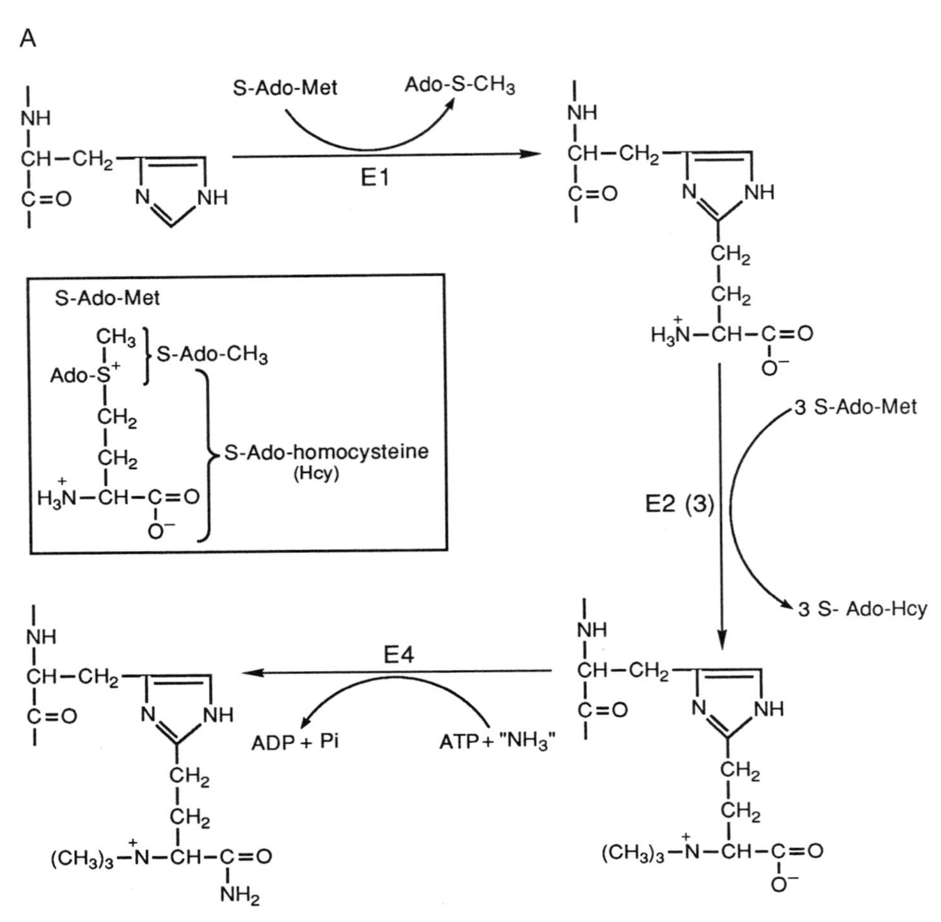

Fig. 7 Examples of fairly complex enzymatic reactions involved in posttranslational modifications. (A) Possible steps in the biosynthesis of diphthamide [redrawn from the representation by Chen and Bodley (1988)]. The commitment step is the one catalyzed by enzyme 1, and it seems fair to assume that the specificity of the reaction, the selection of one His out of an estimated 30,000 available in all the cellular proteins is determined entirely by this enzyme. (B) Possible steps in the biosynthesis of hypusine [redrawn from the reactions proposed by Park *et al.* (1993b)]. In this case the specificity must be determined by the second enzyme in the proposed reaction scheme, and again that specificity is extremely high, selecting a single Lys out of some 50–60,000 estimated to be available in the cellular proteins.

fied forms of the protein are different, and it is consequently proper to think of these reactions as the activation/deactivation switches in living cells. The total biochemical knowledge and new insight that have been derived from the study of members of this group add up to one of the major scientific achievements in the last few decades. The most obvious example, and historically also the first, is the protein kinases that transfer phosphate from ATP to Ser (and Thr) in proteins, and the phosphatases that can remove the phosphate and return the protein to the original form. As mentioned previously, a large number of these protein kinases–protein phosphatases have now been characterized in terms of specificity (Ser, Thr, Tyr in different consensus sequences), cofactor

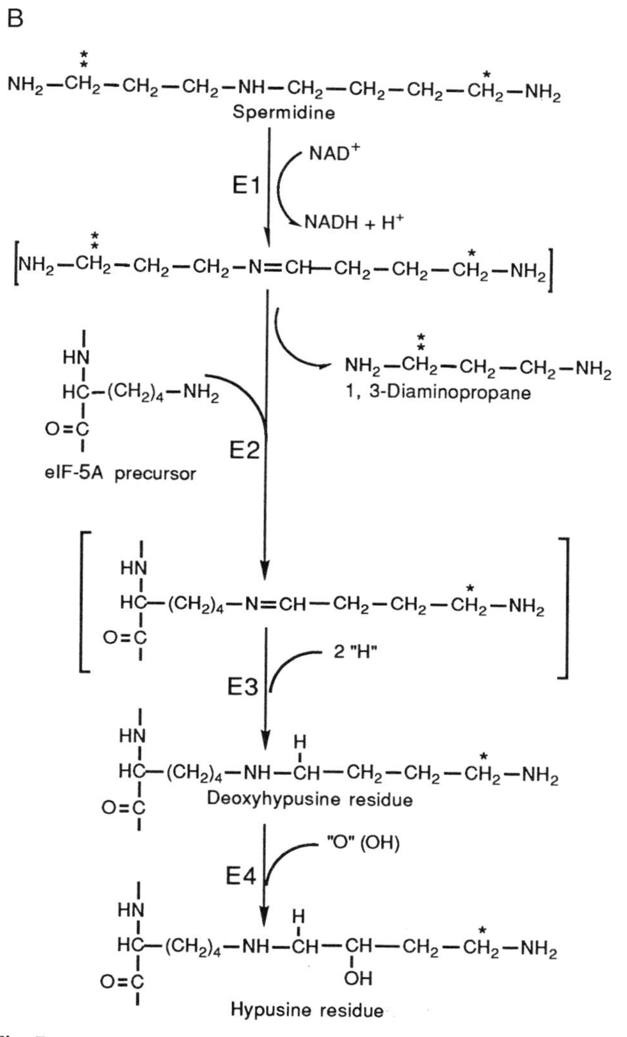

Fig. 7 *(continued)*

and activator requirements (cAMP, Ca, Zn, phosphatidylinositol), cellular location (cytosolic or, for the Tyr kinases, mostly membrane bound), and, most importantly, the biological functions correlated and explained in terms of an increasing volume of detailed structural information. The fact that the kinases and phosphatases are themselves substrates for other kinase–phosphatase combinations establishes an exquisitely sensitive signal amplification system in which a given signal can start a cascade of reactions leading to a large final output effect. The role of various interconvertible enzyme cascades in cellular regulation has been discussed in some detail by Chock *et al.* (1980). The "polarity" of the activation–deactivation reactions is not in any way fixed; the key is that the modified and the unmodified forms have different properties, not which one is the most active. Other regulatory cascades than the phosphorylation/dephosphorylation reactions are also known. The uridylyl and adenylyl transfer in bacterial glutamine synthase is one important example, and the case of reversible ADP-ribosylation in dinitrogenase reductase in *Rhodospirillum rubrum* is yet another one (Pope *et al.*, 1986). The fact that animal cells have both transferases for incorporation and hydrolases for removing ADP-ribose suggests that this regulatory switch may be used more extensively than presently thought.

3. Reversible Reactions

The third type in which a single enzyme catalyzes a reversible reaction is a rather rare one. The only obvious example is protein disulfide-isomerase, catalyzing the reaction

$$RS{-}SR + R'SH \rightleftharpoons R'S{-}SR + RSH$$

and even in this case the readily reversible reaction may become a unidirectional one by coupling the reaction with other processes such as protein folding. In fact, that feature probably needs to be considered for all the posttranslational reactions. If the chemical modification (P to PR) is coupled to a secondary structural modulation (PR to P*R), both reactions determine the equilibrium, and even if the posttranslational reaction itself is reversible, the coupled reactions could have all the earmarks of an "irreversible" unidirectional reaction, if the structural modulation pulls it in that direction.

$$P + R \overset{\log K=0}{\rightleftharpoons} PR \overset{\log K=3}{\rightleftharpoons} P^{*}R$$

C. Specificity

1. General Aspects

Posttranslational reactions are highly specific; some of them may well be the most specific reactions in living cells. The specificity is a characteristic feature of all posttranslational reactions; we know that a given modification will involve one or a few residues of a given amino acid, but never all of them, or even more obviously, will cleave one or a few peptide bonds, but never all of them. What are the structural features that determine this high degree of specificity? Figure 8 illustrates in a general way the different levels of structure that could be the basis for specificity determinants "read" by the various processing enzymes. The concept of the black box that was introduced in the beginning of this chapter emphasizes the difficulties in trying to define the actual structure that exists, perhaps only transiently, during the biosynthetic processing of each individual protein and the problems one has to face in trying to design meaningful experiments to establish what the specificity might be. As indicated in Fig. 8, the specificity determinants for one set of reactions is quite readily defined. That is in the case of those modifications that take place at the level of amino acyl-tRNA. At least four such reactions are well established at this level: the formation in prokaryotes of N-fMet-tRNA^{fMet}, but not of N-fMet-tRNA^{Met}; the formation in all cells of Sec-tRNA^{Sec} from Ser-tRNA^{Sec}, but not from Ser-tRNA^{Ser} in the presence of excess Se; the formation in some prokaryotes and in mitochondria of Gln-tRNA^{Gln} from Glu-tRNA^{Gln}, but not from Glu-

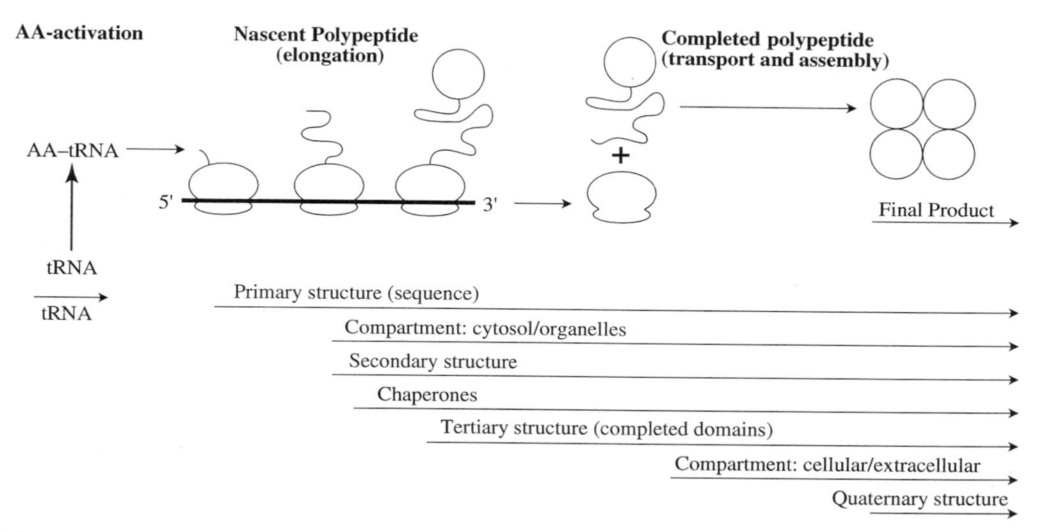

Fig. 8 Specificity determinants for posttranslational modifications. [Redrawn from the scheme proposed by Krishna and Wold (1993a).] This is an attempt to illustrate what aspects of structure, topology, and dynamics may determine the reactivity of certain residues or certain peptide bonds during protein synthesis and processing. Considering the mRNA structure as one known feature at the start of the scheme, and the completed, folded, and assembled protein structure as the other known at the end of the scheme, everything else is essentially unknown, in that the different reactions are likely to occur on transient intermediates of unknown structure and often in unknown compartments. As indicated in the text, it is possible to speculate about the various features of the nascent and completed polypeptide chain that may be "read" by the processing enzymes, and this figure summarizes these speculations. Only the initial step, the amino acid activation, is clear-cut with regard to the specificity determinants involved, namely the individual tRNAs used in the different modification reactions at this level. All the other reactions take place in the "black box" that makes up the rest of this figure. It seems reasonable, however, to propose that the primary sequence must provide a significant set of recognition signals all the way from the first assembly of a short polypeptide to the completed protein product. As indicated in the text, the primary sequence may be read directly or indirectly as part of specific secondary, tertiary, or quaternary structures. After a significant stretch of polypeptide has been assembled, three events may occur either separately or together: The peptide is localized to different compartments (e.g., cytosol and ER), secondary structure may start to form, and the nascent peptide may start its interaction with molecular chaperones. As the nascent polypeptide gets larger, partial domains may become discernible and tertiary structural features may start to direct the specificity of the reactions. When the completed polypeptide chain is released from the polysome (it appears that the presence of chaperones is required for that process in many proteins), new choices can be made about compartment location (extracellular, intracellular, organelle association), and the final folded structure either as a monomer or as a multimer can provide unique signals that reflect the native structure. Drawn to indicate a possible start of the expression of a given specificity determinant, it is assumed that except for the tRNA, each one will remain as an active determinant for the rest of the biosynthetic process.

$tRNA^{Glu}$; and the formation of phosphoseryl-$tRNA^{SerP}$. In all cases specific tRNAs, $tRNA^{fMet}$, $tRNA^{Sec}$, $tRNA^{Gln}$, and $tRNA^{SerP}$, are required for the reactions involved, and the specificity is thus clearly determined by that specific tRNA. The SerP case is similar to that of selenocysteine in that in having its own tRNA and codon (again, the termination codon UGA is used), it could probably be considered as the 22nd primary amino acid. Since the majority of the Ser-phosphates are made through the action of protein kinases, that means that we probably should consider two types of Ser-phosphates, stable, structural ones and regulatory ones. The remainder of the specificity determinants illustrated in Fig. 8 are much less clear-cut to interpret and emphasize the unknown aspects of the structures actually involved at the time of the reaction. In all cases we need to consider the reactivity of a unique amino acid residue or peptide bond, and the issue is to define the environment that gives that particular residue or bond its unique character. It seems reasonable to assume that the linear sequence of amino acids is always the basis of specificity, but how is that primary sequence expressed in the local environment? Is it a simple linear sequence with little secondary, ternary, or quaternary structure imposed? Is it a secondary structure element requiring, for example, a specific turn for the reaction to take place? Are folded structures required for certain displays of primary structures to become manifest? Are linear structures in transit through the membranes separating two compartments the true substrates for modification? Since the unique makeup of processing enzymes in the different compartments will obviously affect the outcome of the modification reactions, the signals that determine compartment residence also become important specificity determinants of posttranslational modification. It is taken for granted that none of the substrates exist as isolated, single proteins, and so it is quite likely that the actual *in vivo* substrates are complexes with chaperones and other helper proteins. All of the structures that are acted on in the posttranslational reactions are probably transient ones; even the products of late reactions involving folded structures are likely to differ in detail from the unmodified precursors, and so the black-box concept, the knowledge of modified structures but not of precursors and intermediates, appears to have general application across the entire field illustrated in Fig. 8, and all the questions just raised are real issues in understanding the specificity of the reactions.

2. *Individual Reactions*

In looking at individual posttranslational modifications, it is clear that different aspects of specificity can be identified and used to illustrate a number of possible interpretations. The cases of amino acyl-tRNA modifications with tRNA determining the specificity have been mentioned. At least two cases of specific primary sequence can also be pointed out. In the case of hypusine, the peptide -STSKTGK*HGHAL-, in which the second K, K*, becomes hypusine (Hpu), is conserved in the eIF-5A (eukaryotic initiation factor 5A) of all eukaryotic species. Moreover, this sequence is found in no other proteins in the database (Park *et al.*, 1993 a, b). Along the same lines, the peptide -VHDVTLHADAIH*R-, in which the second H, H*, becomes a diphthamide, has been observed to be highly conserved in the eEF-2 (eukaryotic elongation factor 2) of eukaryotic species, and also with minor replacements (E-8 instead of A-8) in archaebacteria; apparently it is not found in any other proteins in the database (Gehrmann *et al.*, 1985). In both of these cases the posttranslational modification involves a single residue in a single protein, and the presence of a unique sequence for both reactions suggest that Nature has developed these modification reactions by evolving both the enzymatic machinery required and the unique structures to be modified in terms of specific, nonrepeated amino acid sequences.

The one obvious structural determinant that can be assessed from the data banks is still primary structure, but if one looks for sequence conservancy for posttranslational reactions in general, rather few clear patterns of specificity are found. In a thorough and broad review, Han and Martinage (1992) have surveyed the sequence specificity for a large number of different reactions, and found that the primary sequence observed for a number of peptides that are substrates for a given modification may have no obvious common structural determinants or may reflect a consensus structure where the minimal set of structural features are required, but insufficient for the reaction to take place. The following is a summary of the main findings reported by Han and Martinage (1992). They show that the various *O*-glycosylation sites do not appear to have any consensus structures, whereas the *N*-glycosylation sites are well established to require as a minimum the structure -Asn-Xxx-Ser(Thr)- in which Asn is the residue to be glycosylated and Xxx can be any amino acid. The residue Xxx affects the reactivity of the Asn residue to a certain extent; thus, if Xxx is Pro, no glycosylation takes place, and if Xxx is Trp, Asp, Glu, or Leu, the glycosylation is inefficient (Shakin-Eshleman *et al.*, 1996). The specificity of protein kinases reflect the presence of different families of enzymes acting on Ser(Thr) or Tyr, but also reflect different general consensus structures. Tyrosine sulfation, which is catalyzed by a protein sulfotransferase found in Golgi membranes of animals, does not appear to have any consensus peptide, but the substrates do contain acidic residues (Asp and Glu) in the neighborhood of the reactive Tyr residues. The presence of a C-terminal amide requires, as indicated previously, that a Gly residue be present to be converted to hydroxyGly and in turn to the C-terminal amide and glyoxylic acid, but the product amide does not seem to require specific amino acid residues. Isoprenylation to yield either farnesylated or geranylgeranylated proteins involves the C-terminal peptide. For farnesylation, the required C-terminal structure is Cys-Aaa-Bbb-Xxx, in which Aaa can be a variety of aliphatic amino acids, Bbb must be an uncharged one, and Xxx is preferentially Met, Ser, or Phe. In the final product the C-terminal tripeptide is removed and the generated C-terminal S-prenylated Cys is carboxylmethylated. It is of interest to note that some proteins contain an additional Cys close to the C-terminal Cys-Aaa-Bbb-Xxx sequence. If these proteins are prenylated, proteolytically cleaved, and carboxylmethylated, the additional Cys may also become *S*-esterified by palmitic acid. The specificity for geranylgeranylation is somewhat different in that several C-terminal sequences can be acceptors for the transferase. Proteins terminating with -Xxx-Xxx-Cys-Cys, -Xxx-Cys-Xxx-Cys, or -Cys-Cys-Xxx-Xxx may all be geranylgeranylated at both Cys residues, and except in the case of sequences with terminal -Cys-Cys: They are also methylated (Smeland *et al.*, 1994; Farnsworth *et al.*, 1994). It is of great interest that a specific N-terminal sequence, Xxx-Xxx-Tyr-Xxx-Tyr-Leu-Phe-Lys-, is also required for the geranylgeranylation to take place, and that interactions in the folded structure thus are involved in determining the specificity of the reaction (Sanford *et al.*, 1995). The incorporation of the glycolipid anchors also follow the pattern of removing the C-terminal end to expose the new terminus, to which the ethanolamine end of the anchor is attached (see Fig. 10, Section IV, for structure), but it does not seem that a clear consensus sequence can be derived from the data. Very similar types of specificity determinants and reaction pattern are involved in the incorporation of a diacylglyceryl group at the N-terminal end of a major outer membrane (Braun's) lipoprotein in *E. coli*. This lipoprotein is the prototype of bacterial lipoproteins and can be used to illustrate the reaction properties. The protein is synthesized as a prolipoprotein containing an N-terminal signal sequence. The signal sequence from a large number of lipoproteins show a consensus sequence at the signal peptide cleavage site: Leu-Ala (Ser)-Gly (Ala)-Cys in the −3 to +1 positions. The lipoprotein is synthesized in a multiple step reaction as follows: (1) incorporation of glycerol as a thioether to Cys at position +1; (2) incorporation of the two fatty acids into the bound glycerol; (3) cleavage of the signal peptide by signal peptidase to leave the modified Cys as the new N-terminus; and finally (4), the incorporation of an acyl group at the α-amino group of the Cys. The resulting structure is shown along with other membrane anchors in Fig. 10 in Section IV. In the case of other modifications of the N terminus, a good deal of work has been directed toward the understanding of the most common N-terminal modification, the *N*-acetyl derivative. As discussed previously, certain rules have been elucidated that determine whether acetylation will or will not take place; the majority of the natural acetyl derivatives involve Ala, Ser, Gly, Met as the N-terminal amino acids, and the nature of the amino acids in the second, third, and later positions also determine whether the acetylation will take place (Arfin and Bradshaw, 1988). N-Terminal myristoylation is also specific for the N terminus, and only Gly has been found to be myristoylated. In this case it appears that an N-terminal consensus peptide can be described for the reaction, and again a relatively small second residue is required for efficient reaction. In the case of γ-carboxylation, consensus peptides containing high levels of Glu can be recognized, and all of these Glu residues appear to be substrates for the carboxylation reaction. Similarly, specific domains can be shown to be common to the hydroxylation of Asp and Asn in vitamin K-dependent proteins, suggesting a consensus sequence -Cys-Xxx-Asp/Asn Xxx_4-Phe/Tyr-Xxx-Cys-Xxx-Cys- for the hydroxylation of the Asp/Asn in position 3. Partial specificity needs to be considered for many of these reactions. Thus, transglutaminase shows quite high specificity for the Gln partner of the reaction, but very little specificity for the amino group that is the second substrate of the reaction.

It is clear that all the unknowns considered in the overview in Fig. 8 are back to haunt us when we look for specific structural determinants in terms of common structural features. As already established, the main problem is that we really do not know what the actual structure is, and that primary sequence is too narrow an indicator of structure in this context. A more careful consideration of higher levels of structure is clearly in order, and it is significant that these structural elements can be assigned specific roles in determining the biological activities of certain structures. One such study covers the role of Pro in a number of biologically important peptides (Vanhoof *et al.*, 1995). The idea here is that Pro characteristically disrupts stable secondary structures and that Pro motifs consequently may reflect kinked, bent, and distorted structures with unique amino acid exposures. These proline motifs may thus present special structures to a variety of receptors and processing enzymes, and a number of unique biological properties of such proline motifs have indeed been observed. In some cases proteolytic resistance is induced by the presence of the Pro residue in a given peptide; in other cases the biological activity is strongly enhanced. Thus, the thrombin activity on coagulation cascade peptides requires a Pro in the P2 position next to the P1 Arg, which is cleaved by thrombin, and in the case of the processing of the polyproteins involved in retrovirus replication, an Asp protease appears to be autocatalytically produced by the cleavage of two Phe-Pro bonds. The authors (Vanhoof *et al.*, 1995) also note the high content of Pro in immunopharmacologically active peptides, and that it is the universal residue 2 in the N-terminal sequence of cytokines and growth factors. Clearly these positions have been conserved, but the functional significance is not understood at this time. The important point here is that unique residues like Pro may induce secondary structures or structural variations that affect the specificity of posttranslational reactions well beyond just the primary sequence. Site-specific mutagenesis has become a very powerful tool with which to explore broad structural features that affect how a given primary sequence is "seen" by the modifying enzymes. Only one example will be included here, namely the specificity of the biotinylation of the biotinyl subunit of transcarboxylases. This 123-amino acid subunit is biotinylated at Lys-89 in the conserved sequence -Ala-Met-Lys-Met-. The biotinylation still proceeds normally in a mutant missing the first 18 amino acids and also in one in which the C-terminal Gly has been removed. However, if the penultimate residue, Ile-122, is removed, no biotinylation takes place. A single hydrophobic residue, separated from the active site Lys by 33 residues in the linear sequence, thus dramatically affects the structure of the site to be modified by biotin holoenzyme synthetase (Murtif and Samols, 1987). This long-range effect implying three-dimensional structures as specificity determinants is the same as that already discussed for geranylgeranylation.

A final level of structural analysis that has recently become possible needs to be included to show where this whole area is headed in terms of sophistication and new understanding. When the structure of the insulin receptor Tyr kinase became available, it became possible to compare the detailed structures of the two families of protein kinases, those that phosphorylate Tyr and those that phosphorylate Ser/Thr. More than 400 different enzymes have been identified in the two families, and their comparative specificity and mechanistic properties have been assessed by comparing sequence alignments. In the comparison of the 3D structures (Taylor *et al.*, 1995), a general fold of the backbone is found to be conserved, but unique features superimposed on the common properties suggest how the two families of proteins may differ in catalytic activity and specificity. This is clearly approaching the question of substrate specificity and reaction efficiency at a new level. Future work will undoubtedly proceed in that direction, and to the extent we will be able to interpret the data outside a black box, the definition of the rules that cause a certain reaction to take place should be greatly improved. At the present stage it seems clear that we have a good deal of difficulty in predicting final structures of extensively processed mature proteins from the knowledge of their cDNA structure.

D. Chaperones Involved in Posttranslational Modification

With the elegant experiments of Anfinsen and co-workers, which showed that all the information needed for a protein to fold into a unique three-dimensional structure is encoded in the primary sequence (Anfinsen, 1973), the stage was set for broad studies of the processes involved in the folding and unfolding of proteins. The early experiments with dilute solutions of pure, single-chain proteins generally fit relatively simple mechanistic models, but it rapidly became clear that if the experiments involved multichain proteins in which the refolding required coordination of the folding and the multimolecular association of individual chains, the models became much more complex. Multiple side products would result and the yields of the refolded multichain proteins would be extremely low. These early observations lent significance to the ever-present, nagging question as to the relevance of the observations made of dilute aqueous solutions of pure proteins to the *in vivo* folding process that takes place in a concentrated "soup" of completed and uncompleted proteins. If one attempted to study the folding *in vitro* of a mixture of proteins at the concentrations that apply in the cytosolic or ER compartments, the product would undoubtedly be an insoluble glob of denatured proteins. So how does Nature solve this problem and produce individually folded proteins in excellent yield? Nature mobilizes helper proteins that protect and guide each new polypeptide chain during its assembly and after its completion to keep it from interacting with the wrong partner, from folding in the wrong compartment or with the wrong fold, and to transport it to its proper destination in the cell. Among these various helper proteins are the chaperones. A proper discussion of chaperones is beyond the scope of this chapter, but it is appropriate to briefly consider them in connection with the question about their involvement in posttranslational reactions.

Some reviews have appeared covering the discovery and our current understanding of molecular chaperones (Rassow and Pfanner, 1995; Hendrick and Hartl, 1995; Lorimer, 1996; Buchner, 1996; Ellis and Hartl, 1996). The molecular chaperones are a distinct class of proteins; they are highly conserved and widely distributed among prokaryotes, plants, and animals. They bind transiently and noncovalently to nascent polypeptides and to unfolded or unassembled proteins, and they mediate the folding, assembly of oligomers, and translocation of proteins in different stages of biosynthetic processing (Fenton *et al.*, 1994; Frydman *et al.*, 1994). The major classes of chaperones involved in these reactions are the Hsp60 system (the designation heat shock protein, Hsp, reflects the early discovery of most of the chaperones as stress-induced proteins; the number designates the molecular mass in kilodaltons), the TCP1 system, the Hsp70 family, the Hsp90 family, the Hsp110 family, and small Hsps (Rassow and Pfanner, 1995). The designation "intramolecular chaperone" has been introduced to describe the role of the propeptides in the folding of single-chain proteins prior to proteolytic activation into multichain, folded structures (e.g., insulin, proteases) (Shinde and Inouye, 1994). Although the propeptide portion certainly exhibits a chaperone's influence in directing the correct folding of each of these proteins, the process appears to be sufficiently different from that involving the molecular chaperones to make the value of the "intramolecular chaperone" concept questionable.

The Hsp60 system includes two subclasses, a Hsp60 (heat shock protein 60) family and a 10 kDa-sized cochaperone. The Hsp60 family includes GroEL (protein identified by a *gro* mutant of *E. coli*) in prokaryotes and in the mitochondrial matrix of eukaryotes; Cpn60 (chaperonin60) in yeast; and Hsp58 in mammals. The cochaperones include GroES (protein identified by a *gro* mutant of *E. coli*) in prokaryotes and in the mitochondrial matrix of eukaryotes; Cpn10 in yeast, and Hsp10 in mammals (Fenton *et al.*, 1994; Richarme and Kohiyama, 1994). GroEL is a 14-subunit protein with ATPase activity, and the active protein consists of two stacked rings of seven subunits of M_r 57,259. GroES contains a single ring of six to eight identical subunits of M_r 10,368. In the presence of

ATP, the GroEL and GroES rings bind each other, and this binding inhibits the ATPase activity of GroEL (Hemmingsen *et al.*, 1988; Hendrick and Hartl, 1995). Both these chaperones function cooperatively in preventing the aggregation of newly synthesized polypeptides and then mediating the various steps of translocation and folding to the native state. GroEL folds the substrate peptide/protein in the presence of Mg and ATP. GroES binding stabilizes the complex and regulates its ATPase activity. Hydrolysis of ATP results in the release of GroES and the folded protein. Several of these chaperone/cochaperone reaction cycles are found necessary to achieve complete folding (Martin *et al.*, 1993).

The TCP1 system (t-complex polypeptide 1) consists of the TF55 (thermophilic factor55) complex in prokaryotes, Tcp1p (t-complex polypeptide 1p) in yeast, and cytosolic CCTα, CCTβ, CCTγ, CCTδ, CCTε, and CCTη in mammals. This group of chaperones is similar to Hsp60 chaperones; they are unique in that they do not seem to require the small molecular cochaperones to be active in cytosol, chloroplast, and mitochondria. They form high-molecular-weight complexes consisting of two heptamer rings (Rassow and Pfanner, 1995; Hartl *et al.*, 1994). These Hsp60 and TCP1 systems with their tetradecameric complex barrel-like structures constitute the characteristic protein processing "machines" in different cells and cellular compartments.

The Hsp70 system includes DnaK in prokaryotes; Ssa1p (stress protein of 70 kDa, subfamily A), Ssa2p, Ssa3p, Ssa4p, Ssb1p, Ssb2p in yeast cytosol; Kar2p (karyogamy mutant protein) in yeast ER; Ssc1p, Yge1p (yeast homolog of GrpE protein) in yeast mitochondria; Hsp70, Hsp72, Hsc73 (Hsp70 cognate protein of 73 kDa) in mammalian cytosol; Bip in mammalian ER and Grp75 (glucose-regulated protein75) in mammalian mitochondria; and a group of low-molecular-weight chaperones, namely, DnaJ (J component involved in the replication of DNA in *E. coli*), and GrpE in prokaryotes; yeast cytosolic Sis1p, scj1p (*Saccharomyces cerevisiae* homolog of DnaJ1 protein), Ydj1p (yeast homolog of DnaJ protein), and mammalian cytosolic Hsj1 (*Homo sapiens* homolog of DnaJ), Hdj1 (human homolog of DnaJ), Hdj2, Hsp40; sec63p (secretion mutant 63p) in ER and Mdj1p (mitochondrial homolog of DnaJ) in mitochondria of yeast that cooperate with the Hsp70 (Frydman *et al.*, 1994). Hsp70s interact with incompletely folded proteins, such as nascent peptide chains on ribosomes and proteins in the process of translocation from the cytosol into mitochondria and endoplasmic reticulum. The best-illustrated chaperones in this group, the Dnak, Dnaj, GrpE system of *E. coli* or Bip of the mammalian ER, bind to the hydrophobic stretches of the nascent peptides or denatured proteins during folding (Kim and Arvan, 1995).

The Hsp90 family of proteins is in both ER and cytosol and is the most abundant chaperone family in the cells under normal growth. The family includes HtpG in prokaryotes, Hsc83 and Hsp83 in yeast, and Hsp90α, Hsp90β, Grp94 in mammals (Rassow and Pfanner, 1995). Unlike the Hsp60 chaperones, this group of proteins act in an ATP-independent fashion and participate in several regulatory functions in the cell, including binding to the oncogene products and steroid hormone receptors (Wiech *et al.*, 1993).

Among the other families of chaperones are the Hsp110 family, which includes prokaryotic ClpB, yeast cytosolic Hsp104, yeast mitochondrial Hsp78, and mammalian cytosolic Hsp110 (Rassow and Pfanner, 1995; Parsell *et al.*,

1994). In addition, "small Hsps" are found in yeast, plants, and higher animals. With a monomeric size of 15–30 kDa, these proteins form oligomeric structures of about 800 kDa in mammalian systems. These molecules seem to function as molecular chaperones in an ATP-independent fashion by preventing protein aggregation and promoting refolding *in vitro* (Jacob *et al.*, 1993), but their role in the involvement in protein translocation is yet to be explored. α-Crystallin, a bovine eye lens protein, has been shown to be structurally and functionally similar to Hsp25 of this family (Merck *et al.*, 1993).

In addition to the Hsps, two other classes of highly conserved proteins that play a role in protein folding *in vivo* have been identified, the protein disulfide isomerases and peptidyl-prolyl *cis–trans*-isomerases. The disulfide isomerases are about 57 kDa in size and are localized in the lumen of the ER in eukaryotes and in the periplasm of *E. coli.* They catalyze the formation of disulfide bonds cotranslationally by interacting with the nascent polypeptide chains (Freedman, 1992; Freedman *et al.*, 1994; Noiva and Lennarz, 1992). The prolyl isomerases, which are also called immunophilins, catalyze the *cis–trans* isomerization at X-Pro peptide bonds and accelerate many slow steps in the folding of proteins *in vitro.* However, although many isomerases have been identified in a variety of organisms, their *in vivo* role in protein folding remains unclear (Sudha *et al.*, 1995).

In an oversimplified summary of how the chaperones may work, we can consider first a number of monomeric or dimeric components that bind to specific regions of an unfolded polypeptide chain and prevent folding and association with undesirable partners while facilitating translocation and perhaps also correct folding. In addition to these "workhorse" proteins, of which Hsp70 is a typical example, there are the machines of the Hsp60 family, in which two stacks of heptamers form cylinders, sometimes supplemented by Hsp10 heptamers, into which the unfolded chain can be deposited and the folding process can take place. Based on the X-ray structure of a mycobacterial Cpn10 oligomer, models of the Cpn10-Cpn60 (Hsp10-Hsp60, GroES-GroEL) machines have been proposed (Mande *et al.*, 1996). The complete process is undoubtedly quite complex. Thus, it has been demonstrated in the *in vitro* biosynthesis of firefly luciferase (Frydman *et al.*, 1994) that Hsp70 binds to the nascent polypeptide chain, and that this binding is required for proper folding to take place. Hsp70 must be present during the chain elongation part of the translation; if it is added later, it has no effect and folding does not take place. TCP1 reacts in sequence with the Hsp70. In another case, the *in vitro* synthesis of rhodanese was demonstrated to require the addition of the chaperones to release the full length protein from the ribosome and to yield a biologically active, folded product (Kudlicki *et al.*, 1994). In the absence of chaperones, a full-length ribosome-bound rhodanese protein is the product. In the presence of the chaperones DnaJ, DnaK, GrpE, GroEL and GroES plus ATP, active protein was released. All chaperones are required for optimal yield; DnaJ alone inhibited the release of protein, and GroES or DnaK alone stimulated release of inactive protein. The total system by which individual proteins can be processed in the cell without interference from all the other proteins is certainly complex, but considering the fact that the chaperones can interact with a large number of different polypeptides to facilitate the efficient production of biologically active proteins in their proper compartments of action, we may conclude that the process represents an economical solution to a difficult problem.

So, are chaperones involved in the posttranslation reactions? The answer is obviously yes, but the extent to which they are involved cannot be answered very precisely at this stage. One reason for this is that most of the work to date has focused on translocation and folding, and very little is known about the possible role of chaperones in the covalent modification of amino acids. Another uncertainty in the answer has to do with the definition of chaperone. In the introduction earlier, we used "helper proteins" to describe the whole spectrum of protein involved in the biosynthesis of new proteins; if we agree to call all these proteins "chaperones," we can include processing enzymes such as protein disulfide isomerase, prolyl *cis–trans*-isomerase, and proteases such as signal peptidases, and we can unequivocally state that chaperones are directly involved in covalent modifications both of the peptide bond and of amino acid side chains. It seems likely that many reactions are affected by chaperones, and that the real substrate for the modifying enzymes really is a polypeptide–chaperone complex rather than just the free polypeptide. We have looked for evidence for this type of involvement in the modification of the N terminus of proteins, but have so far been unable to demonstrate any kind of chaperone involvement *in vitro*. Considering the complexity of the *in vivo* system, negative results are probably not very conclusive in this case. It is likely that the search for specific examples of chaperone-mediated posttranslational covalent modification may intensify in the future, and that the documentation of this function will become as compelling as the documentation of the role of chaperones in translocation and folding.

IV. Biological Functions of Posttranslational Modifications: Why Modifications Are Made

This section may well include the most complex aspects of posttranslational modification reactions. It is again possible to assign broad and general functions to specific families of modifications, but one frequently encounters exceptions to the generalities and specific functions that do not fit the general picture. In spite of the limitations, it may be useful to briefly consider some of the general functions of families of derivatives. In the following very few references will be given, since the key references for all the derivatives that will be discussed have been presented in Section II, C.

A. Regulation: Interconvertible Modifications

As discussed previously, the reactions that are characterized as irreversible and bidirectional, those that involve a modification catalyzed by one enzyme and the return to the original unmodified form catalyzed by a different enzyme, are characteristically involved in regulation. The basis for this function is that the modified and the unmodified forms of the protein have different biological activities, and that the modification/demodifications consequently can be viewed as regulatory switches of biological activity. Chock *et al.* (1980) have pointed out that these reactions should not be thought of as on/off switches;

the interconversion of an active and an inactive form of an enzyme is a dynamic process arriving at some steady-state level characterized by the relative activity of the converting enzymes as established by the level of allosteric effectors. The activity is probably never zero or 100%, and on/off is thus not a proper designation. Unidirectional cascades such as the proteolytic activation involved in, for example, blood-clotting enzymes can be considered true on/off switches. In the terminology of Chock *et al.* (1980), the amplification of a signal from an allosteric effector of an enzyme can be considered as a cascade in which a small change in active enzyme may result in large changes in substrate conversion. Since the substrate may itself be an enzyme, the net amplification of the initial small signal really constitutes a cascade effect. For the simplest system, a forward and a reverse reaction make up a single cyclic interconversion of modified and unmodified protein, in a monocyclic cascade, as illustrated in Fig. 9. If an interconvertible enzyme catalyzes the modification of a second interconvertible enzyme, the two cycles are coupled so that the properties of both cycles are expressed in the coupled bicyclic cascade; similarly, if *n* interconvertible enzymes are involved, the cascade is made up of *n* cycles. The cyclic cascades all exhibit strong signal amplification and rapid rate amplification, and they can have both positive and negative cooperative responses to changes in regulatory effectors.

A number of enzymes, such as mammalian pyruvate dehydrogenase, liver pyruvate kinase, and phosphofructokinase, and yeast NAD-dependent glutamate dehydrogenase are regulated through monocyclic cascades involving phosphorylation/dephosphorylation. ADP-ribosylation and polyADP-ribosylation also have been found to have separate on and off enzymes (see under Arg and Glu in Section II, C), and along with the reversible addition of a C-terminal Tyr in tubulin (see Tyr in Section II, C), fit the general pattern of

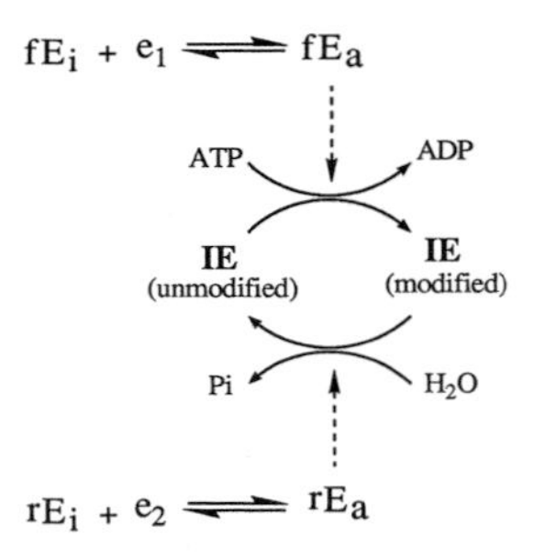

Fig. 9 Illustration of a typical monocyclic cascade. [Redrawn from the scheme presented by Chock *et al.* (1980).] The key reaction is the interconversion of IE (unmodified) and IE (modified) (IE, interconvertible enzyme). The interconversion is carried out by a forward enzyme, fE, which in this case might be assumed to be a protein kinase, and by a reverse enzyme, rE, a protein phosphatase. Each of these enzymes is under allosteric control by the effectors e_1 and e_2 that shift the equilibrium between the inactive forms fE_i and rE_i and the active forms fE_a and rE_a toward the active forms, thus establishing the cascade: A small signal in terms of increased concentration of e_1 will activate fE and thus shift the interconversion of IE toward the modified form, whereas a small signal in terms of e_2 will increase the activity of rE and shift the interconversion back toward the unmodified form. Note that nothing is specified, or need to be specified, about the relative activity of IE (unmodified) and IE (modified). Either one could be the most active form of the enzyme.

monocyclic cascades. It appears that most of these systems respond to multiple effectors; that may well be a characteristic of the cyclic cascades.

Other well-studied systems, such as the coordinated glycogen phosphorylase, glycogen synthase system, 3-hydroxy-3-methylglutaryl-CoA reductase, the *E. coli* glutamine synthase, and more recently the incredible explosion of information about receptor-associated Tyr kinase-coupled cascades and about the regulation of Ca^{2+}, K^{+}, and Cl^{-} channels, fit the bi- or multicyclic cascade models. It should be noted that 40 different metabolites affect the activity of glutamine synthase.

B. Cross-links

Another group of derivatives that may appear easy to categorize in terms of function are the cross-links. The general concept that the cross-links stabilize or "fix" certain folded structures goes back to the extensive early studies of disulfide bonds in proteins and was solidified by studies of the effect of artificial cross-links on protein activity and stability. In general, the discovery of other natural cross-links has provided additional support for their role as "fixatives" in protein structure. There are, however, exceptions to the general rule, and perhaps also unique consequences of the cross-links that need to be considered.

An examination of the derivatives listed in Section II, C shows quite a large number of different cross-links. Cystine, undoubtedly the best studied of the all the cross-links, is unique in that in the presence of protein disulfide isomerase, it can be reversed; this property must in fact be the basis for the important roles of Cys in protein structure and function. Most of the other cross-links are unidirectional reactions from which the products are eventually secreted intact after the rest of the protein has been digested.

One important protein cross-link, the so-called isopeptide, N^{ε}-(γ-glutamyl)-lysine, is produced by replacing the amide NH_2 of protein bound Gln with the ε-amino group of Lys (Loewy, 1984). The transamidation reaction is catalyzed by a family of transglutaminases, enzymes that require a protein-bound Gln as the acyl donor, but can use either protein-bound primary amines (Lys and ornithine) or free amines as the acyl acceptor; in the absence of amines, the enzyme will even use water as an acceptor, catalyzing the hydrolysis of Gln to Glu. The enzyme was originally found as the extracellular factor XIII of the blood-clotting cascade. In this form it catalyzes the introduction of isopeptide bonds at specific sites of fibrin, and this formation of high molecular weight cross-linked fibrin can be considered as the actual coagulation process. Similar extracellular polymerization cross-linking is also observed in seminal fluid, in the insoluble medulla of hair and quill, and in the α-keratin of skin. A second form of the enzyme, the tissue transglutaminase, is responsible for a number of intracellular activities observed in many different cells and tissues, muscle cells, membrane proteins from different cells, eye lens tissue, and fish egg chorion after fertilization, and also in the epidermis, which has been studied in cultured keratinocytes. There is evidence that this keratinocyte enzyme may be the product of a third distinct gene. Several very intriguing functions of have been proposed for the tissue transglutaminase-catalyzed reactions (Fesus and Thomazy, 1988). These include cell proliferation, ageing, certain cases of endocytosis and secretion, differentiation, and apoptosis, programmed cell death. In the case of accidental death, necrosis, the cells swell and burst, and spill their contents into the extracellular space, where an inflammatory response may result. In apoptosis, the cell and the nucleus shrink and sometime fragment, and all the pieces are rapidly phagocytized and cleanly removed. The morphological changes in apoptosis are definitely associated with increased transglutaminase activity, and the rigid, shrunken structures may well reflect the effect of isopeptide cross-linking of the cells (Piacentini *et al.*, 1994). It should be noted that although the enzyme is quite broadly distributed in nature, the isopeptide bond is a rather rare occurrence in most proteins investigated. It seems likely that the cross-links formed between polyamine acceptors (spermine, spermidine, putrescine) and Gln donors are also catalyzed by transglutaminase. It should finally be noted that very small amounts of N^{ε}-(β-aspartyl)lysine have been observed in a few proteins. It is not clear whether these are the products of a different enzyme or are side products formed very slowly by transglutaminase.

Three of the cross-linked derivatives, involving covalent binding of cofactors in the bis-thioether bond to the heme in Cyt *c* and to some phycocyanobilins in cyanobacteria and the 2,4′-bistryptophan-6′,7′-dione, may well represent cross-linking to stabilize the ligand in relation to the protein, making in the process a cross-link that may not have any particular function as a protein crosslink per se. Some other, quite rare derivatives, the thioester bonds between Cys and γ-Glu, and the thioether bonds between Cys and His and Cys and Tyr, have no obvious applications as general stabilizing cross-links. However, recent studies on human α_2-macroglobulin suggest that the S-(γGlu)-Cys esters perhaps are involved as highly sophisticated topological regulators of biological activity (Kolodziej *et al.*, 1996). This protein is a nonspecific protease inhibitor made up of two disulfide-linked dimers, with two *S*-(γ-Glu)-Cys esters per dimer in the native protein. Based on their computer-enhanced EM models, Kolodziej *et al.* (1996) show that the two dimers are arranged into a twisted oval structure (like a figure 8 in a specific view) with the protease binding site exposed. Upon interaction with a protease, the four thioester bonds are broken, and the α_2-macroglobulin oval is twisted in the opposite direction to trap the protease on the inside of the new structure. It is of interest to know that if the native protein is treated with methylamine to cleave the four thioester bonds, the molecule undergoes the same twist reversal. It thus appears that the role of the very small loop thioesters between Cys-949 and Gln-952 (-Cys-Gly-Glu-Gln-) in this molecule (Sottrup-Jensen, 1989) is indeed to stabilize a very specific macromolecular structure. Since the reverse twist is a spontaneous event when the bonds are broken, the initial thiol esters must presumably have been formed during the early folding steps.

All the remainder of the cross-links may well be spontaneous, nonenzymatic reactions. One group arises from dehydroalanine, which in turn is derived mostly from Cys and Ser; another arises from Lys, Hyl and their oxidatively deaminated (enzyme-catalyzed) derivatives, allysine and hydroxyallysine; and a third one involves oxidative association of Tyr residues. It is possible that enzymes are involved for some of these derivatives, but most of them can proceed in the absence of added enzymes. Dehydroalanine can be derived from several amino acids, but mostly from Cys and Ser (or even better, phospho-Ser); it reacts with the SH of Cys, the amino group of lysine, or the imidazole nitrogens of His to give the thioether lanthionine, alaninolysine, and alanino(π- and τ-)histidine, respectively, as the cross-linked derivatives. In some bacterial peptides a dehydrated Thr also reacts with the SH of Cys to give the β-methyllanthionine cross-link. The alaninohystidine derivatives are derived from phosphoSer, forming dehydroalanine by the elimination of P_i (Marsh and Sass, 1984). They are found in small amounts in vertebrate tooth dentin associated with Ca phosphate (hydroxyapatite), and in quite high levels in a phosphoprotein found in the shell lamella and the extrapallial fluid of some bivalve molluscs, associated with Ca carbonate. Both proteins are high in Asp and phospho-Ser; the dentine one has rather low levels of His, whereas the mollusc shell protein contains more than 30% His. The former exists primarily as a monomer of 38 kDa mass; the latter has an unknown monomer mass and is obtained as a phosphoprotein particle of 54×10^6 Da mass, apparently polymerized through the formation of the alaninohistidine cross-links, and with the capacity for binding Ca and thus being part of the mineralization process on the mollusc shells.

The very large Lys family of cross-links has been discussed in Section II, C and is shown in Fig. 3. These cross-links are important components of connective tissue structure and also reflect maturity and age of proteins in this category. The Tyr family of cross-links arising from oxidative phenolic coupling of Tyr rings to yield cross-links such as 3,3′-diyrosine, 3,3′;5′3″ trityrosine, and isodityrosine probably have similar structural roles to play in connective tissue proteins, as well as in many different types of proteins and organisms such as structural proteins of the exoskeleton of insects, the fertilization membrane of sea urchin eggs, adhesives substances of sea mussels, human lens protein, and plant cell walls. Apparently the elastic properties of such proteins reflect the properties of the Tyr cross-links.

C. Covalent Cofactors

An examination of the data in Section II,C shows that several cofactors are covalently linked to proteins. One can make convincing rationalizations of this fact by considering, for example, the efficiency of reactions where the stoichiometric quantity of the cofactor is always associated with the active protein. However, noncovalent cofactors are much more common than covalently bound ones, and so, as is generally true in this area, it becomes problematic to come up with a single model that will adequately justify the existence of covalent cofactors (Rucker and Wold, 1988).

Two derivatives, biotinyllysine (biocytin) in carboxylases and transcarboxylases and lipoyllysine in α-keto acid dehydrogenases, are examples of cofactors that are active only in the covalently bound form. It may also be acceptable to include the Ser-linked pantethein-phosphoryl derivative in acyl carrier proteins in this category, although the free equivalent of this cofactor, coenzyme A, is a much more common acyl carrier in a large variety of reactions. In all three cases the cofactor-containing protein is part of a multisubunit complex in which the cofactor can be shuttled between the specific acceptors and donors involved, and it is quite possible that the covalent attachment may facilitate complex, multistep reactions. In the case of two protein-bound quinone-type cofactors, trihydroxyphenylalanine (TOPA) and 2,4′-bistryptophan-6′,7′-dione, the reason for a covalent derivative is most likely that it represents one way in which cofactors that would be very unstable in aqueous solution can be stabilized by the proper protein environment, and at the same time can be properly located as cofactors for the oxidases and dehydrogenases with which they are associated. The phycobiliproteins with Cys-bound linear tetrapyrroles may well play the double role of stabilization and proper interaction with other proteins in the light-harvesting system of photosynthetic microorganisms. Another two cofactors, phosphopyridoxal and retinal, are both covalently linked to proteins (e.g., transaminases and rhodopsins, respectively) as relatively weak and readily reversible Schiff base derivatives with Lys. In the case of the transaminases, the derivative is only transiently formed in the course of the reactions catalyzed and must consequently be considered unique in this category of derivatives. It is interesting to note that a similar reactivity can be obtained with the N-terminal pyruvoyl derivative given under N-terminal modifications in Section II, B; in this case the carbonyl of the pyruvate replaces that of phophopyridoxal in the reaction.

The derivatives of flavin, FAD linked to His, Cys, or Tyr, and FMN linked to Cys in certain dehydrogenases and oxidases have a large number of equivalent systems with noncovalently bound flavin coenzymes, and similarly, although cytochrome *c* has covalently bound heme, most of the other heme proteins contain noncovalently bound heme. It is interesting to speculate that the covalent incorporation of cofactor in these cases may direct the biosynthetic efficiency and pathway, rather than determining the efficiency of the biological activity. Thus, Cyt *c* is synthesized on soluble polysomes from a nuclear gene and is transported to the mitochondria as the apoprotein. The heme is incorporated in synchrony with the transport into the intermembrane compartment, and the completed heme protein then becomes permanently fixed in that compartment. The fact that the flavin incorporation appears to take place at the level of incompleted, nascent polypeptide chains (Decker, 1982) is also consistent with a role in proper folding and transport for some flavoproteins.

D. Membrane Anchors

In the course of the past couple of decades it has become increasingly clear that a number of modifications appear to have the specific function of anchoring polypeptides to membranes through the introduction of hydrophobic, often membrane-like moieties. The literature is growing rapidly in this area. An entire volume has been devoted to these modification reactions (Casey and Buss, 1995). As indicated in Fig. 10, there are several distinct families of derivatives in eukaryotic proteins, the N-terminal amides, notably the N^{α}-myristoyl proteins, the fatty acyl thioesters of Cys, and, less frequently, esters of Ser and Thr, involving mostly palmitate, but also other fatty acids; the C-terminal glycophospholipids; and several thioethers, including diacylglyceryl and diphytanylglyceryl ethers of N-terminal *N*-acyl-Cys in eubacteria and archaebacteria, respectively, and the prenylated proteins containing farnesyl (C_{15}) and geranylgeranyl (C_{20}) attached to Cys residues at or next to the C terminus in eukaryotes. It seems clear that the important function of the anchors in membrane interaction is that the lipid anchors appear to conserve the fluidity of membranes, whereas the protein anchors seem to give rigid, nonflexible binding regions.

In all cases a major hydrophobic moiety is added to the protein, suggesting unique possibilities for membrane interaction, and the experimental evidence is consistent with this notion. It is not quite that simple, however. In the case of the N-terminal myristoylation, for example, it is clear that several aspects of protein interactions are affected. It should be noted that the introduction of the myristoyl group takes place (cotranslationally) during polymerization of the nascent polypeptide chain, and the 14-carbon acyl group should have some effect on the folding of the myristoylated protein. It should also be noted that several myristoylated proteins are soluble proteins and not associated with membranes at all, and finally that the acylation may be heterogeneous and yield a mixture of saturated and mono- and dienol (*cis*) C_{14} acyl groups. This is the case in transducin, for example, and the acylation in that case does not affect membrane association but protein–protein interaction (Neubert *et al.*, 1992). There are, however, also cases where the N-terminal myristoyl group does cause membrane binding. The proteins are typically protein kinases and phosphatases and are involved in virus assembly and signal transduction. A broader spectrum of different proteins are modified with esterified C_{16} and C_{18} and also some C_{14} and $C_{18:1}$ fatty acids, either to Cys or to Ser(Thr). A large number of viral glycoproteins, including G proteins, are involved, along with receptors and some enzymes. The esterified derivatives can be distinguished from the amide derivatives by their sensitivity to alkaline methanolysis or hydroxylamine. The evidence in this case clearly favors the protein anchoring function of the fatty acid derivatives. Interestingly, there are reports of microsomal esterases in a large number of different cells that have the ability to remove specific acyl groups and thus render membrane-bound proteins soluble (Berger and Schmidt, 1986). Such an enzyme could obviously play a major role in targeting, transport, and protein stability.

The bacterial lipoproteins containing the diacylglyceryl group as a thioether at the N-terminal Cys represent a structure that should fit very well in the membrane bilayer; especially because the α-amino group is acylated in all these derivatives, the N terminus really has the appearance of a somewhat distorted triglyceride. It is highly likely that the principal function of these groups is to anchor the protein to the membrane, using the N-terminal end as the anchoring site. A recent addition to the N-terminal anchors is the diphytanylglyceryl derivative identified in the archaebacteria *Halobacterium halobium* and *Halobacterium cutirubrum*. In this case the lipid is also linked through a thioether link to a Cys residue, which appears to be the N-terminal one. The free OH groups of the glyceryl contain two ether-linked 3′,7′,11′,15′-tetramethylhexadecyl groups instead of fatty acids. This is completely consistent with the fact that the membranes of archaebacteria are made up exclusively of ether-linked diphytyl glyceryl lipids (reviewed by Kates, 1995), and that the diphytyl lipoprotein thus is completely equivalent to the bacterial lipoprotein, both matching the unique composition of their respective membranes. It is a curiosity of evolution that the two classes of lipids are mirror images of each other, the eubacterial ones being *sn*-1,2-diacyl esters, whereas the archaebacterial ones are *sn*-2,3 diphytyl ethers. The conclusions reached from the mass spectrometric data of a copper protein from the haloalkaliphilic arhaeon *Natronobacterium pharaonis* are consistent with a biosynthetic path similar to that of the bacterial lipoprotein. A comparison of the mass of the sequence deduced from cDNA and that obtained

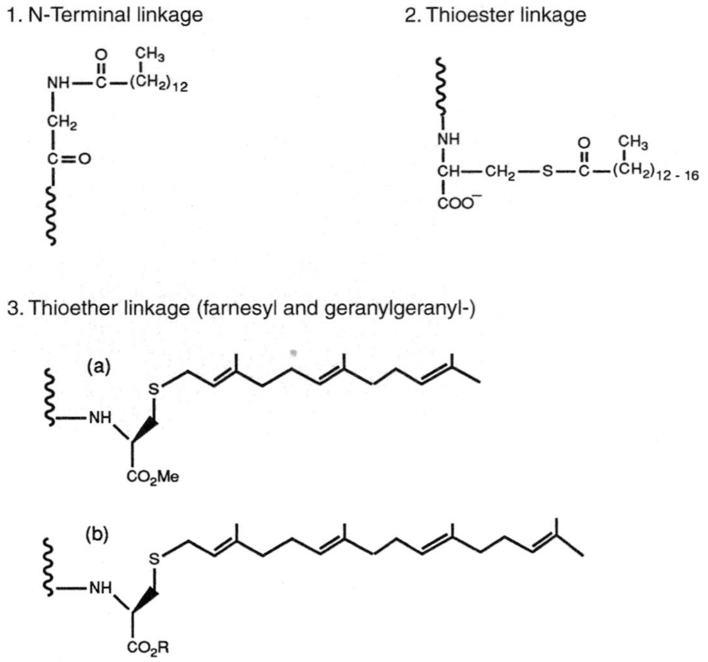

1. N-Terminal linkage

2. Thioester linkage

3. Thioether linkage (farnesyl and geranylgeranyl-)

(a)

(b)

Fig. 10 Protein-bound membrane anchors. These compounds have all been implicated in membrane interactions, either as permanent anchors or as part-time anchors associated with membranes and membrane-bound receptors. Chemically, it may be useful to consider three major families of compounds. One family is made up of *single hydrocarbon chains:* (1) myristoyl and unsaturated C_{14} acyl groups in amide links to the N-terminus, (2) C_{14}–C_{20} fatty acids esterified to Ser/Thr or, as shown in the figure, to Cys, and (3) C_{15} (farnesyl-) and C_{20} (geranylgeranyl-) isoprenyl groups linked by thioether bonds to Cys at or near the C terminus. Another family is made up of *diacyl- or diphytyl-glyceryl derivatives:* (4) 1,2-diacyl-*sn*-glycerol esters linked through a 3-thioether link to Cys at the N-terminal end of the protein (the acyl chain at the α-amino group is shown as a C_{16} chain; it is often shorter), (5) in archaebacteria, 2,3-diphytyl-*sn*-glycerol ethers linked through a 1-thioether link to a Cys (its position is not established, but indirect evidence suggests that it is also located at the N terminus). The third family contains (6) *the complex glycophospholipids* linked to the α-carboxyl of C-terminal amino acids through an amid to ethanolamine, which in turn is linked through a phosphate diester to a glycan; the glucosamine at the reducing end of this glycan forms a glycosidic link to the inositol moiety of phosphatidylinositol. Similar structures in plant and yeast contain ceramide instead of diacylglycerol (Ferguson and Williams, 1988).

from the isolated N-terminally blocked protein is consistent with the modification of Cys-25 with a diphytyl(C_{20})glyceryl group, the removal of the 24-amino acid signal peptide, and the acetylation of the α-amino group of the modified Cys (Mattar *et al.*, 1994).

The glycosylphosphatidylinositol links in proteins also provide a typical phospholipid as the covalently attached component that should interact smoothly with the phospholipid bilayers, and these derivatives again all appear to be membrane-bound exclusively through the lipid anchor. Again, widely different types of proteins are involved: enzymes, cell surface antigens, adhesion molecules, and a variety of others. It is of interest that enzymes exist, often associated with membranes, that will cleave the anchor from the protein. A

4. Thiether linkage (1,2-diacyl-*sn*-glycerl-)

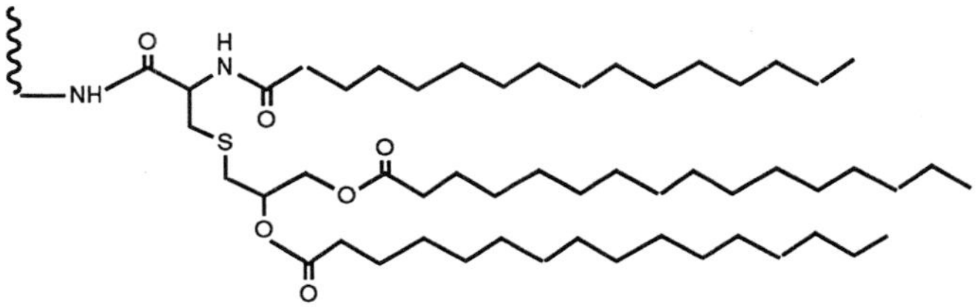

5. Thioether linkage (2,3-diphytyl-*sn*-glyceryl-)

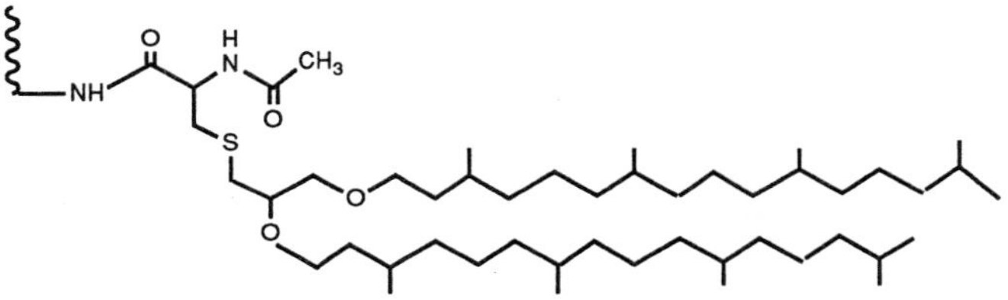

6. COOH- Terminal linkage (glycophospholipids)

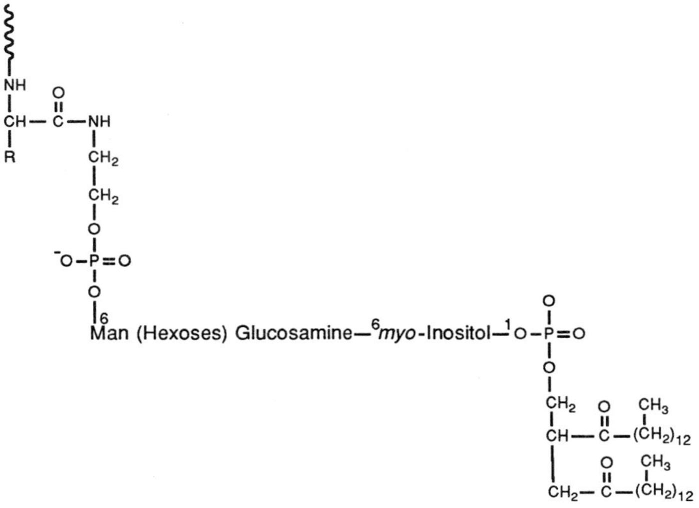

Man (Hexoses) Glucosamine—[6]*myo*-Inositol—[1]O–P=O

Fig. 10 (*continued*)

cell surface phospholipase C, cleaving the glycerol–phosphate bond, is one such activity, permitting release of the protein, but also components such as diacylglycerol and phosphoinositol that may have secondary roles in signal transduction and regulation.

The prenyl proteins in eukaryotic cells represent the final group of possible membrane anchors. When cells are grown in the presence of compactin, which inhibits mevalonic acid biosynthesis, extensive changes in morphology, viability, and protein spectrum have been observed that cannot be reversed by adding cholesterol, but can be reversed by mevalonic acid itself. This implicates nonsteroid products of mevalonic acid as the active principle, and these types of experiments eventually led to the identification of several polyisoprene-derivatized proteins. A large number of labeled proteins can be observed after growth in the presence of [^{14}C]mevalonic acid, but only a few of these have been characterized: yeast mating factors, nuclear lamins, ras proteins, and a protein kinase, as well as other proteins in the retinal phototransduction pathway. In all these cases the isoprenylation and carboxyl methylation at the C-terminal end of the proteins produce a definite hydrophobic site for membrane interaction, and it is presumed that membrane anchoring is the primary function of the modification. As in the case of myristoylation, there appear to be some soluble isoprenylated proteins as well, so the total picture is more complicated. Again it is proposed that interaction with specific receptors may be another function of this modification, and in fact that some membrane associations require the polyisoprenyl group, the methyl group, and a specific amino acid sequence for proper association to take place. Good progress has been made in the understanding of the biosynthesis of these derivatives; the individual specific functional roles will undoubtedly also soon be elucidated.

A final summary point needs to be made regarding the specific role of the many different putative membrane anchors. Accepting the major generalization that integral membrane proteins and protein anchors induce rigidity to the bilayer environment while the lipid anchors simply become parts of the membrane proper and conserve the fluidity and motion of the membrane itself, the obvious questions remain why so many different anchors exist, whether the N-terminal and the C-terminal anchors have unique properties compared to the internal sites that may be involved in fatty acyl ester formation, and how the different hydrocarbons fit in the typical bilayer membrane. One can only speculate about the answers to these questions, but it is intriguing to consider that specific lipid moieties, associated with specific amino acid sequences, for example, may determine which membrane is to be involved in the interaction.

E. Signaling, Recognition, and Structural Amplification: Glycoproteins

The glycoproteins represent another group of posttranslational modification for which we have accumulated a large body of information about biosynthesis and covalent structure, while our understanding of all their biological functions still remains quite incomplete. It has jokingly been remarked that Watson and Crick, by uncovering the universal laws of molecular genetics, have left a terrible scourge on other biochemists, who now naturally have to attempt to show that each new phenomenon has some universal validity. This search for the universality of glycan function has been evident over the last decades; the fact that so many functions have been observed suggests that if there is one fundamental function of glycans in glycoproteins, it has not yet been found. With the increased interest in glycobiology in the past decades (Rademacher *et al.*, 1988), several specific tools have been developed to study the nature and the effects of glycosylation. These tools are summarized in the legend to Fig. 11. Examples of proposed functions for glycoproteins with *N*-linked and *O*-linked glycans will be briefly summarized in the following.

One of the most obvious effects of introducing even a small glycan into the covalent structure of the glycoproteins should be to alter the physical properties of the amino acid residue involved, to grossly affect its packing

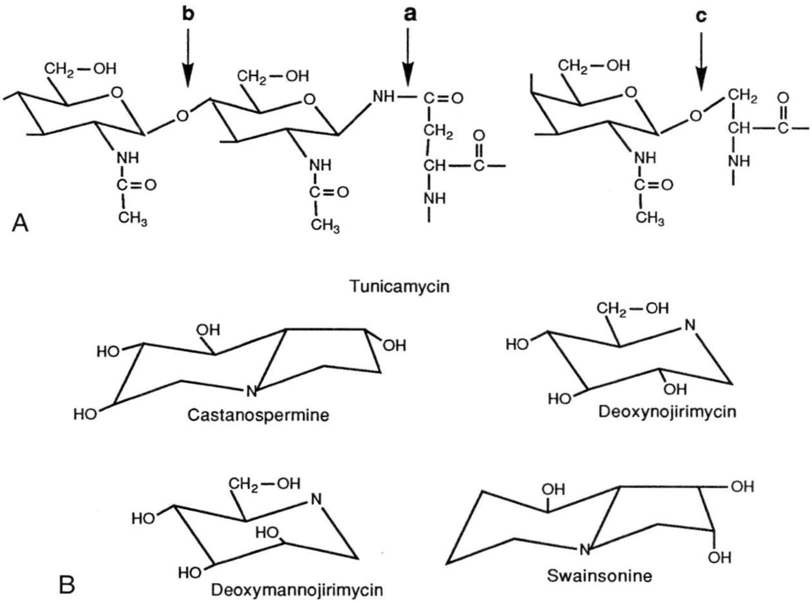

Fig. 11 Some useful tools in the study of glycoprotein structure and function. (A) Endoglycosidases that can be used to release glycans *in vitro* and several specific exo-glycosidases and glycosyltransferases (some of which are shown in Fig. 2) that will alter the covalent structure of the protein-bound glycans have been purified and well characterized, and many are also commercially available. Three endoglycosidases are illustrated in the figure: (a) *N*-Glycanase (peptide-N^4-(*N*-acetyl-β-glucosaminyl)asparagine amidase) cleaves the amide bond releasing the intact glycan, ammonia (released spontaneously), and the protein containing Asp instead of glycosyl-Asn. The enzyme is quite nonspecific and will release different *N*-linked glycans from native glycoproteins. (b) Endoglycosidase H (endo-β-acetylglucosaminidase) cleaves the glycosidic bond between the two inner GlcNAc of *N*-linked glycans. It has a fairly limited specificity, acting on most oligomannose and hybrid glycans, but not on complex ones (see Fig. 2). (c) *O*-Glycanase (endo-α-*N*-acetylgalactosaminidase) catalyzes the release of the common disaccharide Gal-β(1,3)-GalNAc linked to Ser and Thr (see Fig. 5). The sialic acid often attached to the disaccharide must be removed for the enzyme to be active. (B) Several inhibitors that block glycosylation or individual processing steps have been established (Elbein, 1991) and are commercially available. The following ones are illustrated in the figure: Tunicamycin inhibits the first step (dolechyl-P + UDP-GlcNAc $\rightarrow$ dolechyl-P-P-GlcNAc) in the assembly of the complete glycan precursor (shown in the upper part of Fig. 2) on the dolechol donor. Tunicamycin has been of great use in *in vivo* studies in which nonglycosylated forms of glycoproteins are produced for the comparison of their phenotypic expressions to those of the natural glycosylated forms. The glucosidases (I and II) involved in the removal of the three Glc residues in the first step of the glycan processing reactions are not indicated in Fig. 2; the mannosidases inhibited are shown in Fig. 2. The glycosidase inhibitors apparently act as competitive inhibitors, and *in vitro* it has been found that the ratio of substrate and inhibitor must be fairly low to completely block a given reaction. Other examples include: (1) A large number of lectins (defined as proteins other than enzymes and immunoglobulins, that bind sugars and glycans with high a degree of specificity), primarily from plant sources, have been purified and their specificity in recognizing specific glycan structures and changes in glycan structures has been determined. (2) Artificial glycoprotein models have been synthesized and characterized with respect to their interaction with receptors and glycan processing enzymes (Lee and Lee, 1994). (3) Mutants that are missing specific enzymes in the biosynthetic processing have been developed to explore the phenotypic expression of the different steps in glycosylation and glycan processing (Stanley and Ioffe, 1995).

characteristics and intramolecular interactions in the interior of the protein molecule and solvent and both intra- and intermolecular interactions on the exterior of the molecule. In the extreme, the effect of a large number of glycans, such as in the highly *O*-glycosylated stretches in mucins and in some membrane glycoproteins, is to force the polypeptides with 50 residues or more into an extended conformation with very little flexibility, closely approximating a rigid rod. In the case of membrane-bound enzymes and receptors, it has been proposed that the function of such glycosylated stretches is to push the functional domain of the protein away from the crowded cell surface (Jentoft, 1990). Such highly glycosylated polypeptide chains have other altered physical characteristics, such as solubility and hydrodynamic properties (e.g., the high viscosity of mucins), and are generally quite resistant to proteases. The effect of glycosylation is often much less dramatic than this, and the structural and functional impact of introducing a single glycan on the surface of a protein may in fact not be readily observable. Thus, a direct comparison of the structure and function of a well-characterized glycoprotein, bovine pancreatic ribonuclease B, with its nonglycosylated form, ribonuclease A (the nonglycosylated form, A, is the major one, but several glycosylated forms are also produced in the bovine pancreas; B is the major glycosylated form at about 20% of the total enzyme), shows that the *N*-linked glycan moiety is disorganized and does not appear to contribute to structure, since no significant differences in the protein backbone conformation are observed (Williams *et al.*, 1987). However, very subtle differences in enzymatic activity can be detected for the different glycoforms of ribonuclease, and the dynamic stability of the folded structure, as measured by amide proton exchange rates, is found to be greater for ribonuclease B than for ribonuclease A (Rudd *et al.*, 1994). Similarly, IgG antibodies normally have two complex glycans *N*-linked to the asparagine 297 side chains of the C_H2 domains. These glycans are spread over the surfaces of the four-stranded β-pleated sheets in each protomer and thus disrupt normal protomer–protomer interactions in the glycosylated region of the native dimer; the glycan chains appear to interact with each other, however (Guddat *et al.*, 1993). Removal or alteration of the glycans does not alter the basic IgG structure, but significantly alters the effector functions of antibodies (Nose and Wigzell, 1983). Thus, the partial or complete deglycosylation results in reduction of the binding affinity of IgG antibodies for Fc receptors on monocytes and macrophages, of the antibody-dependent cellular cytotoxicity, and of the complement-dependent cytotoxicity. Although most of the rather few glycoproteins that have been subjected to detailed structural analysis are like ribonuclease and show disorganized glycan structures out in the solvent phase with only the inner one or two sugars being restricted by protein interactions, there are examples of the glycan becoming an integral part of the folded completed structure. One of the very few clear-cut examples of this kind of glycan–protein interaction has been observed in the galactose-specific lectin from *Erythrina corallodendron*, which was crystallized as the complex with the lactose ligand (Shaanan *et al.*, 1991). In this glycoprotein, the glycan [the heptasaccharide Man_3-(Xyl-)GlcNAc-(Fuc-)GlcNAc-] is firmly held by intra- and intermolecular hydrogen bonds so that all seven sugars are clearly seen in the electron density map. It is of interest to note that other lectins, such as that from soybeans (Dessen *et al.*, 1995), show the more general properties of permitting observation by X-ray crystallography of only the two GlcNAc of the single N-linked glycan, a homogeneous $Man_9GlcNAc_2$-, attached to each subunit; the rest of the glycan is "invisible" out in the solvent phase, well away from the protein. There is no obvious functional difference between the two lectins, but the strategy for obtaining the lectin–ligand complex differs. In the *Erythrina* lectin, the protein interactions have stabilized a single conformation of the flexible structure of the glycan, and similarly, the glycan has modified protein structure to the extent that the dimer is drastically different from that of the other plant lectins, with which it shares very similar tertiary structures. With the glycan at the surface of what would normally be part of the dimer interface, the dimer must assume a unique structure to permit the participation of the glycan and conserve the ligand-binding properties.

It can be concluded from the preceding considerations that the effect of glycans on the final structure of proteins is not as great as one might expect. Especially for the *N*-linked ones, where the glycans often are attached at relatively nonstructured loops and extend into the solvent phase, the effect, while significant at the site of modification, may be minimal for the total protein. That does not mean that the glycans cannot have dramatic effects on the dynamics of acquiring the folded structure. Since the glycosylation takes place on the nascent chain, the glycan is present during folding of the completed chain and can exert its unique physical and chemical properties on that process. The biosynthesis of insulin-like growth factor provides an excellent illustration of that kind of effect. The protein is synthesized as a glycosylated protomer containing internal disulfide bonds. During the processing, two protomers associate and get linked together by disulfide bonds through the action of disulfide isomerase. Each protomer is next cleaved by proteolysis to give a pair of disulfide-linked (α and β) chains, and the mature, active receptor then consists of a single covalent unit made of two α and two β chains, with the latter two containing the transmembrane sequences linking the receptor to the membrane. If the glycans are removed (enzymatically) from this active receptor, there is no effect on activity. However, if the biosynthesis takes place in the presence of the glycosylation inhibitor tunicamycin, receptor is produced without any biological activity, and it is proposed that the glycans are required to direct the folding and disulfide interchange that lead to the mature, active receptor (Olson and Lane, 1989). This type of transient function directing the biosynthetic processing of glycoproteins may well be an important function of the glycans; it would certainly be consistent with the frequently encountered difficulties in assigning specific functions to the glycans in the final products. The observation made in this case that the alteration of the glycan does not affect biological activity is quite common, but by no means universal. In the case of the glycosylated gonadotropic hormone, the *N*-linked glycans appear to be homogeneous at each glycosylation site, a very unusual feature, since glycoprotein in general show heterogenity at every site. If these glycans are modified enzymatically, the hormone still binds to its receptor, but its biological activity, the coupling of receptor binding to the activation of adenyl cyclase, is abolished (see Rademacher *et al.*, 1988, for a summary).

With so many of the glycans situated on the surface and in the solvent phase of glycoproteins, another anticipated function might be in signalling to and interacting with the environment, and a huge volume of data document such functional properties. As indicated previously, it appears that many of these examples of glycan-directed molecule–molecule, molecule–cell, and cell–cell interactions, if expressed at all, are expressed to a different extent and in different ways for individual glycoproteins, and thus do not seem to demonstrate a universal property of all glycoproteins. In the following, a brief overview of some of the possible functional roles of the glycans in glycoproteins in this area will be presented.

Plant lectins with specificity for monosaccharides, as well as for different glycan structures, have been known for just about 100 years, and have been studied extensively, chiefly as laboratory tools, during the past 30 or so years. They bind their sugar ligands with high affinity and specificity; in legume seeds they are present at extremely high levels, often looking more like storage proteins than functional entities. It has been suggested that the legume lectins in root tissue may play a role in the interaction with specific *Rhizobium* species to establish the symbiotic nitrogen-fixing system. The process is certainly much more complicated than a simple plant–*Rhizobium* interaction, and the suggestion does not provide functions for nonlegumous plant lectins. Although the roles of the lectins in the plants themselves are still uncertain, the plant lectins represent an obvious link in the chain of events that establishes the glycans as participants in recognition processes. The observation about 25 years ago of similar lectin-like activities in vertebrate as well as invertebrate animals in the form of either soluble molecules or membrane-associated, sugar-specific receptors probably provided the impetus for the exciting renaissance of carbohydrate science that is still going on. In contrast to the large number of plant lectins for which no definite function could be elucidated, the animal lectins turned out to have measurable biological effects and could be classified in terms of both structure and function (Drickamer, 1988). Lectins / receptors with different sugar specificities were established first in liver (Ashwell and Harford, 1982) and then in several other vertebrate organs and tissues (see Weigel, 1992, for an overview), and many of these different sugar-recognizing activities could be linked to specific biological functions. Impressive progress has been made in recent years toward our understanding of the structural bases of specific glycan–lectin interactions (Weis and Drickamer, 1996).

One of the early experiments to show that glycans are involved in cell–cell recognition was the demonstration that when cells from mechanically fragmented sponges of different colors were mixed, they reassembled into separate colonies for each color. The species-specific assembly required the presence of a naturally occurring proteoglycan; if the proteoglycan was removed or if free glucuronic acid was added, the aggregation did not take place. Similar observations of specific association of homologous cells have been made after mixing two different embyronic tissues or two samples of the same tissue at different age. In a number of vertebrate tissues, a family of cell adhesion molecules (CAMs) have been observed. They are cell surface glycoproteins and membrane-associated, and the facts that homophilic binding involves the same CAM on the surface of the interacting cells, and that some of the CAMs only appear after certain stages in embryogenesis, strongly suggest that the developmental cell sorting at least in part involves specific glycans (Edelman, 1985). The fact that it is generally found that changes in the glycan components of glycoproteins are associated with cellular development and differentiation, without affecting cell viability, is clearly consistent with the foregoing observations. Interaction of cells with the proteoglycans of the extracellular matrix is mediated by glycoproteins such as fibronectin and laminin and also involve specific glycan interactions, less well understood in this case (Höök *et al.*, 1984). Another cell–cell interaction such as the mutual specific recognition of sperm and egg in sea urchin is well documented as involving an acrosomal lectin, bindin, the specificity of which ensures proper intraspecies matching. Similar studies on the acrosomal protein of abalone show that the gamete recognition process is more complex; the acrosomal protein, lysin, in this case creates a hole in the vitelline envelope through which the sperm can reach the egg. The process is species specific and involves glycoproteins in the vitelline envelope, but the binding is proposed to involve hydrophobic interactions (Vacquier *et al.*, 1990).

The involvement of cell-surface receptors in the clearance of glycoproteins from circulation is also very thoroughly documented in experiments in which injected asialoglycoproteins and asialoagalactoglycoproteins and oligomannose-containing glycoproteins are cleared by different cells containing the appropriate receptors, such as hepatocyte Gal-receptors, macrophage Man/GlcNAc-receptors, and nonparenchymal liver cell Man-receptors. In these cases, the circulating glycoproteins are generally observed to be internalized and destroyed in the lysosomes (Weigel, 1992). It is not at all clear that this clearance is the primary role for all the different receptors *in vivo*, nor is there any direct evidence for the proposal that these receptors are involved in the clearance of antibody–antigen complexes from the circulation. Some of the lectins/receptors are likely to be involved in intracellular sorting and perhaps some very special endocytotic processes. Thus, there are receptors specific for Man 6-phosphate found in most cells. A low molecular weight form is involved in bringing Man 6-P-marked lysosomal enzymes to the lysosomes for residence and action there. If that receptor activity is missing, the lysosomal hydrolases are lost to the extracellular compartment, where they may be recovered by binding to the high molecular weight form, internalized and brought to the lysosomes. It is interesting that this high molecular weight receptor is identical to the insulin-like growth factor receptor. The two ligands, Man 6-P and insulin-like growth factor II, both bind to the receptor, to different sites, since neither one interferes with the other's binding. Another sorting role has been proposed for the Gal receptor in hepatocytes in guiding newly synthesized and processed glycoprotein with exposed Gal residues from the Golgi to the surface, where cell surface sialyltransferases incorporate Sia and thus release the glycoprotein to the extracellular compartment (Weigel, 1992).

One of the puzzles of glycobiology is the fact that the glycans found in glycoproteins often are very heterogeneous, even at a single glycosylation site. How can structures presumably involved in highly specific functions be so "sloppily" expressed? There are several speculative answers to that question. For some possible functions such as solubility, thermal stability, proteolytic resistance, and structural modulations like those in proteoglycans, the actual composition and structure of the glycans may not be important at all. In other cases the glycan "messages" may well be part of dynamic systems in which different structures represent the steady state at the time of measurement or isolation of a mixture of interconvertible forms with different biological properties. If, furthermore, the role of the glycan was expressed during biosynthesis and processing of the glycoprotein long before it was isolated and characterized, the established glycan structure could obviously be totally irrelevant to the function of the finished product. There are clearly a number of questions left to be answered, and the answers may well be different for individual glycoproteins. The fact that the recent structural analyses of carbohydrate-recognizing proteins from animals, plants, and bacteria indicate several groups of unrelated proteins suggests that the trait to recognize and bind carbohydrates has evolved independently in different species at different times.

F. Other Functions

1. Protein Turnover

Spontaneous covalent modification reactions, many of them oxidative alterations of Cys, His, Met, and Tyr, have the effect of marking proteins for destruction (Stadtman, 1990); the actual recognition mechanisms are not well understood, but the modified proteins end up being attacked by individual proteases within or outside the lysosomes or engulfed in high-molecular-weight protease complexes, proteasomes, that catalyze their hydrolysis. One pathway of protein destruction involves ATP-dependent proteases, especially in bacteria. Another well-understood pathway involves the ubiquitination of Lys residues in the protein to be destroyed; it is the ubiquitinated derivative that is the substrate for proteasomes, again in an ATP-dependent reaction (Hershko, 1988, 1991). In the first step of this latter pathway, the C-terminal Gly residue of the 76-amino acid polypeptide ubiquitin is esterified to a Cys SH in enzyme E1 of the ubiquitin–peptide ligase complex; ATP is required for this reaction. In the next step, the ubiquitin moiety is transferred to E2, again as a thioester. The activated ubiquitinyl–E2 complex can be used in two distinct ways: It can be used to ubiquitinate Lys ε-amino groups either in histones in the nucleus, or in cytosolic proteins complexed with E3 of the ligase complex. The step in which E3 binds a protein must be the initiation of the process leading to the destruction of that protein, and must mean that E3 somehow can recognize some feature of the protein that marks it for destruction. The nature of those features is not understood. In the case of the histones a single or at most a few ubiquitin molecules are introduced; in the case of E3–protein complexes the protein is generally polyubiquitinated. The extra ubiquitin units may be added to the protein itself, but most of them are added to ubiquitin units already attached, to form strings of ubiquitin. This polyubiquitinated conjugate is finally broken down by proteasomes (ATP required) to yield peptides and regenerated ubiquitin molecules. It is possible that the ubiquitination of some proteins has other functions than degradation; the observation that ubiquitin in the ubiquitin–histone complexes turns over rapidly, while the histone is quite stable, even suggests that the reaction in this case may be reversible and have some regulatory function. The signals by which E3 recognizes the protein destined for destruction are probably quite complex. One of them has been proposed to involve the N-terminal amino acid residue (Varshavsky *et al.*, 1989). The proposal is based on experiments comparing the *in vivo* half-life of β-galactosidase constructs containing a 40-residue N-terminal extension in which only the N-terminal amino acid is varied. If the residue has a small radius of gyration (see Section II, B), the protein is stable; if it has a large radius of gyration, the half-life of the protein is on the order of 2–30 min. With an N-terminal Glu or Asp, an Arg can be transferred from Arg-tRNA (see II, B) to become the new N terminus, giving a half-life of a few minutes. Acetylated proteins are stable; they should be, since the acetylation involve the same small radius residues. There are exceptions to this general rule in terms of the rate of protein turnover; one possibility is that the 40 residue extension, which has very little structure, is part of the total signal, but it is most likely that the total signal for ubiquitination is more complicated. It is also clear that there must be many other signals that determine the *in vivo* stability of proteins.

2. *Miscellaneous*

Several posttranslational reactions that are quite abundant in nature are even more puzzling in terms of their functional significance than some of those discussed previously. Among these are proteins sulfated on Tyr residues and proteins methylated at a variety of residues. In the case of sulfation it has been estimated that about 1% of protein Tyr are sulfated in multicellular, but apparently not in unicellular organisms; the sulfate transfer takes place in the trans-Golgi, and sulfate is found in many, but not at all, secreted proteins. A variety of functions have been proposed for the sulfation, the introduction of what should be essentially a permanent negative charge on the protein-bound Tyr sulfate; the proposals are very difficult to test because it is so difficult to prepare unsulfated control samples for comparison. By the use of site-directed mutagenesis and the use of the competitive inhibitor chlorate, some sulfate-free proteins have been produced and evaluated. In most of them the activity, whether catalysis or intermolecular interaction, is greatly diminished when the sulfate is absent, so it is clear that the sulfate is important in the biosynthetic processing (folding), in modifying the activity through conformational alterations of the final product, or by direct involvement of the sulfate in the biological function of the protein. The favored proposal at this time appears to be that the key role of sulfate is in stabilizing or destabilizing intermolecular interactions (Niehrs *et al.*, 1994).

The methylation reactions are probably even more abundant in that they occur in all prokaryotic and eukaryotic species and involve several different residues such as the side chains of Asp and Glu (carboxylmethyl esters), Cys and Ser/Thr (*S*- and *O*-ethers), Arg, His and Lys (a variety of *N*-linked mono-, di-, and trimethyl derivatives), and at the free N terminus (N^{α}-methylamines). As mentioned in Section II, C, the carboxylmethyl esters have been implicated in specific functions such as bacterial chemotaxis and ameba movement for Me-Glu and rescue of or indeed the marking for destruction of spontaneously produced D-Asp, or of Asp or Asn spontaneously modified from a normal peptide bond to an isopeptide bond. For the *N*-methylated derivatives, on the other hand, there is virtually no good function documented. Nonmethylated analogs of methylated proteins such as histones, ribosomal proteins, enzymes, binding proteins, and some structural proteins have been produced by mutation or by comparing the same protein produced in different species, sometimes produced with and sometimes without the methyl groups (e.g., N^{ε}-trimethyl-Lys in some cytochrome *c*'s and in calmodulin but not in related proteins), but it has been extremely difficult to assign specific phenotypic traits to the methylation process. There are clear, but rather minor, effects on activity, so the real effects must be quite subtle and perhaps also quite species specific.

The biosynthesis of the active thyroid hormones 3,5,3′-triiodothyronine and 3,5,3′,5′-tetraiodothyronine (thyroxine) represents one set of modification reactions that can be assigned a very specific biological role. As discussed in Section II, C, iodination and ether formation between two iodinated Tyr residues take place after the 180-kDa protein thyroglobulin has been synthesized and released into the lumen of the thyroxine-producing cells, and to produce the actual product of the reaction, the thyroglobulin molecule must be hydrolyzed and the protein-bound hormone released. It is indeed gratifying to understand the biological reason for this set of reactions, but at the same time there is an important lesson to be learned from the thyroxine biosynthesis: Nature often must use circuitous and complex ways to achieve goals, such as, in this case, synthesizing a large protein for the production, storage, and highly regulated release of a small modified amino acid as a hormone. Posttranslational reactions in general exemplify this principle of Nature's often inscrutable ways of developing amazingly versatile responses to life's challenges.

Acknowledgment

Aspects of the preparation of this chapter and some of the work reported were supported by grants from The Robert A. Welch Foundation (AU-0916 and AU-0009).

References

Adamietz, P., and Hilz, H. (1984). Purification and characterization of (ADP-ribosyl)$_n$ proteins. *Meth. Enzymol.* **106,** 461–471.

Adams, J. M., and Capecchi, M. R. (1966). *N*-Formylmethionyl-sRNA as the initiator of protein synthesis. *Proc. Natl. Acad. Sci. USA* **55,** 147–155.

Adler, S. P., Purich, D., and Stadtman, E. R. (1975). Cascade control of *E. coli* glutamine synthetase: Properties of the P11 regulatory protein and the uridylyltransferase–uridylyl removing enzyme. *J. Biol. Chem.* **250,** 6264–6272.

Aeschbach, R., Amado, R., and Neukom, H. (1976). Formation of dityrosine cross-links in proteins by oxidation of tyrosine residues. *Biochim. Biophys. Acta* **439,** 292–301.

Alix, J.-H. (1988). Post-translational methylations of ribosomal proteins. *In:* "Advances in Post-translational Modification of Proteins and Aging" (V. Zappia, P. Galletti, R. Porta, and F. Wold, eds.), pp. 213–228. Plenum Press, New York.

Allfrey, V. G., DiPaola, E. A., and Sterner, R. (1984). Protein side-chain acetylations. *Meth. Enzymol.* **107,** 224–240.

Althaus, F. R., Hilz, H., and Shall, S. (eds.) (1985). "ADP-Ribosylation of Proteins," pp. 1–580. Springer-Verlag, Berlin.

Amado, R., Aeschbach, R., and Neukom, H. (1984). Dityrosine: *In vitro* production and characterization. *Meth. Enzymol.* **107,** 377–388.

Ambros, V., and Baltimore, D. (1978). Protein is linked to the 5′ end of polio virus RNA by a phosphodiester linkage to tyrosine. *J. Biol. Chem.* **253,** 5263–5266.

Andersen, S. O. (1963). Characterization of a new type of cross-linkage in resilin—a rubber-like protein. *Biochim. Biophys. Acta* **69,** 249–262.

Anfinsen, C. B. (1973). Principles that govern the folding of protein chains. *Science* **181,** 223–230.

Arfin, S. M., and Bradshaw, R. A. (1988). Cotranslational processing and protein turnover in eukaryotic cells. *Biochemistry* **27,** 7979–7984.

Ashwell, G., and Harford, J. (1982). Carbohydrate-specific receptors in the liver. *Ann. Rev. Biochem.* **51,** 531–554.

Aswad, D. W. (1984). Stoichiometric methylation of porcine adrenocorticotropin by protein carboxyl methyltransferase requires deamidation of Asn 25: Evidence for methylation at the α-carboxyl group of atypical L-isoaspartyl residues. *J. Biol. Chem.* **259,** 10714–10719.

Augen, J., and Wold, F. (1986). How much sequence information is needed for the regulation of amino-terminal acetylation of eukaryotic proteins? *Trends Biochem. Sci.* **11,** 494–497.

Awade, A., Cleuziate, P., Gonzales, T., and Robert-Baudary, J. (1994). Pyrrolidone carboxyl peptidase (PCP): An enzyme that removes pyroglutamic acid (PGlu) from PGlu-peptides and PGlu-proteins. *Proteins* **20,** 34–51.

Bada, J. L. (1984). *In vivo* racemization in mammalian proteins. *Meth. Enzymol.* **106,** 98–115.

Baldwin, R. L. (1975). Intermediates in protein folding reactions and the mechanism of protein folding. *Ann. Rev. Biochem.* **44,** 453–475.

Bastide, F., Meissner, G., Fleischer, S., and Post, R. L. (1973). Similarity of the active site of phosphorylation of the adenosine triphosphatase for transport of sodium and potassium ions in kidney to that for transport of calcium ions in the sarcoplasmic reticulum of muscle. *J. Biol. Chem.* **248,** 8385–8391.

Baynes, J. W., Thorpe, S. R., and Murtiashaw, M. H. (1984). Nonenzymatic glucosylation of lysine residues in albumin. *Meth. Enzymol.* **106,** 88–98.

Ben-Bassat, A., Bauer, K., Chang, S. Y., Myambo, K., Boosman, A., and Chang, S. (1987). Processing of the initiation methionine from proteins: Properties of the *Escherichia coli* methionine aminopeptidase and its gene structure. *J. Bact.* **169,** 751–757.

Beninati, S., Piacentini, M., Ceru-Argento, M. P., Russo-Caia, S., and Autuori, F. (1985). Presence of di- and polyamines covalently bound to protein in rat liver. *Biochim. Biophys. Acta* **841,** 120–126.

Berger, M., and Schmidt, M. F. G. (1986). Characterization of a protein fatty acylesterase present in microsomal membranes of diverse origin. *J. Biol. Chem.* **261,** 14912–19918.

Bodley, J. W., Dunlop, P. C., and VanNess, B. G. (1984). Diphthamide in elongation factor 2: ADP-ribosylation, purification and properties. *Meth. Enzymol.* **106,** 378–387.

Bogosian, G., Violand, B. N., Dorward-King, E. J., Workman, W. E., Jung, P. E., and Kane, J. F. (1989). Biosynthesis and incorporation into protein of norleucine by *Escherichia coli. J. Biol. Chem.* **264,** 531–539.

Bornstein, P. (1974). The biosynthesis of collagen. *Ann. Rev. Biochem.* **43,** 567–603.

Bowles, A. D. J., and Pappin, D. J. (1988). Traffic and assembly of concanavalin A. *Trends Biochem. Sci.* **13,** 60–64.

Bradbury, A. F., and Smyth, D. G. (1991). Peptide amidation. *Trends Biochem. Sci.* **16,** 212–215.

Brot, N., Fliss, H., Coleman, T., and Weissbach, H. (1984). Enzymatic reduction of methionine sulfoxide residues in proteins and peptides. *Meth. Enzymol.* **107,** 352–360.

Buchner, J. (1996). Supervising the fold: Functional principles of molecular chaperones. *FASEB J.* **10,** 10–19.

Carlisle, T. L., and Suttie, J. W. (1980). Vitamin K-dependent carboxylase: Subcellular location of the carboxylase and enzymes involved in vitamin K metabolism in rat liver. *Biochemistry* **19,** 1161–1167.

Carr, S. A., Biemann, K., Shoji, S., Parmelee, D. C., and Titani, K. (1982). *n*-Tetradecanoyl is the NH_2 terminal blocking group of the catalytic subunit of cyclic AMP-dependent protein kinase from bovine cardiac muscle. *Proc. Natl. Acad. Sci. USA* **79,** 6128–6131.

Casey, P. J., and Buss, J. E. (eds.) (1995). Lipid modifications of proteins. *Meth. Enzymol.* **250,** 1–754. Academic Press, San Diego.

Chen, J.-Y. C., and Bodley, J. W. (1988). Biosynthesis of diphthamide in *Saccharomyces cerevisiae.* Partial purification and characterization of a specific S-Ado Met-EF2 methyl transferase. *J. Biol. Chem.* **263,** 11692–11696.

Chen, R., and Chen-Schmeisser, U. (1977). Isopeptide linkage between N^{α}-monomethylalanine and lysine in ribosomal protein S11 from *E. coli. Proc. Natl. Acad. Sci. USA* **74,** 4905–4908.

Chen, C. C., Bruegger, B. B., Kern, C. W., Lin, Y.-C., Halpern, R. M., and Smith, R. A. (1977a). Phosphorylation of nuclear proteins in rat regenerating liver. *Biochemistry* **16,** 4852–4855.

Chen, R., Brosius, J., Wittmann-Liebold, B., and Schafer, W. (1977b). Occurrence of methylated amino acids as N-termini of proteins from *E. coli* ribosomes. *J. Mol. Biol.* **111,** 173–181.

Chen, J.-Y. C., Bodley, J. W., and Livingston, D. M. (1985). Diphtheria-toxin resistant mutants of *Saccharomyces cerevisiae. Mol. Cell. Biol.* **5,** 3357–3362.

Chock, P. B., Rhee, S. G., and Stadtman, E. R. (1980). Interconvertible enzyme cascades in cellular regulation. *Ann. Rev. Biochem.* **49,** 813–843.

Clarke, S. (1985). Protein carboxyl methyltransferases: Two distinct classes of enzymes. *Ann. Rev. Biochem.* **54,** 479–506.

Clarke, S. (1988). Perspectives on the biological function and enzymology of protein carboxyl methylation reactions in eukaryotic and procaryotic cells. *In:* "Advances in Post-translational Modification of Proteins and Aging" (V. Zappia, P. Galletti, R. Porta, and F. Wold, eds.), pp. 213–228. Plenum Press, New York.

Clarke, S. (1992). Protein isoprenylation and methylation at carboxyterminal cysteine residues. *Annu. Rev. Biochem.* **61,** 355–386.

Cohen, P. (1989). The structure and regulation of protein phosphatases. *Ann. Rev. Biochem.* **58,** 453–508.

Cohen, R. E., Ballou, L., and Ballou, C. E. (1980). *Saccharomyces cerevisiae* mannoprotein mutants: Isolation of the mnn5 mutant and comparison with the mnn3 strain. *J. Biol. Chem.* **255,** 7700–7707.

Creighton, T. E. (1984). Disulfide bond formation in proteins. *Meth. Enzymol.* **107,** 305–329.

Danley, D. E., Strick, C. A., James, L. C., Lanzetti, A. J., Otterness, I. G., Grenett, H. E., and Fuller, G. M. (1991). Identification and characterization of a C-terminally extended form of recombinant murin IL-6. *FEBS Lett.* **283,** 135–139.

Daumy, G. O., Merenda, J. M., McColl, A. S., Andrew, G. C., Franke, A. E., Geoghegan, K. F., and Otterness, I. G. (1989). Isolation and characterization of biologically active murin interleukin-1α derived from expression of a synthetic gene in *E. coli. Biochim. Biophys. Acta* **998,** 32–42.

Davankov, V. A., Bochkov, A. S., Kurganov, A. A., Roumeliotis, P., and Unger, K. K. (1980). Separation of unmodified α-amino acid enantiomers by reverse phase HPLC. *Chromatographia* **13,** 677–685.

de Beer, T., Vliegenthart, J. F. G., Löffler, A., and Hofsteenge, J. (1995). The hexopyranosyl residue that is C-glycosidically linked to the side chain of tryptophan-7 in human RNase U_S is α-mannopyranose. *Biochemistry* **34,** 11785–11789.

Decker, K. (1982). Biosynthesis of covalent flavoproteins. *In:* "Flavins and Flavoproteins" (V. Massey, and C. H. Williams, eds.), pp. 465–472. Elsevier, Amsterdam.

Degani, C., and Boyer, P. D. (1973). A borohydride reduction method for characterization of the acyl phosphate linkage in proteins and its application to sarcoplasmic reticulum adenosine triphosphatase. *J. Biol. Chem.* **248,** 8222–8226.

De Jong, W. W., Mulders, J. W. M., Voorter, C. E. M., Berbers, G. A. M., Hoekman, W. A., and Bloemendal, H. (1988). Post-translational modifications of eye lens crystallins: Cross-linking, phosphorylation and deamidation. *In:* "Advances in Post-translational Modification of Proteins and Aging" (V. Zappia, P. Galletti, R. Porta, and F. Wold, eds.), pp. 95–108. Plenum Press, New York.

Dessen, A., Gupta, D., Sabesan, S., Brewer, C. F., and Sacchettini, J. C. (1995). X-ray crystal structure of the soybean agglutinin cross-linked with a biantennary analog of the blood group I carbohydrate antigen. *Biochemistry* **34,** 4933–4942.

DeVore, D. P., and Gruebel, R. J. (1978). Dityrosine in adhesive formed by the sea mussel, *Mytilus edulis. Biochem. Biophys. Res. Commun.* **80,** 993–999.

Diedrich, D. L., and Schnaitman, C. A. (1978). Lysyl-derived aldehydes in outer membrane proteins of *E. coli. Proc. Natl. Acad. Sci. USA* **75,** 3708–3712.

Docherty, K., and Steiner, D. F. (1982). Post-translational proteolysis in polypeptide hormone biosynthesis. *Ann. Rev. Physiol.* **44,** 625–638.

Drakenberg, T., Fernlund, P., Roepstorff, P., and Stenflo, J. (1983). β-Hydroxyaspartic acid in vitamin K-dependent protein C. *Proc. Natl. Acad. Sci. USA* **80,** 1802–1806.

Drickamer, K. (1988). Two distinct classes of carbohydrate-recognition domains in animal lectins. *J. Biol. Chem.* **263,** 9557–9560.

Durban, E., Nochumson, S., Kim, S., Paik, W. K., and Chan, S.-K. (1978). Cytochrome C-specific protein-lysine methyltransferase from *Neurospora crassa:* Purification, characterization and substrate requirements. *J. Biol. Chem.* **253,** 1427–1435.

Duronio, R. J., Jackson-Machelski, E., Heuckeroth, R. O., Olins, P. O., Devine, C. S., Yonemoto, W., Slice, L. W., Taylor, S. S., and Gordon, J. I. (1990). Protein *N*-myristoylation in *Escherichia coli:* Reconstitution of a eukaryotic protein modification in bacteria. *Proc. Natl. Acad. Sci. USA* **87,** 1506–1510.

Eddé, B., Rossier, J., Le Caer, J. P., Desbruyeres, E., Gros, F., and Denoulet, P. (1990). Posttranslational glutamylation of α-tubulin. *Science* **247,** 83–87.

Edelman, G. M. (1985). Cell adhesion and the molecular processes of morphogenesis. *Ann. Rev. Biochem.* **54,** 135–169.

Elbein, A. D. (1991). Glycosidase inhibitors: Inhibitors of *N*-linked oligosaccharide processing. *FASEB J.* **5,** 3055–3063.

Ellis, R. H., and Hartl, F. U. (1996). Protein folding in the cell: Competing models of chaperonin function. *FASEB J.* **10,** 20–26.

Englund, P. T. (1993). The structure and biosynthesis of glycosylphosphatidylinositol protein anchors. *Ann. Rev. Biochem.* **62,** 121–138.

Eyre, D. (1987). Collagen cross-linking amino acids. *Meth. Enzymol.* **144,** 115–139.

Eyre, D., Paz, M. A., and Gallop, P. M. (1984). Crosslinking in collagen and elastin. *Ann. Rev. Biochem.* **53,** 717–748.

Farnsworth, C. C., Seabra, M. C., Ericsson, L. H., Gelb, M. H., and Glomset, J. A. (1994). Rab geranylgeranyl transferase catalyzes the geranylgeranylation of adjacent cysteines in the small GTPases Rab1A, Rab3A, and Rab5A. *Proc. Natl. Acad. Sci. USA* **91,** 11963–11967.

Faulstich, H., Buku, A., Bodenmuller, H., and Wieland, T. (1980). Virotoxins: Actin binding cyclic peptides of *Amarita virosa* mushrooms. *Biochemistry* **19,** 3334–3343.

Fenton, W. A., Kashl, Y., Furtak, K., and Horwich, A. L. (1994). Residues in chaperonin GroEL required for polypeptide binding and release. *Nature (London)* **371,** 614–619.

Ferguson, M. A. J., and Williams, A. F. (1988). Cell-surface anchoring of proteins via glycosyl–phosphatidylinositol structures. *Ann. Rev. Biochem.* **57,** 285–320.

Fesus, L., and Thomazy, V. (1988). Searching for the function of tissue transglutaminase: Its possible involvement in the biochemical pathway of programmed cell death. *In:* "Advances in Post-translational Modification of Proteins and Aging" (V. Zappia, P. Galletti, R. Porta, and F. Wold, eds.), pp. 119–134. Plenum Press, New York.

Flavin, M., and Murofushi, H. (1984). Tyrosine incorporation in tubulin. *Meth. Enzymol.* **106,** 223–237.

Foerder, C. A., and Shapiro, B. M. (1977). Release of ovoperoxidase from sea urchin eggs hardens the fertilization membrane with tyrosine cross-links. *Proc. Natl. Acad. Sci. USA* **74,** 4214–4218.

Folk, J. E. (1980). Transglutaminases. *Ann. Rev. Biochem.* **49,** 517–531.

Folk, J. E. (1983). Mechanism and basis for specificity of transglutaminase-catalyzed N^{ε}-(γ-glutamyl) lysine bond formation. *Adv. Enzymol.* **54,** 1–54.

Freedman, R. B. (1992). Protein folding in the cell. *In:* "Protein Folding" (T. E. Creighton, ed.), pp. 455–539. Freeman Press, New York.

Freedman, R. B., Hirst, T. R., and Tuite, M. F. (1994). Protein disulfide isomerase: Building bridges in protein folding. *Trends Biochem. Sci.* **19,** 331–336.

Fry, S. C. (1984). Isodityrosine, a diphenyl ether cross-link plant cell wall glycoprotein: Identification, assay, and chemical synthesis. *Meth. Enzymol.* **107,** 388–397.

Frydman, J., Nimmesgern, E., Ohtsuka, K., and Hartl, F. U. (1994). Folding of nascent peptide chains in a high molecular mass assembly with molecular chaperones. *Nature (London)* **370,** 111–117.

Fujimoto, D., Moriguchi, T., Ishida, T., and Hayashi, H. (1978). The structure of pyridinoline, a collagen cross-link. *Biochem. Biophys. Res. Commun.* **84,** 52–57.

Fujitaki, J. M., and Smith, R. A. (1984). Techniques in the detection and characterization of phosphoramidate-containing proteins. *Meth. Enzymol.* **107,** 23–36.

Fukae, M., and Mechanic, G. L. (1980). Maturation of collageneous tissue: Temporal sequence of formation of peptidyl lysine-derived cross-linking aldehydes and cross-links in collagen. *J. Biol. Chem.* **255,** 6511–6518.

Furfine, E. S., Leban, J. J., Landavazo, A., Moomaw, J. F., and Casey, P. J. (1995). Protein farnesyltransferase: Kinetics of farnesyl pyrophosphate binding and product release. *Biochemistry* **34,** 6857–6862.

Gallop, P. M., and Paz, M. A. (1975). Posttranslational protein modifications, with special attention to collagen and elastin. *Physiol. Rev.* **55,** 418–472.

Garcia-Castineiras, S., Dillon, J., and Spector, A. (1978). Detection of bityrosine in cataractous human lens protein. *Science* **199,** 897–899.

Gavaret, J.-M., Cahnmann, H. J., and Nunez, J. (1979). The fate of the "lost side chain" during thyroid hormonogenesis. *J. Biol. Chem.* **254,** 11218–11222.

Gehrmann, R., Henschen, A., and Klink, F. (1985). Primary structure of elongation factor 2 around the site of ADP-ribosylation is highly conserved from archaebacteria to eukaryotes. *FEBS Lett.* **185,** 37–42.

Geiger, T., and Clarke, S. (1987). Deamidation, isomerization, and racemization at asparaginyl and aspartyl residues in peptides. Succinimide-linked reactions that contribute to protein degradation. *J. Biol. Chem.* **262,** 785–794.

Gershey, E. L., Haslett, G. W., Vidali, G., and Allfrey, V. G. (1969). Chemical studies of histidine methylation: Evidence for the occurrence of 3-methylhistidine in avian erythrocyte histone fractions. *J. Biol. Chem.* **244,** 4871–4876.

Ghisla, S., Kenney, W. C., Knappe, W. R., McIntire, W., and Singer, T. P. (1980). Chemical synthesis and some properties of 6-substituted flavins. *Biochemistry* **19,** 2537–2544.

Glaser, G. B., and Karic, L. (1976). α-1-Antitrypsin studies on the native and reduced-alkylated protein. *Fed. Proc.* **35,** 1465.

Glazer, A. N. (1984). S^{β}-(Bilin) cysteine derivatives: Structure, spectroscopic properties, and quantitation. *Meth. Enzymol.* **106,** 359–364.

Glomset, J. A., Gelb, M. H., and Farnsworth, C. C. (1990). Prenyl proteins in eukaryotic cells: A new type of membrane anchor. *Trends. Biochem. Sci.* **15,** 139–142.

Goss, N. H., and Wood, H. G. (1984). Formation of N^{ε}-(biotinyl)lysine in biotin enzymes. *Meth. Enzymol.* **107,** 261–278.

Grandhee, S. K., and Monnier, V. M. (1991). Mechanism of formation of the Maillard protein crosslink pentosidine: Glucose, fructose, and ascorbate as pentosidine precursors. *J. Biol. Chem.* **266,** 11649–11653.

Graves, D. J., Martin, B. L., and Wang, J. H. (eds.) (1994). "Co- and Post-translational Modifications of Proteins: Chemical Principles and Biological Effects," pp. 1–339. Oxford University Press, New York.

Greenberg, C. S., Birckbichler, P. J., and Rice, R. H. (1991). Transglutaminases: Multifunctional cross-linking enzymes that stabilize tissues. *FASEB J.* **5,** 3071–3077.

Grinnell, B. W., Hermann, R. B., and Yan, S. B. (1994). Human Protein C inhibits selectin-mediated cell adhesion: role of unique fucosylated oligosaccharide. *Glycobiology* **4,** 221–225.

Guay, M., and Lamy, F. (1979). The troublesome crosslinks of elastin. *Trends Biochem. Sci.* **4,** 160–164.

Guddat, L. W., Herron, J. N., and Edmundson, A. B. (1993). Three-dimensional structure of a human immunoglobulin with a hinge deletion. *Proc. Natl. Acad. Sci. USA* **90,** 4271–4275.

Guptasharma, P., and Balasubramanyam, D. (1992). Dityrosine formation in the proteins of the eye lens. *Current Eye Res.* **11,** 1121–1125.

Gustafson, G. L., and Gander, J. E. (1984). O^{β}-(*N*-Acetyl-α-glucosamine-1-phosphoryl)serine in proteinase I from *Dictyostelium discoideum*. *Meth. Enzymol.* **107,** 172–183.

Hale, G., and Perham, R. N. (1980). Amino acid sequence around lipoic acid residues in the pyruvate dehydrogenase multienzyme complex of *E. coli*. *Biochem. J.* **187,** 905–908.

Hamm, H. H., and Decker, K. (1980). Cell-free synthesis of a flavoprotein containing the 8α-(N^3-histidyl)-riboflavin linkage. *Eur. J. Biochem.* **104,** 391–395.

Hampsey, D. M., Das, G., and Sherman, F. (1986). Amino acid replacements in yeast iso-1-cytochrome *c*: Comparison with the phylogenetic series and the tertiary structure of related cytochrome *c*. *J. Biol. Chem.* **261,** 3259–3271.

Han, K. K., and Martinage, A. (1992). Possible relationship between coding recognition amino acid sequence motif or residue(s) and post-translational chemical modification of proteins. *Intl. J. Biochem.* **24,** 1349–1363.

Hancock, J. F., Magee, A. I., Childs, J. E., and Marshall, C. J. (1989). All *ras* proteins are polyisoprenylated but only some are palmitoylated. *Cell* **57,** 1167–1177.

Hantke, K., and Braun, V. (1973). Covalent binding of lipid to protein: Diglyceride and amid-linked fatty acid at the N-terminal end of the murein-lipoprotein of the *E. coli* outer membrane. *Eur. J. Biochem.* **34,** 284–296.

Hardingham, T. E., and Fosang, A. J. (1992). Proteoglycans: Many forms and many functions. *FASEB J.* **6,** 861–870.

Hart, G. W., Haltiwanger, R. S., Holt, G. D., and Kelly, W. G. (1989). Glycosylation in the nucleus and cytoplasm. *Ann. Rev. Biochem.* **58,** 841–874.

Hartl, F-U., Hlodan, R., and Langer, T. (1994). Molecular chaperones in protein folding: The art of avoiding sticky situations. *Trends Biochem. Sci.* **19,** 20–25.

Hauschka, P. V., Lian, J. B., and Gallop, P. M. (1975). Direct identification of the calcium binding amino acid, γ-carboxyglutamate, in mineralized tissue. *Proc. Natl. Acad. Sci. USA* **72,** 3925–3929.

Hayaishi, O., and Ueda, K. (1982). Poly- and mono-(ADP)-ribosylation reactions: Their significance in molecular biology. *In:* "ADP-Ribosylation reactions: Biology and Medicine" (O. Hayaishi and K. Ueda, eds.), pp. 3–16. Academic Press, New York.

Hayaishi, O., and Ueda, K. (1984). Evidence for poly(ADP-ribosyl) derivatives of carboxylates in histone. *Meth. Enzymol.* **106,** 450–461.

Hayashi, S., and Wu, H. C. (1990). Lipoproteins in bacteria. *J. Bioenerg. Biomembr.* **22,** 451–471.

Hedner, U., and Davie, E. W. (1989). Introduction to hemostasis and the vitamin K-dependent coagulation factors. *In:* "The Metabolic Basis of Inherited Diseases" (C. R. Scriver, A. L. Beaudet, W. S. Sly, and D. Valle, eds.), pp. 2107–2134. McGraw-Hill, Toronto.

Helfman, P. M., and Bada, J. L. (1976). Aspartic acid racemization in dentine as a measure of ageing. *Nature (London)* **262,** 279–281.

Hemmingsen, S. M., Woolford, C., van der Vies, S. M., Tilly, K., Dennis, D. T., Georgopolous, C., Hendrix, R. W., and Ellis, R. J. (1988). Homologous plant and bacterial proteins chaperone oligomeric protein assembly. *Nature (London)* **333,** 330–334.

Hendrick, J. P., and Hartl, F.-U. (1995). The role of molecular chaperones in protein folding. *FASEB J.* **9,** 1559–1569.

Hendrix, R. W. (1979). Purification and properties of groE, a host protein involved in bacteriophage assembly. *J. Mol. Biol.* **129,** 375–392.

Hershko, A. (1988). Ubiquitin-mediated protein degradation. *J. Biol. Chem.* **263,** 15237–15240.

Hershko, A. (1991). The ubiquitin pathway for protein degradation. *Trends Biochem. Sci.* **16,** 265–268.

Hoff, W. D., Dux, P., Hard, K., Devreese, B., Nugteren-Roodzant, I. M., Crielaard, W., Boelens, R., Kaptein, R., Van Beeumen, J., and Hellingwerf, K. J. (1994). Thiol ester-linked *p*-coumaric acid as a new photoactive prosthetic group in a protein with rhodopsin-like photochemistry. *Biochemistry* **33,** 13960–13962.

Hofsteenge, J., Muller, D. R., Beer, T., Loffler, A., Richter, W. J., and Vliegenthart, J. F. G. (1994). New type of linkage between a carbohydrate and a protein: C-Glycosylation of a specific tryptophan residue in human Rnase Us. *Biochemistry* **33,** 13524–13530.

Höök, M., Kjellén, L., Johansson, S., and Robinson, J. (1984). Cell surface glycosaminoglycans. *Ann. Rev. Biochem.* **53,** 847–869.

Hunt, S. (1984). Halogenated tyrosine derivatives in invertebrate scleroproteins: Isolation and identification. *Meth. Enzymol.* **107,** 413–438.

Hunter, T., and Cooper, J. A. (1985). Protein-tyrosine kinases. *Ann. Rev. Biochem.* **54,** 897–930.

Huszar, G. (1984). Methylated lysines and 3-methyl histidine in myosine: Tissue and developmental differences. *Meth. Enzymol.* **106,** 287–295.

Iglewski, W. J., Lee, H., and Muller, P. (1985). A cellular mono(ADP-ribosyl) transferase which modifies the diphthamide residue of elongation factor-2. *In:* "ADP-Ribosylation of Proteins" (F. R. Althaus, H. Hilz, and S. Shall, eds.), pp. 536–543. Springer-Verlag, Berlin.

Inglese, J., Glickman, J. F., Lorenz, W., Caron, M. G., and Lefkowitz, R. J. (1992). Isoprenylation of a protein kinase, requirement of farnesylation/α-carboxyl methylation for full enzymatic activity of rhodopsin kinase. *J. Biol. Chem.* **267,** 1422–1425.

Ito, N., Phillips, S. E. V., Stevens, C., Ogel, Z. B., McPherson, M. J., Keen, J. N., Yadav, K. D. S., and Knowles, P. F. (1991). Novel thioether bond revealed by a 1.7 Å crystal structure of galactose oxidase. *Nature (London)* **350,** 87–90.

Jacob, U., Gaestel, M., Engel, K., and Buchner, J. (1993). Small heat shock proteins are molecular chaperones. *J. Biol. Chem.* **268,** 1517–1520.

James, D. C., Goldman, M. H., Hoare, M., Jenkins, N., Oliver, R. W. A., Green, B. N., and Freedman, R. B. (1996). Posttranslational processing of recombinant human interferon-γ in animal expression systems. *Prot. Sci.* **5,** 331–340.

Janes, S. M., Mu, D., Wemmer, D., Smith, A. J., Kaur, S., Maltby, D., Burlingame, A. L., and Klinman, J. P. (1990). A new redox cofactor in eukaryotic organisms: 6-Hydroxydopa at the active site of bovine serum amine oxidase. *Science* **248,** 981–987.

Jentoft, N. (1990). Why are proteins *O*-glycosylated? *Trends Biochem. Sci.* **15,** 291–294.

Jespersen, A. M., Christensen, T., Klausen, N. K., Nielsen, P. F., and Sorensen, H. H. (1994). Characterization of a trisulphide derivative of biosynthetic human growth hormone produced in *Escherichia coli*. *Eur. J. Biochem.* **219,** 365–373.

Jimenez, E. C., Craig, A. G., Watkins, M., Hillyard, D. R., Gray, W. R., Gulyas, J., Rivier, J. E., Cruz, L. J., and Olivera, B. M. (1997). Bromocontryptophan: Post-translational bromination of tryptophan. *Biochemistry* **36,** 989–994.

Johnson, D. R., Bhatnagar, R. S., Knoll, L. J., and Gordon, J. I. (1994). Genetic and biochemical studies of protein *N*-myristoylation. *Ann. Rev. Biochem.* **63,** 869–914.

Kaji, H. (1976). Amino-terminal arginylation of chromosomal proteins of arginyl-tRNA. *Biochemistry* **15,** 5121–5125.

Kaji, A., Kaji, H., and Novelli, G. D. (1965). Soluble amino acid-incarporating system: 1. Preparation of the system and nature of the reaction. *J. Biol. Chem.* **240,** 1185–1191.

Kaletta, C., Entian, K-D., and Jung, G. (1991). Peptide sequence of cinnamycin (Ro 09-0198): The first structural gene of a duramycin-type 1 antibiotic. *Eur. J. Biochem.* **199,** 411–415.

Kates, M. (1995). Adventures with membrane lipids. *Biochem. Soc. Trans.* **23,** 697–709.

Keiler, K. C., Waller, P. R. H., and Sauer, R. T. (1996). Role of a peptide tagging system in degradation of proteins synthesized from damaged messenger RNA. *Science* **271,** 990–993.

Kennedy, L., and Baynes, J. W. (1984). Non-enzymatic glycosylation and the chronic complications of diabetes: An overview. *Diabetologia* **26,** 93–98.

Khorana, H. G. (1988). Bacteriorhodopsin, a membrane protein that uses light to translocate proteins. *J. Biol. Chem.* **263,** 7439–7442.

Kikuchi, Y., and Tamiya, N. (1992). The use of amino acid composition data for the chemical taxonomy of the hinge-ligament proteins of molluscan bivalve species. *In:* "Frontiers and New Horizons in Amino Acid Research" (K. Takai, ed.), pp. 309–314. Elsevier, Amsterdam.

Kim, P. S., and Arvan, P. (1995). Calnexin and BIP act as sequential molecular chaperones during thyroglobulin folding in the endoplasmic reticulum. *J. Cell. Biol.* **128,** 29–38.

Kivirikko, K., Myllyla, R., and Philajaniemi, T. (1989). Protein hydroxylation: Prolyl 4 hydroxylase, an enzyme with four cosubstrates and a multifunctional subunit. *FASEB J.* **3,** 1609–1617.

Kivirikko, K. I., Myllylä, R., and Philajaniemi, T. (1992). Hydroxylation of proline and lysine residues in collagen and other animal and plant proteins. *In:* "Post-translational Modification of Proteins" (J. J. Harding and M. J. C. Crabbe, eds.). CRC Press, Boca Raton, FL, pp. 1–51.

Klinman, J. P., and Mu, D. (1994). Quinoenzymes in biology. *Ann. Rev. Biochem.* **63,** 299–344.

Klostermeyer, H. (1984). N^{ε}-(β-Aspartyl)lysine. *Meth. Enzymol.* **107,** 258–261.

Klotz, A. V., and Glazer, A. N. (1985). γ-*N*-Methylasparagine in phycobiliproteins: Occurrence, location and biosynthesis. *J. Biol. Chem.* **262,** 17350–17355.

Koch, T. H., Christy, M. R., Barkley, R. M., Sluski, R., Bohemier, D., VanBuskirk, J. A., and Kirsch, W. M. (1984). β-Carboxyaspartic acid. *Meth. Enzymol.* **107,** 563–575.

Kolattukudy, P. E. (1984). Detection of an N-terminal glucuronamide linkage in proteins. *Meth. Enzymol.* **106,** 210–217.

Kolodziej, S. J., Schroeter, J. P., Strickland, D. K., and Stoops, J. K. (1996). The Novel three-dimensional structure of native human α_2-macroglobulin and comparisons with the structure of the methylamine derivative. *J. Struct. Biol., 116,* 366–376.

Kornfeld, R., and Kornfeld, S. (1985). Assembly of asparagine-linked oligosaccharides. *Ann. Rev. Biochem.* **54,** 631–664.

Koshland, D. E., Jr. (1981). Biochemistry of sensing and adaptation in a simple bacterial system. *Ann. Rev. Biochem.* **50,** 765–782.

Kowalak, J. A., and Walsh, K. A. (1996). β-Methylthio-aspartic acid: Identification of a novel posttranslational modification in ribosomal protein S12 from *Escherichia coli. Protein Science* **5,** 1625–1632.

Kräusslich, H.-G., and Wimmer, E. (1988). Viral proteinases. *Ann. Rev. Biochem.* **57,** 701–754.

Kreil, G. (1984). Occurrence, detection and biosynthesis of carboxyterminal amides. *Meth. Enzymol.* **106,** 218–223.

Kreil, G. (1994). Peptides containing a D-amino acid from frogs and molluscs. *J. Biol. Chem.* **269,** 10967–10970.

Kreil, G., and Kreil-Kiss, G. (1967). The isolation of *N*-formylglycine from a polypeptide present in bee venom. *Biochem. Biophys. Res. Commun.* **27,** 275–280.

Krishna, R. G., and Wold, F. (1993a). Post-translational modification of proteins. *Adv. Enzymol.* **67,** 265–298.

Krishna, R. G., and Wold, F. (1993b). Post-translational modifications: mass spectrometry. *In:* "Methods in Protein Sequence Analysis" (K. Imahori, and F. Sakiyama, eds.), pp. 167–172. Plenum Press, New York.

Krishna, R. G., and Wold, F. (1997). Identification of common post-translational modifications. *In:* "Protein Structure: A Practical Approach," 2nd ed. (T. E. Creighton, ed.). pp. 91–116. IRL Press at Oxford University Press, Oxford.

Krishna, R. G., Chin, C. C. Q., and Wold, F. (1991). N-Terminal sequence analysis of N^{α}-acetylated proteins after unblocking with *N*-acylaminoacyl-peptide hydrolase. *Anal. Biochem.* **199,** 45–50.

Kudlicki, W., Odom, O. W., Kramer, G., and Hardesty, B. (1994). Activation and release of enzymatically inactive, full-length rhodanese that is bound to ribosomes as peptidyl-tRNA. *J. Biol. Chem.* **269,** 16549–16553.

Lagarias, J. C., Glazer, A. N., and Rapoport, H. (1979). Chromopeptides from C-phycocyanin: Structure and linkage of a phycocyanobilin bound to the subunit. *J. Amer. Chem. Soc.* **101,** 5030–5037.

Lampen, J. O., and Nielsen, J. B. K. (1984). N-Terminal glyceride–cysteine modification of membrane penicillinases in gram-positive bacteria. *Meth. Enzymol.* **106,** 365–368.

Lamport, D. T. A. (1984). Hydroxyproline glycosides in the plant kingdom. *Meth. Enzymol.* **106,** 523–528.

Lederer, F., Alix, J. H., and Hayes, D. (1977). *N*-Trimethylalanine, a novel blocking group found in *E. coli* ribosomal protein L11. *Biochem. Biophys. Res. Commun.* **77,** 470–480.

Lee, Y. C., and Lee, R. T. (eds.) (1994). "Neoglycoconjugates. Preparation and Applications," pp. 1–549. Academic Press, New York.

Lerch, K. (1984). S^{β}-(2-Histidyl) cysteine: Properties, assay and occurrence. *Meth. Enzymol.* **106,** 355–359.

Levine, M. J., and Spiro, R. G. (1979). Isolation from glomerula basement membrane of a glycopeptide containing both asparagine-linked and hydroxylysine-linked carbohydrate units. *J. Biol. Chem.* **254,** 8121–8124.

Lhoest, J., and Colson, C. (1977). Genetics of ribosomal protein methylation in *E. coli:* II. A mutant lacking a new type of methylated amino acid N^5-methylglutamine in protein L3. *Mol. Gen. Genet.* **154,** 175–180.

Lin, T. S., and Kolattukudy, P. E. (1980). Structural studies on cutinase, a glycoprotein containing novel amino acids and glucuronic acid amide at the N-terminus. *Eur. J. Biochem.* **106,** 341–351.

Live, D. H., and Edmondson, D. E. (1988). Studies of phosphorylated sites in proteins using ^1H–^{31}P two dimensional NMR: Further evidence for a phosphodiester link between a seryl and threonyl residue in *Azotobacter flavodoxin. J. Am. Chem. Soc.* **110,** 4468–4470.

Loewy, A. G. (1984). The N^{ε}-(γ-glutamic)lysine crosslink: Method of analysis, occurrence in extracellular and cellular proteins. *Meth. Enzymol.* **107,** 241–257.

Lommerse, J. P. M., Kroon-Batenburg, L. M. J., Kamerling, J. P., and Vliegenthart, J. F. G. (1995). Conformational analysis of the xylose-containing *N*-glycan of pinapple stem bromelain as part of the intact glycoprotein. *Biochemistry* **34,** 8196–8206.

Lorand, L. (1988). Transglutaminase-mediated cross-linking of proteins and cell ageing: the erythrocyte and lens model. *In:* "Advances in Post-translational Modification of Proteins and Aging" (V. Appia, P. Galletti, R. Porta, and F. Wold, eds.), pp. 79–94. Plenum Press, New York.

Lorimer, G. H. (1996). A quantitative assessment of the role of the chaperonin proteins in protein folding *in vivo. FASEB J.* **10,** 5–9.

Lote, C. J., and Weiss, J. B. (1971). Identification of digalactosylcysteine in a glycopeptide isolated from urine by a new preparative technique. *FEBS Lett.* **16,** 81–85.

Lou, M. F. (1975). Isolation and identification of L-β-aspartyl-L-lysine and L-γ-glutamyl-L-ornithine from normal human urine. *Biochemistry* **14,** 3503–3508.

Lowry, P. J., and Chadwick, A. (1970). Interrelations of some pituitary hormones. *Nature* (*London*) **226,** 219–224.

Lukens, L. N. (1976). Time of occurrence of disulfide linking between procollagen chains. *J. Biol. Chem.* **251,** 3530–3538.

Lynch, D. R., and Snyder, S. H. (1986). Neuropeptides: Multiple molecular forms, metabolic pathways, and receptors. *Ann. Rev. Biochem.* **55,** 773–799.

Mande, S. C., Mehra, V., Bloom, B. R., and Hol, W. G. J. (1996). Structure of the heat shock protein chaperonin-10 of *Mycobacterium leprae. Science* **271,** 203–207.

Margoliash, E., and Schejter, A. (1966). Cytochrome C. *Adv. Prot. Chem.* **21,** 113–286.

Marsh, M. E., and Sass, R. L. (1984). Phosphoprotein particles: Calcium and inorganic phosphate binding structures. *Biochemistry* **23,** 1448–1456.

Martensen, T. M. (1984). Chemical properties, isolation and analysis of *O*-phosphates in proteins. *Meth. Enzymol.* **107,** 3–23.

Martin, N. C., Rabinowitz, M., and Fukuhara, H. (1977). Yeast mitochondrial DNA specifies tRNA for 19 amino acids: Deletion mapping of the tRNA genes. *Biochemistry* **16,** 4672–4677.

Martin, J., Mayhew, M., Langer, T., and Hartl, F. U. (1993). The reaction cycle of GroEL and GroES in chaperonin-assisted protein folding. *Nature* (*London*) **366,** 228–233.

Masters, P. M., Bada, J. L., and Zigler, J. S. Jr. (1977). Aspartic acid racemization in the human lens during aging and in cataract formation. *Nature* (*London*) **268,** 71–73.

Mato, J. M., and Marin-Cao, D. (1979). Protein and phospholipid methylation during chemotaxis in *Dictyostelium discoideum* and its relationship to calcium movement. *Proc. Natl. Acad. Sci. USA* **76,** 6106–6109.

Mattar, S., Scharf, B., Kent, S. B. H., Rodewald, K., Oesterhelt, D., and Engelhard, M. (1994). The primary structure of halocyanin, an archaeal blue copper protein, predicts a lipid anchor for membrane function. *J. Biol. Chem.* **269,** 14939–14945.

McIlhinney, R. A. J. (1990). The fats of life: The importance and function of protein acylation. *Trends Biochem. Sci.* **15,** 387–391.

McIntire, W. S., Wemmer, D. E. Chistoserdov, A., and Lidstrom, M. E. (1991). A new cofactor in a procaryotic enzyme: Tryptophan tryptophylquinone as the redox prosthetic group in methylamine dehydrogenase. *Science* **252,** 817–824.

Mecham, R. P., and Davis, E. C. (1994). Elastic fiber structure and assembly. *In:* "Extracellular Matrix Assembly and Structure" (P. D. Yurchenco, D. E. Birk, and R. P. Mecham, eds.), pp. 281–314. Academic Press, San Diego.

Merck, K. B., Groenen, P. J., Voorter, C. E. M., de Haard-Hoekman, W. A., Horwitz, J., Bloemendal, H., and de Jong, W. W. (1993). Structural and functional similarities of bovine α-crystallin and mouse small heat-shock protein: A family of chaperones. *J. Biol. Chem.* **268,** 1046–1052.

Midelfort, C. F., and Mehler, A. H. (1972). Deamidation *in vivo* of an asparagine residue of rabbit muscle aldolase. *Proc. Natl. Acad. Sci. USA* **69,** 1816–1819.

Mirelman, D., and Siegel, R. C. (1979). Oxidative deamination of ε-aminolysine residues and formation of Schiff base cross-linkages in cell envelops of *E. coli*. *J. Biol. Chem.* **254,** 571–574.

Moehring, T. M., and Moehring, T. J. (1984). Diphthamide: *In vitro* biosynthesis. *Meth. Enzymol.* **106,** 388–395.

Moerschell, R. P., Hosokawa, Y., Tsunasawa, S., and Sherman, F. (1990). The specificities of yeast methionine amino peptidase and acetylation of amino-terminal methionine *in vivo*. *J. Biol. Chem.* **265,** 19638–19643.

Monnier, V. M. (1989). Toward a Maillard reaction theory of aging. *In:* "The Maillard Reaction in Aging, Diabetes and Nutrition" (J. W. Baynes and V. M. Monnier, eds.), pp. 1–22. Alan R. Liss, New York.

Moscarello, M. A., Pang, H., Pace-Asciak, C. R., and Wood, D. D. (1992). The N-terminus of human myelin basic protein consists of C_2, C_4, C_6, and C_8 alkyl carboxylic acids. *J. Biol. Chem.* **267,** 9779–9782.

Müller-Eberhard, H. J. (1988). Molecular organization and function of the complement system. *Ann. Rev. Biochem.* **57,** 321–347.

Murtif, V. L., and Samols, D. (1987). Mutagenesis affecting the carboxyl terminus of the biotinyl subunit of transcarboxylase. Effects on biotinylation. *J. Biol. Chem.* **262,** 11813–11816.

Nakajima, T., and Ballou, C. E. (1974). Structure of the linkage region between the polysaccharide and protein parts of *Saccharomyces cerevisiae* mannan. *J. Biol. Chem.* **249,** 7685–7694.

Nakajima, T., and Volcani, B. (1970). ε-*N*-Trimethyl-L-δ-hydroxylysine phosphate and its non-phosphorylated compound in diatom cell walls. *Biochem. Biophys. Res. Commun.* **39,** 28–33.

Nakamura, F., and Suyama, K. (1994). Analysis of aldosine, an amino acid derived from aldol crosslink of elastin and collagen by high-pressure liquid chromatography. *Anal. Biochem.* **223,** 21–25.

Narayana, A. S., and Page, R. C. (1976). Demonstration of a precursor–product relationship between soluble and cross-linked elastin and the biosynthesis of the desmosines *in vitro*. *J. Biol. Chem.* **251,** 1125–1130.

Neubert, T. A., Johnson, R. S., Hurley, J. B., and Walsh, K. A. (1992). The rod transducin α subunit amino terminus is heterogeneously fatty acylated. *J. Biol. Chem.* **267,** 18274–18277.

Neurath, H., and Walsh, K. A. (1976). Role of proteolytic enzymes in biological regulation. *Proc. Natl. Acad. Sci. USA* **73,** 3825–3832.

Niehrs, C., Beibwanger, R., and Huttner, W. B. (1994). Protein sulfation, 1993—An update. *Chemico-Biological Interactions* **92,** 257–271.

Noiva, R., and Lennarz, W. J. (1992). Protein disulfide isomerase: A multifunctional protein resides in the lumen of the endoplasmic reticulum. *J. Biol. Chem.* **267,** 3553–3556.

Nordwig, A., and Pfab, F. K. (1969). Specificities of protocollagen hydroxylases from different sources. *Biochim. Biophys. Acta* **181,** 52–58.

Nose, M., and Wigzell, H. (1983). Biological significance of carbohydrate chains on monoclonal antibodies. *Proc. Natl. Acad. Sci. USA* **80,** 6632–6636.

Nunez, J. (1984). Thyroid hormones: Mechanism of phenoxy ether formation. *Meth. Enzymol.* **107,** 476–488.

Ogata, N., Ueda, K., and Hayaishi, O. (1980a). ADP-Ribosylation of histone H2B: Identification of glutamic acid residue 2 as the modification site. *J. Biol. Chem.* **255,** 7610–7615.

Ogata, N., Ueda, K., Kagamiyama, H., and Hayaishi, O. (1980b). ADP-Ribosylation of histone H1: Identification of glutamic acid residues 2, 14 and the COOH-terminal lysine residue as modification sites. *J. Biol. Chem.* **255,** 7616–7620.

Olson, T. S., and Lane, M. D. (1989). A common mechanism for posttranslational activation of plasma membrane receptors. *FASEB J.* **3,** 1618–1624.

Oppenheimer, N. J. (1984). ADP-Ribosylarginine. *Meth. Enzymol.* **106,** 399–403.

Orlowski, M., and Meister, A. (1971). Enzymology of pyrrolidone carboxylic acid. *In:* "The Enzymes" (P. D. Boyer, ed.), Vol. 4, pp. 123–151. Aacdemic Press, New York.

Paborsky, L. R., Tate, K. M., Harris, R. J., Yansura, D. G., Band, L., McCray, G., Gorman, C. M., O'Brien, D. P., Chang, J. Y., Swartz, J. R., Fung, V. P., Thoman, J. N., and Vehar, G. A. (1989). Purification of recombinant human tissue factor. *Biochemistry* **28,** 8072–8077.

Paik, W. K., and Dimaria, P. (1984). Enzymatic methylation and demethylation of protein-bound lysine residues. *Meth. Enzymol.* **106,** 274–286.

Paik, W. K., and Kim, S. (eds.) (1990). "Protein Methylation," pp. 1–434. CRC Press, Boca Raton, FL.

Park, M. H., Wolff, C. E., and Folk, J. J. (1993a). Is hypusine essential for eukaryotic cell proliferation? *Trends Biochem. Sci.* **18,** 475–479.

Park, M. H., Wolff, C. E., and Folk, J. J. (1993b). Hypusine: Its post-translational formation in eukaryotic initiation factor 5A and its potential role in cellular regulation. *Biofactors* **4,** 95–104.

Parsell, D. A., Kowal, A. S., Singer, M. A., and Lindquist, S. (1994). Protein disaggregation mediated by heat-shock protein HSP104. *Nature (London)* **372,** 475–478.

Paulson, J. C. (1989). Glycoproteins: What are the sugar chains for? *Trends Biochem. Sci.* **14,** 272–275.

Pettigrew, G. W., and Smith, G. M. (1977). Novel *N*-terminal protein blocking group identified as dimethylproline. *Nature (London)* **265,** 661–662.

Pfeifer, E., Pavela-Vrancic, M., von Döhren, H., and Kleinkauf, H. (1995). Characterization of tyrocidine synthetase 1 (TY1): Requirements of posttranslational modification for peptide biosynthesis. *Biochemistry* **34,** 7450–7459.

Piacentini, M., Ceru-Argento, M. P., Farrace, M. G., and Autuori, F. (1988). Post-translational modifications of cellular proteins by polyamines and polyamine-derivatives. *In:* "Advances in Post-translational Modification of Proteins and Aging" (V. Zappia, P. Galletti, R. Porta, and F. Wold, eds.), pp. 185–198. Plenum Press, New York.

Piacentini, M., Davies, P. J. A., and Fesus, L. (1994). Tissue transglutaminase in cells undergoing apoptosis. *In:* "Apoptosis II: The Molecular Basis of Apoptosis in Disease" (L. D. Tomei and F. O. Cope, eds.), pp. 143–163. Cold Spring Harbor Laboratory Press, New York.

Pigiet, V., and Conley, R. P. (1978). Isolation and characterization of phosphothioredoxin from *E. coli*. *J. Biol. Chem.* **253,** 1910–1920.

Pokharna, H., Monnier, V., Boja, B., and Moskowitz, R. W. (1995). Lysyl oxidase and Maillard reaction-mediated crosslinks in aging and osteoarthritic rabbit cartilage. *J. Orthopaedic Res.* **13,** 13–21.

Pope, M. R., Saari, L. L., and Ludden, P. W. (1986). *N*-Glycohydrolysis of adenosine diphosphoribosylarginine linkages by dinitrogenase reductase activating glycohydrolase (activating enzyme) from *Rhodospirillum rubrum*. *J. Biol. Chem.* **261,** 10104–10111.

Portridge, S. M. (1989). Cross-linking in elastin. *In:* "Elastin and Elastases" (L. Robert and W. Hornebeck, eds.), Vol. 1, pp. 127–140. CRC Press, Boca Raton, FL.

Rademacher, T. W., Parekh, R. B., and Dwek, R. A. (1988). Glycobiology. *Ann. Rev. Biochem.* **57,** 785–838.

Rassow, J., and Pfanner, N. (1995). Molecular chaperones and intracellular protein translocation. *Rev. Physiol. Biochem. & Pharmacol.* **126,** 199–264.

Redeker, V., Levilliers, N., Schmitter, J-M., Le Caer, J-P., Rossier, J., Adoutte, A., and Bre, M.-H. (1994). Polyglycylation of tubulin: A posttranslational modification in axonemal microtubules. *Science* **266,** 1688–1691.

Rhee, S. G. (1984). 5'-Nucleotidyl-*O*-tyrosine bond in glutamine synthetase. *Meth. Enzymol.* **107,** 183–200.

Richarme, G., and Kohiyama, M. (1994). Amino acid specificity of the *E. coli* chaperone GroEL (heat shock protein 60). *J. Biol. Chem.* **269,** 7095–7098.

Riquelme, P. T., Buzzio, L. O., and Koide, S. S. (1979). ADP-Ribosylation of rat liver lysine-rich histone *in vitro*. *J. Biol. Chem.* **254,** 3018–3028.

Robinson, A. B., Scotchler, J. W., and McKerrow, J. H. (1973). Rates of nonenzymatic deamidation of glutamyl and asparaginyl residues in pentapeptides. *J. Am. Chem. Soc.* **95,** 8156–8159.

Rosenquist, G. L., and Nicholas, H. B. Jr. (1993). Analysis of sequence requirements for protein tyrosine sulfation. *Prot. Sci.* **2,** 215–222.

Rothnagel, J. A., and Rogers, G. E. (1984). Citrulline in proteins from the enzymatic deimination of arginine residues. *Meth. Enzymol.* **107,** 624–631.

Rucker, R. B., and Wold, F. (1988). Cofactors in and as posttranslational protein modifications. *FASEB J.* **2,** 2252–2261.

Rudd, P. M., Joao, H. C., Coghill, E., Fiten, P., Saunders, M. R., Opdenakker, G., and Dwek, R. A. (1994). Glycoforms modify the dynamic stability and functional activity of an enzyme. *Biochemistry* **33,** 17–22.

Rudman, D., Chawla, R. K., and Hollins, B. M. (1979). *N,O*-Diacetylserine$_1$ α-melanocyte-stimulating hormone, a naturally occurring melanotropic peptide. *J. Biol. Chem.* **254,** 10102–10108.

Rudnick, D. A., McWherter, C. A., Gokel, G. W., and Gordon, J. I. (1993). Myristoyl-coA: Protein *N*-myristoyltransferase. *Adv. Enzymol.* **67,** 375–430.

Sagami, H., Kikuchi, A., and Ogura, K. (1995). A novel type of protein modification by isoprenoid-derived materials: Diphytanylglycerylated proteins in *Halobacteria. J. Biol. Chem.* **270,** 14851–14854.

Sanders, S. L., and Sheckman, R. (1992). Polypeptide translocation across the endoplasmic reticulum membrane. *J. Biol. Chem.* **267,** 13791–13794.

Sanford, J. C., Pan, Y., and Wessling-Resnick, M. (1995). Properties of Rab5 N-terminal domain dictate prenylation of C-terminal cysteines. *Mol. Biol. Cell* **6,** 71–85.

Sass, R. L., and Marsh, M. E. (1983). N^{τ}- And N^{π}-histidinoalanine: Naturally occurring crosslinking amino acids in calcium binding phosphoproteins. *Biochem. Biophys. Res. Commun.* **114,** 304–309.

Sass, R. L., and Marsh, M. E. (1984). Histidinoalanine: A naturally occurring cross-linking amino acid. *Meth. Enzymol.* **106,** 351–355.

Scaloni, A., Simmaco, M., and Bossa, F. (1991). Determination of the chirality of amino acid residues in the course of subtractive Edman degradation of peptides. *Anal. Biochem.* **197,** 305–310.

Scaloni, A., Barra, D., and Bossa, F. (1994). Analysis of dehydroamino acids after β-elimination, followed by thiol addition. *Anal. Biochem.* **218,** 226–228.

Schmid, F. X. (1992). Kinetics of unfolding and refolding of single-domain proteins. *In:* "Protein Folding" (T. E. Creighton, ed.), pp. 197–242. Freeman Press, New York.

Seabra, M., C., Reiss, Y., Casey, P. J., Brown, M. S., and Goldstein, J. L. (1991). Protein farnesyltransferase and geranylgeranyltransferase share a common α-subunit. *Cell* **65,** 429–434.

Seetharam, H., Heeren, R. A., Wong, E. Y., Bradfors, S. R., Klein, B. K., Aykent, S., Kotts, C. E., Mathis, K. J., Bishop, B. F., Jennings, M. J., Smith, C. E., and Seigel, N. R. (1988). Mistranslation in IGF-1 during over-expression of the protein in *Escherichia coli* using a synthetic gene containing low frequency codons. *Biochem. Biophys. Res. Commun.* **155,** 518–523.

Sekine, A., Fujiwara, M., and Narumiya, S. (1989). Asparagine residue in the rho gene product in the modification site for botulinum ADP-ribosyltransferase. *J. Biol. Chem.* **264,** 8602–8605.

Sell, D. R., and Monnier, V. M. (1989). Structure elucidation of a senescence cross-link from human extracellular matrix: Implication of pentoses in the aging process. *J. Biol. Chem.* **264,** 21597–21602.

Shaanan, B., Lis, H., and Sharon, N. (1991). Structure of a legume lectin with an ordered *N*-linked carbohydrate in complex with lactose. *Science* **254,** 862–866.

Shakin-Eshleman, S. H., Spitalnik, S. L., and Kasturi, L. (1996). The amino acid at the X position of an Asn-X-Ser sequon is an important determinant of *N*-linked core-glycosylation efficiency. *J. Biol. Chem.* **271,** 6363–6666.

Shall, S. (1988). ADP-Ribosylation of proteins: A ubiquitous cellular control mechanism. *In:* "Advances in Post-translational Modification of Proteins and Aging" (V. Zappia, P. Galletti, R. Porta, and F. Wold, eds.), pp. 597–611. Plenum Press, New York.

Shao, M.-C., Sokolik, C. W., and Wold, F. (1994). Noncovalent neoglycoproteins. *In:* "Neoglycoconjugates: Preparation and Applications" (Y. C. Lee and R. T. Lee, eds.), pp. 225–249. Academic Press, San Diego.

Shao, Y., Xu, M.-Q., and Paulus, H. (1995). Protein splicing: Characterization of the aminosuccinimide residue at the carboxyl terminus of the excised intervening sequence. *Biochemistry* **34,** 10844–10850.

Shapiro, B. M., and Stadtman, E. R. (1968). 5'-Adenylyl-*O*-tyrosine: The novel phosphodiester residues of adenylylated glutamine synthetase from *E. coli. J. Biol. Chem.* **243,** 3769–3771.

Sheid, B., and Pedrinan, L. (1975). DNA-dependent protein methylation activity in bull seminal plasma. *Biochemistry* **14,** 4357–4361.

Shinde, U., and Inouye, M. (1994). The structural and functional organization of intramolecular chaperones: The N-terminal properties which mediate protein folding. *J. Biochem.* **115,** 629–636.

Siegel, F. L. (1988). Enzymatic *N*-methylation of calmodulin. *In:* "Advances in Post-translational Modification of Proteins and Aging" (V. Zappia, P. Galletti, R. Porta, and F. Wold, eds.), pp. 341–352. Plenum Press, New York.

Siegel, R. C., and Fu, J. C. C. (1976). Collagen cross-linking: Purification and substrate specificity of lysyl oxidase. *J. Biol. Chem.* **251,** 5779–5785.

Silberstein, S., and Gilmore, R. (1996). Biochemistry, molecular biology and genetics of the oligosaccharyl transferase. *FASEB J.* **10,** 849–858.

Singer, T. P., and McIntire, W. S. (1984). Covalent attachment of flavin to flavoproteins: occurrence, assay and synthesis. *Meth. Enzymol.* **106,** 369–378.

Sletten, K., and Aakesson, I. (1971). Presence of ornithine in the urate-binding α1-α2 globulin. *Nature New Biol.* **231,** 118–119.

Smeland, T. E., Seabra, M. C., Goldstein, J. L., and Brown, M. S. (1994). Geranylgeranylated Rab proteins terminating in Cys-Ala-Cys, but not Cys-Cys, are carboxy-methylated by bovine brain membranes *in vitro. Proc. Natl. Acad. Sci. USA* **91,** 10712–10716.

Smith, L. S., Kern, C. W., Halpern, R. M., and Smith, R. A. (1976). Phosphorylation on basic amino acids in myelin basic protein. *Biochem. Biophys. Res. Commun.* **71,** 459–465.

Sokolik, C. W., Liang, T. C., and Wold, F. (1994). Studies on the specificity of acetylaminoacyl-peptide hydrolase. *Prot. Sci.* **2,** 126–131.

Sottrup-Jensen, L. (1989). α-Macroglobulins: structure, shape, and mechanism of proteinase complex formation. *J. Biol. Chem.* **264,** 11539–11542.

Stadtman, E. R. (1990). Covalent modifications are marking steps in protein turnover. *Biochemistry* **29,** 6323–6331.

Stadtman, T. C. (1996). Selenocysteine. *Ann. Rev. Biochem.* **65,** 83–100.

Stanley, P., and Ioffe, E. (1995). Glycosyltransferase mutants: Key to new insights in glycobiology. *FASEB J.* **9,** 1436–1444.

Steinert, P. M., and Idler, W. W. (1979). Postsynthetic modifications of mammalian epidermal α-keratin. *Biochemistry* **18,** 5664–5669.

Steinert, P. M., North, A. C., and Parry, D. A. (1994). Structural features of keratin intermediate filaments. *J. Inv. Dermatol.* **103,** 19S–24S.

Stenflo, J., and Suttie, J. W. (1977). Vitamin K-dependent formation of γ-carboxyglutamic acid. *Ann. Rev. Biochem.* **46,** 157–172.

Stenflo, J., Lundwall, A., and Dahlbäck, B. (1987). β-Hydroxyasparagine in domains homologous to the epidermal growth factor precursor in vitamin K-dependent protein S. *Proc. Natl. Acad. Sci. USA* **84,** 368–372.

Stenflo, J., Holme, E., Lindstedt, S., Chandramouli, N., Huang, L. H. T., Tam, J. P., and Merrifield, R. B. (1989). Hydroxylation of aspartic acid in domains homologous to the epidermal growth factor precursor is catalyzed by a dioxoglutarate-dependent dioxygenase. *Proc. Natl. Acad. Sci. USA* **86,** 444–447.

Stewart, T. S., and Sharp, S. (1984). Characterizing the function of O^{β}-phosphoseryl tRNA. *Meth. Enzymol.* **106,** 157–161.

Stimson, E., Virji, M., Barker, S., Panico, M., Blench, I., Saunders, J., Paine, G., Moxon, E. R., Dell, A., and Morris, H. R. (1996). Discovery of a novel protein modification: α-Glycerophosphate is a substituent of meningococcal pilin. *Biochem. J.* **316,** 29–33.

Stock, J., and Simms, S. (1988). Methylation, demethylation, and deamidation at glutamate residues in membrane chemoreceptor proteins. *In:* "Advances in Post-translational Modification of Proteins and Aging" (V. Zappia, P. Galletti, R. Porta, and F. Wold, eds.), pp. 201–212. Plenum Press, New York.

Struck, D. K., and Lennarz, W. J. (1980). The function of saccharide-lipids in synthesis of glycoproteins. *In:* "The Biochemistry of Glycoproteins and Proteoglycans" (W. J. Lennarz, ed.), pp. 35–83. Plenum Press, New York.

Suchanek, G., Kreil, G., and Hermodson, M. A. (1980). Amino acid sequence of honey bee prepromelittin synthesized *in vitro. Proc. Natl. Acad. Sci. USA* **75,** 701–704.

Sudha, V., Rodriguez-Ghidarpour, S., MacKinnon, C., McGee, W. A., Pierce, M. M., and Nall, B. T. (1995). Prolyl isomerase as a probe of stability of slow-folding intermediates. *Biochemistry* **34,** 12892–12902.

Sugino, Y., Tsunasawa, S., Yutani, K., Ogasahara, K., and Suzuki, M. (1980). Amino-terminally formylated tryptophan synthase α-subunit produced by the *trp* operon cloned in a plasmid vector. *J. Biochem.* **87,** 351–355.

Suttie, J. W. (1985). Vitamin K-dependent carboxylase. *Ann. Rev. Biochem.* **54,** 459–477.

Suzuki, T., Seko, A., Kitajima, K., Inoue, Y., and Inoue, S. (1994). Purification and enzymatic properties of peptide: *N*-glycanase from C3H mouse-derived L-929 fibroblast cells: Possible widespread occurrence of posttranslational remodification of proteins by *N*-deglycosylation. *J. Biol. Chem.* **269,** 17611–17618.

Swanson, R. J., and Applebury, M. (1983). Methylation of proteins in photo-receptor rod outer segments. *J. Biol. Chem.* **258,** 10599–10605.

Swenson, R. P., and Howard, J. B. (1979). Characterization of alkylamine-sensitive site in α_2-macroglobulin. *Proc. Natl. Acad. Sci. USA* **76,** 4313–4316.

Tabas, I., and Kornfeld, S. (1980). Biosynthetic intermediates of β-glucuronidase containing high mannose oligosaccharides with blocked phosphate residues. *J. Biol. Chem.* **255,** 6633–6639.

Tack, B. F., Harrison, R. A., Janatova, J., Thomas, M. L., and Prahl, J. W. (1980). Evidence for presence of an internal thiolester bond in third component of human complement. *Proc. Natl. Acad. Sci. USA* **77,** 5764–5768.

Tanase, S., Kojima, H., and Morino, Y. (1979). Pyridoxal 5′-phosphate binding site of pig heart alanine aminotransferase. *Biochemistry* **18,** 3002–3007.

Tanner, W., and Lehle, L. (1987). Protein glycosylation in yeast. *Biochim. Biophys. Acta* **906,** 81–99.

Taylor, S. S., Buechler, J. A., and Yonemoto, W. (1990). cAMP-Dependent protein kinase: Framework for a diverse family of regulatory enzymes. *Ann. Rev. Biochem.* **59,** 970–1005.

Taylor, S. S., Radzio-Andzelm, E., and Hunter, T. (1995). Protein kinases 8: How do protein kinases discriminate between serine/threonine and tyrosine? *FASEB J.* **9,** 1255–1266.

Thorneley, R. N. F., Abell, C., Ashby, G. A., Drummond, M. H., Eady, R. R., Huff, S., Macdonald, C. J., and Shneier, A. (1992). Posttranslational modification of *Klebsiella pneumoniae* flavodoxin by covalent attachment of coenzyme A, shown by ^{31}P NMR and electrospray mass spectrometry, prevents electron transfer from the *nifJ* protein to nitrogenase. A possible new regulatory mechanism for biological nitrogen fixation. *Biochemistry* **31,** 1216–1224.

Towler, D. A., Gordon, J. I., Adams, S. P., and Glaser, L. (1988). The biology and enzymology of eukaryotic protein acylation. *Ann. Rev. Biochem.* **57,** 69–99.

Tsang, M. L., and Schiff, J. A. (1976). Sulfate-reducing pathway in *E. coli* involving bound intermediates. *J. Bacteriol.* **125,** 923–933.

Tse, Y.-C., Kirkegaard, K., and Wang, J. C. (1980). Covalent bonds between protein and DNA: Formation of phosphotyrosine linkage between certain DNA topoisomerases and DNA. *J. Biol. Chem.* **255,** 5560–5565.

Tsunasawa, S., and Sakiyama, F. (1984). Amino-terminal acetylation of proteins: An overview. *Meth. Enzymol.* **106,** 165–170.

Tu, G.-F., Reid, G. E., Zhang, J.-G., Moritz, R. L., and Simpson, R. J. (1995). C-terminal extension of truncated recombinant proteins in *Escherichia coli* with a 10Sa RNA decapeptide. *J. Biol. Chem.* **270,** 9322–9326.

Tuhy, P. M., Bloom, J. W., and Mann, K. G. (1979). Decarboxylation of bovine prothrombin fragment 1 and prothrombin. *Biochemistry* **18,** 5842–5848.

Udenfriend, S., and Kodukula, K. (1995). How glycosylphosphatidylinositol-anchored membrane proteins are made. *Ann. Rev. Biochem.* **64,** 563–591.

Ueda, K., and Hayaishi, O. (1985). ADP-Ribosylation. *Ann. Rev. Biochem.* **54,** 73–100.

Uy, R., and Wold, F. (1977). Posttranslational covalent modification of proteins. *Science* **198,** 890–896.

Vacquier, J. D., Carner, K. R., and Stout, C. D. (1990). Species-specific sequences of abalone lysin, the sperm protein that creates a hole in the egg envelope. *Proc. Natl. Acad. Sci. USA* **87,** 5792–5796.

Vagelos, P. R. (1973). Acyl group transfer (acyl carrier protein). *In:* "The Enzymes" (P. D. Boyer, ed.), Vol. VIII, pp. 155–199. Academic Press, New York.

Van Buskirk, J. J., and Kirsch, W. M. (1978). γ-Carboxyglutamic acid in eukaryotic and procaryotic ribosomes. *Biochem. Biophys. Res. Commun.* **82,** 1329–1331.

van Heynigen, S. (1992). Mono-ADP-ribosylation of proteins. *In:* "Post-translational Modification of Proteins" (J. J. Harding and M. J. C. Crabbe, eds.), pp. 153–183. CRC Press, Boca Raton, FL.

Vanhoof, G., Gossens, F., De Meester, I., Hendriks, D., and Scharpé, S. (1995). Proline motifs in peptides and their biological processing. *FASEB. J.* **9,** 736–744.

Van Kleef, F. S. M., DeJong, W. W., and Hoenders, H. J. (1975). Stepwise degradations and deamidation of the eye lens protein α-crystallin in aging. *Nature* (*London*) **258,** 264–266.

Van Ness, B. G., Howard, J. B., and Bodley, J. W. (1978). Isolation and properties of the trypsin-derived ADP-ribosyl peptides from diphtheria toxin-modified yeast elongation factor. *J. Biol. Chem.* **258,** 8687–8690.

Van Ness, B. G., Howard, J. B., and Bodley, J. W. (1980a). ADP-Ribosylation of elongation factor 2 by diphtheria toxin: NMR spectra and proposed structures of ribosyl-diphthamide and its hydrolysis products. *J. Biol. Chem.* **255,** 10710–10716.

Van Ness, B. G., Howard, J. B., and Bodley, J. W. (1980b). ADP-Ribosylation of elongation factor 2 by diphtheria toxin: Isolation and properties of the novel ribosyl-amino acid and its hydrolysis products. *J. Biol. Chem.* **255,** 10717–10720.

van Poelje, P. D., and Snell, E. E. (1990). Pyruvoyl-dependent enzymes. *Ann. Rev. Biochem.* **59,** 29–59.

Varshavsky, A., Bachmair, A., and Finley, D. (1989). Targeting of proteins for degradation. *Biotechnology* **13,** 109–143.

Vidali, G., Ferrari, N., and Pfeffer, U. (1988). Histone acetylation: A step in gene activation. *In:* "Advances in Post-translational Modification of Proteins and Aging" (V. Zappia, P. Galletti, R. Porta, and F. Wold, eds.), pp. 583–596. Plenum Press, New York.

Waite, J. H., and Benedict, C. V. (1984). Assay of dihydroxyphenylalanine (DOPA) in invertebrate structural proteins. *Meth. Enzymol.* **107,** 397–413.

Waite, J. H., and Rice-Ficht, A. C. (1987). Presclerotized eggshell protein from the liver fluke *Fasciola hepatica*. *Biochemistry* **26,** 7819–7825.

Wallace, C. J. A. (1993). The curious case of protein splicing: Mechanistic insights suggested by protein semisynthesis. *Prot. Sci.* **2,** 697–705.

Wang, S. X., Mure, M., Medzihradszky, K. F., Burlingame, A. L., Brown, D. E., Dooley, D. M., Smith, A. J., Kagan, H. M., and Klinman, J. P. (1996). A crosslinked cofactor in lysyl oxidase: Redox function for amino acid side chains. *Science* **273,** 1078–1084.

Webster, D. R., Gundersen, G. G., Bulinski, J. C., and Borisy, G. G. (1987). Differential turnover of tyrosinated and detyrosinated microtubules. *Proc. Natl. Acad. Sci. USA* **84,** 9040–9044.

Weigel, P. H. (1992). Mechanism and control of glycoconjugate turnover. *In:* "Glycoconjugates. Composition, Structure, and Function" (H. J. Allen, and E. C. Kisailus, eds.), pp. 421–497. Marcel Dekker, New York.

Weis, W. I., and Drickamer, K. (1996). Structural basis of lectin-carbohydrate recognition. *Ann. Rev. Biochem.* **65,** 441–473.

Weiss, J. B., Lote, C. J., and Bobinski, H. (1971). New low molecular weight glycopeptide containing triglucosylcysteine in human erythrocyte membrane. *Nature New Biol.* **234,** 25–26.

West, R. A. Jr., Moss, J., Vaughan, M., Liu, T., and Liu, T.-Y. (1985). Pertussis toxin-catalyzed ADP-ribosylation of transducin. Cysteine 347 is the ADP-ribose acceptor site. *J. Biol. Chem.* **260,** 14428–14430.

Whitby, A. J., Stone, P. R., and Whish, W. J. D. (1979). Effect of polyamines and Mg^{++} on poly(ADP-ribose) and ADP-ribosylation of histones in wheat. *Biochem. Biophys. Res. Commun.* **90,** 1295–1304.

Whiteheart, S. W., Shenbagamurthi, P., Chen, L., Cotter, R. J., and Hart, G. W. (1989). Murine elongation factor 1α (EF-1α) is posttranslationally modified by novel amide-linked ethanolamine–phosphoglycerol moieties. *J. Biol. Chem.* **264,** 14334–14341.

Wiech, H., Buchner, J., Zimmermann, M., and Zimmermann, R. (1993). Hsc70, heavy chain binding protein, and hsp90 differ in their ability to stimulate transport of precursor protein into mammalian microsomes. *J. Biol. Chem.* **268,** 7414–7422.

Williams, V. P., and Glazer, A. N. (1978). Structural studies on phycobiliproteins: Bilin-containing peptides of C-phycocyanin. *J. Biol. Chem.* **253,** 202–211.

Williams, R. L., Greene, S. M., and McPherson, A. (1987). The structure of ribonuclease B at 2.5 Å resolution. *J. Biol. Chem.* **262,** 16020–16031.

Willie, A., Edmondson, D. E., and Jorns, M. S. (1996). Sarcosine oxidase contains a novel covalently bound FMN. *Biochemistry* **35,** 5292–5299.

Wold, F. (1981). *In vivo* chemical modification of proteins. *Ann. Rev. Biochem.* **50,** 783–814.

Wold, F. (1983). Posttranslational protein modifications: Perspectives and prospectives. *In:* "Posttranslational Covalent Modifications of Proteins for Function" (C. B. Johnson, ed.), pp. 1–17. Academic Press, New York.

Wold, F. (1985). Reactions of the amide side-chains of glutamine and asparagine *in vivo*. *Trends Biochem. Sci.* **10,** 4–6.

Wolff, J., and Covelli, I. (1969). Factors in the iodination of histidine in proteins. *Eur. J. Biochem.* **9,** 371–377.

Yamane, H. Y., Farnsworth, C. C., Xie, H., Howald, W., Fung, B. K. K., Clarke, S., Gelb, M. H., and Glomset, J. A. (1990). Brain G protein γ subunits contain *all-trans*-geranylgeranyl-cysteine methyl ester at their carboxyl termini. *Proc. Natl. Acad. Sci. USA* **87,** 5868–5872.

Yamauchi, D., and Minamikawa, T. (1990). Structure of the gene encoding concanavalin A from *Canavalia gladiata* and its expression in *E. coli* cells. *FEBS Lett.* **260,** 127–130.

Yan, S. C. B., Razzano, P., Chao, Y. B., Walls, J. D., Berg, D. T., McClure, D. B., and Grinnell, B. W. (1990). Characterization and novel purification of recombinant human protein C from three mammalian cell lines. *Bio/Technology* **8,** 655–661.

Yan, S. B., Chao, Y. B., and van Halbeek, H. (1993). Novel Asn-linked oligosaccharides terminating in GalNAcβ(1-4)[Fucα(1-3)]GlcNAcβ(1-•) are present in recombinant human protein C expressed in human kidney 293 cells. *Glycobiology* **3,** 597–608.

Yang, Y. R., and Schachman, H. K. (1993). Aspartate transcarbamylase containing circularly permuted catalytic polypeptide chains. *Proc. Natl. Acad. Sci. USA* **90,** 11980–11984.

Young, V. R., and Munro, H. N. (1978). N^{τ}-methylhistidine (3-methylhistidine) and muscle protein turnover: An overview. *Fed. Proc.* **37,** 2291–2300.

Zhang, P., and Schachman, H. K. (1996). *In vivo* formation of allosteric aspartate transcarbamoylase containing circularly permuted catalytic polypeptide chains: Implications for protein folding and assembly. *Protein Science* **5,** 1290–1300.

Zhang, J. G., Moritz, R. L., Reid, G. E., Ward, L. D., and Simpson, R. J. (1992). Purification and characterization of a recombinant murine interleukin-6: isolation of N- and C-terminally truncated forms. *Eur. J. Biochem.* **207,** 903–913.

Chapter 3

Design and Use of Synthetic Peptides as Biological Models

Janelle L. Lauer and Gregg B. Fields

The development of accurate model systems is important in all aspects of scientific research. A common approach for dissecting protein structure and function is to utilize appropriate synthetic peptide models. Peptides have been constructed as mimics of protein primary, secondary, and tertiary structures and have been applied to studies of protein interactions with membranes, surfaces, and other proteins, including cell surface receptors, intracellular kinases, and numerous other enzymes. In this chapter we discuss the various ways in which peptides have been used as models for understanding protein behavior. Selected examples highlight the versatility, strengths, and weaknesses of peptide-based approaches.

I. Peptides as Structural Models of Proteins

A. Primary Structure of Peptides, Including Incorporation of Protein Posttranslational Modifications

Proteins are composed of amino acid building blocks and are usually >50 residues in length. The power of peptide chemistry, particularly the solid-phase method, allows for virtually exact reconstruction of protein primary structure (Fig. 1). Not only are the 20 common amino acids readily accommodated, but protocols exist for the assembly of peptides containing such posttranslational modifications as phosphorylated, glycosylated, hydroxylated, and sulfated residues and disulfide bonds (Fields *et al.*, 1992). On the level of primary structure, peptides can be used as models of proteins for study of biological function. For example, synthetic peptides containing Tyr(PO_3H_2) have been used to study the kinetics of dephosphorylation by several protein tyrosine phosphatases (Ottinger *et al.*, 1993). Glycosylated peptides have been constructed to create analogs of the immunodominant T-cell antigen located at α1(II)256–270 in type II collagen (Broddeflak *et al.*, 1996). Overall, peptides can be highly accurate models of protein primary structure.

B. Secondary and Tertiary Structures of Peptides

It has long been assumed that peptides exist in highly fluctuating states, with few defined secondary or tertiary structural elements. This is an inherant weakness for the use of peptides as models of proteins. Early attempts to identify secondary structure formation within peptide fragments of proteins were generally unsuccessful (see, for example, Epand and Scheraga, 1968). This led to the conclusion that nonlocal interactions occurring within proteins were essential for the formation of secondary structure and that short linear peptides were largely unstructured in aqueous solution. A major reason for the dearth of observations of secondary structure in short peptides in an aqueous environment was the absence of a sensitive probe for small populations of structured conformers within a conformational ensemble. However, one exception did exist early on: the C-peptide of ribonuclease A (residues 1–13). This peptide was found to populate helical conformations to approximately 30% [as assayed using circular dichroism (CD) spectroscopy] in aqueous solution at 1.7°C (Brown and Klee, 1971). Since that time, many peptide fragments of proteins have been found to populate a variety of folded conformations in aqueous solution, in dynamic equilibrium with fully unfolded forms (Dyson *et al.*, 1991). The utility of nuclear magnetic resonance (NMR) spectroscopy to examine protein structure has demonstrated that many peptides do indeed have distinct secondary structural elements, and that these elements may in some cases

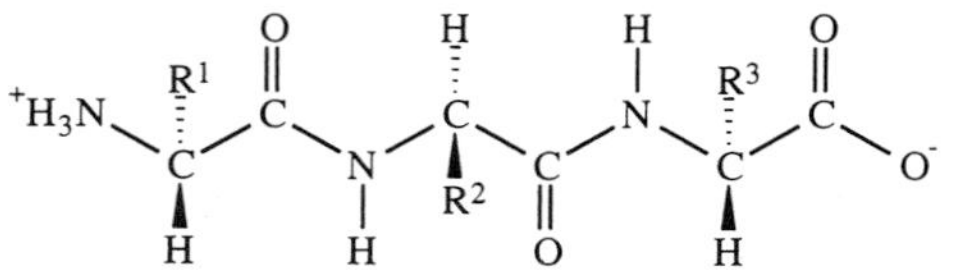

Fig. 1 Structure of the peptide backbone. Side-chain functionalities are represented by R_1, R_2, and R_3.

model the protein from which they are derived. By investigating short linear peptides derived from specific structural units in the native protein, that is, helix, turn, or β-sheet-turn regions, conformational preferences in early folding intermediates may be identified under equilibrium conditions. These intermediates may or may not remain as part of the native protein structure.

The most common secondary structural elements are α helices and β sheets (Fig. 2). An α helix is essentially formed of a series of multiple type-I β turns. In this respect, both helices and turns define secondary structural elements. Since a β sheet represents more nonlocal, tertiary structural interactions, it is likely that unfolded proteins and peptides have a higher tendency to form more local interactions such as turns and helices, which fold faster (Gruenewald *et al.*, 1979) than β sheets (Finkelstein, 1991). Most short linear peptides studied in aqueous solution, while showing some turn (e.g., Rose *et al.*, 1985), multiple-turn (e.g., Mayo *et al.*, 1991), or helical tendency (e.g., Burke *et al.*, 1991; Mayo

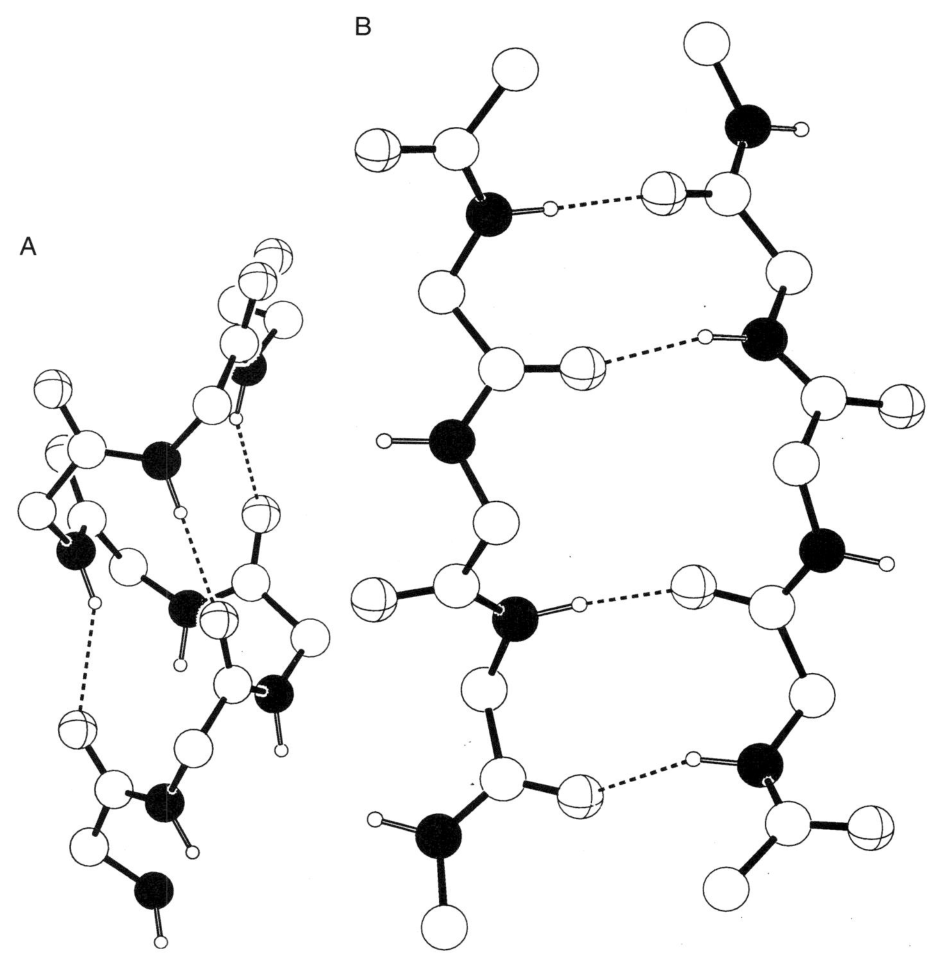

Fig. 2 Common protein secondary structural motifs. (A) α Helix, showing i, $i + 4$ hydrogen bonding. (B) Antiparallel β sheet, with hydrogen bonding between strands.

et al., 1990), are generally populated with random coil-like structures exhibiting some nascent helix conformation (Dyson *et al.*, 1988). Exceptions to this "rule of thumb" do exist and are discussed later. In most cases, conformation has been stabilized by the presence (either by design or by nature) of side-chain to side-chain or side-chain to backbone electrostatic interactions, hydrogen bonding, and hydrophobic interactions.

Aside from stabilizing forces arising from backbone to backbone hydrogen bonding interactions in helices and turns, *N*- and *C*-terminal hydrogen bonding capping interactions, generally between polar side-chain and backbone groups, have been observed (Zhou *et al.*, 1994, and references therein). These types of interactions have been noted to play an important role in helix stabilization even in short linear peptides. Although hydrophobic interactions play some general role in intramolecular conformational stabilization of short peptides, their more important role lies in intermolecular conformational stabilization between peptide fragments.

Electrostatic forces have been noted for some time as playing a crucial role in helix stability (Basharov *et al.*, 1986; Lyu *et al.*, 1989). Marqusee and Baldwin (1987) showed that helix stabilization is promoted by Glu^- . . . Lys^+ salt bridges in short peptides. Four Ala-based peptides (of the kind acetyl-Tyr-Glu-Ala-Ala-Ala-Lys-Glu-Ala-Ala-Ala-Lys-Glu-Ala-Ala-Ala-Lys-Ala-NH_2) were designed, synthesized, and tested by CD spectroscopy for α-helix formation in H_2O. Each terminally blocked peptide, 16 or 17 amino acids long, contained three Glu/Lys residue pairs. In one set of peptides ($i + 4$), the Glu and Lys residues were spaced 4 residues apart. In the other set ($i + 3$), the spacing was 3 residues. Within each of these sets, a pair of peptides was made in which the positions of the Glu and Lys residues were reversed [Glu, Lys vs Lys, Glu] in order to assess the interaction of the charged side chains with the helix dipole. Since the amino acid compositions of these peptides differed at most by a single Ala residue, differences in helicity were caused chiefly by the spacing and positions of the charged residues. The basic aim of this study was to test for helix stabilization by ion pairs (Glu^-, Lys^+) or salt bridges (H-bonded ion pairs). The results indicated as follows: (i) All four peptides showed significant helix formation, and the stability of the α helix did not depend on peptide concentration in the range studied. The best helix former was $i + 4$ Glu, Lys, which showed approximately 80% helicity in 0.01 *M* NaCl at pH 7 and 0°C. (ii) The two $i + 4$ peptides showed more helix formation than the $i + 3$ peptides. pH titration gave no evidence for helix stabilization by $i + 3$ ion pairs. (iii) Surprisingly, the $i + 4$ peptides formed more stable helices than the $i + 3$ peptides at extremes of pH (pH 2 and pH 12) as well as at pH 7. Their results (Marqusee and Baldwin, 1987) were explained by helix stabilization through Glu^- . . . Lys^+ salt bridges at pH 7 and singly charged H bonds at pH 2 (Glu0 . . . Lys^+) and pH 12 (Glu^- . . . Lys0). The reason why these links stabilized the α-helix more effectively in the $i + 4$ than in the $i + 3$ peptides remains unknown. (iv) Reversal of the positions of Glu and Lys residues usually affected helix stability in the manner expected for interaction of these charged groups with the helix dipole (Fairman *et al.*, 1989). (v) α-Helix formation in these Ala-based peptides was enthalpy-driven, as was helix formation by the C-peptide of ribonuclease A. Park *et al.* (1993) showed that the ionization of an amino acid residue can also alter the helix $\rightleftharpoons$ coil equilibrium of model peptides in the presence of high-ionic-strength solvents, which presumably screen out such electrostatic interactions. This observation suggests that ionization of a residue may change its inherent helical propensity.

Studies using electron-spin resonance labels have shown that short (16 residue) Ala-containing peptides form 3_{10} helices (i to $i + 3$ hydrogen bonding) instead of α helices (i to $i + 4$ hydrogen bonding) (Miick *et al.*, 1992). Transitions from the 3_{10} helix to an α helix occur with increased chain length (Fiori *et al.*, 1993).

Several model systems have been used to evaluate the α-helical propensities of different amino acids (Blaber *et al.*, 1993; Cox *et al.*, 1993; Dyson *et al.*, 1990; Horovits *et al.*, 1992; Lyu *et al.*, 1990; O'Neill and DeGrado, 1990; Padmanabha *et al.*, 1990). Chakrabartty *et al.* (1994) measured helix propensities of amino acids in 58 Ala-based peptides in the absence of helix-stabilizing side-chain interactions. A modified form of the Lifson–Roig (Lifson and Roig, 1961) theory for the helix $\rightleftharpoons$ coil transition, which includes helix capping (Doig *et al.*, 1994), was used to analyze the results. Substitutions were made at various positions of homologous helical peptides. Helix-capping interactions were found to contribute to helix stability, even when the substitution site was not at the end of the peptide. Analysis with the original Lifson–Roig theory, which neglects capping effects, does not produce as good a fit to the experimental data. At 0°C, Ala is a strong helix former, Leu and Arg are helix-indifferent, and all other amino acids are helix breakers of varying severity. Because Ala has a small side chain that cannot interact significantly with other side chains, helix formation by Ala is stabilized predominantly by the backbone ("peptide H-bonds"). The implication for peptide folding is that formation of peptide H-bonds can largely offset the unfavorable entropy change caused by fixing the peptide backbone. They conclude that helix propensities of most amino acids oppose folding, such that the majority of protein-derived helices are unstable unless specific side-chain interactions stabilize them.

Marqusee *et al.* (1989) also found unusually stable helix formation in short 16-residue, Ala-based peptides in H_2O. Their results were surprising when contrasted with the classical view that regards the α helix as a marginally stable structure in H_2O and considers short helices unstable. Although these Ala-based peptides were solubilized by insertion of three or more residues of a single charge type, Lys^+ or Glu^-, results could not be explained by helix stabilization resulting from concentration-dependent association or by the interaction of charged residues with the helix dipole. The host–guest method would predict that such a 16-residue peptide should not show measurable α-helix formation. Analysis of the role of the hydrophobic interaction in α-helix formation does not show an unusually strong hydrophobic interaction in a helical block of Ala residues. The likely explanation for these results is, therefore, that individual Ala residues have a high helical potential.

Krstenansky *et al.* (1989) have utilized all known factors important for α-helical stabilization in the design of model α-helical peptides based on the repeating 11 residues Glu-Leu-Leu-Glu-Lys-Leu-Leu-Glu-Lys-Leu-Leu-Lys. The CD spectra of these peptides provided evidence for more α-helical content than had been reported for any other short peptide of less than 18 amino acids. Moreover, α-helical tendency did not require the presence of cosolvents or reduced temperatures. For example, one 17-residue peptide showed 100% and

80% α-helical contents at concentrations of 1.7×10^{-4} *M* and 1.7×10^{-5} *M*, respectively. Another 10-residue peptide was 51% α-helical at 1.7×10^{-4} *M* in 0.1 *M* phosphate buffer at room temperature.

Wright and co-workers (Shin *et al.*, 1993a, Waltho *et al.*, 1993) have undertaken an extensive study of peptide fragments corresponding to regions in myoglobin that have been implicated in early folding events. Waltho *et al.* (1993) and Shin *et al.* (1993a) studied the conformational preferences of peptides corresponding to the G and H helices and of the intervening sequence, termed the G-H turn. The earlier studies mentioned previously on peptides derived from myoglobin concentrated on the *C*-terminal cyanogen bromide fragment (residues 132–153, termed Mb-F) which contains three extra nonhelical residues at the *C* terminus. Optical methods failed to detect helical structure in that peptide (Epand and Scheraga, 1968; Hermans and Puett, 1981), but small conformational preferences for helical structure were found by NMR (Waltho *et al.*, 1989). Shin *et al.* (1993a, b) and Waltho *et al.* (1993) demonstrated that a peptide corresponding to the entire H helix contains a considerably greater population of helical structure in solution than Mb-F1, as a consequence of the removal of nonhelical residues at the *C* terminus and the additional residues at the *N* terminus that form part of the H helix in the native protein.

Apocytochrome *c* (cyt *c*) has hydrodynamic and spectroscopic properties characteristic of a disordered coil (Damaschun *et al.*, 1991; Fisher *et al.*, 1973; Stellwagen *et al.*, 1972), indicating that formation of the native cyt *c* structure relies on the covalently attached heme group. Involvement of the heme group in the contact between *N*- and *C*-terminal regions of the folded state suggests it might play a role in helix–helix association during folding (Bushnell *et al.*, 1990; Takano and Dickerson, 1981). Wu *et al.* (1993) investigated a peptide model for the early cyt *c* folding intermediate, consisting of a noncovalent complex between a heme-containing *N*-terminal fragment (residues 1–38) and a synthetic peptide corresponding to the *C*-terminal helix (residues 87–104). Those studies indicated that a partially folded intermediate with interacting *N*- and *C*-terminal helices is formed at an early stage of folding when most of the chain is still disordered. Far-UV CD and ^{1}H NMR indicate that the isolated peptides are largely disordered, but when combined, they form a flexible, yet tightly bound complex with enhanced helical structure. These results emphasize the importance of interactions between marginally stable elements of secondary structure in forming tertiary subdomains in protein folding.

Kippen *et al.* (1994b) found that the region of barnase containing its two major helices, fragment 1–36, binds rapidly and tightly to the region containing its β sheet, fragment 37–110, to reconstitute an active enzyme. CD showed that the structure of each fragment was largely random coil. Individual fragments should not have significant catalytic activity because groups important for catalysis are shared between the fragments. In trifluoroethanol (TFE) solutions, the α-helical region adopts helical structure. Moreover, the fraction of helix present in water could be estimated from titration with TFE. It was postulated that one possible explanation for the rapid rate of association of the two fragments is that a small amount of the larger fragment also adopted a nativelike structure. Moult and Unger (1991) have predicted that the helical region, residues 10–18 in barnase, could act as one initiation site and the β hairpin, residues 85–100 in the center of the β sheet, as another. Both regions are formed in the folding intermediate of barnase, and their docking appears to be part of the rate-determining step for folding (Matousheck *et al.*, 1992). Using a second set of fragments, Kippen *et al.* (1994a) examined the association of fragments 1–22 and 5–21, which contain the major α helix (residues 6–18), with the fragment (23–110) that contains all of the residues involved in catalysis and all strands of the β sheet. It appears that barnase can fold by the association of independently folded parts, consistent with reports on previous aspects of barnase folding (Matouschek *et al.*, 1990, 1992; Serrano *et al.*, 1992a,b).

In terms of studying short peptides with predictive β-sheet folding tendencies, the situation is different, as attempts to observe isolated β hairpins/β sheets have usually been hampered by the strong tendency of these peptides to aggregate (Bazzi *et al.*, 1987). Osterman and Kaiser (1985) designed and characterized peptides with amphiphilic β-strand structures, by using CD and infrared spectroscopy. In a β-strand conformation, the synthetic peptides will possess a hydrophobic face composed of Val side chains and a hydrophilic face composed of alternating acidic (Glu) and basic (ornithine or Lys) residues. The peptides studied had a variety of chain lengths (5, 9, and 13 residues) and had the amino groups either free or protected with the trifluoroacetyl group. Although the peptides did not possess a high potential for β-sheet formation based on the Chou–Fasman parameters, they possessed significant β-sheet content, with up to 90% β-sheet calculated for the 13-residue protected peptide. The driving force for β-sheet formation is the potential amphiphilicity of this conformation. The β-strand conformation of the 13-residue deprotected peptide was stable in 50% TFE, 6 *M* guanidine hydrochloride (GdnHCl), and octanol. The peptides were strongly self-associating in H_2O, which would reduce the unfavorable contacts of the hydrophobic residues with H_2O.

The self-association of short peptides clearly can stabilize β-sheet formation in aqueous solution. The 33 amino acid β-sheet domain from PF4 (residues 23–55) has been studied by CD and NMR (Ilyina and Mayo, 1995). At 10°C and low concentration, peptide 23–55 appears to exist in aqueous solution in a "random coil" distribution of highly flexible conformational states. Some preferred conformation, however, is observed, particularly within a relatively stable chain reversal from Leu^{45} to Arg^{49}. As the peptide concentration and/or temperature is increased, one or more new conformational states appear and intensify as slowly exchanging (600 MHz ^1H NMR chemical shift time scale) random coil resonances disappear. Hill plots of the concentration dependence indicate mostly tetramer formation as found in native PF4. Although apparent resonance line widths in aggregate state(s) are on the order of 100 Hz, sequence-specific assignments for most resonances could be made. NMR/NOE structural analysis indicates the formation of multiple native-like antiparallel β-sheet conformations, kinetically trapped via subunit association-induced hydrophobic collapse and stabilized by low-dielectric electrostatic interactions among/between Glu^{28} and Lys^{50} in opposing subunits.

Terzi *et al.* (1994) had similar findings with the aggregation of fragment 25–35 of the β-amyloid protein [β-AP(25–35)OH] studied under a variety of conditions. β-AP(25–35)OH in solution at pH 4.0 or 5.5 exhibits a concentration-dependent random coil → β sheet transition. The equilibrium is characterized spectroscopically by an isodichroic point and can be described quantitatively by a simple association model with association constants between 1.8×10^4

M^{-1} (noncooperative model, nucleation parameter $\sigma = 1$) and $2.9 \times 10^4\ M^{-1}$ (cooperative model, $\sigma = 0.2$). The enthalpy of association is $\Delta H \sim -3$ kcal/mol as determined by titration calorimetry. The equilibrium is shifted completely toward β-structured fibrils at pH 7.4 where the Met^{35} carboxyl group is fully charged. In contrast, removal of the charged C terminus by amidation locks the equilibrium in the random coil conformation. Model calculations suggest an antiparallel β-sheet structure involving residues 28–35 that is stabilized at both ends of the β sheet by ion pairs formed between Lys^{28} and Met^{35}. Removal of fibrils via Millipore filtration leads to solutions with random coil monomers only. Seeding these solutions with a few fibrils establishes a new random coil $\rightleftharpoons$ β sheet equilibrium.

Two well characterized examples of β-sheet formation in short peptides are found in a disulfide-bridged heterodimeric model of a bovine pancreatic trypsin inhibitor (BPTI) folding domain (Kwon and Kim, 1994; Oas and Kim, 1988) and a zinc-finger peptide (Lee *et al.*, 1989). In both cases, the presence of a short length of β sheet was established by interstrand NOEs. On the other hand, a well-defined β-sheet structure stabilized by salt bridges has been identified using Fourier transform infrared spectroscopy (FTIR) and CD for an antigenic peptide (Muga *et al.*, 1990).

There is at present little evidence for the formation of monomeric, short linear peptide β sheets in aqueous solution. Two cases have, however, been reported, from the B1 domain of immunoglobulin G binding protein (protein G) (Blanco *et al.*, 1994a; Gronenborn *et al.*, 1991) and platelet factor-4 (Ilyina and Mayo, 1995; Ilyina *et al.*, 1994). Protein G consists of 56 amino acids arranged in a fold with a central α helix packed against two β hairpins. Despite its small size and absence of disulfide bridges, the thermodynamic properties of protein G are similar to those of larger proteins, and it is quite resistant to denaturation by heat or urea (Alexander *et al.*, 1992). Structural characterization of a fragment corresponding to the first β hairpin in aqueous solution has indicated that the peptide does not adopt the β-hairpin conformation to any significant degree, although in 30% aqueous TFE there were clear signs of β-hairpin formations. Blanco *et al.* (1993, 1994a,b) have investigated the conformational properties of a 16-residue peptide, corresponding to the second β hairpin of protein G, by NMR. This fragment, monomeric in pure H_2O, adopts a conformational population containing up to 40% native-like β-hairpin structure. The detection by NMR of a native-like β hairpin in aqueous solution indicates that these structural elements may have an important role in the early steps of protein G folding.

A peptide corresponding to the amino acid region 71–93 of the plasma protein transthyretin (TTR) has been synthesized to investigate its role in the native folding of the molecule and the possible relationship between mutations in this region and amyloid formation of TTR (Jarvis *et al.*, 1994). In the native structure this fragment includes a β strand followed by a short helix and turns back on itself to form part of an antiparallel β sheet. NMR spectroscopy has shown that minor populations of turn-like character were apparent in deuterated dimethyl sulfoxide (DMSO-d_6), while some indication of nascent helix between residues 5 and 12 was observed in water, and upon the addition of 20% TFE. The intrinsic tendency to form a helical structure between residues 5 and 12 in solution suggests that the helical region, also present in the native crystallographically determined TTR structure at corresponding residues 75–82, is an important folding initiation site. In contrast, the β-sheet motif observed in the native structure was not detected in solution.

Platelet factor-4 (PF4) is a 70-residue protein that contains a three-stranded antiparallel β-sheet domain onto which is folded a *C*-terminal α helix and an aperiodic *N*-terminal region. Ilyina *et al.* (1994) have studied three synthetic PF4-derived peptides from β-sheet (residues 24–46 and 38–57) and helix (residues 57–70) domains in aqueous solution by using CD and NMR. Although peptides 24–46 and 56–70 demonstrate some weak conformational preferences, peptide 38–57 maintains the relatively well-defined, NOE-rich chain reversal sequence Leu^{45}-Lys^{46}-Asn^{47}-Gly^{48}-Arg^{49}-Lys^{50}, which apparently is stabilized by hydrophobic side-chain interactions from flanking sequences Leu^{41}-Leu^{45} and Ile^{51}-Leu^{53}. Some helix-like conformational populations are noted in the native PF4 Ile^{42}-Ala^{43}-Thr^{44}-Leu^{45} β-strand segment. NOE-based distance geometry calculations yield native-liek conformations within the Leu^{45}-Lys^{50} sequence. Among 40 structures, backbone root mean square (RMS) deviations range from 0.5 to 1.2 Å, and compared to the same sequence in native PF4, the average RMS deviation is 1.1 Å. These results suggest that β-sheet/turn residues Leu^{41}-Leu^{53} present a folding initiation site on the PF4 folding pathway. Both the B1 domain protein G-derived peptide and the PF4-derived β-sheet peptides provide good models to study in detail the sequence determinants of β-hairpin structure stability, as has been done with α-helices. A *de novo* designed α-helical hairpin peptide has also been proposed as a model to study protein folding "intermediates" (Fezoui *et al.*, 1994).

Several laboratories have reported on the design of small peptides with distinct structures using zinc finger motifs. Berg and colleagues used 7 conserved residues characteristic of zinc finger domains to construct a 26-residue peptide that bound zinc and had a three-dimensional structure comparable to zinc finger motifs (Michael *et al.*, 1992). Alternatively, the laboratory of Imperiali used an iterative design process to reproduce the $\beta\beta\alpha$ architecture of zinc fingers in a 23-residue peptide (Struthers *et al.*, 1996a,b).

Peptides have long been used as structural models of collagenous proteins. Collagen is the major structural protein of all connective tissues, including skin, bone, basement membrane, and blood vessels. Vertebrates contain at least 19 different collagen types (Linsenmayer, 1991; Prockop and Kivirikko, 1995). Collagens are composed of three peptide strands (α chains) of primarily repeating Gly-X-Y triplets, which induce each α chain to adopt a left-handed polyPro II helix (Brodsky and Shah, 1995). Three left-handed chains then intertwine to form a right-handed triple-helical coiled coil. The X and Y positions of collagen sequences are occupied frequently by imino acids (Pro and Hyp), which favor triple-helix formation and stability. Collagens can be distinguished from other proteins based on their triple-helical structure. In 1968, Sakakibara and colleagues used the solid-phase procedure to assemble the first two triple-helical peptides of a defined length, $(Pro\text{-}Pro\text{-}Gly)_{10}$ and $(Pro\text{-}Pro\text{-}Gly)_{20}$ (Sakakibara *et al.*, 1968). These sequences were designed based on the collagen X-Y-Gly repeat and the high content of imino acids (Pro and Hyp) found in the X and Y positions of native collagens. In the early 1990s, the two-dimensional NMR techniques TOCSY (totally correlated spectroscopy) and NOESY (nuclear Overhauser enhancement spectroscopy) were used to assign Pro $C_{\delta}H$ proton and Gly $C_{\alpha}H$ proton resonances in triple-helical and non-triple-helical (melted) $(Pro\text{-}Hyp\text{-}Gly)_{20}$ (Brodsky *et al.*, 1992). One-dimensional ^{1}H NMR spectra showed a quantitative relationship beween the Pro $C_{\delta}H$ and Gly $C_{\alpha}H$ proton resonances and triple helicity (Brodsky *et al.*, 1992; Long *et al.*, 1992). Specifically, the Pro $C_{\delta}H$ proton resonance at 3.2 ppm was present only in the triple-helical molecule, whereas the Gly $C_{\alpha}H$ proton resonance at 4.2 ppm was present only in the non-triple-helical molecule, with the total area of these two peaks remaining constant during a triple-helical melt of $(Pro\text{-}Hyp\text{-}Gly)_{10}$ (Brodsky *et al.*, 1992; Long *et al.*, 1992). The extent of triple helicity could thus be quantitated by dividing the area of the 3.2-ppm peak by the area of the 3.2-ppm + 4.2-ppm peaks. X-ray crystallographic studies further confirmed the triple-helical structures of $(Pro\text{-}Pro\text{-}Gly)_{10}$ (Okuyama *et al.*, 1981, 1976) and $(Pro\text{-}Hyp\text{-}Gly)_{10}$ (Bella *et al.*, 1994).

Associated triple-helical peptides have been studied for (i) individual chain orientation (parallel versus antiparallel) and registar and (ii) the effects of chain length and amino acid composition on triple helicity. Orientation and registar were first studied by Berg and associates (1970), who examined changes in the T_m of the $(Pro\text{-}Pro\text{-}Gly)_{10}$ triple helix as a function of pH. The T_m of the triple helix was greatest at either low pH (<3) or high pH (>9), that is, fully protonated or fully unprotonated (Berg, 1970). Similar results were obtained using $(Pro\text{-}Hyp\text{-}Gly)_{10}$ (Venugopal *et al.*, 1994). Based on electrostatic effects, it was proposed that triple-helical peptides such as $(Pro\text{-}Pro\text{-}Gly)_{10}$ have all three chains in registar and packed in parallel fashion, similar to collagen (Berg *et al.*, 1970). Electron microscopy studies of $(Pro\text{-}Pro\text{-}Gly)_{10}$ and $(Pro\text{-}Pro\text{-}Gly)_{20}$, either negatively stained with sodium silicotungstate or uranyl acetate or positively stained with silicotungstic acid, also indicated that the three individual chains were parallel and in registar (Olsen *et al.*, 1971).

Replacement of Pro in either the X or Y position by Hyp was found to have significant effects on triple helicity. For example, optical rotatory dispersion (ORD) measurements in 10% aqueous acetic acid showed that $(Pro\text{-}Hyp\text{-}Gly)_{10}$ had a $T_m = 58°C$ (T_m/triplet = 1.93°C), whereas $(Pro\text{-}Pro\text{-}Gly)_{10}$ had a $T_m = 24°C$ (T_m/triplet = 0.83°C) (Sakakibara *et al.*, 1973). In 50% aqueous ethanol, $(Pro\text{-}Hyp\text{-}Gly)_5$ had a $T_m = 5°C$, whereas $(Pro\text{-}Pro\text{-}Gly)_5$ did not appear to form a triple helix. This initial study established the stabilizing influence of Hyp in the Y position. Acetylation of Hyp reduced the T_m of $(Pro\text{-}Hyp\text{-}Gly)_{10}$ from 58 to 25°C in 1.0 *M* NaCl (Weber and Nitschmann, 1978). $(Hyp\text{-}Pro\text{-}Gly)_{10}$ did not form a triple helix, indicating that although Hyp in the Y position was triple-helix stabilizing compared with Pro, Hyp in the X position was triple-helix destabilizing compared with Pro (Inouye *et al.*, 1982). It was suggested that Hyp in the X position may help stabilize single-stranded peptide structures, whereas Hyp in the Y position stabilizes triple helices. X-ray crystallographic studies indicated that the additional thermal stability resulting from Hyp in the Y position is due to enhanced water bridging provided by the 4-hydroxyl group (Bella *et al.*, 1995, 1994).

One- and two-dimensional NMR studies of a native collagen sequence "sandwiched" between Pro-Hyp-Gly repeats [$(Pro\text{-}Hyp\text{-}Gly)_3$-Ile-Thr-Gly-Ala-Arg-Gly-Leu-Ala-Gly-$(Pro\text{-}Hyp\text{-}Gly)_4$] indicated that (i) Gly residues within the collagen sequence are dynamically different (i.e., have faster hydrogen exchange rates) than Gly residues within Pro-Hyp-Gly repeats, (ii) an equilibrium between trimeric and monomeric species existed, and (iii) an alternate conformation was not present within the collagen sequence (Fan *et al.*, 1993; Li *et al.*, 1993). These results are consistent with the concept that significant cooperativity occurs between triple-helical stabilizing residues. Electrostatic effects can also contribute favorably to the stability of collagen-model triple-helical peptide. (Balakrishnan *et al.*, 1995; Venugopal *et al.*, 1994), whereas hydrophobic interactions do not contribute to triple-helical stability (Shah *et al.*, 1996).

To study the nucleation step for the folding of collagen, a protocol was developed for liquid-phase synthesis by which three peptide strands were covalently linked via a *C*-terminal branch (Germann and Heidermann, 1988; Roth *et al.*, 1979; Roth and Heidemann, 1980; Thakur *et al.*, 1986). The C-terminal branch was expected to enhance triple-helical thermal stability, and to provide a model of the disulfide-linked *C* terminus of type III collagen (Roth and Heidemann, 1980). The folding kinetics of triple helices have also been evaluated using NMR measurements of the association of collagen-model peptides (Liu *et al.*, 1996). In addition, the association of heterotrimeric triple helices has been studied using peptide models of the *C*-terminal region of type IX collagen (Mechling *et al.*, 1996). It was found that the NC1 domain contained all of the information necessary for proper heterotrimer chain selection and assembly.

Fields and co-workers have systematically evaluated several general solid-phase methods for synthesizing covalently branched triple-helical peptides (THPs) that incorporate native collagen sequences (Fields and Fields, 1992; Fields *et al.*, 1993a). Other branching methods using *N*-terminal Lys-Lys structures (Tanaka *et al.*, 1996, 1993) or Kemp triacids (*cis*, *cis*-1,3,5-trimethylcyclohexane-1,3,5-tricarboxylic acid) (Feng *et al.*, 1996; Goodman *et al.*, 1996a) have been developed. All have been shown to enhance triple-helical stability. Work with the Kemp triacid template has shown that peptoid residues such as *N*-isobutylglycine can be incorporated into triple helices (Goodman *et al.*, 1996b; Melacini *et al.*, 1996).

Several receptors expressed by macrophages contain collagen-like triple-helical regions, such as scavenger receptors types I and II (Ashkenas *et al.*, 1993; Kodama *et al.*, 1990; Rohrer *et al.*, 1990) and the bacteria-binding receptor MARCO (Elomaa *et al.*, 1995). A triple-helical model of the bovine type I scavenger receptor incorporating residues 332–343 has been assembled by a chemoselective ligation strategy using a peptide branch that contains bromoacetylated *N* termini and a collagen sequence that contains an *N*-terminal Cys (Tanaka *et al.*, 1996, 1993). Of all of the peptide structural models discussed here, the widest range of applications has been for the collagen triple-helical supersecondary structure.

C. Surface Behavior of Peptides

The design of biomaterials has examined the surface behavior of proteins with the goal of improving biocompatibility. Peptides containing protein active sites could be attached to biomaterials to improve biocompatibility. As in the case of protein folding, peptide models have been used to examine protein interactions with surfaces. Lattice statistical mechanics predicts that surfaces would enhance peptide secondary structure formation by minimizing unfolded states

(Chan *et al.*, 1991; Wattenbarger *et al.*, 1990). For very simple peptide models, surfaces have been shown to induce secondary structures, either α helical or β sheet (DeGrado and Lear, 1985; Osapay and Taylor, 1990, 1992). Specific examples exist of where peptides retain biological function upon attachment to surfaces. Mooradian and co-workers prepared a photoreactive analog of a fibronectin-derived peptide, FN-C/H-V (Huebsch *et al.*, 1996). FN-C/H-V has been shown to support cell adhesion and spreading *in vitro* (Mooradian *et al.*, 1993). The photoreactive peptide was attached to polystyrene or polyethylene terephthalate film to assess its usefulness in preparing biomaterials that would support re-endothelialization. The FN-C/H-V peptide analog retained its biological function when attached to either biomaterial, suggesting a generally effective and convenient means of modifying surfaces to provide cell adhesion-promoting properties (Huebsch *et al.*, 1996).

D. Peptide Insertion into Membranes

Peptides have long been used as models to understand how proteins span cell membranes and how ion-channel proteins function across biological membranes. In some cases, sequences from naturally occurring membrane-spanning proteins have been used. For example, to study the configurational selectivity of biological membranes, several membrane-spanning proteins were synthesized in their native all-L-amino acid forms and in the mirror-image all-D-amino acid forms (Wade *et al.*, 1990). The D enantiomers of cecropin A, magainin 2 amide, and melittin had equivalent activity compared with the L-peptides in regards to conductivity and antibacterial activity (Wade *et al.*, 1990). In each case, the peptide activity involved the formation of ion-channel pores spanning the membranes, without specific interaction with receptors or enzymes (Wade *et al.*, 1990). Thus, peptide models were used to demonstrate that membranes did not have chiral preferences in terms of interaction with proteins.

A synthetic peptide from the mas oncogene receptor corresponding to residues 253–266 was made with and without a spin label on the *N* terminus. Peptide binding to bilayers and micelles was monitored by electron spin resonance (ESR) imaging and CD spectroscopy. Conformational changes are detected both at the peptide and at the lipid level when the mas 253–266 peptide interacts with lipid structures. The lipid environment becomes more tightly packed as a result of peptide binding (Pertinhez *et al.*, 1995). There is also an increase in peptide secondary structure (an increased α-helix content) upon binding to micelles and lipid bilayers (Pertinhez *et al.*, 1995). The ESR and CD data are in agreement with the predicted conformation for the entire mas oncogene receptor where amino acids 253–266 are located, and thus the peptide is capable of displaying a behavior similar to that of the intact protein (Pertinhez *et al.*, 1995). In addition, it was shown that peptide binding to bilayers occurred only when peptide and phospholipid bore opposite charges, whereas binding to micelles took place irrespective of charge (Pertinhez *et al.*, 1995). These results suggest that changes in the lipid packing of a membrane could modulate conformational changes in receptor domains related to the triggering of signal transduction.

Designed peptides were used to study the relationships among peptide helicity, model membrane permeability, and biological activity by dye release from liposomes and antibacterial and hemolytic activity. Accumulation of Lys-Leu-Ala-Leu-Lys-Leu-Ala-Leu-Lys-Ala-Leu-Lys-Ala-Ala-Lys-Leu-Ala-NH_2 peptides at lipid bilayers and the membrane-disturbing effect on bilayers of high negative surface charge were found to be dominated by charge interactions. Cationic peptide side chains bind to anionic phosphatidylglycerol moieties, where they disturb lipid headgroup organization by forming peptide–lipid clusters. Peptide interaction with bilayers of low negative surface charge was highly dependent on peptide helicity. Less helical peptides exhibit reduced bilayer-disturbing activity; thus, the hydrophobic helix domain is decisive for binding at and inducing permeability in membranes of low negative surface charge. Hydrophobic interactions drove the penetration of the amphipathic peptide structure into the inner membrane region, thus disturbing the arrangement of the lipid acyl chains and causing local disruption (Dathe *et al.*, 1996).

One specific example of peptide models for understanding ion transport is that of gramicidin A. The high-resolution structure of gramicidin A in lipid bilayers was determined using solid-state NMR and synthetic, isotopically labeled gramicidins (Ketchem *et al.*, 1993). Gramicidin A (Formyl-Val-Gly-Ala-D-Leu-Ala-D-Val-Val-D-Val-Trp-D-Leu-Trp-D-Leu-Trp-D-Leu-Trp-NH-CH_2-CH_2-OH) forms cation-selective channels in membranes. Uniformly aligned samples of isotopically labeled gramicidin A in dimyristoylphosphatidylcholine were shown to form single-stranded right-handed β helices, with six to seven residues per turn. The carbonyl oxygens are directed toward the channel lumen and available for solvating cations. The incorporation of isotopic labels into synthetic peptides allowed for the gramicidin channel structure to be solved.

Alternatively, *de novo* designed peptides have been used to study ion channels. (Leu-Ser-Ser-Leu-Leu-Ser-Leu)$_3$-NH_2 and (Leu-Ser-Leu-Leu-Leu-Ser-Leu)$_3$-NH_2 were designed to be membrane-spanning amphiphilic α-helices with ion channel activity (Lear *et al.*, 1988). Both peptides were active, with the latter showing proton selectivity. Further studies with (Leu-Ser-Ser-Leu-Leu-Ser-Leu)$_3$-NH_2 indicated that the peptide is induced by the transmembrane voltage gradient to switch from a surface to a transmembrane orientation (Chung *et al.*, 1992). Modification of the (Leu-Ser-Leu-Leu-Leu-Ser-Leu)$_3$-NH_2 peptide with α-helical inducing α-aminoisobutyric acid (Aib) residues produced (Leu-Ser-Leu-Aib-Leu-Ser-Leu)$_3$-NH_2, which had structural and conductive properties similar to those of the parent peptide (DeGrado and Lear, 1990).

II. Cell/Peptide Interactions

A. Peptides and Cell Adhesion, Spreading, Migration, and Invasion

Normal and pathogenic cell behaviors involve interactions with extracellular matrix (ECM) glycoproteins, such as laminin, fibronectin, collagen, vitronectin, entactin/nidogen, and thrombospondin. An approach to better define the mechanisms of cell adhesion, spreading, and motility is to utilize synthetic peptides to "map" regions of ECM proteins that promote cellular activities (reviewed in Humphries, 1990; Yamada, 1991). Peptides have been derived from laminin, fibronectin, and collagen, that promote normal and tumor cell adhesion, spreading, and migration and/or inhibit or promote tumor cell metastasis (Table I).

Table I
Cell Adhesion-Promoting Peptides Derived from Extracellular Matrix Proteins

Protein	Location	Code	Sequence
Fibronectin	1492–1497		GRGDSP
Fibronectin	1906–1924	FN-C/H-I	YEKPGSPPPREWPRPRPGV
Fibronectin	1946–1960	FN-C/H-II	KNNQKSEPLIGRKKT
Fibronectin	1721–1736	FN-C/H-III	YRVTVTPKEKTGMKE
Fibronectin	1784–1792	FN-C/H-IV	SPPRRARVT
Fibronectin	1892–1899	FN-C/H-V	WQPPRARI
Fibronectin	1961–1985	CS-1	DELPQLVTLPHPNLHGPEILDVPST
Laminin	A Chain 43–63	TG-1/R18	RPVRHAQCRVCDGNSTNPRERH
Laminin	A Chain 1115–1129	PA21	CQAGTFALRGDNPQG
Laminin	A Chain 2087–2101	AD-1	KLLISRARKQAASIK
Laminin	A Chain 2091–2108	PA22-2	CSRARKQAASIKVAVSADR
Laminin	A Chain 2362–2382	GD-7	TDRRYNNGTWYKIAFQRNRKQ
Laminin	A Chain 2443–2463	GD-3	KNLEISRSTFDLLRNSYGVRK
Laminin	A Chain 2547–2565	GD-5	TSLRKALLHAPTGSYSDGQ
Laminin	A Chain 2615–2631	GD-1	KATPMLKMRTSFHGCIK
Laminin	A Chain 2779–2795	GD-4	DGKWHTVKTEYIKRKAF
Laminin	A Chain 2890–2910	GD-2	KEGYKVRLDLNITLEFRTTSK
Laminin	A Chain 3011–3032	GD-6/R30	KQNCLSSRASFRGCVRNLRLSR
Laminin	B-1 Chain 202–218	AC15	RIQNLLKITNLRIKFVK
Laminin	B-1 Chain 641–660	F-9	RYVVLPRPBCFEKGMNYTVR
Laminin	B-1 Chain 925–933	Peptide 11	CDPGYIGSR-NH_2
Laminin	B-1 Chain 960–978	F-11	NIDTTDPEACDKDTGRCLK
Laminin	B-1 Chain 1133–1148	F-12	VEGVEGPRCDKCTRGY
Laminin	B-1 Chain 1171–1188	F-13	ELTNRTHKFLEKAKALKI
Type I Collagen	α1(I)430–442		GPAGKDGEAGAQG
Type I Collagen	α1(I)772–786		GPQGIAGQRGVVGLP*
Type III Collagen	α1(III)53–64		PSGKDGEP*GRP*G
Type III Collagen	α1(III)65–76		RP*GERGLP*GPP*G
Type III Collagen	α1(III)77–88		IKGPAGIP*GFP*G
Type III Collagen	α1(III)478–489	CB4	GKP*/PGEP*/PGPKGEA
Type IV Collagen	α1(IV)531–543	HEP-III	GEFYFDLRLKGDK
Type IV Collagen	α1(IV)1263–1277	IV-H1	GVKGDKGNPGWPGAP

Fibronectin is a dimeric 440- to 550-kDa protein composed of highly homologous A and B chains (reviewed in Ruoslahti, 1988). Both chains are composed of repeating modules referred to as type I (~45 residues), type II (~60 residues), and type III (~90 residues) domains (Ruoslahti, 1988). The C-terminal domain of fibronectin contains many cellular binding sites (Drake *et al.*, 1992, 1993; Haugen *et al.*, 1992; Pierschbacher and Ruoslahti, 1984; Mohri and Ohkubo, 1993; Mooradian *et al.*, 1993). The well-known "RGD" cell adhesion sequence is found in the type III domain on a loop between β strands (Baron *et al.*, 1992). The Arg-Gly-Asp sequence occupies residues 1493–1495 in the human plasma fibronectin A chain. Arg-Gly-Asp-Ser was initially established as promoting adhesion of rat kidney fibroblasts and inhibiting adhesion of these cells to fibronectin (Pierschbacher and Ruoslahti, 1984). Pretreatment of cells *ex vivo* with Gly-Arg-Gly-Asp-Ser was subsequently shown to reduce the metastatic spread of mouse melanoma cell colonies to the lungs, whereas Gly-Arg-Gly-Glu-Ser was not effective (Humphries *et al.*, 1986). Cell surface integrin receptors for fibronectin that recognize Arg-Gly-Asp sequences include $\alpha 5\beta 1$, $\alpha v\beta 3$, $\alpha IIb\beta 3$, and $\alpha v\beta 6$ (reviewed in Ruoslahti, 1991; Hynes, 1992). Integrin recognition of Arg-Gly-Asp has been proposed to be conformation dependent (Ruoslahti 1991). X-ray crystallographic studies of Arg-Gly-Asp show an extended conformation with a kink at Gly (Eggleston and Feldman, 1990).

Arg-Gly-Asp-containing synthetic peptide analogs were made to determine the structural requirements of biological activity, as assayed by inhibition of cell adhesion to fibronectin or vitronectin. Substitution of L-Arg with D-Arg showed no difference in activity, but substituting Gly with D-Ala, or L-Asp with D-Asp, resulted in completely inactive peptides (Pierschbacher and Ruoslahti, 1987). Substitution of L-Ser with D-Ser resulted in a peptide that showed good inhibition of adhesion to vitronectin, but no inhibition to fibronectin-mediated adhesion. When the Ser was substituted with L-Asn, the peptide showed an increased preference for the fibronectin receptor as compared to the vitronectin receptor. Substituting this same residue with Thr had the opposite effect. Cyclization of Arg-Gly-Asp in the form of Gly-Pen-Gly-Arg-Gly-Asp-Ser-Pro-Cys-Ala (where Pen is penicillamine) produced a more potent inhibitor of cell adhesion (receptor gpIIb-IIIa) to vitronectin. This peptide had almost no effect on cell adhesion to fibronectin (Pierschbacher and Ruoslahti, 1987). However, the head-to-tail cyclic Arg-Gly-Asp peptide cyclo[Gly-Arg-Gly-Asp-Ser-Pro-Ala] was a more effective inhibitor of platelet adhesion to fibronectin than the linear form (Mohri *et al.*, 1991). Cyclic Gly-Arg-Gly-Asp-Ser-Pro-Ala is 20-fold more effective at inhibiting mouse melanoma cell adhesion to fibronectin than the linear peptide and 10-fold more effective at reducing *in vivo* mouse melanoma cell colony formation in the lungs (Kumagai *et al.*, 1991). A peptide model of the Tyr-Gly-Arg-Gly-Asp-Ser-Pro sequence found in fibronectin forms a type II β turn, whereas the integrin inactive sequence Tyr-Gly-Arg-Gly-Glu-Ser-Pro assumes a type I or III β turn (Johnson *et al.*,1993). These results suggest the possibility that peptides could be tailored to become more selective for individual receptors than the peptides modeled after natural ligands.

Several sites within fibronectin have been characterized for Arg-Gly-Asp-independent cell adhesion. One of these, a peptide model of residues 1946–1960 from the C terminus of a type III domain of the fibronectin A chain designated "FN-C/H-II," supported cellular adhesion, spreading, and migration when covalently coupled to ovalbumin (McCarthy *et al.*, 1988). Adhesion of FN-C/H-II is initiated by a cell surface phosphatidylinositol-anchored heparan sulfate proteoglycan in neuronal, melanoma, and epithelial cells (Drake *et al.*, 1992; Haugen *et al.*, 1992; Drake *et al.*, 1993; Mooradian *et al.*, 1993). The C-terminal Ser-Glu-Pro-Leu-Ile-Gly-Arg-Lys-Lys-Thr-Tyr region has the same level of mouse melanoma cell adhesion activity as intact FN-C/H-II and nearly equal ability to bind heparan sulfate proteoglycan (Drake *et al.*, 1993). Each basic residue of the Arg-Lys-Lys region is important for mouse melanoma cell adhesion and binding heparan sulfate proteoglycan (Drake *et al.*, 1993). FN-C/H-II has been implicated as one of several sites in fibronectin that may induce "clustering" of melanoma cell surface receptors and influence cellular behavior (Iida *et al.*, 1992). In fact, the use of fibronectin-derived peptides led to the development of model in which two cell surface receptors, the $\alpha 4\beta 1$ integrin and chondroitin sulfate proteoglycan, have a coordinate role in mediating melanoma cell adhesion to fibronectin (Iida *et al.*, 1992).

The process of focal adhesion and stress fiber formation requires downstream signaling in addition to Arg-Gly-Asp promotion of cell adhesion. The heparin-binding fragments of fibronectin provide this signal. Five fibronectin-derived synthetic peptides, FN-C/H-I through FN-C/H-V, supported fibroblast cell attachment when coupled to ovalbumin. Three of the ovalbumin-coupled peptides (FN-C/H-I, III, and V) promoted focal adhesion formation, while only one of these (FN-C/H-V) retained its activity when used as an uncoupled peptide (Woods *et al.*, 1993). FN-C/H-V also promotes corneal epithelial cell adhesion, spreading, and motility (Mooradian *et al.*, 1993).

Leukocytes play an important role in the development of ischemia-reperfusion injury. The fibronectin-derived peptide FN-C/H-V was tested for its ability to protect brain tissue from injury due to transient artery occlusion. In theory, the peptide's protective function would occur by blocking leukocyte adhesion and infiltration into the damaged tissue. There was a significant reduction in infarct size and leukocyte infiltration in the ischemic tissue in rats when peptide FN-C/H-V was added prior to, during, and after cerebral artery occlusion (Yanaka *et al.*, 1996).

Mice homozygous for loss of transforming growth factor-β1 (TGF-β1) undergo mononuclear leukocyte (MNL) infiltration followed by death. In TGF-β1 knockout mice, MNL cells exhibit enhanced adhesion to ECM proteins and endothelial cells. Synthetic peptides corresponding to cell and heparin-binding domains of fibronectin significantly reduced this upregulated MNL adhesion. These peptides were able to rescue the mice from tissue inflammation, cardiopulmonary failure, and death when administered intraveinously (Hines *et al.*, 1994).

Laminin is a cruciform-shaped protein composed of three polypeptide chains, A (300–400 kDa), B-1 (180–200 kDa), and B-2 (180–200 kDa) (reviewed in Engel, 1992). The "long arm" region where all three chains associate is an α-helical coiled conformation (Engel, 1992). The "short arm" regions contain globular domains and less well defined secondary structures (Engel, 1992). A peptide incorporating residues 925–933 from the "short arm" of the mouse Engelbreth–Holm–Swarm (EHS) tumor B-1 chain, commonly referred to as peptide 11, was the first laminin peptide demonstrated to inhibit experimental metastasis following *ex vivo* treatment of cells prior to tail-vein injection (Iwamoto *et al.*, 1987). Peptide 11 supports adhesion and spreading of mouse melanoma cells. In human fibrosarcoma cells, peptide 11 supports adhesion alone (Tashiro *et al.*, 1989). This peptide also inhibits growth of solid tumor tissue of sarcoma and lung tumor colonies, but not ascitic tumor of sarcoma (Sakamoto *et al.*, 1991). More malignant lines of mouse melanoma cells showed greater affinity for peptide 11. Cellular recognition of this sequence is believed to affect both tumor colony formation and growth (Yamamura *et al.*, 1993).

A laminin peptide incorporating residues 641–660 from the globular domain of the mouse EHS tumor B-1 short arm referred to as F-9 promotes the adhesion of mouse melanoma, mouse fibrosarcoma, and rat glioma cells and inhibits the adhesion of mouse melanoma cells to laminin (Charonis *et al.*, 1988). Metastatic mouse fibrosarcoma and human renal carcinoma cells adhere to and spread on peptide F-9. This peptide also inhibits adhesion and migration of these cell lines toward laminin (Skubitz *et al.*, 1990).

A peptide incorporating residues 2091–2108 from the C-terminal region of the mouse EHS tumor laminin A chain "long arm," referred to as PA22-2, promotes adhesion of mouse melanoma and human fibrosarcoma cells, and migration of mouse melanoma cells (Tashiro *et al.*, 1989). Interestingly, human adenocarcinoma cells showed no adhesion or spreading activity on this sequence (Tashiro *et al.*, 1989). PA 22-2 increases experimental metastasis of the lungs by mouse melanoma cells (Kanemoto *et al.*, 1990). All-L and all-D versions of the Ala-Ala-Ser-Ile-Lys-Val-Ala-Val-Ser-Ala-Asp-Arg region of PA22-2 increased *in vivo* mouse melanoma tumor growth equally well, suggesting that the cell surface receptor or receptors do not discriminate for chirality (Nomizu *et al.*, 1992).

An $\alpha 3\beta 1$ integrin binding site was identified in laminin within the C-terminal region of the molecule (Gehlsen *et al.*, 1992). This region was determined first by screening laminin-derived synthetic peptides for the ability to support cell adhesion. The site, identified as residues 3011–3032 of laminin, was then attached to a solid support and incubated with cell lysates. Since the $\alpha 3\beta 1$ integrin was the only receptor isolated from the peptide affinity column, it was determined to be the receptor mediating cell binding to this region within laminin (Gehlsen *et al.*, 1992).

Types I, II, III, and IV collagen are among the most important of structural proteins. Collagens are distinguished from other ECM proteins by their triple-helical conformation. The interaction of cells with triple-helical collagens influences normal physiological processes, such as endothelial cell initiation of sheet migration and restoration of cellular continuity when blood vessel walls are damaged, and pathological conditions, such as tumor cell metastasis and atherosclerosis. Several regions or sequences within the triple-helical regions of types I, II, III, and IV collagen have been identified as cell adhesion sites (Table I). Triple-helical structure can be important for preserving high-affinity cell binding sites (Aumailley and Timpl, 1986; Eble *et al.*, 1993; Gullberg *et al.*, 1992; Morton *et al.*, 1994; Perris *et al.*, 1993; Pfaff *et al.*, 1993; Tuckwell *et al.*, 1994; Vandenberg *et al.*, 1991). Cellular adhesion to the triple-helix of native collagens can be mediated by the $\alpha 1\beta 1$, $\alpha 2\beta 1$, and $\alpha 3\beta 1$ integrins (Kuhn and Eble, 1994), as well as cell surface proteoglycans (Faassen *et al.*, 1992a). Adhesion to denatured collagen can involve cell surface receptors, such as the $\alpha 5\beta 1$ and $\alpha v\beta 3$ integrins (Davis, 1992; Gullberg *et al.*, 1992; Pfaff *et al.*, 1993; Tuckwell *et al.*, 1994), as well as fibronectin bridges (Aumailley and Timpl, 1986; Tuckwell *et al.*, 1994) that are not involved in adhesion to triple-helical collagen.

Type IV collagen has a heterotrimeric structure, where the major isoform consists of two $\alpha 1$(IV) chains and one $\alpha 2$(IV) chain (Yamada *et al.*, 1990; Yurchenco and O'Rear, 1994). Type IV collagen is structurally divided into three domains, one an interrupted triple helix and two (7S and NC1) containing non-triple-helical structural elements. Type IV collagen can directly promote the adhesion and migration of diverse cell types, including melanoma, corneal epithelial, endothelial, and neural crest cells (Chelberg *et al.*, 1989; Etoh *et al.*, 1993; Herbst *et al.*, 1988; Olivero and Furcht, 1993; Padmanabha *et al.*, 1990; Perris *et al.*, 1993). The type IV collagen triple helix can serve as a recognition element, as heat denaturation of type IV collagen results in a 10-fold reduction of fibrosarcoma cell binding (Aumailley and Timpl, 1986).

Several synthetic peptides derived from the triple-helical region of type IV collagen have been tested for the ability to support cell adhesion and motility.

These sites may be differentiated based on the types of cellular activities that they promote. A heparin-binding synthetic peptide, incorporating $\alpha1$(IV) residues 531–543 and designated HEP-III (Koliakos *et al.*, 1989), promotes the adhesion of melanoma cells, ovarian carcinoma cells, keratinocytes, and epithelial cells and inhibits ovarian carcinoma cell adhesion to type IV collagen (Cameron *et al.*, 1991; Miles *et al.*, 1994; Wilke and Furcht, 1990). An all-D version of HEP-III has activities identical to those of the normal, all-L version (Miles *et al.*, 1994). Antibodies to various integrin subunits have been used to demonstrate $\alpha2\beta1$ and $\alpha3\beta1$ integrin involvement for corneal epithelial cell adhesion to HEP-III (Maldonado and Furcht, 1995). Subsequent affinity chromatography and immunoprecipitation experiments have indicated that the $\alpha3\beta1$ integrin is the primary binding receptor for melanoma and ovarian carcinoma cell adhesion to HEP-III (Miles *et al.*, 1995). Although the $\alpha1$(IV)531–543 sequence appears to be a general cell adhesion-promoting site, there are differences in the levels of activity promoted based on cell type (Miles *et al.*, 1994). Higher levels of adhesion to the $\alpha1$(IV)531–543 sequence were achieved with human keratinocytes (Wilke and Furcht, 1990), melanoma, and ovarian carcinoma cells compared to lymphocytes (Jurkats) at similar HEP-III concentrations (Miles *et al.*, 1994). In addition to cell adhesion, HEP-III promotes the motility of keratinocytes and inhibits keratinocyte migration on type IV collagen via the $\alpha2\beta1$ integrin (Kim *et al.*, 1994a). Another peptide, incorporating $\alpha1$(IV) residues 1263–1277 and designated IV-H1, has been demonstrated to support melanoma cell adhesion (Chelberg *et al.*, 1990; Fields *et al.*, 1993b; Mayo *et al.*, 1991). IV-H1 also supports melanoma cell motility and selectively inhibits cell adhesion to type IV collagen (Chelberg *et al.*, 1990). Melanoma cell motility is dependent upon IV-H1 conformation (Chelberg *et al.*, 1990; Mayo *et al.*, 1991).

Studies with HEP-III and IV-H1 have utilized linear peptides, although both sequences are from the triple-helical region of type IV collagen. Several general solid-phase methods for synthesizing covalently branched THPs that incorporate native collagen sequences have been developed (also see Section I,B) (Fields *et al.*, 1993a,b). The $\alpha1$(IV)531–543 sequence exhibited cell adhesion-promoting activity when incorporated into a THP (Miles *et al.*, 1994), suggesting that this region is biologically active in type IV collagen. However, triple helicity did not substantially enhance activity of the $\alpha1$(IV)531–543 sequence compared with the linear sequence. Since the $\alpha1$(IV)531–543 sequence contains a triple-helical interruption, it is not surprising that biological activity could be based on the non-triple-helical structure. Conversely, a THP incorporating the $\alpha1$(IV)1263–1277 sequence had 100-fold greater melanoma cell adhesion activity than the linear sequence (Fields *et al.*, 1993b). Thus, "conformationally dependent" and "conformationally independent" cellular recognition sites exist within the triple-helical domain of type IV collagen. The THP activities further support the concept that tumor-cell adhesion to type IV collagen involves multiple, distinct domains, as at least three domains within type IV collagen in triple-helical conformation are tumor cell adhesion sites, and it would not be surprising if others were found (Chelberg *et al.*, 1989; Fields *et al.*, 1993b; Vandenberg *et al.*, 1991).

Type I collagen has a heterotrimeric structure consisting of two $\alpha1$(I) chains and one $\alpha2$(I) chain. Type I collagen can directly promote the adhesion and migration of numerous cell types, including hepatocytes, keratinocytes, fibroblasts, melanoma, and neural crest cells (Chen *et al.*, 1993a; Faassen *et al.*, 1992b; Grzesiak *et al.*, 1992; Rubin *et al.*, 1981; Perris *et al.*, 1993; Scharffetter-Kochanek *et al.*, 1992). Denaturation of the type I collagen triple-helix reduces the binding of hepatocytes (Rubin *et al.*, 1981), neural crest cells (Perris *et al.*, 1993), and platelets (Morton *et al.*, 1994). In similar fashion to type IV collagen, several distinct regions within the triple-helical domain of type I collagen have been identified as cell adhesion sites. $\alpha1\beta1$ and $\alpha2\beta1$ integrin binding to the triple-helical region of type I collagen is, in general, conformationally dependent (Gehlsen *et al.*, 1992; Morton *et al.*, 1994; Santoro, 1986; Gullberg *et al.*, 1992). However, the $\alpha2\beta1$ integrin from human fibroblasts and human platelets adheres to a non-triple-helical peptide incorporating residues 430–442 of the rat $\alpha1$(I) chain in a Mg^{2+} dependent fashion (Santoro, 1986; Staatz *et al.*, 1990). This peptide inhibited platelet and breast adenocarcinoma cell adhesion to type I collagen (Staatz *et al.*, 1991) but was not effective at inhibiting $\alpha2\beta1$-mediated chondrosarcoma cell adhesion to type II collagen (Tuckwell *et al.*, 1994). It is not known if these different inhibitory effects are due to binding of different regions within the $\alpha2\beta1$ integrin by the respective collagens or due to the lack of peptide conformation.

A peptide incorporating residues $\alpha1$(I)769–783 supports human fibroblast adhesion and migration and inhibits human fibroblast and human T-lymphocyte attachment to type I collagen (Bhatnagar *et al.*, 1992; Qian and Bhatnagar, 1996; Winkler *et al.*, 1991). This peptide also induces the production of interstitial collagenase (MMP-1) in fibroblasts (Qian *et al.*, 1993). A THP which incorporates the $\alpha1$(I)772–786 sequence has greatly enhanced fibroblast cellular adhesion activity compared with the linear sequence (Grab *et al.*, 1996).

The examples given have been for ECM-derived peptides. The cellular activities of many other proteins, including many extracellular receptors, have been mapped in similar fashion. By synthesizing peptide analogs of known extracellular regions of a given receptor and screening them for inhibition of ligand binding, the specific site of ligand–receptor interaction can be elucidated. Human immunodeficiency virus type-1 (HIV-1) uses the CD4 protein as a receptor for cellular infection. Epitope mapping with a family of CD4 monoclonal antibodies (MAb) in conjunction with a panel of CD4-derived synthetic peptides identified the binding site of HIV-1. A synthetic peptide analog of this region was found to inhibit HIV-1 fusion to cells (Jameson *et al.*, 1988).

Synthetic peptide mapping of biological function has raised questions as to whether or not the identified sites are truly recognized in the native protein. For example, the focal contact formation activity of three fibronectin-derived peptides was enhanced when coupled to ovalbumin (Woods *et al.*, 1993). Coupling to ovalbumin is believed to enhance peptide secondary structure (Woods *et al.*, 1993). Other possibilities exist, such as the creation of a multivalent ligand. However, due to the nonspecific nature in which peptides are coupled to ovalbumin, large complexes of peptide bound to peptide, ovalbumin bound to ovalbumin, and "dumbbells" (peptide forming a bridge between ovalbumin molecules) are possible. Also, ovalbumin, when added in a soluble form, increases cell adhesion to laminin (Wilson and Weiser, 1992). The proposed mechanism involves ovalbumin enhancing cell–cell interactions via binding to the cell surface enzyme $\beta(1 \rightarrow 4)$-galactosyltransferase. A more direct method for producing multivalent ligands is via peptide coupling to magnetic beads. The use of tosyl-activated magnetic beads allows for specific attachment of peptides and proteins via free primary amino groups. The binding of integrin ligands on magnetic beads has been used to cluster integrins and form focal adhesion complexes (FACs) (Lauer *et al.*, submitted; Plopper *et al.*, 1995). Substrate clustering and subsequent receptor clustering promotes cytoskeletal reorganization that can affect signaling pathways (Miyamoto *et al.*, 1995a,b).

B. Peptides and Induction of Signaling

The previous section has dealt with peptide models of proteins that promote cell adhesion. Numerous peptide models have been shown to increase cellular signaling via adhesive and nonadhesive cellular receptors. In addition, synthetic peptides have been shown to modulate signal transduction when incorporated intracellularly. The laminin-derived synthetic peptide Tyr-Ile-Gly-Ser-Arg-NH_2 has been shown to decrease tumor metastasis and growth in experimental animals. A multiple-antigen peptide system (MAPS) incorporating the sequence Tyr-Ile-Gly-Ser-Arg-NH_2 induced the apoptosis of HT-1080 cells *in vitro*, as assayed by acridine orange staining, DNA degradation (ladder pattern formation), and increased TGF-β1 mRNA production. Tyr-Ile-Gly-Ser-Arg-NH_2-induced apoptosis was inhibited with an antibody raised against the peptide (Kim *et al.*, 1994b).

An Arg-Gly-Asp-related synthetic peptide, Pro-Arg-Gly-Asp-Asn, can induce both Ca^{2+} signaling and reduction of osteoclast cell spread area (retraction). Structural changes were made to the peptide to determine what features were responsible for its activity. Pro-Arg-Gly-Asp-Asn elicited a transient increase in $[Ca^{2+}]_i$ without promoting retractile events. Pro-D-Arg-Gly-Asp-Asn and benzoyl-Arg-Gly-Asp-Asn, which affect osteoclast retraction, have minimal Ca^{2+} signaling capabilities. *C*-Terminal charge was shown to be an important requirement for both retraction and calcium signaling, whereas the *N*-terminal charge is not required for either signaling or retraction (Shankar *et al.*, 1995).

Random phage display peptide libraries were used to isolate small peptides that activate the erythropoietin (EPO) receptor. The consensus sequence that was determined was not found in the EPO primary sequence (Wrighton *et al.*, 1996). It was shown that the signaling pathways initiated by the peptide substrates were identical to those initiated by EPO (Wrighton *et al.*, 1996). The activity of the peptide is probably due to its ability to dimerize the EPO receptor, thus initiating signal transduction (Livnah *et al.*, 1996).

Binding of Tyr(PO_3H_2) peptides to phosphatidylinositol 3-kinase (PI-3K) p85 src homology (SH) 2 domain results in conformational changes that could be transmitted to the p110 catalytic subunit, causing an increase in enzymatic activity. Maximal activity was obtained when both SH2 domains were engaged. Similar stimulation has been reported for other SH2 domain enzymes (Beattie, 1996). Zvelebil *et al.* examined interactions between solutions of SH2 domains with various Tyr(PO_3H_2)-containing peptides. Based on biotinylated Tyr(PO_3H_2) peptides immobilized on an avidin-coated sensor chip, they determined K_d values on the order of 10^{-5} to 10^{-7} M (Zvelebil *et al.*, 1995).

The main problems with the use of Tyr(PO_3H_2)-containing peptides are the instability of the phosphate group and the inability of the peptide to cross cell membranes. The use of nonhydrolyzable Tyr(PO_3H_2) analogs can circumvent one of these problems. These analogs are discussed in a later section. The latter problem can be overcome with the use of cyclized peptides. Decreased charge helps the peptide cross membranes. Some studies have examined *N*-acetylated or *C*-amidated peptides as an approach to decrease charge, but the resultant peptides do not always retain biological activity. Alternatively, fatty acyl chains can be attached to the peptide to increase membrane permeability (Beattie, 1996).

C. Peptide Induction of Platelet Aggregation

The mechanisms by which platelet activation and aggregation occur are of great interest for understanding thrombus formation during the wound healing process. Interstitial collagens are well-known inducers of platelet aggregation, and thus peptide-based approaches have been used to identify platelet recognition sites in collagen. Type III collagen-induced human platelet aggregation is inhibited by a peptide incorporating bovine α1(III)482–489 (Legrand *et al.*, 1980). An analog of this peptide only weakly inhibits platelet adhesion to type III collagen, suggesting that it affects platelet activation rather than cellular recognition of substrate (Karniguian *et al.*, 1983). Consistent with this hypothesis is the result that the α1(III)482–489 peptide does not inhibit migration of human fibrosarcoma or bladder carcinoma cells on type I collagen gels (Yamada *et al.*, 1990). The linear peptide itself does not promote platelet activation or aggregation (Karniguian *et al.*, 1983; Legrand, *et al.*, 1980). However, a sandwiched, disulfide cross-linked peptide containing the α1(III)479–489 sequence, Cys-(Pro-Hyp-Gly)$_4$-Lys-Pro-Gly-Glu-Pro-Gly-Pro-Lys-Gly-Glu-Ala-Gly-(Pro-Hyp-Gly)$_4$, does promote platelet aggregation (Morton *et al.*, 1993).

The α1(IV)1263–1277 THP promoted the adhesion and aggregation of platelets (Rao *et al.*, 1994). Rotary shadowing images indicated that aggregates of the THP could form distinct quaternary structures. These results were among the first demonstrations of a synthetic peptide promoting platelet adhesion, activation, and aggregation, and suggest that the combination of THP primary, secondary, tertiary, and quaternary structural features is required for platelet aggregation.

The triple-helical peptides Gly-Lys-Hyp-(Gly-Pro-Hyp)$_{10}$-Gly-Lys-Hyp-Gly and Gly-Cys-Hyp-(Gly-Pro-Hyp)$_{10}$-Gly-Cys-Hyp-Gly, when cross-linked, are able to induce platelet activation of the αIIbβ3 integrin. Downstream events linked to receptor binding include Ca^{2+} mobilization, protein kinase C activation, release of arachidonic acid, and ultimately platelet aggregation (Achison *et al.*, 1996; Morton *et al.*, 1995). These events were independent of $\alpha2\beta1$ integrin activation, and the mechanism appeared to differ from that of platelet aggregation induced by the α1(IV)1263–1277 THP (Rao *et al.*, 1994).

Thrombin activates platelets by cleaving the thrombin receptor, creating a new *N* terminus and a "tethered" ligand. A synthetic peptide composed of residues 42–55 from the thrombin receptor mimicks this ligand and can initiate platelet activity. The minimal sequence initiating the full range of activity is Ser-Phe-Leu-Leu-Arg-Asn (residues 42–47), which is fivefold more potent than the parent peptide at initiating platelet aggregation, increasing $[Ca^{2+}]_i$, increasing tyrosine phosphorylation, decreasing cAMP production, and up-regulating serotonin secretion (Nanevicz *et al.*, 1995). Other data suggest that Ser-Phe-Leu-Leu (Hui *et al.*, 1992) or Ser-Phe-Leu-Leu-Arg (Sabo *et al.*, 1992) are the minimal sequences required for activity. These truncated sequences either do not produce the full range of activities or do so only at much higher doses (300 μM as compared to 1.3 μM) (Hui *et al.*, 1992; Sabo *et al.*, 1992). In addition, the synthetic peptide Phe-Leu-Leu-Arg-Asn was shown to inhibit thrombin-induced platelet activation (Vassallo *et al.*, 1992).

Thrombin-stimulated platelet adhesion to fibronectin can be inhibited by linear and cyclic analogs of the synthetic peptide Gly-Arg-Gly-Asp-Ser-Pro-Ala. Native fibronectin and both peptide analogs blocked thrombin-induced platelet aggregation (Mohri and Ohkubo, 1993). Thus, any of the above-mentioned molecules could be utilized to modulate platelet activity.

III. Protein/Peptide Interactions

A. Stereochemistry and Protein Recognition

In an attempt to design peptides than not only model proteins but have enhanced *in vivo* stabilities, researchers have explored the effects of ligand stereochemistry on protein recognition. A straightforward design approach is the complete D-amino acid substitution of the peptide sequence, creating a D-enantiomeric peptide. The all-D peptide is a conformational mirror image of the all-L peptide (see Fig. 3). However, this mirror image may not retain the biological activity of the parent compound.

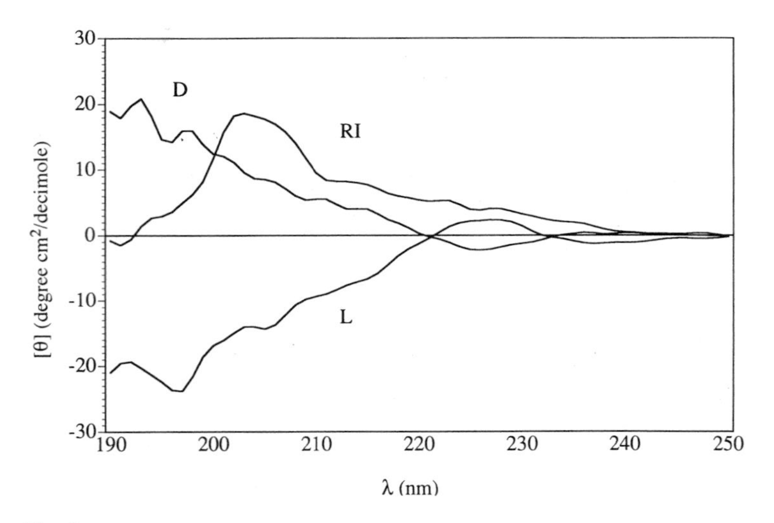

Fig. 3 CD spectrum of peptide IV-H1 (Gly-Val-Lys-Gly-Asp-Lys-Gly-Asn-Pro-Gly-Trp-Pro-Gly-Ala-Pro-Tyr) in all-L (L), all-D (D), and retro-inverso (RI) forms. Peptide concentrations were ~0.5 mM in water.

Several early experiments showed that all-D-amino acid peptides did not retain any of the activity of the parent all-L peptide. However, recent studies have shown that all-D-peptides can retain biological activity in certain systems. A laminin-derived synthetic peptide (A chain residues 2097–2108) and its all-D enantiomer had similar activity for promotion of cell attachment and support of tumor growth in mice (Nomizu *et al.*, 1992). In similar fashion, both all-L and all-D versions of the collagen-derived synthetic peptide HEP-III [α1(IV) residues 531–543] promote the adhesion of melanoma and ovarian carcinoma cells and inhibit ovarian carcinoma cell adhesion to type IV collagen (Miles *et al.*, 1994). Calmodulin binds both the all-L and all-D forms of melittin with apparently similar affinities (Fischer *et al.*, 1994).

A synthetic combinatorial library was used to select an all-D-peptide (acetyl-Arg-Phe-Trp-Ile-Asn-Lys-NH_2) as a potent new ligand for the μ-opioid receptor. The peptide was shown to be a full agonist, which means it not only binds to the μ receptor, but also induces a subsequent conformational change that allows for signal transduction. It also crossed the blood–brain barrier and produced long-lasting analgesia in mice (Dooley *et al.*, 1994). In this case, the all-L peptide was not active.

Retro-inverso peptides, while containing all-D-amino acids, are not mirror images of the native peptide (Fig. 3). These compounds also contain the opposite sequence order of the parent peptide, so they should show the same three-dimensional spatial orientation of amino acid side chains. The retro-inverso analog of a synthetic inhibitor of protein kinase C (PKC) was shown to have activity similar to that of the parent peptide. Substitution of a D-Ala with D-Ser in this peptide produced a compound with greater inhibitory activity (Ricouart *et al.*, 1989). The retro-inverso analog of tuftsin (Thr-Lys-Pro-Arg, derived from the heavy chain of γ-globulin) had the same anti-inflammatory activity, with a 10-fold greater potency, as natural tuftsin (Becherucci *et al.*, 1992). Synthetic retro-inverso analogs of the CD4 protein surface inhibit the incidence and severity of experimental autoimmune encephalomyelitis (EAE), an animal model of multiple sclerosis (MS), and do so without producing any deleterious side effects. EAE is known to be a CD4-mediated disease. Antibodies to CD4 also inhibit the formation of the disease, but do so by depleting $CD4^+$ cells as well as producing an immune response in the host (Jameson *et al.*, 1994).

B. Protease Interactions with D-Amino Acid Substrates

The use of all-D and retro-inverso peptides to enhance *in vivo* activities is based on the premise that proteases have reduced activity toward these potential substrates. For many years, single D-amino acids have been used to protect synthetic substrates from exopeptidase activity. Extensive studies have been conducted in recent years to examine enzyme specificity in terms of L- and D-substrates. The most definitive has been work with the HIV-1 protease. Kent and coworkers synthesized all-L and all-D HIV-1 proteases, then examined the chiral specificity of the two enzymes using the substrate 2-aminobenzoyl-Thr-Ile-Nle-Nph-Gln-Arg-NH_2 (where Nph is nitrophenylalanine) (Schnolzer and Kent, 1992). The synthetic all-L enzyme cleaved only the all-L, not the all-D, version of 2-aminobenzoyl-Thr-Ile-Nle-Nph-Gln-Arg-NH_2, whereas the synthetic all-D enzyme cleaved only the all-D substrate (Schnolzer and Kent, 1992). The chiral specificity of enzymes was established by these results.

The results of other enzyme studies are consistent with those from the HIV-1 study, in that native (all-L) enzymes cleave only all-L substrates, not all-D or retro-inverso substrates. For example, thermolysin cleaves all-L atrial natriuretic factor $(ANF)_{8\text{-}15}$ peptide analogs (Phe-Gly-Gly-Arg-Ile-Asp-Arg-Ile-Asp-Arg-Ile-NH_2), but does not cleave retro-inverso ANF peptide analogs (Berman *et al.*, 1989). Trypsin and pronase cleave the all-L PKC substrate Ala-Phe-Ala-Arg-Lys-Gly-Ala-Leu-Arg-Gln-Lys-Asn-Val, but do not cleave the retro-inverso PKC substrate (Ricouart *et al.*, 1989). Trypsin cleaves all-L cecropin A but does not cleave all-D cecropin A (Wade *et al.*, 1990). Although they are not cleaved by enzymes, there are indications that retro-inverso peptides may still be bound in the enzyme active site. Enkephalinase activity is inhibited by the dipeptide L-Phe-Gly ($IC_{50} \sim 3\ \mu M$) and by the retro-inverso dipeptide ($IC_{50} \sim 12\ \mu M$) (Roques *et al.*, 1983). Renin cleaves angiotensinogen at a Phe-Phe bond and a Phe-Val bond to form angiotensin I. Retro-inverso modification of the scissile bonds does not affect the binding of synthetic peptide analogs to the renin active site (Pinori and Verdini, 1987). Thiorphan is an inhibitor of both enkephalinase and angiotensin converting enzyme (ACE). The retro-inverso form of thiorphan retains its activity as an enkephalinase inhibitor while losing the inhibition of ACE (Roques *et al.*, 1983). Its increased specificity makes it more desirable for use as a therapeutic agent. Overall, all-D peptides may not be suitable as substrates for enzymes, but can be useful inhibitors of enzymatic activity.

C. Antibody Production and Specificity

Synthetic peptides have commonly been used for epitope identification via antibody production. For example, sequential overlapping synthetic peptides were made from human TGF-α and used in conjunction with a panel of murine monoclonal antibodies (MAbs) and rabbit polyclonal antibodies to identify epitopes within the TGF-α molecule. MAbs that did not inhibit TGF-α in mitogenic assays recognized two immunodominant regions of the *N*-terminal region (residues 1–9) and a β-sheet forming region (residues 22–31). These regions were also recognized by the polyclonal antibodies along with another *C*-terminal minor epitope. None of the neutralizing MAbs recognized any synthetic peptides. The receptor-binding surface of TGF-α appears to involve residues 12–20 and 34–43, which occur of the opposite face of the molecule as the *N*-terminal and β-sheet forming regions (Hoeprich *et al.*, 1989). The *C*-terminal region may also be important for TGF-α receptor binding (Tam *et al.*, 1991).

Individuals latently infected with the HIV-1 virus, as characterized by seronegativity to structural HIV-1 proteins, often have an antibody response against the regulatory HIV-1 nonstructural protein NEF. Three HIV-1 isolates were tested against a series of overlapping synthetic peptides representing the entire NEF sequence. Nine distinct epitopes were recognized by the sera samples. These peptides could be used to diagnose early or latent HIV-1 infection (Gombert *et al.*, 1990).

D-Peptide analogs have been used for antibody production. Antibodies of IgG_3 isotype reacted with retro-inverso and all D-peptide analogs of Ile-Arg-

Gly-Glu-Arg-Ala in a similar fashion as to the homologous L-peptide, whereas antibodies of the IgG_1, IgG_{2a}, and IgG_{2b} isotypes showed configurational preference (Guichard *et al.*, 1994). Substitution of Arg^{131} to Lys or D-Arg was found to change neither the antigenic nor the immunogenic properties of the resulting peptides. Substitution of Glu, Arg^{134}, and Ala by D-enantiomers altered the antigenicity of the peptides (Benkirane *et al.*, 1993). The K_a of antibodies to the retro-inverso peptide was 7- to 75-fold higher than to the parent Ile-Arg-Gly-Glu-Arg-Ala peptide (Benkirane *et al.*, 1995). Antibodies to L- and retro-inverso peptides cross-reacted with the parent molecule (Benkirane *et al.*, 1995).

D. Interruption of Signal Transduction Pathways

Signal transduction pathways rely on protein–protein interactions mediated by various domains within signaling molecules (Table II). Some of these domains include Src homology domains (SH2 and SH3) (Bachelot *et al.*, 1996; Beattie, 1996; Kavanaugh and Williams, 1994; Smithgall, 1995; Ward *et al.*, 1996; Zhou *et al.*, 1995; Zvelebil *et al.*, 1995), phosphotyrosine binding domains (PTBs), which have also been termed phosphotyrosine interaction domains (PIDs) (Blaikie *et al.*, 1994; Bork and Margolis, 1995; Eck, 1995; Kavanaugh and Williams, 1994), LIM domains (Agulnick *et al.*, 1996; Wu *et al.*, 1996), the PDZ domain (Cowburn, 1996; Fanning and Anderson, 1996; Harrison, 1996), and WW domains (Sudol, 1996a,b). There are also important examples of protein–protein interactions that are not mediated by such generalized domain sequences. Some of these include the binding of receptors with intracellular signaling molecules, such as the $\beta1$ integrin subunit cytoplasmic domain interacting with p125 focal adhesion kinase (FAK) and talin as well as other cytoskeletal components. The ability to understand and predict how these proteins interact with each other will make it possible to design specific synthetic peptide ligands to interfere with signaling pathways in a selective manner. To date, many peptide analogs have been designed based on binding domain sequences. These peptides can disrupt subcellular localization of proteins by physically interfering with binding domains. They can also directly affect enzymatic activity, both positively and negatively, by interfering with protein–protein interac-

Table II
Protein Binding Domains

Domain	Consensus binding site[a]
Src Homology 2 (SH2)	pYXXZ
Phosphotyrosine binding domain (PTB)	NPXpY
LIM (General)	NNAYF
LIM-2	NKLY
LIM-3	GPP*GPP*YA
WW	XPPXB
Src Homology 3 (SH3)	PXXP
PDZ	X(S/T)XV

[a] X, Any residue; Z, any hydrophobic residue; pY, phosphotyrosine; P*, 4-hydroxyproline; B, any aromatic residue.

tions. In all cases, the outcome is the modulation of signaling pathways. Specific examples of each case are presented next.

The best-studied of the above-mentioned binding domains are the SH domains. They mediate protein–protein interactions in a wide variety of signaling molecules. SH2 domains bind with high affinity to peptide sequences within target proteins that contain Tyr(PO_3H_3) residues, but have little or no affinity for the analogous nonphosphorylated sequences. Key substrate amino acids for binding SH2 domains are the +1, +2, and +3 positions on the *C*-terminal side of the Tyr(PO_3H_2) residue (Beattie, 1996; Smithgall, 1995; Songyang and Cantley, 1995; Zvelebil 1995). These residues vary widely between different signaling proteins, and thus there is no known SH2 binding consensus sequence. The domain contains a central β sheet that divides the domain into two functionally distinct sides. One half forms a binding pocket for the Tyr(PO_3H_2) moiety (providing high-affinity binding interactions involving ionic, amino aromatic, and hydrophobic contacts), while the second half binds the side chain of the +3 amino acid (providing specificity) (Bachelot *et al.*, 1996; Beattie, 1996; Smithgall, 1995; Songyang and Cantley, 1995; Ward *et al.*, 1996; Zhou *et al.*, 1995; Zvelebil *et al.*, 1995). Both SH2 and SH3 domains retain their biological activity when expressed independently of their source proteins (Smithgall, 1995).

Target sequences for SH3 domains are rich in proline and hydrophobic amino acids, but do not require phosphorylation. The *N* and *C* termini of SH3 domains are located in close proximity to one another, suggesting that this domain can extend from the surface of a protein without perturbing the adjacent structure (Smithgall, 1995). The SH3 consensus binding sequence is X-Pro-X-X-Pro-Pro-Pro-Z-X-Pro, where Z is a hydrophobic amino acid and X is any amino acid. The Pro-X-X-Pro core is absolutely conserved. Peptides containing Arg-X-Leu-Pro-Pro-Leu-Pro-Arg-Z bind to Src, and those containing Arg-X-Leu-Pro-Pro-Arg-Pro-X-X bind to p85 of PI-3K, with a K_d in the range of 10 μM (Smithgall, 1995), comparable to natural SH3-binding motifs (Chen *et al.*, 1993b).

The other binding domains have been less intensively studied, so their consensus binding sequences are less well characterized. The phosphotyrosine binding domain (PTB) was first discovered in Shc, an adaptor protein that binds to Grb-2, coupling it to the *ras* activation pathway (Blaikie *et al.*, 1994; Kavanaugh and Williams, 1994). The PTB, approximately 160 amino acids in length, was shown to bind tyrosine-phosphorylated proteins of a consensus sequence Asn-Pro-X-Tyr(PO_3H_2) (Bork and Margolis, 1995). LIM domains are cysteine-rich zinc-binding motifs that bind tyrosine-containing proteins. Some examples of sequences bound by LIM include Asn-Asn-Ala-Tyr-Phe, Asn-Lys-Leu-Tyr, and Gly-Pro-Hyp-Gly-Pro-Hyp-Tyr (Wu *et al.*, 1996). No clear consensus binding sequence has been determined for this domain. The PDZ domain recognizes the consensus sequence X-Ser/Thr-X-Val (Cowburn, 1996; Fanning and Anderson, 1996; Harrison, 1996). The WW domain is named for two conserved tryptophan residues located 20–22 residues apart in the binding motif. The consensus binding site for the WW domain is X-Pro-Pro-X-B where B is any aromatic amino acid. It has been proposed that this domain might help regulate SH2, SH3, or PTB domain interactions, due to the similarity of consensus binding sequences (Sudol, 1996a). For example, the WW domain could bind the sequence X-Pro-Pro-X-Tyr when it is not phosphorylated, and an SH2 domain could bind this same sequence when phosphorylated on the Tyr residue.

Tyr(PO_3H_2)-containing synthetic peptides were used to determine the relative importance of two separate SH2 domains of the protein tyrosine phosphatase SH-PTP1. One peptide, whose sequence was derived from the EPO receptor, was shown to bind to the *N*-terminal SH2 domain of SH-PTP1 and activate the phosphatase 60-fold (Pei *et al.*, 1996). The other peptide, containing a sequence from the IgG Fcγ receptor, was shown to bind both the *N*-terminal and *C*-terminal SH2 domains, and activate the phosphatase 12-fold (Pei *et al.*, 1996). A mutant SH-PTP1 was produced that did not contain the *C*-terminal SH2 domain. When this phosphatase was tested with both Tyr(PO_3H_2)-containing peptides, it was determined that both peptides could bind to and stimulate phosphatase activity (Pei *et al.*, 1996). In this study, the use of synthetic peptides allowed the researcher to determine that the *N*-terminal SH2 domain of SH-PTP1 was necessary and sufficient for catalytic activity, whereas the role of the *C*-terminal SH2 domain probably lies in recruiting substrates to the enzyme (Pei *et al.*, 1996).

Synthetic peptides have been used to affect cellular signaling via inhibition of Lyn kinase activity (Ruzza *et al.*, 1995). The parent peptide, Glu-Asp-Asn-Glu-Tyr-Thr-Ala, models the main autophosphorylation site of Lyn, a Src family enzyme. Determination of a sequence that is effectively phosphorylated by Lyn could be used to obtain substrates suitable for the detection of Src-like tyrosine kinase activity, or active site directed inhibitors specific for this class of enzymes. It was shown that removal of the *N*-terminal charge by acetylation, removal of the *C*-terminal charge by amidation, or cyclization of the peptide induces a dramatic loss of Lyn enzyme activity (Ruzza *et al.*, 1995). Dimerization of the peptide, in linear or cyclic form, created substrates especially suitable for Lyn, as indicated by their K_m values (20 μM each, compared to 120 μM for the parent peptide) (Ruzza *et al.*, 1995). When the parent peptide was compared to the best derivative, the cyclic dimer, the IC_{50} values were 340 μM and 43 μM, respectively (Ruzza *et al.*, 1995).

The use of synthetic peptides to interfere with cellular processes has a few drawbacks. Peptide stability in a living system can be too low to sustain adequate cellular responses. The lack of transport of synthetic peptide-based drugs across cellular membranes can also interfere with their activity. Optimization of activity is reached via reduction of the core peptide bonds to improve stability against proteolysis, as well as esterification or other modification of the *C*- and/or *N*-terminal charges to facilitate penetration of some of these compounds into cells. Cyclic peptides can also be used to improve binding affinity as well as increase resistance to proteolysis (Smithgall, 1995).

The development of high-affinity analogs of Tyr(PO_3H_2) residues that are resistant to cellular phosphatases is under investigation (Fig. 4). Phosphonate-based Tyr(PO_3H_2) mimetics such as phosphonomethyl phenylalanine (PMP) introduce fluorines to the phosphorus, which has provided higher affinity. This strategy has also been applied to malonyltyrosine (OMT) mimetics (Wange *et al.*, 1995; Burke *et al.*, 1996). The nonhydrolyzable Tyr(PO_3H_2) analog can be incorporated into synthetic peptides and used in cellular systems. The inhibition of PTP1B with FOMT was 10-fold higher than for OMT (Burke *et al.*, 1996). Increased affinity may be attributable to new H-bonding interactions between

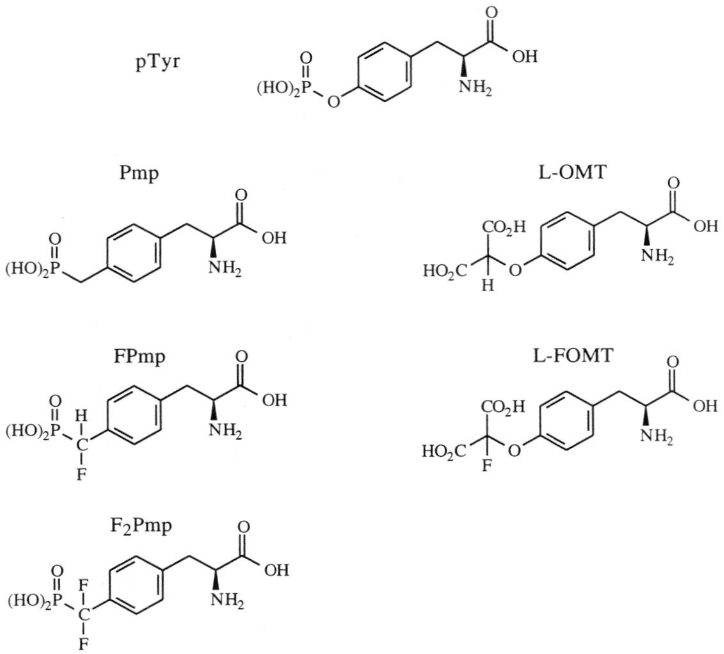

Fig. 4 Structures of phosphotyrosine (pTyr) and nonhydrolyzable analogs based on phosphonomethyl phenylalanine (Pmp) or *O*-malonyltyrosine (OMT).

the fluorine atom and the enzyme catalytic site and not due to lowering of pK_a values. This theory is based on data derived from a competition binding assay in which an FOMT peptide had the same affinity for an SH2 domain as the OMT peptide (Burke *et al.*, 1996). In permeabilized T lymphocytes, the intracellular incorporation of synthetic peptides containing F_2PMP residues affected the association of the T-cell receptor for a signaling molecule known as ZAP-70 (Wange *et al.*, 1995). Binding of the T-cell receptor after cellular activation is known to occur via the SH2 domain of ZAP-70. When the synthetic peptide incorporating the nonhydrolyzable Tyr(PO_3H_2) analog was inserted into the T cells, the T-cell receptor did not associate with ZAP-70. The downstream effects of this inhibition included a lack of phosphorylation of the ZAP-70 protein as well as a lack of up-regulation of its kinase activity (Wange *et al.*, 1995).

Protein kinase C is suspected to play a role in axonal development. This proposal is based on the use of drugs that are known to be rather nonspecific. A peptide was designed to include a highly specific PKC inhibitory region attached to a signal sequence that would translocate this peptide through the extracellular membrane. The synthetic peptide was shown to accumulate in the cytoplasm of neuronal cells in culture and induce a rapid modification in growth cone morphology related to PKC inhibition (Theodore *et al.*, 1995).

Synthetic peptides containing a sequence derived from the transcription factor NF-κB were synthesized with an attached signal sequence to allow the peptides to cross cellular membranes. These peptides were shown to inhibit the nuclear translocation of NF-κB, thus inhibiting transcription (Lin *et al.*, 1995).

Synthetic Tyr(PO_3H_2)-containing peptides representing a portion of the cytoplasmic domain of the $\beta1$ integrin subunit were utilized for a study of tyrosine-phosphorylated integrins. The cytoplasmic domain of the $\beta1$ subunit was chosen because it is known to be the target of oncogenic tyrosine kinases. Following peptide synthesis, antibodies were produced that would recognize the phosphorylated $\beta1$ integrin subunit. It was shown that cellular localization of the phosphorylated $\beta1$ subunit was distinct from that of the nonphosphorylated $\beta1$ subunit (Johansson *et al.*, 1994). One of the SH2 domains of PI-3K was shown to bind the phosphroylated $\beta1$ peptide, but not to the nonphosphorylated analog (Johansson *et al.*, 1994).

Peptide models of the $\alpha1$ and $\alpha2$ integrin subunit cytoplasmic domains have been used for affinity chromatography. Purified actin was shown to bind to the $\alpha2$ peptide but not the $\alpha1$ peptide. The selective binding of cytoskeletal proteins may help to explain why both $\alpha2\beta1$ and $\alpha1\beta1$ integrins bind to collagen, but only $\alpha2\beta1$ can mediate contraction of extracellular collagen matrices (Kieffer *et al.*, 1995).

Subunit-specific peptides from a G protein receptor can inhibit different responses related to receptor occupancy. The sequence Lys-Glu-Asn-Leu-Lys-Asp-Cys-Gly-Leu-Phe was shown to inhibit IL-8/G protein interaction, while Leu-Glu-Arg-Ile-Ala-Gln-Ser-Asp-Tyr-Ile was shown to specifically inhibit downstream signal transduction (Damaj *et al.*, 1996).

Synthetic peptides derived from intracellular segments of the *N*-formyl peptide receptor were used to identify which regions within the receptor were necessary for interaction with the G protein responsible for its signal transduction. The second intracellular loop and the *C*-terminal tail were identified as being important for effective G protein coupling (Schreiber *et al.*, 1994).

The intracellular protein cell adhesion regulator (CAR) has been proposed to influence integrin-mediated binding to extracellular matrix proteins, including basement membrane (type IV) collagen (Pullman and Bodmer, 1992). The effects of phosphorylated and nonphosphorylated peptide models of CAR on human melanoma cell adhesion to type IV collagen have been examined (Lauer and Fields, 1997; Lauer, *et al.*, 1997). Three analogs of the $CAR_{138-142}$ (Val-Glu-Ile-Leu-Tyr) were tested for inhibitory activity. The first contained the 138–142 sequence ($CAR_{138-142}$); the second contained the 138–142 sequence with a Tyr(PO_3H_2) for residue 142 ($pCAR_{138-142}$); and the third contained the reversed 138–142 sequence ($rCAR_{138-142}$). In order to quantitate the intracellular effect of CAR peptide analogs, a reversible cell permeabilization procedure was required. The procedure used a water-soluble lipid derivative, thus being relatively mild and allowing peptides to be incorporated into the cell without affecting cell viability or receptor function subsequent to permeabilization. The cells could then be tested for normal activities. When added extracellularly, none of the analogs inhibited cell adhesion to type IV collagen. Intracellular incorporation of both $CAR_{138-142}$ and $pCAR_{138-142}$ resulted in inhibition of cell adhesion in a dose-dependent fashion. The IC_{50} values were ~80 and ~10 μM for $CAR_{138-142}$ and $pCAR_{138-142}$, respectively. Intracellular incorporation of the $rCAR_{138-142}$ peptide was ineffective. Fluorescence microscopy of a fluorescein-labeled $CAR_{138-142}$ peptide revealed that the reversible permeabilization procedure resulted in the peptides crossing the cellular membrane. The results indicate that the *C*-terminal region of CAR, spanning residues 138–142, appears to influence integrin recognition of type IV collagen. The specific sequence of CAR is responsible for this behavior, as the reverse sequence of residues 138–142 had no activity. The behaviors of $CAR_{138-142}$ and $pCAR_{138-142}$ were different, with $pCAR_{138-142}$ providing greater inhibition at similar concentrations. This indicates that Tyr phosphorylation, although not essential, enhances the function of the *C*-terminal region of CAR. It is possible that intracellular kinases can phosphorylate $CAR_{138-142}$, allowing for cellular modulation of activity. The differences in activity also provide evidence that the phosphate group was not removed entirely by protein Tyr phosphatases during the permeabilization/adhesion experiment. A possible therapeutic approach for inhibition of melanoma cell adhesion to extracellular matrix proteins is the use of CAR peptide analogs intracellularly.

The aforementioned studies illustrate cases where peptide probes provided greater specificity than compounds previously used. Taken together, these examples indicate many different ways in which peptides and peptide analogs can be used to interrupt signal transduction at defined points. The data derived from experiments such as these will help delineate signal transduction pathways, increasing our understanding of cellular processes.

IV. Peptides and Modulation of T-Cell Behavior

In general, there are several possible sites at which to interrupt T-cell related diseases. These include: (i) human leukocyte antigen (HLA)–peptide interactions; (ii) T-cell receptor (TCR) recognition of HLA peptides; (iii) TCR biochemical signals; (iv) cellular adhesion or signaling pathways; (v) second messengers; or (vi) transcription factors or effector molecules (Krensky, 1994). Since the TCR responds to internally derived peptides in the natural system, synthetic peptides can be readily used to modulate T-cell behavior. More specifically, synthetic peptides have been used to create TCR antagonists, modulate helper T-cell activity, control cytotoxic T-cell behavior, and suppress undesired or excessive immune responses.

Synthetic peptides have been a valuable tool in the determination of T-cell epitopes, both natural (those determined from protein sequences) and nonnatural (those determined from random peptide libraries). T-cell epitope determination plays an important role in the development of synthetic vaccines, T-cell receptor antagonists, and major histocompatibility complex (MHC) blockers. Strategies for the determination of natural T-cell epitopes are well established and have contributed substantially to our understanding of the nature of T-cell responses. Usually, the source proteins are identified using molecular genetic techniques. Portions of the sequence that correspond to the T-cell epitopes are determined via the use of (i) truncation mutants of the original genes and/or (ii) synthetic peptides corresponding to overlapping sections of the protein sequences (Gundlach *et al.*, 1996). With the recent advances in peptide synthesis, the second method of epitope determination has proven to be easily performed. Synthetic techniques have also opened up the possibility of nonnatural epitope determination and the modification of natural epitopes. Both avenues will help advance the development of T-cell research beyond what is available using natural peptide sequences.

The TCR recognizes antigenic peptides presented by MHC molecules. Synthetic peptide analogs produced by amino acid substitution of natural epitopes can generate T-cell responses that are qualitatively different from those produced by the original antigenic peptide. Some examples include the stimulation of interleukin production without the stimulation of proliferation (Evavold and Allen, 1991), cytotoxicity with neither proliferation nor interleukin production (Evavold *et al.*, 1993), alteration of the pattern of cytokine production (Windhagen *et al.*, 1995), and the production of anergy (Sloan-Lancaster *et al.*, 1993). Analogs of antigenic peptides have also been shown to inhibit antigen-specific T-cell responses. This phenomenon has been called TCR antagonism. A synthetic peptide and its naturally occurring TCR antagonist analog were used with mature T cells and immature thymocytes from an $\alpha\beta$ TCR transgenic mouse. Both peptides were shown to antagonize T-cell cytolytic response without being able to directly stimulate mature T cells (Williams *et al.*, 1996). The development of these specific TCR antagonists that do not stimulate T-cell proliferation could be beneficial in the treatment of T-cell related diseases.

Interaction of the TCR with an immunogenic peptide bound to MHC class II molecules leads to the activation of $CD4^+$ helper T-cells. Helper T-cells are subdivided into two subsets on the basis of the cytokines secreted. TH1 cells secrete IL-2, IFN-γ, and TNF-β and induce delayed-type hypersensitivity. TH2 cells secrete IL-4, IL-5, IL-6, and IL-10 and induce the humoral immune response. Three known immunogenic peptides were compared for their ability to induce selective activation of TH1 and/or TH2 cells. Peptide K1A2 [Glu-Tyr-Lys-Glu-Tyr-Ala-Ala-Tyr-Ala-$(\text{Glu-Tyr-Ala})_2$] induced TH1 cells and was shown to bind the cells with a high affinity (Chaturvedi *et al.*, 1996). The second peptide, K3 [Glu-Tyr-Lys-$(\text{Glu-Tyr-Ala})_3$], induced TH2 cells and had a moderate affinity for the cell type. K4 [Glu-Tyr-Lys-$(\text{Glu-Tyr-Ala})_4$] had a low binding affinity and induced both TH1 and TH2 derived cytokines (Chaturvedi *et al.*, 1996). This study also showed that cells primed with high-affinity peptide are committed to differentiate into the TH1 subset irrespective of the priming dose and affinity of the challenge antigen (Chaturvedi *et al.*, 1996). The differentiation of cells primed with low-affinity peptide depends upon the immunization dose and binding affinity of the challenge antigen. T-cell differentiation can be altered in a controlled way, with the use of specific peptide sequences chosen for the desired effect.

Another example of synthetic peptides being used to modulate helper T-cell activity involves the ras protein, a small G protein known to play a role in cell growth and differentiation. Modification of the ras p21 protein is associated with transformation. Mutated ras p21 proteins may bear unique tumor-specific T-cell epitopes not found on nontransformed ras. Thus, tumor expressing mutant ras p21 may become targets for host-derived T lymphocytes. Synthetic peptides that model segments of mutated ras p21 were shown to be immunogenic in mice (Abrams *et al.*, 1995). The cell type mediating the immunological effect was $CD4^+$ helper T cells of the TH1 subset (Abrams *et al.*, 1995). These results show that peptides can induce specific T-cell mediated lysis if an epitope can be determined to exist on the undesired cells differentiating them from natural cells, as was the case with this ras mutation.

Synthetic peptide analogs can also be used to modulate T-cell activation following infection with HIV-1. Subsequent to HIV-1 infection, humans exhibit lymphocyte dysfunction prior to the loss of $CD4^+$ T cells. The major HIV-1 surface glycoprotein, gp120, can modulate lymphocyte function *in vitro*. A synthetic peptide derived from the conserved CD4 binding region of gp120 transiently reduced the levels of protein tyrosine phosphorylation that occurs in response to T-cell activation (Phipps *et al.*, 1995). The reduction of tyrosine phosphorylation was associated with the inhibition of p56-lck kinase activity. The peptide also induced an increase in Src kinase activity, even though total cellular tyrosine phosphorylation was decreased (Phipps *et al.*, 1995). These results show that T-lymphocyte infection by the HIV-1 virus is mediated to some degree by gp120, and that gp120-derived peptides could be used to modulate T-cell activation due to HIV-1 infection.

Another interesting example of the use of peptides to inhibit undesired T-cell behavior involves the rejection of organs following transplantation. Synthetic peptides corresponding to a particular conserved region of the HLA class I α_1 α-helix can significantly prolong rat heterotropic heart transplants (Krensky, 1994). In this study the peptide was given in conjunction with subtherapeutic doses of cyclosporin A, an 11-residue cyclic peptide that blocks IL-2 production. Neither compound alone produced significant responses.

T-cell mediated autoimmune diseases tend to develop focused antigen-specific responses that overutilize certain TCR variable region segments, causing the induction of anti-TCR specific T cells and antibodies that can inhibit the pathogenic T cells. This natural regulatory pathway can be manipulated through the use of synthetic peptide vaccines that correspond to segments of the overexpressed variable regions. The use of peptides in this way offers a unique and relatively selective approach for regulating the immune system. One such example is found in experimental autoimmune encephalomyelitis, which is the animal model for multiple sclerosis. In this system, the pathogenic T cells are directed at myelin components, which cause a slow deterioration of the nervous system. *In vivo* administration of peptides corresponding to the particular V regions overexpressed in EAE (Vβ8.2) can prevent and treat EAE by boosting the regulatory anti-Vβ8.2 T cells (Vandenbark *et al.*, 1996). Parallel studies have taken place in MS patients, using Vβ5.2 and Vβ6.1 peptides. These peptides boosted the frequency of TCR peptide-specific T cells in some cases with clinical benefit. However, not all autoimmune pathogenic responses involve strongly biased TCR variable gene expression, so not all patients respond to the variable region peptides. A panel of peptides may be developed that will include the most frequently overexpressed genes, so that maximal clinical benefit can be achieved. At the present time, though, no consensus has been reached as to which variable genes are commonly overutilized in MS patients. Of the 11 patients included in this study, 7 responded to the Vβ5.2 peptide, and 6 responded to the Vβ6.1 peptide (as indicated by antibody screening). Four patients failed to respond to either peptide. Among the responders, 2 showed improvements after peptide treatment, 2 were stable, and 3 deteriorated after peptide treatment. All of the nonresponders showed deterioration during the study (Vandenbark *et al.*, 1996). The 2 patients that showed the most benefit were the ones that were treated earliest after the onset of their disease. These results give some hope that if the correct peptides could be identified and used early in treatment, the benefits of peptide therapy to combat MS could be maximized (Vandenbark *et al.*, 1996).

Another example of peptide modulation of an autoimmune disease involves murine AIDS, which is caused by infection with a murine leukemia retrovirus mixture. This disease is strikingly similar to human AIDS, and thus may prove a useful model for study. Murine retrovirus infection produces an inhibition of B- and T-lymphocyte stimulation as well as the excessive stimulation of the TH2 subset of helper T cells. The TH1 to TH2 cell conversion may be the determining factor in inducing the fatal outcome of the disease, because of the different cytokines produced by each subset of helper T cells. Autoantibody binding to the peptide determinant CDR1 of the TCR is elevated during murine retrovirus infection. These abnormalities eventually lead to profound immunodeficiency. Treatment of mice with multiple peptides corresponding to the CDR1 epitopes has been shown to slow the loss of B-cells in infected animals (Watson *et al.*, 1995). The retrovirally infected mice produced antibodies to some of the peptides, and these antibodies were able to reverse some of the cytokine abnormalities induced by the virus. Thus, the peptides prevented the retrovirus-induced reduction in B and T lymphocyte proliferation and TH1 cytokine production while suppressing excessive production of TH2 cytokines (Liang *et al.*, 1996).

Experimental arthritis can be induced in animal models and subsequently treated with the administration of type II collagen. The disease is produced by the stimulation of aberrant T-cell mediated breakdown of joint tissue, which leads to the production of arthritic symptoms. Type II collagen has been shown to have both conformationally dependent and independent epitopes (Cremer and Kang, 1988; Glattauer *et al.*, 1991; Morgan *et al.*, 1992). At this point in time, only conformationally independent sequences from the type II collagen triple helix have been identified that can directly mediate T-cell responses (Table III). For example, the α1(II)316-333 peptide inhibits T-cell proliferation induced by type II collagen (Burkhardt *et al.*, 1992). The α1(II)58-73 peptide stimulates T-cell hybridomas and inhibits induction of experimental arthritis in rats (Ku *et al.*, 1993). The α1(II)190-200 or α1(II)245-270 peptide and analogs, or the α1(II)607-621 peptide, inhibits experimental arthritis in mice (Myers *et*

Table III
Peptide Sequences Derived from Type II Collagen That Modulate T-Cell Function

Location	Sequence	Biological activity
α1(II) 181–196	GARGPEGAQGPRGEPG	Suppressed collagen-induced arthritis (CIA)
α1(II) 252–266	KGQTGELGIAGFKGE	Suppressed CIA, but to lesser degree than above
α1(II) 74–93	ARGFPGTPGLPGVKGHRGYP	Produced T-cell proliferation and suppressed CIA, minimal immunostimulatory residues underlined
α1(II) 181–209	GARGPEGAQGPRGESGTP*GP*GPAGAP*GN	Suppressed CIA, critical T-cell determinant is underlined
α1(II) 250–270	GPKGQTGKPGIAGFKGEQGPK	Diminished T-cell proliferation, suppressed CIA
α1(II) 257–270	ELGIAGFKGEQGPK	Suppressed CIA, major T-cell determinant is underlined
α1(II) 254–273	TGGKPGIAGFKGEQGPKGEP	Produced T-cell proliferation but did not suppress CIA
α1(II) 271–285	GEPGPAGPQGAPGPA	Supported T-cell proliferation
α1(II) 316–333	GFPGQDGLAGPKGAPGER	Inhibited T-cell proliferation
α1(II) 316–333	CFPGQDGLAGPKGAPGER	No effect (control)
α1(II) 324–333	AGPKGAPGER	Inhibited T-cell proliferation
α1(II) 337–350	GLAGPKGANGDPGR	No effect (control)
α1(II) 607–621	GPAGPAGTAGARGAP	Suppressed CIA, minimal effective sequence is underlined
α1(II) 924–943	PGERGLKGHRGFTGLQGLPG	Produced T-cell proliferation and suppressed CIA, minimal immunostimulatory residues underlined

al., 1989, 1993; Khare *et al.*, 1995; Miyahara *et al.*, 1995; Myers *et al.*, 1995). Eventual identification of conformationally dependent epitopes is most desirable in light of the positive effects of oral doses of intact type II collagen on rheumatoid arthritis (Trentham *et al.*, 1993).

The preceding examples of T-cell related disease treatments are representative of a few of the possible ways in which synthetic peptides can be used to modulate immunological systems. Because the immune system uses peptides to regulate responses, it has long been recognized as a viable target for peptide-related research. The continued application of peptide and peptidomimetic approaches will allow for the development of new drugs and vaccines for the treatment of T-cell related diseases.

V. Enzyme/Peptide Interactions: Peptides as Enzyme Substrates

There exists an extensive history of peptide use as both models of enzyme binding sites in proteins and as tools to dissect enzyme specificity. A few representative examples are discussed next.

The incorporation of phosphorylated residues has allowed for the study of protein phosphatases. Several Tyr(PO_3H_2)-containing peptides were used as substrates for rat brain protein tyrosine phosphatase and human adipocyte acid phosphatase (Ottinger *et al.*, 1993). Kinetic analysis revealed that the two enzymes catalyzed dephosphorylation with different rates and affinities. This work demonstrated the utility of phosphopeptides for assessing phosphatase activities.

Catabolism of ECM components has been ascribed to a family of Zn^{2+} metalloenzymes. These matrix metalloproteinases (MMPs; also termed matrixins) are believed to be important in connective tissue remodeling during development and wound healing. MMPs have also been implicated in a variety of disease states, including arthritis, glomerulonephritis, periodontal disease, tissue ulcerations, and tumor cell invasion and metastasis (Woessner, 1991; Birkedal-Hansen *et al.*, 1993; Stetler-Stevenson *et al.*, 1993; Nagase, 1996). Primary structures of human enzymes have been determined for MMP-1 (interstitial collagenase) (Goldberg *et al.*, 1986), MMP-2 (gelatinase A) (Collier *et al.*, 1988), MMP-3 (stromelysin 1) (Whitham *et al.*, 1986), MMP-7 (pump 1, matrilysin) (Muller *et al.*, 1988), MMP-8 (neutrophil collagenase) (Hasty *et al.*, 1990), MMP-9 (gelatinase B) (Wilhelm *et al.*, 1989), MMP-10 (stromelysin 2) (Muller *et al.*, 1988), MMP-11 (stromelysin 3) (Basset *et al.*, 1990), MMP-12 (metalloelastase) (Shapiro *et al.*, 1992), MMP-13 (collagenase 3) (Freije *et al.*, 1994), and three membrane-type MMPs (MT-MMPs) (MMP-14, MMP-15, MMP-16) (Sato *et al.*, 1994; Takino *et al.*, 1995; Will and Hinzmann, 1995). Most MMPs contain a propeptide domain, a catalytic domain, and a hemopexin/vitronectin-like domain. One approach for defining the specificity of each MMP utilizes synthetic peptides (Nagase and Fields, 1996). Results of these studies have been used to define the differences in MMP behaviors and to design "optimal" substrates for the MMPs.

Initial MMP peptidase studies focused on substrates designed from the MMP-1/MMP-8 cleavage site in the $\alpha 1$ chain of types I or III collagen. The sequence utilized, Pro-Gln/Leu-Gly-Ile-Ala-Gly, encompasses the substrate P_3

through P_3' subsites according to the nomenclature of Schechter and Berger (1967). The first mammalian MMP peptidase activities demonstrated were the cleavage of Dnp-Pro-Gln-Gly-Ile-Ala-Gly-Gln-D-Arg and Dnp-Pro-Leu-Gly~Ile-Ala-Gly-D-Arg by MMP-2 (Seltzer *et al.*, 1981), Dnp-Pro-Leu-Gly~Ile-Ala-Gly-Arg-NH_2 by MMP-3 (Galloway *et al.*, 1983), Dnp-Pro-Leu-Gly~Ile-Ala-Gly and Dnp-Pro-Leu-Gly~Ile-Ala-Gly~Arg-NH_2 by MMP-8 (Williams and Lin, 1984), and several peptides containing Pro-Gln-Gly~Ile-Ala by MMP-9 (Williams and Lin, 1984). Weingarten and associates used a series of peptides, peptolides and peptide esters to explore certain aspects of MMP-1 sequence specificity (Weingarten and Feder, 1986; Seltzer *et al.*, 1989, 1990). The α1(I) collagen sequence Gly-Pro-Gln-Gly~Ile-Ala-Gly-Gln was used as the starting substrate for a comprehensive study of MMP-1, MMP-3, and MMP-8 sequence specificities (Fields *et al.*, 1987; Fields, 1988). Comparison of the interstitial collagenases MMP-1 and MMP-8 showed similar, but not identical, specificities. The MMP-3 comprehensive sequence specificity showed similar patterns to MMP-1 and MMP-8, which include preferences for (i) a long chain (Leu, Met) or bulky, uncharged residue in subsite P_2, (ii) Ala in subsite P_1, (iii) Leu and aromatics (Phe, Trp, Tyr) in subsite P_1', and (iv) aromatics (Phe and Trp), Leu, or Arg in subsite P_2'. The P_1' subsite was very sensitive to substitution.

The activities of the full-length and the C-terminal-truncated MMP-1 and MMP-3 were compared using the substrate Dnp-Pro-Leu-Gly~Leu-Trp-Ala-D-Arg-NH_2 (Murphy *et al.*, 1992). The k_{cat}/K_m values for MMP-1 hydrolysis of Dnp-Pro-Leu-Gly~Leu-Trp-Ala-D-Arg-NH_2 do not change significantly even if the enzyme lacks the C-terminal domain. C-terminal truncation also has little effect on MMP-3 hydrolysis of this peptide. MMP-1 and MMP-3 peptidase specificity appears to be determined by the catalytic domain, with little, if any, contribution from the C-terminal hemopexin-like domain.

The collagen sequence-based substrates reported by Fields (1988) were subsequently used to study the sequence specificities of MMP-2, MMP-9, and MMP-7 (Netzel-Arnett *et al.*, 1993). The general patterns for subsite preferences were more similar between the two gelatinases (MMP-2 and MMP-9) than those between the two collagenases (MMP-1 and MMP-8). The MMP-7 sequence specificity had many deviations from that of MMP-2 and MMP-9 and appeared to have peptidase activities closer to MMP-1 and MMP-8.

Relative substrate specificities of MMP-1 and MMP-9 have been evaluated by "substrate mapping" in which a mixture of substrates is hydrolyzed by MMP-1 or MMP-9 and the products analyzed by reversed-phase HPLC coupled to fast-atom-bombardment mass spectrometry (Berman *et al.*, 1992; McGeehan *et al.*, 1994).

One of the obvious goals of peptide-based specificity studies is to better understand the behaviors of enzymes toward their natural substrates. Peptide specificities of MMPs, however, do not necessarily match protein specificities. For example, peptide studies to not reveal how triple-helical collagen can be cleaved by MMPs. The type I collagen-derived sequence Gly-Pro-Gln-Gly~Ile-Ala-Gly-Gln is cleaved at a similar rate (within one order of magnitude) by MMP-1, MMP-2, MMP-3, MMP-7, MMP-8, and MMP-9, even though only MMP-1, MMP-2, and MMP-8 have been shown to cleave type I collagen at this site. Similarly, MMP-1 cannot cleave the Asn^{80}-Tyr^{81} bond of MMP-2 during activation (Crabbe *et al.*, 1994), whereas MMP-1 can cleave peptides containing

Asn in the P_1 position or Tyr in the P_1' position (Fields, 1988; McGeehan *et al.*, 1994). In an indirect fashion, peptidase activities have revealed the importance of substrate secondary and tertiary structures in directing MMP specificities.

A caveat to peptide-based enzyme specificity studies is that the starting peptide "template" can influence the individual substitution results. Single substitutions within a substrate are often assumed to be independent (noninteractive), and hence the sum of the free energy changes from single substitutions is equal to that for a substrate possessing the multiple substitutions. Sequence specificity studies of MMP-1 and MMP-3 have resulted in designed substrates where the assumption of additive free energy changes appeared reasonable (Fields *et al.*, 1987; Niedzwiecki *et al.*, 1992). However, in a series of substrates designed to discriminate between MMP-3 and other MMPs, a discrepancy on the order of 2000–5000 was found between the predicted rate and the actual rate of hydrolysis. For example, the substrate NFF-3 [Mca-Arg-Pro-Lys-Pro-Val-Glu~Nva-Trp-Arg-Lys(Dnp)-NH_2] should have been hydrolyzed by MMP-2 at 8.9×10^{-2} times the rate of substrate NFF-2 [Mca-Arg-Pro-Lys-Pro-Tyr-Ala~Nva-Trp-Met-Lys(Dnp)-NH_2] based on sequence specificity studies (Netzel-Arnett *et al.*, 1993), which would have resulted in $k_{cat}/K_m = 4800\ \sec^{-1}M^{-1}$. However, MMP-2 exhibited little hydrolysis of NFF-3, and thus the prediction made by assuming independent sites was in error.

Exceptions to the assumption of mutation additivity for enzymatic activity, and protein–protein interactions in general, have been recognized for some time (Wells, 1990). The work of McGeehan *et al.* (1994) demonstrated definitively that sequence specificity of MMP substrates can be "template dependent." One example, for MMP-1, was a substitution of Gly with Ala in subsite P_1. For the substrate Dnp-Pro-Leu-Gly~Cys(CH_3)-His-Ala-D-Arg-NH_2, this substitution resulted in a 5-fold decrease in susceptibility to MMP-1, whereas only a 1.6-fold increase was seen for the same substitution of Dnp-Pro-Leu-Gly~Leu-Trp-Ala-D-Arg-NH_2 (McGeehan *et al.*, 1994). Substitution of the P_1 subsite Gly with Ala in Gly-Pro-Gln-Gly~Ile-Ala-Gly-Gln enhanced activity 6.6-fold (Fields, 1988). Thus, depending upon the template, substitution of the P_1 subsite Gly with Ala can significantly increase or decrease MMP-1 activity. Similar template-dependent results were shown for MMP-9 specificity. Using the template Dnp-Pro-Leu-Gly~Xaa-Trp-Ala-D-Arg-NH_2, the MMP-9 preference for subsite P_1' was Cys(CH_3) > Cys(CH_2CH_3) > Cys(S-mercaptoethyl). The template Dnp-Pro-Leu-Gly~Xaa-His-Ala-D-Arg-NH_2 gave MMP-9 preferences of Cys(CH_2CH_3) > Cys(CH_3) ~ Cys(*S*-mercaptoethyl). Template dependency has a significant effect on specificity studies and the design of optimized substrates and inhibitors.

Peptide-based sequence specificity studies have proved useful for understanding HIV-1 protease behavior. Once the HIV-1 virus enters a cell, the HIV-1 reverse transcriptase converts viral single-stranded RNA into an RNA–DNA hybrid. This hybrid molecule loses the RNA strand and replicates into a double-stranded DNA molecule, which becomes integrated into the host-cell genome. The host cells then transcribe the viral DNA into mRNA for which the primary translation products are the gag and pol polyproteins along with envelope glycoproteins. Proteolytic cleavage of the gag and pol polyproteins by HIV-1 protease is necessary for the production of the individual structural and enzymatic components of HIV-1. The HIV-1 protease is a member of the aspartic protease family. It differs from other members of this family in that it has approximately half the number of the amino acids as other aspartic proteases. It is thus dependent on dimerization to produce a functional enzyme. Peptides were used to effectively map the cleavage site specificity of HIV-1. It was noted that the enzyme appeared to have two different preferences for cleavage, either at a hydrophobic–hydrophobic site or an aromatic–Pro site (Griffiths, *et al.*, 1992). The rate of cleavage of these sites was affected in different ways by the nature of the neighboring substrate residues. The enzyme was believed to utilize nonlocal interactions with substrates and have flexible loops (Griffiths *et al.*, 1992).

Synthetic peptides have been used to extensively map the specificity of a unique serine protease, the fiddler crab collagenolytic serine protease. Crab collagenolytic serine protease has an interesting dual functionality, in that it hydrolyzes both serine protease substrates (small peptides) (Grant and Eisen, 1980) and zinc metalloprotease substrates (types I–V collagen) (Welgus and Grant, 1983). Synthetic substrates were used to study the P_1', P_2', and P_3' specificity of this serine protease (Grant and Eisen, 1980; Tsu *et al.*, 1994). The peptide specificity studies closely matched the collagenolytic activity of the enzyme, with particular preference for Arg residues in the P_1' subsite.

VI. Summary

Peptides have long served as effective models of proteins. Applications of peptide-related research include a wide variety of biological systems, conformation studies of protein domains, identification of cellular recognition sites, examination of protein–protein associations including signal transduction, and definition of enzyme specificity. The use of peptide models of proteins does have certain shortcomings. First and foremost, the conformation of peptide models as compared to proteins is a concern. Second, peptide applications *in vivo* are subject to problems of stability and membrane permeability. Third, assumptions of independent binding sites with respect to peptide–protein interactions may not always be valid. Taking these caveats into consideration, peptides will continue to be useful models applied for the dissection of protein structure and function. In addition, peptide models of proteins will also serve as lead compounds for the development of peptidomimetic and nonpeptide drugs and therapeutic agents.

References

Abrams, S. I., Dobrzanski, M. J., Wells, D. T., Stanziale, S. F., Zaremba, S., Masuelle, L., Knator, J. A., and Schlom, J. (1995). Peptide-specific activation of cytolytic $CD4^+$ T lymphocytes against tumor cells bearing mutated epitopes of k-*ras* p21. *Eur. J. Immunol.* **25,** 2588–2597.

Achison, M., Joel, C., Hargreaves, P. G., Sage, S. O., Barnes, M. J., and Farndale, R. W. (1996). Signals elicited from human platelets by synthetic, triple helical, collagen-like peptides. *Blood Coagulation Fibrinolysis* **7,** 149–152.

Agulnick, A. D., Taira, M., Breen, J. J., Tanaka, T., Dawid, I. B., and Westphal, H. (1996). Interactions of the LIM-domain-binding factor Ldb1 with LIM homeodomain proteins. *Nature* **384,** 270–272.

Alexander, P., Faneshtock, S., Lee, T., Orban, J., and Bryan, P. (1992). Thermodynamic analysis of the folding of streptococcal protein G IgG-binding domains B1 and B2: Why small proteins tend to have high denaturation temperatures. *Biochemistry* **31,** 3597–3603.

Ashkenas, J., Penman, M., Vasile, E., Acton, S., Freeman, M., and Krieger, M. (1993). Structures and high and low affinity ligand binding properties of murine type I and type II macrophage scavenger receptors. *J. Lipid Res.* **34,** 983–1000.

Aumailley, M., and Timpl, R. (1986). Attachment of cells to basement membrane collagen type IV. *J. Cell Biol.* **103,** 1569–1575.

Bachelot, C., Rameh, L., Parsons, T., and Cantley, L. C. (1996). Association of phosphatidylinositol 3-kinase, via the SH2 domains of p85, with focal adhesion kinase in polyoma middle t-transformed fibroblasts. *Biochim. Biophys. Acta* **1311,** 45–52.

Balakrishnan, R., Siegel, D. L., Baum, J., and Brodsky, B. (1995). Acid destabilization of a triple-helical peptide model of the macrophage scavenger receptor. *FEBS Lett.* **368,** 551–555.

Baron, C., Mayo, K., Skubitz, A. P. N., and Furcht, L. T. (1992). NMR assignment and secondary structure of the cell adhesion type III module of fibronectin. *Biochemistry* **31,** 2068–2073.

Basharov, M. A., Vol'kenshtein, M. V., Golovanov, I. B., Nauchitel, V. V., and Sobolev, V. M. (1986). The role of electrostatic interactions in the stabilization of the α-helix peptide. *Doklady Akademii Nauk. SSSR* **286,** 1261–1264.

Basset, P., Bellocq, J. P., Wolf, C., Stoll, I., Hutin, P., Limacher, J. M., Podhajcer, O. L., Chenard, M. P., Rio, M. C., and Chambon, P. (1990). A novel metalloproteinase gene specifically expressed in stromal cells of breast carcinomas. *Nature* **348,** 699–704.

Bazzi, M. D., Woody, R. W., and Brack, A. (1987). Interaction of amphipathic polypeptides with phospholipids: Characterization of conformations and the CD of oriented α-sheets. *Biopolymers* **26,** 1115–1124.

Beattie, J. (1996). SH2 domain protein interaction and possibilities for pharmacological intervention. *Cell. Signal.* **8,** 75–86.

Becherucci, C., Perretti, M., Nencioni, L., Silvertri, S., and Parente, L. (1992). Anti-inflammatory effect of tuftsin and its retro-inverso analogue in rat adjuvant arthritis. *Agents Action* (Special Conference Issue), C115–C117.

Bella, J., Eaton, M., Brodsky, B., and Berman, H. M. (1994). Crystal and molecular structure of a collagen-like peptide at 1.9 Å resolution. *Science* **226,** 75–81.

Bella, J., Brodsky, B., and Berman. H. M. (1995). Hydration structure of a collagen peptide. *Structure* **3,** 893–906.

Benkirane, N., Friede, M., Guichard, G., Briand, J. P., Van Regenmortel, M. H. V., and Muller, S. M. (1993). Antigenicity and immunogenicity of modified synthetic peptides containing D-amino acid residues. *J. Biol. Chem.* **268,** 26279–26285.

Benkirane, N., Guichard, G., Van Regenmortel, M. H. V., Briand, J. P., and Muller, S. (1995). Cross-reactivity of antibodies to retro-inverso peptidomimetics with the parent protein histone H3 and chromatin core particle. *J. Biol. Chem.* **270,** 11921–11926.

Berg, R. A., Olsen, B. R., and Prockip, D. J. (1970). Titration and melting curves of the collagen-like triple helices formed from $(Pro\text{-}Pro\text{-}Gly)_{10}$ in aqueous solution. *J. Biol. Chem.* **245,** 5759–5763.

Berman, J. M., Hassman, C. F., Buch, S. H., and Chen, T. M. (1989). Receptor binding affinity and thermolysin degradation of truncated and retro-inverso-isomeric ANF analogs. *Life Sci.* **44,** 1267–1270.

Berman, J., Green, M., Sugg, E., Anderegg, R., Millington, D. S., Norwood, D. L., McGeehan, G., and Wiseman, J. (1992). Rapid optimization of enzyme substrates using defined substrate mixtures. *J. Biol. Chem.* **267,** 1434–1437.

Bhatnagar, R. S., Sorensen, K. R., Scaria, P. V., and Qian, J. J. (1992). Interactions of collagen may be explained by molecular behavior at interfaces. *FASEB J.* **6,** A288.

Birkedal-Hansen, H., Moore, W. G. I., Bodden, M. K., Windsor, L. J., Dirkedal-Hansen, B., DeCarlo, A., and Engler, J. A. (1993). Matrix metalloproteinases: A review. *Crit. Rev. Oral. Biol. Med.* **4,** 197–250.

Blaber, M., Zhang, X., and Matthews, B. W. (1993). Structural basis of amino acid α helix propensity. *Science* **260,** 1637–1640.

Blaikie, P., Immanuel, D., Wu, J., Li, N., Yajnik, V., and Margolis, B. (1994). A region in Shc distinct from the SH2 domain can bind tyrosine-phosphorylated growth factor receptors. *J. Biol. Chem.* **269,** 32031–32034.

Blanco, F. J., Jimenez, M. A., Herrany, J., Rico, M., Santoro, J., and Nieto, J. L. (1993). NMR evidence of a short linear peptide that folds into a β-hairpin in aqueous solution. *J. Am. Chem. Soc.* **115,** 5887–5888.

Blanco, F. J., Jimenez, M. A., Pineds, A., Rico, M., Santoro, J., and Nieto, J. L. (1994a). NMR solution structure of the isolated N-terminal fragment of protein-GB1 domain: Evidence of trifluoroethanol induced native-like β-hairpin formation. *Biochemistry* **33,** 6009–6014.

Blanco, R., Rivas, G., and Serrano, L. (1994b). A short linear peptide that folds into a native stable β-hairpin in aqueous solution. *Nature Struct. Biol.* **1,** 584–590.

Bork, P., and Margolis, B. (1995). A phosphotyrosine interaction domain. *Cell* **80,** 693–694.

Broddeflak, J., Bergquist, K. E., and Kihlberg, J. (1996). Preparation of a glycopeptide analogue of type II collagen: Use of acid labile protective groups for carbohydrate moieties in solid phase synthesis of *O*-linked glycopeptides. *Tet. Letters* **37,** 3011–3014.

Brodsky, B., and Shah, N. K. (1995). The triple-helix motif in proteins. *FASEB J.* **9,** 1537–1546.

Brodsky, B., Li, N. H., Long C. G., Apigo, J., and Baum, J. (1992). NMR and CD studies of triple-helical peptides. *Biopolymers* **32,** 447–451.

Brown, J. E., and Klee, W. A. (1971). Helix–coil transition of the isolated amino terminus of ribonuclease. *Biochemistry* **10,** 470–476.

Burke, C., Mayo, K. H., Skubitz, A., and Furcht, L. T. (1991). ^{1}H-NMR and CD secondary structure analysis of cell adhesion promoting peptide F9 from laminin. *J. Biol. Chem.* **266,** 19407–19412.

Burke, T. R., Ye, B., Akamatsu, M., Ford, H., Yan, X., Kole, H. K., Wolf, G., Shoelson, S. E., and Roller, P. P. (1996). 4′*O*-[2-2(-fluoromalonyl)]-L-tyrosine: A phosphotyrosyl mimic for the preparation of signal transduction inhibitory peptides. *J. Med. Chem.* **39,** 1021–1027.

Burkhardt, H., Yan, T., Broker, B., Beck-Sikinger, A., Holmdahl, R., Von er Mark, K., and Emmrich, F. (1992). Antibody binding to a collagen type II epitope gives rise to an inhibitory peptide for autoreactive T cells. *Eur. J. Immunol.* **22,** 1063–1067.

Bushnell, G. W., Louis, G. V., and Brayer, G. D. (1990). High-resolution three-dimensional structure of horse heart cytochrome *c. J. Mol. Biol.* **214,** 585–595.

Cameron, J. D., Skubitz, A. P. N., and Furcht, L. T. (1991). Type IV collagen and corneal epithelial adhesion and migration. *Invest. Ophthalmol. Vis. Sci.* **32,** 2766–2773.

Chakrabartty, A., Kortemme, T., and Baldwin, R. L. (1994). Helix propensities of the amino acids measured in alanine-based peptides without helix-stabilizing side-chain interactions. *Protein Sci.* **3,** 843–852.

Chan, H. S., Wattenbarger, M. R., Evans, D. J., Bloomfield, V. A., and Dill, K. A. (1991). Enhanced structure in polymers at interfaces. *J. Chem. Phys.* **94,** 8542–8556.

Charonis, A. S., Skubitz, A. P. N., Koliakos, G. G., Dege, J., Vogel, A. M., Wolhuerter, R., Reger, L., and Furcht, L. T. (1988). A novel synthetic peptide from the β1 chain of laminin with heparin-binding and cell adhesion-promoting activities. *J. Cell Biol.* **107,** 1253–1260.

Chaturvedi, P., Yu, Q., Southwood, S., Sette, A., and Singh, B. (1996). Peptide analogs with different affinities for MHC alter the cytokine profile of helper T cells. *International Immunol.* **8,** 745–755.

Chelberg, M. K., Tsilibary, E. C., Hauser, A. R., and McCarthy, J. B. (1989). Type IV collagen-mediated melanoma cell adhesion and migration: Involvement of multiple, distinct domains of the collagen molecule. *Cancer Res.* **49,** 4796–4802.

Chelberg, M. K., McCarthy, J. B., Skubitz, A. P. N., Furcht, L. T., and Tsilibary, E. C. (1990). Characterization of a synthetic peptide from type IV collagen that promotes melanoma cell adhesion, spreading, and motility. *J. Cell Biol.* **111,** 261–270.

Chen, J. D., Kim, J. P., Zhang, K., Sarret, Y., Wynn, K. C., Kramer, R. H., and Woodley, D. T. (1993a). Epidermal growth factor (EGF) promotes human keratinocyte locomotion of collagen by increasing the α2 integrin subunit. *Exp. Cell. Res.* **209,** 216–223.

Chen, J. K., Lane, W. S., Brauer, A. W., Tanaka, A., and Schreiber, S. L. (1993b). Biased combinatorial libraries: Novel ligands for the SH3 domain of phosphatidylinositol 3-kinase. *J. Am. Chem. Soc.* **115,** 12591–12592.

Chung, L. A., Lear, J. D., and DeGrado, W. F. (1992). Fluorescence studies of the secondary structure and orientation of a model ion channel peptide in phospholipid vesicles. *Biochemistry* **31,** 6608–6616.

Collier, I. E., Whilhelm, S. M., Eisen, A. Z., Marmer, B. L., Grant, G. A., Seltzer, J. L., Kronberger, A., He, C., Bauer, E. A., and Goldberg, G. I. (1988). H-ras oncogene-transformed human bronchial epithelial cells (TBE-1) secrete a single metalloprotease capable of degrading basement membrane collagen. *J. Biol. Chem.* **263,** 6579–6587.

Cowburn, D. (1996). Adaptors and integrators. *Structure* **4,** 1005–1008.

Cox, J. P. L., Evans, P. A., Packman, L. C., Williams, D. H., and Woolfson, D. N. (1993). Dissecting the structure of a partially folded protein: Circular dischroism and nuclear magnetic resonance studies of peptides from ubiquitin. *J. Mol. Biol.* **234,** 483–492.

Crabbe, T., O'Connell, J. P., Smith, B. J., and Docherty, A. J. P. (1994). Reciprocated matrix metalloproteinase activation: A process performed by interstitial collagenase and progelatinase A. *Biochemistry* **33,** 11419–14425.

Cremer, M. A., and Kang, A. H. (1988). Collagen-induced arthritis in rodents: A review of immunity to type II collagen with emphasis on the importance of molecular conformation and structure. *International Rev. Immunol.* **4,** 65–81.

Damaj, B. B., McColl, S. R., Mahana, W., Crouch, M. F., and Naccache, P. H. (1996). Physical association of $G_{i2}\alpha$ with interleukin-8 receptors. *J. Biol. Chem.* **271,** 12783–12789.

Damaschun, G., Damaschun, H., Gast, K., Gernat, C., and Zirwer, D. (1991). Acid denatured apocytochrome *c* is a random coil: Evidence from small-angle X-ray scattering and dynamic light scattering. *Biochim. Biophys. Acta* **1078,** 289–295.

Dathe, M., Schumann, M., Wieprecht, T., Winkler, A., Beyermann, M., Krause, E., Matsuzaki, K., Murase, O., and Bienert, M. (1996). Peptide helicity and membrane surface charge modulate the balance of electrostatic and hydrophobic interactions with lipid bilayers and biological membranes. *Biochemistry* **35,** 12612–12622.

Davis, G. E. (1992). Affinity of integrins for damaged extracellular matrix $\alpha v\beta 1$ binds to denatured collagen type I through RGD sites. *Biochem. Biophys. Res. Commun.* **182,** 1025–1031.

DeGrado, W. F., and Lear, J. D. (1985). Induction of peptide conformation at apolar/water interfaces: A study with model peptides of defined hydrophobic periodicity. *J. Am. Chem. Soc.* **107,** 7684–7689.

DeGrado, W. F., and Lear, J. D. (1990). Conformationally constrained α-helical peptide models for protein ion channels. *Biopolymers* **29,** 205–213.

Doig, A. J., Chakrabartty, A., Klingler, T. M., and Baldwin, R. L. (1994). Determination of free energies of N-capping in α-helices by modification of the Lifson–Roig helix–coil theory to include N- and C-capping. *Biochemistry* **33,** 3396–3403.

Dooley, C. T., Chung, N. N., Wilkes, B. C., Schiller, P. W., Bidlack, J. M., Pasternak, G. W., and Houghten, R. A. (1994). An all D-amino acid opioid peptide with central analgesic activity from a combinatorial library. *Science* **266,** 2019–2022.

Drake, S. L., Klein, D. J., Mickelson, D. J., Oegema, T. R., and Furcht, L. T. (1992). Cell surface phosphatidylinositol-anchored heparan sulfate proteoglycan initiates mouse melanoma cell adhesion to a fibronectin-derived heparin-binding synthetic peptide. *J. Cell. Biol.* **117,** 1331–1341.

Drake, S. L., Varnum, J., Mayo, K. H., Letourneau, P. C., Furcht, L. T., and McCarthy, J. B. (1993). Identification of the active site of fibronectin synthetic peptide FN-C/H II, responsible for cell adhesion and heparan sulfate binding. *J. Biol. Chem.* **268,** 15859–15867.

Dyson, H. J., Rance, M., Houghten, R. A., Lerner, R. A., and Wright, P. E. (1988). Folding of immunogenic peptide fragments of proteins in water solution I: Sequence requirements for the formation of a reverse turn. *J. Mol. Biol.* **201,** 161–200.

Dyson, H. J., Satterthwait, A. C., Lerner, R. A., and Wright, P. E. (1990). Conformational preferences of synthetic peptides derived from the immunodominant site of the circumsporozoite protein of *Plasmodium falciparum* by ^{1}H-NMR. *Biochemistry* **29,** 7828–7837.

Dyson, H. J., Wright, P. E., Houghten, R. A., Wilson, I. A., Wright, P. E., and Lerner, R. A. (1991). Defining solution conformations of small linear peptides. *Annu. Rev. Biophys. Chem.* **20,** 519–538.

Eble, J., Golbik, R., Mann, K., and Kuhn, K. (1993). The $\alpha 1\beta 1$ integrin recognition site of the basement membrane collagen molecule $\alpha 1(IV)_2\alpha 2(IV)$. *EMBO J,* **12,** 4795–4802.

Eck, M. J. (1995). A new flavor in phosphotyrosine recognition. *Structure* **3,** 421–424.

Eggleston, D. S., and Feldman, S. H. (1990). Structure of the fibrinogen binding sequence: Arginylglycylaspartic acid (RGD). *Int. J. Peptide Protein Res.* **36,** 161–166.

Elomaa, O., Kangas, M., Sahlberg, C., Tuukkanen, J., Sormunen, R., Liakka, A., Thesleff, I., Kraal, G., and Tryggvason, K. (1995). Cloning of a novel bacteria-binding receptor structurally related to scavenger receptors and expressed in a subset of macrophages. *Cell* **80,** 603–609.

Engel, J. (1992). Laminins and other strange proteins. *Biochemistry* **31,** 10643–10651.

Epand, R. M., and Scheraga, H. A. (1968). The influence of long-range interactions on the structure of myoglobin. *Biochemistry* **7,** 2864–2871.

Etoh, T., Thomas, L., Pastel-Levy, C., Colvin, R. B., Mihm, M. C., Jr., and Byers, H. R. (1993). Role of integrin $\alpha 2\beta 1$ (VLA-2) in the migration of human melanoma cells on laminin and type IV collagen. *J. Invest. Dermatol.* **100,** 640–647.

Evavold, B. D., and Allen, P. M. (1991). Separation of IL-4 production from Th cell proliferation by an altered T cell receptor ligand. *Science* **363,** 1308–1309.

Evavold, B. D., Sloan-Lancaster, J., Hsu, B. L., and Allen, P. M. (1993). Separation of T helper 1 clone cytolysis from proliferation and lymphokine production using analog peptides. *J. Immunol.* **150,** 3131–3134.

Faassen, A. E., Drake, S. L., Iida, J., Knutson, J. R., and McCarthy, J. B. (1992a). Mechanisms of normal cell adhesion to the extracellular matrix and alterations associated with tumor invasion and metastasis. *Adv. Pathol. Lab. Med* **5,** 229–259.

Faassen, A. E., Schrager, J. A., Klein, D. J., Oegema, T. R., Couchman, J. R., and McCarthy, J. B. (1992b). A cell surface chondroitin sulfate proteoglycan, immunologically related to CD44, is involved in type I collagen-mediated melanoma cell motility and invasion. *J. Cell Biol.* **116,** 521–531.

Fairman, R., Shoemaker, K. R., York, E. J., Stewart, J. M., and Baldwin, R. L. (1989). Further studies of the helix dipole model: Effects of a free NH_3^+ or COO^- group on helix stability. *Proteins* **5,** 1–7.

Fan, P., Li, M. H., Brodsky, B., and Baum, J. (1993). Backbone dynamics of $(\text{Pro-Hyp-Gly})_{10}$ and a designed collagen-like triple-helical peptide by ^{15}N-NMR relaxation and hydrogen-exchange measurements. *Biochemistry* **32,** 13299–13309.

Fanning, A. S., and Anderson, J. M. (1996). Protein–protein interactions: PDZ domain networks. *Curr. Biol.* **6,** 1385–1388.

Feng, Y., Melacini, G., Taulane, J. P., and Goodman, M. (1996). Acetyl-terminated and template-assembled collagen-based polypeptides composed of Gly-Pro-Hyp sequences. 2. Synthesis and conformational analysis by circular dichroism, ultraviolet absorbance, and optical rotation. *J. Am. Chem. Soc.* **118,** 10351–10358.

Fezoui, Y., Weaver, D. L., and Osterhout, J. J. (1994). *De novo* design and structural characterization of an α-helical hairpin peptide: A model system for the study of protein folding intermediates. *Proc. Natl. Acad. Sci. USA* **91,** 3675–3679.

Fields, G. B. (1988). The application of solid phase peptide synthesis to the study of structure-function relationships in the collagen-collagenase system. Ph.D. Thesis, Tallahassee, Florida State University: 213 pages.

Fields, G. B., and Fields, C. G. (1992). Optimization strategies for FMOC solid phase peptide synthesis: Synthesis of triple-helical collagen-model peptides. *In* "Innovation and Perspectives in Solid Phase Synthesis: Peptides, Polypeptides, and Oligonucleotides" (R. Epton, ed.), pp. 153–162. Intercept, Andover, U.K.

Fields, G. B., VanWart, H. E., and Birkedal-Hansen, H. (1987). Sequence specificity of human skin fibroblast collagenase: Evidence for the role of collagen structure in determining the collagenase cleavage site. *J. Biol. Chem.* **262,** 6221–6226.

Fields., G. B., Tian, Z., and Barany, G. (1992). Principles and practice of solid-phase peptide synthesis. *In* "Synthetic Peptides: A User's Guide" (G. A. Grant, ed.), pp. 77–183. W. H. Freeman, New York.

Fields, C. G., Lovdahl, C. M., Miles, A. J., Matthias-Hagen, V. L., and Fields, G. B. (1993a). Solid-phase synthesis and stability of triple-helical peptides incorporating native collagen sequences. *Biopolymers* **33,** 1695–1707.

Fields, C. G., Mickelson, D. J., Drake, S. L., McCarthy, J. B., and Fields, G. B. (1993b). Melanoma cell adhesion and spreading activities of a synthetic 124-residue triple-helical "mini-collagen." *J. Biol. Chem.* **268,** 14153–14160.

Finkelstein, A. V. (1991). Rate of β-structure formation in polypeptides. *Proteins* **9,** 23–27.

Fiori, W. R., Miick, S. M., and Millhauser, G. L. (1993). Increasing sequence length favors α-helix over 3_{10}-helix in analine-based peptides: Evidence for a length-dependent structural transition. *Biochemistry* **32,** 11957–11962.

Fischer, P. J., Prendergast, F. G., Ehrhardt, M. R., Urbauer, J. L., Wand, A. J., Sedarous, S. S., McCormick, D. J., and Buckley, P. J. (1994). Calmodulin interacts with amphiphilic peptides composed of all D-amino acids. *Nature* **368,** 651–653.

Fisher, A., Tuniuchi, H., and Anfinsen, C. B. (1973). On the role of heme in the formation of the structure of cytochrome *c*. *J. Biol. Chem.* **248,** 3188–3195.

Freije, J. M. P., Diez-Itza, T., Balbin, M., Sanchez, L. M. Blasco, R., Tolivia, J., and Lopez-Otin, C. (1994). Molecular cloning and expression of collagenase-3, a novel human matrix metalloproteinase produced by breast carcinomas. *J. Biol. Chem.* **269,** 16766–16773.

Galloway, W. A., Murphy, G., Sandy, J. D., Gavrilovic, J., Crawston, T. E., and Reynolds, J. J. (1983). Purification and characterization of a rabbit bone metalloproteinase that degrades proteoglycan and other connective-tissue components. *Biochem. J.* **209,** 741–752.

Gehlsen, K. R., Sriramarao, P., Furcht, L. T., and Skubitz, A. P. N. (1992). A synthetic peptide derived from the carboxy terminus of the laminin A chain represents a binding site for the $\alpha3\beta1$ integrin. *J. Cell Biol.* **117,** 449–459.

Germann, H. P., and Heidermann, E. (1988). A synthetic model of collagen: An experimental investigation of the triple-helix stability, *Biopolymers* **27,** 157–163.

Galttauer, V., Ramshaw, J. A. M., Tebb, T. A., and Werkmeister, J. A. (1991). Conformational epitopes on interstitial collagens. *Int. J. Biol. Macromol.* **12,** 140–146.

Goldberg, G. I., Wilhelm, S. M., Kronberger, A., Bauer, E. A., Grant, G. A., and Eisen, Z. A. (1986). Human fibroblast collagenase. Complete primary structure and homology to an oncogene transformation-induced rat protein. *J. Biol. Chem.* **261,** 6600–6605.

Gombert, F. O., Blecha, W., Tahtinen, M., Ranki, A., Pfeifer S., Troger, W., Braun, R., Muller-Lantzsch, N., Jung, G., Rubsamen-Waigmann, H., and Krohn, K. (1990). Antigenic epitopes of NEF proteins from different HIV-1 trains as recognized by sera from patients with manifest and latent HIV infection. *Virology* **176,** 458–466.

Goodman, M., Feng, Y., Melacini, G., and Taulane, J. P. (1996a). A template-induced incipient collagen-like triple-helical structure. *J. Am. Chem. Soc.* **118,** 5156–5157.

Goodman, M., Melacini, G., and Feng, Y. (1996b). Collagen-like triple helices incorporating peptoid residues. *J. Am. Chem. Soc.* **118,** 10928–10929.

Grab, B., Miles, A. J., Furcht, L. T., and Fields, G. B. (1996). Promotion of fibroblast adhesion by triple-helical peptide models of type I collagen-derived sequences. *J. Biol. Chem.* **271,** 12234–12240.

Grant, G. A., and Eisen, A. Z. (1980). Substrate specificity of the collagenolytic serine protease from *Uca pugilator:* Studies with noncollagenous substrates. *Biochemistry* **19,** 6089–6095.

Griffiths, J. T., Phylip, L. H., Konvalinka, J., Strop, P., Gustchina, A., Wlodawer, A., Davenport, R. J., Briggs, R., Dunn, B. M., and Kay, J. (1992). Different requirements for productive interaction between the active site of HIV-1 proteinase and substrates containing -hydrophobic*hydrophobic- or -aromatic*Pro- cleavage sites. *Biochemistry* **31,** 5193–5200.

Gronenborn, A. M., Filpula, D. R., Essig, N. Z., Achari, A., Whitlow, M., Wingfield, P. T., and Clore, G. M. (1991). A novel, highly stable fold of the immunoglobulin binding domain on streptococcal protein G. *Science* **253,** 657–661.

Gruenewald, B., Nicola, C. U., Lustig, A., Schwarz, G., and Klump, H. (1979). Kinetics of the helix–coil transition of a polypeptide with non-ionic side groups, derived from ultrasonic relaxation measurements. *Biophys. Chem.* **9,** 137–147.

Grzesiak, J. J., Davis, G. E., Kirchhofer, D., and Pierschbacher, M. D. (1992). Regulation of $\alpha2\beta1$-mediated fibroblast migration of type I collgen by shifts in the concentrations of extracellular Mg^{2+} and Ca^{2+}. *J. Cell Biol.* **117,** 1109–1117.

Guichard, G., Benkirane, N., Zeder-Lutz, G., Van Regenmortel, M. H. V., Briand, J. P., and Muller, S. (1994). Antigenic mimicry of natural L-peptides with retro-inverso-peptidomimetics. *Proc. Natl. Acad. Sci. USA* **91,** 9765–9769.

Gullberg, D., Gehlsen, K., Turner, D., Ahlen, K., Zijenah, L., Barnes, M., and Rubin, K. (1992). Analysis of the $\alpha1\beta1$, $\alpha2\beta1$, and $\alpha3\beta1$ integrins in cell-collagen interactions: Identification of conformation dependent $\alpha1\beta1$ binding sites in collagen type I. *EMBO J.* **11,** 3865–3873.

Gundlach, B. R., Wiesmuller, K. H., Junt, T., Kienle, S., Jung, G., and Walden, P. (1996). Determination of T cell epitopes with random peptide libraries. *J. Immunol. Meth.* **192,** 149–155.

Harrison, S. C. (1996). Peptide–surface association: The case of PDZ and PTB domains. *Cell* **86,** 341–343.

Hasty, K. A., Pourmotabbed, T. F., Goldberg, G. I., Thompson, J. P., Spinella, D. G., Stevens, R. M., and Mainardi, C. L. (1990). Human neutrophil collagenase. *J. Biol. Chem.* **265,** 11421–11424.

Haugen, P. K., Letourneau, P. C., Drake, S. L., Furcht, L. T., and McCarthy, J. B. (1992). A cell-surface heparan sulfate proteoglycan mediates neural cell adhesion and spreading on a defined sequence from the C-terminal cell and heparin binding domain of fibronectin, FN-C/H II. *J. Neurosci.* **12,** 2597–2608.

Herbst, T. J., McCarthy, J. B., Tsilibary, E. C., and Furcht, L. T. (1988). Differential effects of laminin, intact type IV collagen, and specific domains of type IV collagen on endothelial cell adhesion and migration. *J. Cell Biol.* **106,** 1365–1373.

Hermans, J., and Puett, D. (1981). Relative effects of primary and tertiary structure on helix formation on myoglobin and α-lactalbumin. *Biopolymers* **10,** 895–914.
Hines, K. L., Kulkarni, A. B., McCarthy, J. B., Tian, H., Ward, J. M., Christ, M., McCartney-Francis, N. L., Turcht, L. T., Karlsson, S., and Wahl, S. M. (1994). Synthetic fibronectin peptides interrupt inflammatory cell infiltration in transforming growth factor β1 knockout mice. *Proc. Natl. Acad. Sci. USA* **91,** 5187–5191.
Hoeprich, P. D., Langton, B. C., Zhang, J., and Tam, J. P. (1989). Identification of immunodominant regions of transforming growth factor α. *J. Biol. Chem.* **264,** 19086–19091.
Horovits, A., Matthews, J. M., and Fersht, A. R. (1992). α-Helix stability in proteins II: Factors that influence stability at an internal position. *J. Mol. Biol.* **227,** 560–568.
Huebsch, J. B., Fields, G. B., Triebes, T. G., and Mooradian, D. L. (1996). Photoreactive analog of peptide FN-C/H-V from the carboxy-terminal heparin-binding domains of fibronectin supports endothelial cell adhesion and spreading on biomaterial surfaces. *J. Biomed. Mat. Res.* **31,** 555–567.
Hui, K. Y., Jakubowski, J. A., Wyss, V. L., and Angelton, E. L. (1992). Minimal sequence requirement of thrombin receptor agaonist peptide. *Biochem. Biophys. Res. Commun.* **184,** 790–796.
Humphries, M. J. (1990). The molecular basis and specificity of integrin–ligand interactions. *J. Cell Sci.* **97,** 585–592.
Humphries, M. J., Olden, K., and Yamada, K. M. (1986). A synthetic peptide from fibronectin inhibits experimental metastasis of murine melanoma cells. *Science* **233,** 467–470.
Hynes, R. O. (1992). Integrins: Versatility, modulation, and signaling in cell adhesion. *Cell* **69,** 11–25.
Iida, J., Skubitz, A. P. N., Furcht, L. T., Wayner, E., and McCarthy, J. B. (1992). Coordinate role for cell surface chondroitin sulfate proteoglycan and α4β1 integrin in mediating melanoma cell adhesion to fibronectin. *J. Cell Biol.* **118,** 431–444.
Ilyina, E., and Mayo, K. H. (1995). Multiple native-like conformations trapped via self-association-induced hydrophobic collapse of the 33-residue β-sheet domain from platelet factor-4. *Biochem. J.* **306,** 407–419.
Ilyina, E., Milius, R., and Mayo, K. H. (1994). Synthetic peptides probe folding initiation sites in platelet factor-4: Stable chain reversal found within the hydrophobic sequence LIATLKNGRKISL. *Biochemistry* **33,** 13436–13444.
Inouye, K., Kobayashi, Y., Kishida, Y., Sakakibara, S., and Prockip, D. J. (1982). Synthesis and physical properties of $(Hyp\text{-}Pro\text{-}Gly)_{10}$: Hydroxyproline in the X-position decreases the melting temperature of the collagen triple helix. *Arch. Biochem. Biophys.* **219,** 198–203.
Iwamoto, Y., Robey, F. A., Graf, J., Sasaki, M., Kleinman, H. K., Yamada, Y., and Martin, G. (1987). YIGSR, a synthetic laminin pentapeptide, inhibits experimental metastasis formation. *Science* **238,** 1132–1134.
Jameson, B. A., Rao, P. E., Hing, L. I., Shaw, G. M., Hood, L. E., and Kent, S. B. H. (1988). Location and chemical synthesis of a binding site for HIV-1 on the CD4 protein. *Science* **240,** 1335–1339.
Jameson, B. A., McDonnell, J. M., Marini, J. C., and Korngold, R. (1994). A rationally designed CD4 analogue inhibits experimental allergic encephalomyelitis. *Nature* **368,** 744–746.
Jarvis, J., Munro, S., and Craik, D. (1994). Structural analysis of peptide fragment 71–93 of transthyretin by NMR spectroscopy and electron microscopy. Insight into amyloid fibril formation. *Biochemistry* **33,** 33–41.
Johansson, M. W., Larsson, E., Luning, B., Pasquale, E. B., and Ruoslahti, E. (1994). Altered localization and cytoplasmic domain-binding properties of tyrosine-phosphorylated β1 integrin. *J. Cell Biol.* **126,** 1299–1309.
Johnson, W. C., Pagano, T. G., Basson, C. T., Madri, J. A., Gooley, P., and Armitage, I. M. (1993). Biologically active Arg-Gly-Asp oligopeptides assume a type II β-turn in solution. *Biochemistry* **32,** 268–273.
Kanemoto, R., Reich, R., Royce, L., Greatorex, D., Adler, S. H., Shiraishi, N., Martin, G. R., Yamada, Y., and Kleinman, H. K. (1990). Identification of an amino acid sequence from the laminin A chain that stimulates metastasis and collagenase IV production. *Proc. Natl. Acad. Sci. USA* **87,** 2279–2283.
Karniguian, A., Legrand, Y. J., Lefrancier, P., and Caen, J. P. (1983). Effect of collagen derived octapeptide on different steps of the platelet/collagen interaction. *Thrombosis Res.* **32,** 593–604.
Kavanaugh, W. M., and Williams, L. T. (1994). An alternative to SH2 domains for binding tyrosine-phosphorylated proteins. *Science* **266,** 1862–1864.
Ketchem, R. R., Hu, W., and Cross, T. A. (1993). High-resolution conformation of gramicidin A in a lipid bilayer by solid-state NMR. *Science* **261,** 1457–1460.

Khare, S. D., Krco, C. J., Griffiths, M. M., Luthra, H. S., and David, C. S. (1995). Oral administration of an immunodominant human collagen peptide modulates collagen-induced arthritis. *J. Immunol.* **155,** 3653–3659.

Kieffer, J. D., Plopper, G., Ingber, D. E., Hartwig, J. H., and Kupper, T. S. (1995). Direct binding of f-actin to the cytoplasmic domain of the α2 integrin chain *in vitro*. *Biochem. Biophys. Res. Commun.* **217,** 466–474.

Kim, J. P. Chen, J. D., Wilke, M. S., Schall, T. J., and Woodley, D. T. (1994a). Human keratinocyte migration on type IV collagen: Roles of heparin-binding site and $\alpha 2\beta 1$ integrin. *Lab. Invest.* **71,** 401–408.

Kim, W. H., Schnaper, W., Nomizu, M., Yamada, Y., and Kleiman, H. K. (1994b). Apoptosis in human fibrosarcoma cells is induced by a multimeric synthetic Tyr-Ile-Gly-Ser-Arg (YIGSR)-containing polypeptide from laminin. *Cancer Res.* **54,** 5005–5010.

Kippen, A., Sancho, J., and Fersht, A. (1994a). Folding of barnase in parts. *Biochemistry* **33,** 3778–3786.

Kippen, A. D., Arcus, V. L., and Fersht, A. R. (1994b). Structural studies on peptides corresponding to mutants of the major α-helix of barnase. *Biochemistry* **33,** 10013–10021.

Kodama, T., Freeman, M., Rohrer, L., Zabrecky, J., Matsudaira, P., and Krieger, M. (1990). Type I macrophage scavenger receptor contains α-helical and collagen-like coiled coils. *Nature* **343,** 531–535.

Koliakos, G. G., Kouzi-Koliakos, K., Furcht, L. T., Reger, L. A., and Tsilibary, E. C. (1989). The binding of heparin to type IV collagen: Domain specificity with identification of peptide sequences from the α1(IV) and α2(IV) which preferentially bind heparin. *J. Biol. Chem.* **264,** 2313–2323.

Krensky, A. M. (1994). T cells in autoimmunity and allograft rejection. *Kidney Int.* **45,** S50–S56.

Krstenansky, J. L., Owen, T. J., Hagaman, K. A., and McLean, L. R. (1989). Short model peptides having a high α-helical tendency: Design and solution properties. *FEBS Lett.* **242,** 409–413.

Ku, G., Kronenberg, M., Peacock, D. J., Tempst, P., Banquerigo, M. L., Braun, B. S., Reeve, J. R., Jr., and Brahn, E. (1993). Prevention of experimental autoimmune arthritis with a peptide fragment of type II collagen. *Eur. J. Immunol.* **23,** 591–599.

Kuhn, K., and Eble, J. (1994). The structural basis of integrin–ligand interactions. *Trends Cell Biol.* **4,** 256–261.

Kumagai, H., Tajima, M., Ueno, Y., Giga-Hama, Y., and Ohba, M. (1991). Effect of cyclic RGD peptide on cell adhesion and tumor metastasis. *Biochem. Biophys. Res. Commun.* **177,** 74–82.

Kwon, D. Y., and Kim, P. S. (1994). The stabilizing effects of hydrophobic cores on peptide folding of bovine pancreatic trypsin inhibitor folding intermediate model. *Eur. J. Biochem.* **223,** 631–636.

Lauer, J. L., and Fields, G. B. (1997). *In vitro* incorporation of synthetic peptides into cells. *In* "Methods in Enzymology" **289** (G. B. Fields, ed.), pp. 564–571. Academic Press, Orlando, FL.

Lauer, J. L., Furcht, L. T., and Fields, G. B. (1997). Inhibition of melanoma cell binding to type IV collagen by analogs of cell adhesion regulator. *J. Med. Chem.* **40,** 3077–3084.

Lear, J. D., Wasserman, Z. R., and DeGrado, W. F. (1988). Synthetic amphiphilic peptide models for protein ion channels. *Science* **240,** 1177–1181.

Lee, M. S., Gippert, G., Soman, K. Y., Case, D. A., and Wright, P. E. (1989). Three dimensional solution structure of a single zinc finger DNA-binding domain. *Science* **245,** 635–637.

Legrand, Y. J., Karniguian, A., Le Francier, P., Fauvel, F., and Caen, J. P. (1980). Evidence that a collagen-derived nonapeptide is a specific inhibitor of platelet-collagen interaction. *Biochem. Biophys. Res. Commun.* **96,** 1579–1585.

Li, M. H., Fan, P., Brodshy, B., and Baum, J. (1993). Two-dimensional NMR assignments and conformation of $(Pro\text{-}Hyp\text{-}Gly)_{10}$ and a designed collagen triple-helical peptide. *Biochemistry* **32,** 7377–7387.

Liang, B., Marchalonis, J. J., Zhang, Z., and Watson, R. R. (1996). Effects of vaccination against different T cell receptors on maintenance of immune function during murine retrovirus infection. *Cell. Immunol.* **172,** 126–134.

Lifson, S., and Roig, A. (1961). On the theory of helix–coil transition in polypeptides. *J. Chem. Phys.* **34,** 1963–1974.

Lin, Y. Z., Yao, S. Y., Yeach, R. A., Torgerson, T. R., and Hawiger, J. (1995). Inhibition of nuclear translocation of transciption factor NF-κB by a synthetic peptide containing a cell membrane-permeable motif and nuclear localization sequence. *J. Biol. Chem.* **270,** 14255–14258.

Linsenmayer, T. F. (1991). Collagen. *In* "Cell Biology of Extracellular Matrix" (E. D. Hay, ed.), pp. 7–44. Plenum Press, New York.

Liu, X., Siegel, D. L., Fan, P., Brodsky, B., and Baum, J. (1996). Direct NMR measurement of the folding kinetics of a trimeric peptide. *Biochemistry* **35,** 4306–4313.

Livnah, O., Stura, E. A., Johnson, D. L., Middleton, S. A., Mulcahy, L. S., Wrighton, N. C., Dower, W. J., Jolliffe, L. K., and Wilson, I. A. (1996). Functional mimicry of a protein hormone by a peptide agonist: The EPO receptor complex at 2.8 Å. *Science* **273,** 464–471.

Long, C. G., Li, M. H., Baum, J., and Brodsky, B. (1992). Nuclear magnetic resonance and circular dichroism studies of a triple-helical peptide with a glycine substitution. *J. Mol. Biol.* **225,** 1–4.

Lyu, P. C., Marky, L. A., and Kallenbach, N. R. (1989). The role of ion pairs in α-helix stability: Two designed helical peptides. *J. Am. Chem. Soc.* **111,** 2733–2734.

Lyu, P. C., Liff, M. I., Marky, L. A., and Kallenbach, N. R. (1990). Side chain contributions to the stability of α-helical structure in peptides. *Science* **250,** 669–673.

Maldonado, B. A., and Furcht, L. T. (1995). Involvement of integrins with adhesion promoting, heparin-binding peptides of type IV collagen in cultured human corneal epithelial cells. *Invest. Ophthalmol. Vis. Sci.* **36,** 364–372.

Marqusee, S., and Baldwin, R. L. (1987). Helix stabilization by Glu^-. . . Lys^+ salt bridges: Short peptides of *de novo* design. *Proc. Natl. Acad. Sci. USA* **84,** 8898–8902.

Marqusee, S., Robbins, V. H., and Baldwin, R. L. (1989). Unusually stable helix formation in short alanine-based peptides. *Proc. Natl. Acad. Sci. USA* **86,** 5286–5290.

Matouschek, A., Kellis, J. J., Serrano, L., and Fersht, A. R. (1990). Mapping the transient state-pathway of protein folding by protein engineering. *Nature* **340,** 122–126.

Matousheck, A., Serrano, L., and Fersht, A. R. (1992). The folding of an enzyme IV: Structure of an intermediate in the refolding of barnase analyzed by a protein engineering procedure. *J. Mol. Biol.* **224,** 819–835.

Mayo, K. H., Burke, C., Lindon, T., and Kloczewiak, M. (1990). 1H-NMR sequential assignments and secondary structure analysis of human fibrinogen chain C-terminal residues 385-411. *Biochemistry* **29,** 3277–3286.

Mayo, K. H., Parra-Diaz, D., McCarthy, J. B., and Chelberg, M. (1991). Cell adhesion promoting peptide GVKGDKGNPGWPGAP from the collagen type IV triple helix. *Biochemistry* **30,** 8251–8267.

McCarthy, J. B., Chelberg, M. K., Mickelson, D. J., and Furcht, L. T. (1988). Localization and chemical synthesis of fibronectin peptides with melanoma adhesion and heparin binding activities. *Biochemistry* **27,** 1380–1388.

McGeehan, G. M., Bickett, D. M., Green, M., Kassel, D., Wiseman, J. S., and Berman, J. (1994). Characterization of the peptide substrate specificities of interstitial collagenase and 92-kDa gelatinase. *J. Biol. Chem.* **269,** 32814–32820.

Mechling, D. E., Gambee, J. E., Morris, N. P., Sakai, L. Y., Keene, D. R., Mayne, R., and Bachinger, H. P. (1996). Type IX collagen NC1 domain peptides can trimerize *in vitro* without forming a triple helix. *J. Biol. Chem.* **271,** 13781–13785.

Melacini, G., Feng, Y., and Goodman, M. (1996). Collagen-based structures containing the peptoid residue *N*-isobutylglycine (Nleu). 6. Conformational analysis of Gly-Pro-Nleu sequences by 1H-NMR, CD, and molecular modeling. *J. Am. Chem. Soc.* **118,** 10725–10732.

Michael, S. F., Kilfoil, B. J., Schmidt, M. H., Amann, B. T., and Berg, J. M. (1992). Metal binding and folding properties of a minimalist Cys_2His_2 zinc finger peptide. *Proc. Natl. Acad. Sci. USA* **89,** 4796–4800.

Miick, S. M., Martinez, G. V., Fiori, W. R., Todd, A. P., and Millhauser, G. L. (1992). Short alanine-based peptides may form 3_{10}-helices and not α-helices in aqueous solution. *Nature* **359,** 653–655.

Miles, A. J., Skubitz, A. P. N., Furcht, L. T., and Fields, G. B. (1994). Cell adhesion activities of single-stranded and triple-helical peptide models of basement membrane collagen α1(IV)531-543: Evidence for conformationally dependent and conformationally independent type IV collagen cell adhesion sites. *J. Biol. Chem.* **269,** 30939–30945.

Miles, A. J., Knutson, J. R., Skubitz, A. P. N., Furcht, L. T., McCarthy, J. B., and Fields, G. B. (1995). A peptide model of basement membrane collagen α1(IV)531-543 binds the $\alpha 3\beta 1$ integrin. *J. Biol. Chem.* **270,** 29047–29050.

Miyahara, H., Myers, L. K., Rosloniec, E. F., Brand, D. D., Seyer, J. M., Stuart, J. M., and Kang, A. H. (1995). Identification and characterization of a major tolerogenic T cell epitope of type II collagen that suppresses arthritis in B10.RIII mice. *Immunology* **86,** 110–115.

Miyamoto, S., Akiyama, S. K., and Yamada, K. M. (1995a). Synergistic roles for receptor occupancy and aggregation in integrin transmembrane function. *Science* **267,** 883–885.

Miyamoto, S., Teramoto, H., Coso, O. A., Gutkind, J. S., Burbelo, P. D., Akiyama, S. K., and Yamada, K. M. (1995b). Integrin function: Molecular hierarchies of cytoskeletal and signaling molecules. *J. Cell Biol.* **131,** 791–805.

Mohri, H., and Ohkubo, T. (1993). Effect of cyclic Arg-Gly-Asp-containing peptide on fibronectin binding to activated platelets; role of fibronectin on platelet aggregation. *Peptides* **14,** 861–865.

Mohri, H., Hashimoto, Y., Ohba, M., Kumagai, H., and Ohkubo, T. (1991). Novel effect of cyclization of the Arg-Gly-Asp-containing peptide on vitronectin binding to platelets. *Am. J. Hematol.* **37,** 14–19.

Mooradian, D. L., McCarthy, J. B., Skubitz, A. P., Cameron, J. D., and Furcht, L. T. (1993). Characterization of FN-C/H-V, a novel synthetic peptide from fibronectin that promotes rabbit corneal epithelial cell adhesion, spreading, and motility. *Invest. Ophthalmol. Vis. Sci.* **34,** 153–164.

Morgan, K., Turner, S. L., Reynolds, I., Hajeer, A. H., Brass, A., and Worthington, J. (1992). Identification of an immunodominant B-cell eptope in bovine type II collagen and the production of antibodies to type II collagen by immunization with a synthetic peptide representing this epitope. *Immunol.* **77,** 609–616.

Morton, L. F., McCulloch, I. Y., and Barnes, M. J. (1993). Platelet aggregation by a collagen-like synthetic peptide. *Thrombosis Res.* **72,** 367–372.

Morton, L., Peachy, A., Zijenah, L., Goodall, A., Humphries, M., and Varnes, M. (1994). Conformation-dependent platelet adhesion to collagen involving integrin $\alpha2\beta1$-mediated and other mechanisms: Multiple $\alpha2\beta1$-recognition sites in collagen type I. *Biochem. J.* **299,** 791–797.

Morton, L. F., Hargreaves, P. G., Farndale, R. W., Young, R. D., and Barnes, M. J. (1995). Integrin $\alpha2\beta1$-independent activation of platelets by simple collagen-like peptides. Collagen tertiary (triple-helical) and quaternary (polymeric) structures are sufficient alone for $\alpha2\beta1$-independent platelet reactivity. *Biochem. J.* **306,** 337–344.

Moult, J., and Unger, R. (1991). An analysis of protein folding pathways. *Biochemistry* **30,** 3816–3824.

Muga, A., Surewicz, W. K., Wong, P. T. T., Mantsch, H. H., Singh, V. K., and Shinohara, T. (1990). Structural studies with the oveopathogenic peptide M derived from retinal S-antigen. *Biochemistry* **29,** 2925–2934.

Muller, D., Quantin, B., Gesnel, M. C., Millon-Collared, R., Abecassis, J., and Breathnach, R. (1988). The collagenase gene family in humans consists of at least four members. *Biochem. J.* **253,** 187–192.

Murphy, G., Allan, J. A., Willenbrock, F., Cockett, M. I., O'Connell, J. P., and Docherty, A. J. P. (1992). The role of the C-terminal domain in collagenase and stromelysin specificity. *J. Biol. Chem.* **267,** 9612–9618.

Myers, L. K., Stuart, J. M., Seyer, J. M., and Kang, A. H. (1989). Identification of an immunosuppressive epitope of type II collagen that confers protection against collagen-induced arthritis. *J. Exp. Med.* **170,** 1999–2010.

Myers, L. K., Rosloniec, E. F., Seyer, J. M., Stuart, J. M., and Kang, A. H. (1993). A synthetic peptide analogue of a determinant of type II collagen prevents the onset of collagen-induced arthritis. *J. Immunol.* **150,** 4652–4658.

Myers, L. K., Cooper, S. W., Terato, K., Seyer, J. M., Stuart, J. M., and Kang, A. H. (1995). Identification an characterization of a tolerogenic T cell determinant within residues 181–209 of chick type II collagen. *Clin. Immunol. Immunopath.* **75,** 33–38.

Nagase, H. (1996). Matrix metalloproteinases. *In* "Zinc Metalloproteases In Health and Disease" (N. M. Hooper, ed.), pp. 153–204. Taylor & Francis, London.

Nagase, H., and Fields, G. B. (1996). Human matrix metalloproteinase specificity studies using collagen sequence-based synthetic peptides. *Biopolymers* **40,** 399–416.

Nanevicz, T., Ishii, M., Wang, L., Chen, M., Turck, C. W., Cohen, F. E., and Coughlin, S. R. (1995). Mechanisms of thrombin receptor agonist specificity. *J. Biol. Chem.* **270,** 21619–21625.

Netzel-Arnett, S., Sang, Q. X., Moore, W. G. I., Navre, M., Birkedal-Hansen, B., and VanWart, H. E. (1993). Comparative sequence specificities of human 72- and 92-kDa gelatinases (type IV collagenases) and PUMP (matrilysin). *Biochemistry* **32,** 6427–6432.

Niedzwiecki, L., Teahan, J., Harrison, R. K., and Kim, P. S. (1992). Substrate specificity of the human matrix metalloproteinase stromelysin and the development of continuous fluorometric assays. *Biochemistry* **31,** 12618–12623.

Nomizu, M., Utani, A., Shiraishi, N., Kibbey, M. C., Yamada, Y., and Roller, P. P. (1992). The all-D-configuration segment containing the IKVAV sequence of laminin A chain has similar activities to the all-L-peptide *in vitro* and *in vivo*. *J. Biol. Chem.* **267,** 14118–14121.

Oas, T. G., and Kim, P. S. (1988). A peptide model of a protein folding intermediate. *Nature* **336,** 42–48.

Okuyama, K., Tanaka, N., Ashida, T., and Kakudo, M. (1976). Structure analysis of a collagen model polypeptide, $(\text{Pro-Pro-Gly})_{10}$. *Bull. Chem. Soc. Jpn.* **49,** 1805–1810.

Okuyama, K., Arnott, S., Takayanagi, M., and Kakudo, M. (1981). Crystal and molecular structure of a collagen-like polypeptide $(\text{Pro-Pro-Gly})_{10}$. *J. Mol. Biol.* **152,** 427–443.

Olivero, D. K., and Furcht, L. T. (1993). Type IV collagen, laminin and fibronectin promote the adhesion and migration of rabbit lens epithelial cells *in vitro*. *Invest. Opthalmol. Vis. Sci.* **34,** 2825–2834.

Olsen, B. R., Berg, R. A., Sakakibara, S., Kishida, Y., and Prockop, D. J. (1971). The synthetic polytripeptides $(\text{Pro-Pro-Gly})_{10}$ and $(\text{Pro-Pro-Gly})_{20}$ form micro-crystalline structures similar to segmental structures formed by collagen. *J. Mol. Biol.* **57,** 589–596.

O'Neill, K. T., and DeGrado, W. F. (1990). A thermodynamic scale for the helix-forming tendencies of the commonly occurring amino acids. *Science,* **250,** 646–651.

Osapay, G., and Taylor, J. W. (1990). Multicyclic polypeptide model compounds 1: Synthesis of a tricyclic amphiphilic α-helical peptide using an oxime resin, segment-condensation approach. *J. Am Chem. Soc.* **112,** 6046–6051.

Osapay, G., and Taylor, J. W. (1992). Multicyclic polypeptide model compounds 2: Synthesis and conformational properties of a highly α-helical uncosapeptide constrained by three side-chain to side-chain lactam bridges. *J. Am. Chem. Soc.* **114,** 6966–6973.

Osterman, D. G., and Kaiser, E. T. (1985). Design and characterization of peptides with amphiphilic β-strand structures. *J. Cell. Biochem.* **29,** 57–72.

Ottinger, E. A., Shekels, L. L., Bernlohr, D. A., and Barany, G. (1993). Synthesis of phosphotyrosine-containing peptides and their use as substrates for protein tyrosine phosphorylation. *Biochemistry* **32,** 4354–4361.

Padmanabha, S., Marqusee, S., Ridgeway, R., Laue, T. M., and Baldwin, R. L. (1990). Relative helix-forming tendencies of non-polar amino acids. *Nature* **344,** 268–270.

Park, S., Shalongo, W., and Stellwagen, E. (1993). Modulation of the helical stability model peptide by ionic residues. *Biochemistry* **32,** 12901–12905.

Pei, D., Wang, J., and Walsh, C. T. (1996). Differential functions of the two Src homology 2 domains in protein tyrosine phosphatase SH-PTP1. *Proc. Natl. Acad. Sci. USA* **93,** 1141–1145.

Perris, S., Syfrig, J., Paulsson, M., and Bronner-Praser, M. (1993). Molecular mechanisms of neural crest cell attachment and migration on types I and IV collagen. *J. Cell Sci.* **106,** 1357–1368.

Pertinhez, T. A., Nakaie, C. R., Carbalho, R. S. H., Paiva, A. C. M., Tabak, M., Toma, F., and Schreier, S. (1995). Conformational changes upon binding of a receptor loop to lipid structures: Possible role in signal transduction. *FEBS Lett.* **375,** 239–242.

Pfaff, M., Aumailley, M., Specks, U., Knolle, J., Zermes, H., and Timpl, R. (1993). Integrin and Arg-Gly-Asp dependence of cell ashesion to the native and unfolded triple helix of collagen type IV. *Exp. Cell Res.* **206,** 167–176.

Phipps, D. J., Reed-Doob, P. MacFadden, D. K., Piovesan, J. P., Mills, G. B., and Branch, D. R. (1995). An octapeptide analogue of HIV gp120 modulates protein tyrosine kinase activity in activated peripheral blood T lymphocytes. *Clin. Exp. Immunol.* **100,** 412–418.

Pierschbacher, M. D., and Ruoslahti, E. (1984). Cell attachment activity of fibronectin can be duplicated by small synthetic fragments of the molecule. *Nature* **309,** 30–33.

Pierschbacher, M. D., and Ruoslahti, E. (1987). Influence of sterochemistry of the sequence Arg-Gly-Asp-Xaa on binding specificity in cell adhesion. *J. Biol. Chem.* **262,** 17297–17298.

Pinori, M., and Verdini, A. S. (1987). Renin inhibition by partially modified retro-inverso analogues of the renin inhibitory peptide (RIP). *Peptide Chem.,* 645–648.

Plopper, G. E., McNamee, H. P., Dike, L. E., Bojanowski, K., and Ingber, D. E. (1995). Convergence of integrin and growth factor receptor signaling pathways within the focal adhesion complex. *Mol. Biol. Cell* **6,** 1349–1365.

Prockop, D. J., and Kivirikko, K. I. (1995). Collagens: Molecular biology, diseases, and potentials for therapy. *Ann. Rev. Biochem.* **64,** 403–434.

Pullman, W. E., and Bodmer, W. F. (1992). Cloning and characterization of a gene that regulates cell adhesion. *Nature* **356,** 529–532.

Qian, J. J., and Bhatnagar, R. S. (1996). Enhanced cell attachment to anorganic bone mineral in the presence of a synthetic peptide related to collagen. *J. Biomed. Materials Res.* **31,** 545–554.

Qian, J. J., Rosen, M. A., and Bhatnagar, R. (1993). A collagen-mimetic peptide induces collagenase in dermal fibroblasts. *FASEB J.* **7,** A1306.

Rao, G. H. R., Fields, C. G., White, J. G., and Fields, G. B. (1994). Promotion of human platelet adhesion and aggregation by a synthetic, triple-helical "mini-collagen." *J. Biol. Chem.* **269,** 13899–13903.

Ricouart, A., Tartar, A., and Sergheraert, C. (1989). Inhibition of protein kinase *c* by retro-inverso pseudosubstrate analogues. *Biochem. Biophys. Res. Comm.* **165,** 1382–1390.

Rohrer, L., Freeman, M., Kodama, T., Penman, M., and Krieger, M. (1990). Coiled-coil fibrous domains mediate ligand binding by macrophage scavenger receptor type II. *Nature* **343,** 570–572.

Roques, B. P., Lucassoroca, E., Chaillet, P., Costentin, J., and Fournie-Zaluski, M. C. (1983). Complete differentiation between enkephalinase and angiotensin-converting enzyme inhibition by retro-thiorphan. *Proc. Natl. Acad. Sci. USA* **80,** 3178–3182.

Rose, G. D., Gierasch, L. M., and Smith, J. A. (1985). Turns in peptides and proteins. *Adv. Prot. Chem.* **37,** 1–109.

Roth, W., and Heidemann, E. (1980). Triple helix–coil transition of covalently bridged collagen-like peptides. *Biopolymers* **19,** 1909–1917.

Roth, W., Heppenheimer, K., and Heidemann, E. R. (1979). Die Struktur kollagenahnlicher Homo- und Heteropolytripeptide 4: Polytripeptide durch repetitive Peptidsynthese und Verbruckung von Oligopeptiden. *Makromol. Chem.* **180,** 905–917.

Rubin, K., Hook, M., Obrink, B., and Timpl, R. (1981). Substrate adhesion of rat hepatocytes: Mechanism of attachment to collagen substrates. *Cell* **24,** 463–470.

Ruoslahti, E. (1988). Fibronectin and its receptors. *Annu. Rev. Biochem.* **57,** 375–413.

Ruoslahti, E. (1991). Integrins as receptors for extracellular matrix. *In* "Cell Biology of the Extracellular Matrix" (E. D. May, ed.), pp. 343–363. Plenum Press, New York.

Ruzza, P., Calderan, A., Filippi, B., Biondi, B., Deana, A. D., Cesaro, L., Pinna, L. A., and Borin, G. (1995). Synthetic Tyr-phospho and non-hydrolyzable phosphonopeptides as PTKs and TC-PTP inhibitors. *Int. J. Peptide Protein Res.* **45,** 529–539.

Sabo, T., Gurwitz, D., Motola, L., Brodt, P., Barak, R., and Elhanaty, E. (1992). Structure-activity studies of the thrombin receptor activating peptide. *Biochem. Biophys. Res. Comm.* **188,** 604–610.

Sakakibara, S., Kishida, Y., Kikuchi, Y., Sakai, R., and Kakiuchi, K. (1968). Synthesis of poly-(L- prolyl-L-prolylglycyl) of defined molecular weights. *Bull. Chem. Soc. Jpn,* **41,** 1273.

Sakakibara, S., Inouye, K., Shudo, K., Kishida, Y., Kobayashi, Y., and Prockop, D. J. (1973). Synthesis of (Pro-Hyp-Gly)$_n$ of defined molecular weights: Evidence for the stabilization of collagen triple helix by hydroxyproline. *Biochim. Biophys. Acta* **303,** 198–202.

Sakamoto, N., Iwahana, M., Tanaka, N. G., and Osada, Y. (1991). Inhibition of angiogenesis and tumor growth by a synthetic laminin peptide CDPGYIGSR-NH_2. *Cancer Res.* **51,** 903–906.

Santoro, S. A. (1986). Identification of a 160,000 dalton platelet membrane protein that mediates the initial divalent cation-dependent adhesion of platelets to collagen. *Cell* **46,** 913–920.

Sato, H., Takino, T., Okada, Y., Cao, J., Shinagawa, A., Yamamoto E., and Seiki, M. (1994). A matrix metalloproteinase expressed on the surface of invasive tumour cells. *Nature* **370,** 61–65.

Scharffetter-Kochanek, K. Klein, C. E., Heinen, G., Mauch, C., Schaefer, T., Adelmann-Grill, B. C., Goerz, G., Fusenig, N. E., Krieg, T. M., and Plewig, G. (1992). Migration of a human keratinocyte cell line (HACAT) to interstitial collagen type I is mediated by the $\alpha 2\beta 1$-integrin receptor. *J. Invest. Dermatol.* **98,** 3–11.

Schechter, I., and Berger, A. (1967). On the size of the active site in proteases. I. Papain. *Biochem. Biophys. Res. Commun.* **27,** 159–162.

Schnolzer, M., and Kent, S. B. H. (1992). Constructing proteins by dovetailing unprotected synthetic peptides. *Science* **256,** 221–225.

Schreiber, R. E., Prossnitz, E. R., Ye, R. D., Cochrane, C. G., and Bokoch, G. M. (1994). Domains of the human neutrophil *N*-formyl peptide receptor involved in G protein coupling. *J. Biol. Chem.* **269,** 326–331.

Seltzer, J. L., Adams, S. A., Grant, G. A., and Eisen, A. Z. (1981). Purification and properties of a gelatin-specific neutral protease from human skin. *J. Biol. Chem.* **256,** 4662–4668.

Seltzer, J. L., Weingarten, H., Akers, K. T., Eschbach, M. L., Grant, G. A., and Eisen, A. S. (1989). Cleavage specificity of type IV collagenase (gelatinase) from human skin. *J. Biol. Chem.* **264,** 19583–19586.

Seltzer, J. L., Akers, K. T., Weingarten, H., Grant, G. A., McCourt, D. W., and Eisen, A. Z. (1990). Cleavage specificity of human skin type IV collagenase (gelatinase). *J. Biol. Chem.* **265,** 20409–20413.

Serrano, L., Kellis, J. R., Cann, P., Matouschek, A., and Fersht, A. R. (1992a). The folding of an enzyme II: Substructure of barnase and the contribution of different interactions to protein stability. *J. Mol. Biol.* **224,** 783–804.

Serrano, L., Matouschek, A., and Fersht, A. R. (1992b). The folding of an enzyme VI: The folding pathway of barnase: Comparison with theoretical models. *J. Mol. Biol.* **224,** 847–859.

Shah, N. K., Ramshaw, J. A. M., Kirkpatrick, A., Shah, C., and Brodsky, B. (1996). A host–guest set of triple-helical peptides: stability of Gly-X-Y triplets containing common nonpolar residues. *Biochemistry* **35,** 10262–10268.

Shankar, G., Gadek, T. R., Burdick, D. J., Davison, I., Mason, W. T., and Horton, M. A. (1995). Structural determinants of calcium signaling by RGD peptides in rat osteoclasts: Integrin-dependent and -independent actions. *Exp. Cell Res.* **219,** 364–371.

Shapiro, S. D., Griffin, G. L., Gilbert, D. J., Jenkins, N. A., Copeland, N. G., Welgus, H. G., Senior, R. M., and Ley, T. J. (1992). Molecular cloning, chromosomal localization, and bacterial expression of a murine macrophage metalloelastase. *J. Biol. Chem.* **267,** 4664–4671.

Shin, H., Merutka, G., Waltho, J., Wright, P., and Dyson, H. (1993a). Peptide models of protein folding initiation sites 2: The G-H turn region of myoglobulin acts as a helix stop signal. *Biochemistry* **32,** 6348–6355.

Shin, H., Merutka, G., Waltho, J., Tennant, L., Dyson, J., and Wright, P. (1993b). Peptide models of protein folding initiation sites 3: The G-H helical hairpin myoglobulin. *Biochemistry* **32,** 6356–6364.

Skubitz, A. P. N., McCarthy, J. B., Zhao, Q., Yi, X., and Furcht, L. T. (1990). Definition of a sequence, RYVVLPR, within laminin peptide F-9 that mediates metastatic fibrosarcoma cell ashesion and spreading, *Cancer Res.* **50,** 7612–7622.

Sloan-Lancaster, J., Evavold, B. D., and Allen, P. M. (1993). Induction of T-cell anergy by altered T-cell-receptor ligand on live antigen-presenting cells. *Nature* **363,** 156–158.

Smithgall, T. E. (1995). SH2 and SH3 domains. Potential targets for anti-cancer drug design. *J. Pharm. Toxicol. Meth.* **34,** 125–132.

Songyang, Z., and Cantley, L. C. (1995). Recognition and specificity in protein tyrosine kinase-mediated signalling. *Trends Biochem. Sci.* **20,** 85–88.

Staatz, W. D., Walsh, J. J., Pexton, T., and Santoro, S. A. (1990). The $\alpha 2\beta 1$ integrin cell surface collagen receptor binds to the $\alpha 1$(I)-CB3 peptide of collagen. *J. Biol. Chem.* **265,** 4778–4781.

Staatz, W. D., Fok, K. F., Zutter, M. M., Adams, S. P., Rodriguez, B. A., and Santoro, S. A. (1991). Identification of a tetrapeptide recognition sequence for the $\alpha 2\beta 1$ integrin in collgen. *J. Biol. Chem.* **266,** 7363–7367.

Stellwagen, E., Rysavy, R. J., and Babul, G. (1972). The conformation of horse heart apocytochrome *c*. *J. Biol. Chem.* **274,** 8074–8077.

Stetler-Stevenson, W. G., Aznavoorian, S., and Liotta, L. A. (1993). Tumor cell interaction with the extracellular matrix during invasion and metastasis. *Ann. Rev. Cell Biol.* **9,** 541–573.

Struthers, M. D., Cheng, R. P., and Imperiali, B. (1996a). Design of a monomeric 23-residue polpeptide with defined tertiary structure. *Science* **271,** 342–345.

Struthers, M. D., Cheng, R. P., and Imperiali, B. (1996b). Economy in protein design: Evolution of a metal-independent $\beta\beta\alpha$ motif based on the zinc finger domains. *J. Am. Chem. Soc.* **118,** 3073–3081.

Sudol, M. (1996a). The WW modules competes with the SH3 domain. *Trends Biochem. Sci.* 21, 161–163.

Sudol, M. (1996b). Structure and function of the WW domain. *In* "Progress in Biophysics and Molecular Bioilogy" (Blundell, T., Noble, D., Pawson, T. eds.), in press.

Takano, T., and Dickerson, R. E. (1981). Conformation change of cytochrome *c* 1: Ferrocytochrome *c* struture refined at 1.5 Å resolution. *J. Mol. Biol.* **153,** 79–94.

Takino, T., Sato, H., Shingawa, A., and Seiki, M. (1995). Identification of the second membrane-type matrix metalloproteinase (MT-MMP-2) gene from a human placenta cDNA library. *J. Biol. Chem.* **270,** 23013–23020.

Tam, J. P., Lin, Y., Wang, D., Ke, X., and Zhang, J. (1991). Mapping the receptor-recognition site of human transforming growth factor-α. *Int. J. Peptide Protein Res.* **38,** 204–211.

Tanaka, T., Wada, Y., Nakamura, H., Doi, T., Imanishi, T., and Kodama, T. (1993). A synthetic model of collagen structure taken from bovine macrophage scavenger receptor. *FEBS Lett.* **334,** 272–276.

Tanakak, T., Nishikawa, A., Tanaka, Y., Nakamura, H., Kodama, T., Imanishi, T., and Doi, T.

(1996). Synthetic collagen-like domain derived from the macrophage scavenger receptor binds acetylated low-density lipoprotein *in vitro*. *Protein Eng.* **9,** 307–313.

Tashiro, K., Sephel, G. C., Weeks, B., Sasaki, M., Martin, G. R., Kleinman, H. K., and Yamada, Y. (1989). A synthetic peptide containing the IKVAV sequence from the A chain of laminin mediates cell attachment, migration, and neurite ourgrowth. *J. Biol. Chem.* **264,** 16174–16182.

Terzi, E., Holzemann, G., and Seelig, J. (1994). Reversible random coil β-sheet transition of the alzheimer β-amyloid fragment (25-35). *Biochemistry* **33,** 1345–1350.

Thakur, S., Vadolas, D., Germann, H. P., and Heidemann, E. (1986). Influence of different tripeptides on the stability of the collagen triple helix II: An experimental approach with appropriate variations of a trimer model oligotripeptide. *Biopolymers* **25,** 1081–1086.

Theodore, L., Derossi, D., Chassaing, G., Llirbat, B., Kubes, M., Jordan, P., Chneiweiss, H., Godement, P., and Prochiantz, A. (1995). Intraneuronal delivery of protein kinase C pseudosubstrate leads to growth cone collapse. *J. Neurosci.* **15,** 7158–7167.

Trentham, D. E., Dynesius-Trentham R. A., Orav, E. J., Combitchi, D., Lorenzo, C., Sewell, K. L., Hafler, D. A., and Weiner, H. L. (1993). Effects of oral administration of type II collagen on rheumatoid arthritis. *Science* **261,** 1727–1730.

Tsu, C. A., Persona, J. J., Schellenberger, V., Turck, C. W., and Craik, C. S. (1994). The substrate specificity of *Uca pugilator* collagenolytic serine protease 1 correlates with the bovine type I collagen cleavage sites. *J. Biol. Chem.* **269,** 19565–19572.

Tuckwell, D. S., Ayad, S., Grant, M. E., Takigawa, M., and Humphries, M. J. (1994). Conformation dependence of integrin-type II collagen binding. *J. Cell Sci.* **107,** 993–1005.

Vandenbark, A. A., Hashim, G. A., and Offner, H. (1996). T cell receptor peptides in treatment of autoimmune disease: Rationale and potential. *J. Neurosci. Res.* **43,** 391–402.

Vandenberg, P., Kerm, A., Ries, A., Lukenbill-Edds, L., Mann, K., and Kuhn, K. (1991). Characterization of a type IV collagen major cell binding site with affinity to the $\alpha1\beta1$ and the $\alpha2\beta1$ integrins. *J. Cell Biol.* **113,** 1475–1483.

Vassallo, R. R., Kieber-Emmons, T., Cichowski, K., and Brass, L. F. (1992). Structure–function relationships in the activation of platelet thrombin receptors by receptor-derived peptides. *J. Biol. Chem.* **267,** 6081–6085.

Venugopal, M. G., Ramshaw, J. A. M., Braswell, E., Zhu, D., and Brodsky, B. (1994). Electrostatic interactions in collagen-like triple-helical peptides. *Biochemistry* **33,** 7948–7956.

Wade, D., Boman, A., Wahlin, B., Drain, C. M., Andreu, D., Boman, H. G., and Merifield, R. B. (1990). All-D amino acid-containing channel-forming antibiotic peptides. *Proc. Natl. Acad. Sci. USA* **87,** 4761–4765.

Waltho, J. P., Feher, V. A., Lerner, R. A., and Wright, P. E. (1989). Conformation of a T cell stimulating peptide in aqueous solution. *FEBS Lett.* **250,** 400–404.

Waltho, J., Feher, V., Merutka, G., Dyson, H., and Wright, P. (1993). Peptide models of protein folding initiation sites 1: Secondary structure formation by peptides corresponding to the G- and H-helices of myoglobin. *Biochemistry* **32,** 6337–6347.

Wange, R. L., Isakov, N., Burke, T. R., Otaka, A., Roller, P. P., Watts, J. D., Aebersold, R., and Samelson, L. E. (1995). $F_2(Pmp)_2$-TAMζ_3, a novel competitive inhibitor of the binding of ZAP-70 to the T cell antigen receptor, blocks early T cell signaling. *J. Biol. Chem.* **270,** 944–948.

Ward, C. W., Gough, K. H., Rashke, M., Wan, S. S., Tribbick, G., and Wang, J. (1996). Systematic mapping of potential binding sites for Shc and Grb2 SH2 domains on insulin receptor substrate-1 and the receptors for insulin, epidermal growth factor, platelet-derived growth factor, and fibroblast growth factor. *J. Biol. Chem.* **271,** 5603–5609.

Watson, R. R., Dehghanpisheh, K., Huang, D. S., Wood, S., Ardestani, S. K., Liang, B., Marchalonis, J. J., and Wang, Y. (1995). T cell receptor V beta complementarity-determining region 1 peptide administration moderates immune dysfunction and cytokine dysregulation induced by murine retrovirus infection. *J. Immunol.* **155,** 2282–2285.

Wattenbarger, M. R., Chan, H. S., Evans, D. F., Bloomfield, V. A., and Dill, K. A. (1990). Surface-induced enhancement of internal structure in polymers and proteins. *J. Chem. Phys.* **93,** 8343–8351.

Weber, R. W., and Nitschmann, H. (1978). The influence of *O*-acetylation upon the conformational behavior of the collagen model peptide $(L\text{-}Pro\text{-}L\text{-}Hyp\text{-}Gly)_{10}$ and of gelatin. *Helv. Chim. Acta* **61,** 701–708.

Weingarten, H., and Feder, J. (1986). Cleavage site specificity of vertebrate collagenase. *Biochem. Biophys. Res. Commun.* **139,** 1184–1187.

Welgus, H. G., and Grant, G. A. (1983). Degradation of collagen substrates by a trypsin-like serine protease from the fiddler crab *Uca pugilator*. *Biochemistry* **22,** 2228–2233.
Wells, J. A. (1990). Additivity of mutational effects in proteins. *Biochemistry* **29,** 8509–8517.
Whitham, S. E., Murphy, G., Angel, P., Rahmsdorf, H. J., Smith, B. J., Lyons, A., Harris, T. J. R., Reynolds, J. J., Herlich, P., and Docherty, A. J. P. (1986). Comparison of human stromelysin and collagenase by cloning and sequence analysis. *Biochem. J.* **240,** 913–916.
Wilhelm, S. M., Collier, I. E., Marmer, B. L., Eisen, A. Z., Grant, G. A., and Goldberg, G. I. (1989). SV-40-transformed human lung fibroblasts secrete a 92-kDa type IV collagenase which is identical to that secreted by normal human macrophages. *J. Biol. Chem.* **264,** 17213–17211.
Wilke, M. S., and Furcht, L. T. (1990). Human keratinocytes adhere to a unique heparin-binding sequence within the triple helical region of type IV collagen. *J. Invest. Dermatol.* **95,** 264–270.
Will, H., and Hinzmann, B. (1995). cDNA sequence and mRNA tissue distribution of a novel human matrix metalloproteinase with a potential transmembrane sequence. *Eur. J. Biochem.* **231,** 602–608.
Williams, H. R., and Lin, T. Y. (1984). Human polymorphonuclear leukocyte collagenase and gelatinase. *Int. J. Biochem.* **16,** 1321–1329.
Williams, O., Tanaka, Y., Bix, M., Murdjeva, M., Littman, D. R., and Kioussis, D. (1996). Inhibition of thymocyte negative selection by T cell receptor antagonist peptides. *Eur. J. Immunol.* **26,** 532–538.
Wilson, J. R., and Weiser, M. M. (1992). Colonic cancer cell (HT29) adhesion to laminin is altered by differentiation: Adhesion may involve galactosyltransferase. *Exp. Cell Res.* **201,** 330–334.
Windhagen, A., Scholz, C., Hollsberg, P., Fukaura, H., Sette, A., and Hafler, D. A. (1995). Modulation of cytokine patterns of human autoreactive T cell clones by a single amino acid substitution of their peptide ligand. *Immunity* **2,** 373–376.
Winkler, J. R., Qian, J. J., and Bhatnagar, R. S. (1991). Synthetic peptides inhibit T-cell attachment to type I collagen. *J. Cell. Biol.* **115,** 442a.
Woessner, J. F. (1991). Matrix metalloproteinases and their inhibitors in connective tissue remodeling. *FASEB J.* **5,** 2145–2154.
Woods, A., McCarthy, J. B., Furcht, L. T., and Couchman, J. R. (1993). A synthetic peptide from the COOH-terminal haparin-binding domain of fibronectin promotes focal adhesion formation. *Mol. Biol. Cell* **4,** 605–613.
Wrighton, N. C., Farrell, F. X., Chang, R., Kashyap, A. K., Barbone, F. P., Mulcahy, L. S., Johnson, D. L., Barrett, R. W., Jolliffe, L. K., and Dower, W. J. (1996). Small peptides as potent mimetics of the protein hormone erythropoietin. *Science* **273,** 458–463.
Wu, L., Laub, P., Elove, G., Carey, J., and Roder, H. (1993). A noncovalent peptide complex as a model for an early folding intermediate of cytochrome c. *Biochemistry* **32,** 10271–10276.
Wu, R., Durick, K., Songyang, Z., Cantley, L. C., Taylor, S. S., and Gill, G. N. (1996). Specificity of LIM domain interactions with receptor tyrosine kinases. *J. Biol. Chem.* **271,** 15934–15941.
Yamada, K. M. (1991). Adhesive recognition sequences. *J. Biol. Chem.* **266,** 12809–12812.
Yamada, K. M., Kennedy, D. W., Yamada, S. S., Gralnick, H. Chen, W. T., and Akiyama, S. K. (1990). Monoclonal antibody and synthetic peptide inhibitors of human tumor cell migration. *Cancer Res.* **50,** 4485–4496.
Yamamura, K., Kibbey, M. C., and Kleinman, H. K. (1993). Melanoma cells selected for adhesion to laminin peptides have different malignant properties. *Cancer Res.* **53,** 423–428.
Yanaka, K., Camarata, P. J., Spellman, S. R., McCarthy, J. B., Furcht, L. T., Low, W. C., and Heros, R. C. (1996). Synthetic fibronectin peptides and ischemic brain injury after transient middle cerebral artery occlusion in rats. *J. Neurosurg,* **85,** 125–130.
Yurchenco, P. D., and O'Rear, J. J. (1994). Basal lamina assembly. *Current Opin. Cell Biol.* **6,** 674–681.
Zhou, H. X., Lyu, P. C., Wemmer, D. E., and Kallenbach, N. R. (1994). Structure of the C-terminal α-helix cap in a synthetic peptide. *J. Am. Chem. Soc.* **116,** 1139–1140.
Zhou, M. M., Meadows, R. P., Logan, T. M., Yoon, H. S., Wade, W. S., Ravichandran, K. S., Burakoff, S. J., and Fesik, S. W. (1995). Solution structure of the Shc SH2 domain complexed with a tyrosine-phosphorylated peptide from the T-cell receptor. *Proc. Natl. Acad. Sci. USA* **92,** 7784–7788.
Zvelebil, M. J., Panayotou, G., Linacre, J., and Waterfield, M. D. (1995). Prediction and analysis of SH2 domain–phosphopeptide interactions. *Protein Eng.* **8,** 527–533.

Chapter 4

Site-Directed Mutagenesis and Protein Engineering

Gary S. Coombs and David R. Corey

I. Introduction

The routine application of protein engineering is a recent phenomenon. Prior to the development of oligonucleotide-directed mutagenesis in the early 1980s, practical considerations stringently limited the manipulation of proteins. Proteins needed to be reactive toward specific chemical modification and to be obtainable directly from biological sources. Since these humble beginnings, advances in gene cloning and expression have exponentially increased the number of proteins available for manipulation. In parallel with this expansion, the development of strategies for site-directed and random mutagenesis has multiplied the techniques suitable for protein modification and has transformed protein engineering from a curiosity into a routine practice. Laboratories in diverse disciplines can now attempt ambitious alterations, and such projects will become increasingly common as genome sequencing adds to the pool of accessible proteins.

The purpose of this chapter is to introduce the reader to strategies, philosophies, and potential pitfalls of protein engineering through site-specific modification. How does one evaluate the potential of a given protein scaffold for alteration? What information guides the choice of strategies for protein engineering? Rather than attempt the near-impossible task of summarizing the large number of experiments that have been performed, this chapter will present important experimental approaches and case studies of successful applications. Examples emphasize the unique potential of protein engineering to supply critical insights that cannot be obtained readily through other approaches. Two books that focus on protein engineering may provide the reader with additional insight into this topic (1,2).

II. Why Alter Proteins?

Protein engineering is defined as the rational manipulation of protein chemistry to elicit novel functional and structural properties. As the name implies, protein engineering is a fundamentally different approach to the investigation of biological interactions. Traditionally, experimenters have studied function through direct examination of native proteins. These investigations are intrinsically justified because the information that they generate increases our knowledge of the natural world. On alteration, however, macromolecules are no longer native, removing this justification. Furthermore, a single alteration may change multiple properties of the protein (including structure, folding pathways, flexibility, stability, and enzymatic and / or interactive properties), making it difficult to interpret knowledge gained through the study of engineered proteins or apply it to further understanding of their native counterparts. Is a mutant protein inactive because a key function has been located and removed, or is inactivity caused by a global disruption in protein structure? Each mutant protein is a new entity, and the researcher can examine each in depth. Will the knowledge gained through these efforts have the potential to provide useful insights about how native proteins work?

The need to address this question before beginning a protein engineering project is especially urgent because of the simplicity and variety of techniques for mutagenesis and chemical modification. Starting from a single protein, mutagenesis allows the generation of an infinite number of new proteins. Many will have novel properties, but only a small fraction of these will be interesting beyond the experimenter's own laboratory. The challenge is to use available knowledge to choose alterations that will provide insights into native function, or introduce novel function for laboratory, medical, or industrial applications. If this theoretical groundwork is not laid carefully, the ease of technical manipulations can draw the experimenter into the pointless production of numerous new proteins that collectively supply little or no useful data.

Given these caveats, why devote resources to protein engineering which could be devoted to direct study of native proteins? As with any scientific endeavor, the importance of the potential experimental outcome must be balanced against the investment required to obtain useful results, and this balance will vary from protein to protein. However, there are general reasons for using protein engineering:

1. The direct study of the rules governing protein structure and function through mutagenesis. This is probably the most common use of protein engineering. It is always an excellent option for the study of proteins. As noted previously, however, the experimenter should realize that alterations can have unintended consequences and that in comparing mutants to their wild-type progenitors, interpretations can be extremely complex.
2. The introduction of novel function possessing practical value. This goal covers a variety of possibilities, including improvement of purification efficiency, adaptation of altered enzyme specificities to yield new experimental tools, and optimization of therapeutically valuable proteins. Striking success in this area is rarely achieved—primarily because of the difficulty of engineering proteins to be both noval and efficient—and should be treated as a worthwhile but difficult goal.
3. The development of new and general strategies for protein engineering. This is analogous to the organic chemist's discovery of a useful new synthetic reaction during synthesis of a specific chemical target.
4. The acquisition of a novel perspective on native function. The experimenter is obliged to ask different questions when altering a protein than during direct study of the native macromolecule. Creativity and insight can often turn the answers to these questions into important lessons about native structure and function that would have gone unappreciated otherwise. We stress, however, that, unless an experiment is well designed, useful answers will probably not be obtained.

III. Techniques for Oligonucleotide-Directed Mutagenesis

Mutagenesis by exposure to chemical mutagens or ultraviolet light has long been an important tool in genetics. The ability of these procedures to create proteins that possess altered properties is undisputed. Mutagenesis by these methods is, however, random, preventing the direct manipulation of specified amino acids or the designed alteration of protein function. This prevents the experimenter from exploiting existing knowledge about the protein in the design of experiments. To accomplish specific modification, Smith and colleagues exploited the advent of efficient methods for the chemical synthesis of oligonucleotides to develop techniques for oligonucleotide-directed mutagenesis (3). This strategy has proven to be widely useful and has become a standard technique for gene-based variation of protein structure and function (4–7).

With the exception of cassette mutagenesis (Section III,D), currently available techniques for oligonucleotide-directed mutagenesis require accomplishment of three processes: (i) annealing of an oligonucleotide primer containing the change to be introduced to the gene of interest (ii) replication of the template using the mutated oligonucleotide as primer, and (iii) some means of selection that favors the newly synthesized mutant strand relative to the wild-type template strand. There are three commonly utilized approaches for accomplishing the replication of template, each of which may be followed by multiple selection strategies. Sections III, A–C will briefly describe each approach and discuss inherent advantages and disadvantages.

A. Mutagenesis Using Single-Stranded DNA as Template

Many mutagenesis protocols employ single-stranded DNA as a template. Experimenters can obtain single-stranded DNA using bacteriophage M13 and its derivatives. M13 is a filamentous bacteriophage that exists in both double-stranded and single-stranded forms (Fig. 1). The double-stranded form, also known as the replicative form (RF), can be introduced into bacteria via transformation, whereas phage which contain the single-stranded form can infect bacteria during transfections. Double-stranded M13 DNA can be obtained by lysis of an infected bacterial culture through standard procedures for the preparation of plasmid DNA (8). Either all or part of the gene to be mutated can be ligated into double-stranded M13 and the product can then be utilized to transform a fresh bacterial culture. The single-stranded form of M13 is packaged into phage particles and secreted by the cells into the supernatant. Unlike λ phage, the bacteria are not lysed. The single-stranded phage DNA can then be isolated and used as template for synthesis of the mutagenic strand.

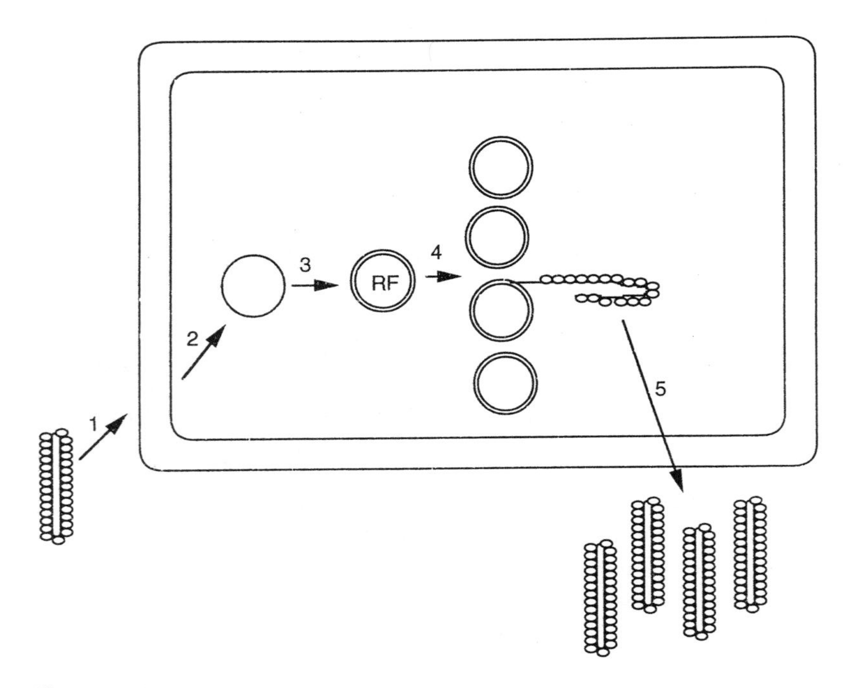

Fig. 1 M13 life cycle.

Single-stranded DNA is a convenient template for mutagenesis because the absence of interstrand base pairing allows primer annealing in one simple step, and M13 is a convenient vector for its production. Disadvantages include the fact that cloning of the gene of interest into M13 introduces an additional procedure, as does cloning of the newly made mutant back into the expression vector. In addition, M13 is noted for its propensity to delete all or part of inserted DNA sequences, especially when the insertions into M13 are larger than 5 kB. Because of the tendency of M13 to delete longer sequences, and for convenience in sequencing mutants, it is advisable to keep the inserted regions short. Genomic instability also makes M13 a risky choice as an expression vector. The likelihood of deletion can be reduced by carefully limiting the growth of M13-bearing cells, as deletions will be more likely in cultures maintained in the plateau phase for extended periods.

Phagemid vectors provide an alternative to the use of M13 (Fig. 2). Phagemids are plasmids that contain an f1 origin of single-stranded replication. This origin allows them to be replicated in single-stranded form and secreted

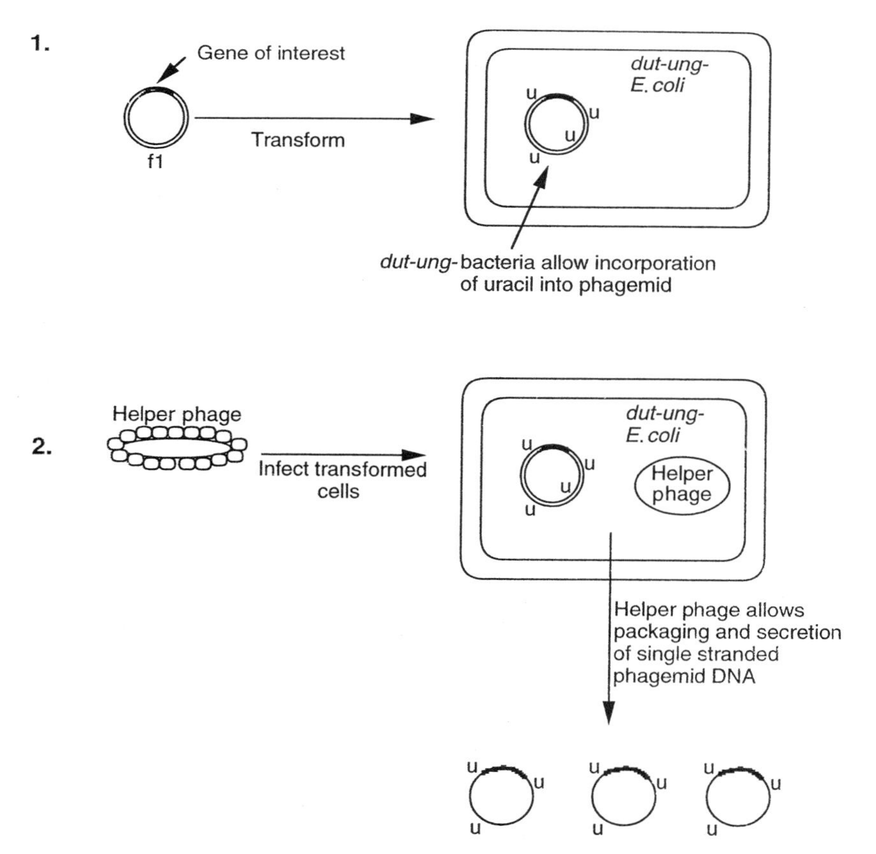

Fig. 2 Typical scheme for obtaining uracil-laden phagemid DNA. In the first step plasmid DNA is transformed in a *dut⁻ung⁻* strain of *Escherichia coli*. In the second step, helper phage are added to allow the synthesis and packaging of uracil-laden single-stranded DNA.

as phage particles when their host cells are superinfected with helper phage that encode the proteins necessary for packaging. Helper phage are mutants of M13 in which the origin of replication is replaced with one that is less efficiently recognized so that synthesis of phagemid-containing particles will be favored relative to that of helper phage. Selectable markers are usually present within the helper phage genome to facilitate selection for only those cells that have been infected by both phagemid and helper phage. Use of either M13 or phagemid plus helper phage requires that the host cell express the F′ episome that codes for pili formation, as these provide the means by which phage particles recognize and gain entry into their target cells.

There are three advantages of using phagemid rather than direct cloning into M13. First, the large size of M13 DNA can make the vector relatively prone to spontaneous loss of portions of the inserted foreign DNA. Phagemid vectors are generally smaller and are not noted to experience problems with deletions, allowing larger genes to be mutated. Second, the smaller size of phagemids enhances transformation efficiency relative to M13. This factor can be critical if many clones need to be obtained, as is often the case for construction of mutant-containing gene libraries. Finally, because phagemid vectors can better maintain the full-length gene, both mutagenesis and expression can be done in the same construct, eliminating cloning steps. These advantages must be balanced against the fact that obtaining single-stranded DNA from phagemid is more complex, and often more difficult, than obtaining single-stranded M13 DNA.

Once single-stranded template is obtained, a variety of methods are available for mutagenesis. Details of these methods vary, but they all represent strategies for synthesizing a mutagenized strand *in vitro* and subsequently selecting for clones containing the newly synthesized mutant strand (Table I). The Kunkel method for mutagenesis is a paradigm for selecting the mutagenized strand over the nonmutagenized strand (9,10). This method exploits incorporation of uracil in the unmutagenized template strand in place of thymine by producing the single stranded phagemid or M13 template in a *dut*$^-$*ung*$^-$ double mutant strain of bacteria. The *dut* mutation inactivates the enzyme dUTPase and allows the cells to accumulate high levels of dUTP. The ung mutation inactivates another enzyme, uracil *N*-glycosylase, which removes uracil residues from DNA. After the *in vitro* synthesis reaction, the DNA is transformed into a bacterium that expresses active uracil *N*-glycosylase, and the mutagenic strand is replicated while the uracil-containing template strand is inactivated by the removal of uracil. This boosts the efficiency of mutagenesis enough so that screening can be conveniently performed by DNA sequencing. Alternatively, the presence of a mutation can be detected by restriction endonuclease digestion if the mutation has introduced or removed a restriction site. Fortunately, the large number of restriction enzyme specificities allows considerable latitude in choosing mutations that will introduce a new restriction site but that will also retain codons for the desired amino acids.

Eckstein and colleagues have developed a similar selection strategy that involves the incorporation of a phosphorothioate deoxyribonucleotide during strand elongation (11). A phosphorothioate in the DNA backbone renders the DNA resistant to cleavage by restriction endonucleases. After incorporation of a phosphorothioate-containing nucleotide during elongation, selection for the

Table I
Mutagenesis Procedures and Kit Suppliers

Kit	Template	Selection strategy	Vendor/Location
Mutagene M13 *in vitro* Mutagenesis	Single-stranded M13	Uracil laden template	Bio-Rad/Richmond, CA
Mutagene Phagemid *in vitro* Mutagenesis	Any plasmid with an f1 origin of single strand replication	Uracil laden template	Bio-Rad
Sculptor Mutagenesis	Single-stranded M13 or Phagemid	Uracil laden template	Amersham/Arlington Heights, IL
T7-GEN *in vitro* Mutagenesis	Any plasmid	Mutant strand incorporates 5-methyl-dCTP allowing selective nicking of template by various restriction enzymes	USBiochemicals/Cleveland, OH
Transformer Site-Directed Mutagenesis	Any plasmid	Second primer removes unique restriction site	Clontech/Palo Alto, CA
U. S. E. Mutagenesis	Any plasmid	Second primer removes unique restriction site	Pharmacia/Piscataway, NJ
Chameleon Double-Stranded Site-Directed Mutagenesis	Any plasmid	Second primer removes unique restriction site	Stratagene/La Jolla, CA
Altered Sites II *in vitro* Mutagenesis Systems	A set of plasmids carrying mutated, inactive antibiotic resistance genes	Second primer restores antibiotic resistance	Stratagene
MORPH Site Specific Plasmid DNA Mutagenesis	Any plasmid	Methylation dependent restriction enzyme used to cleave template DNA	5 Prime→3 Prime, Inc./Boulder, CO
ExSite PCR-Based Site Directed Mutagenesis	Small plasmids	Amplification of mutant sequence by whole plasmid PCR and template inactivation by methylation dependent restriction enzyme	Stratagene

mutagenic strand is achieved by selectively nicking the parent strand with a restriction enzyme that is unable to cleave the newly synthesized, phosphorothioate-containing strand. The nonmutant parental strand is then destroyed by treatment with an exonuclease. The synthesis reaction is then repeated using the phosphorothioate strand as template yielding a mutant homoduplex.

A third method of selection for the mutagenic strand involves the use of multiple mutagenic primers in a single *in vitro* synthesis reaction (12). One primer contains the desired mutation to be made in the gene of interest, and another primer contains a mutation that repairs a mutant antibiotic resistance gene. This method allows for strong selection against the template strand and provides a simple initial screen for potential mutant clones. It avoids the use of phosphorothioate- or uracil-containing strands, thus avoiding their greater propensity for the introduction of unwanted non-site-directed mutations.

Finally, selection for the mutant strand can be achieved by selective amplification using PCR (13) (see Fig. 4). This is achieved by first adding only the primer complementary to the mutant strand and allowing about 10 cycles before adding the primer for the other strand. If the primers are situated at unique restriction sites on either side of the site of mutagenesis, this strategy should result in a fragment that can be cloned into the original expression vector and that should contain greater than 90% mutant DNA.

If the primers are synthesized to randomize nucleotide composition at given positions, libraries of mutant proteins can be obtained. The annealing of primer to discontinuous portions of single-stranded template can also efficiently produce deletions. These can be quite large (up to 950 bp deletions have been reported) (14). Insertions are also possible although efficiency is rapidly reduced with size—the largest insertion to our knowledge is 27 bases (15). Also, as previously mentioned, it is possible to obtain multiple mutations in a single synthesis reaction by the addition of multiple primers. Although there is no guarantee that the resultant variant DNAs will contain every mutation, this approach can save time if multiple widely spaced mutations are required. It is also possible to produce multiple mutations using a single mutagenesis primer, and we have successfully used primers as long as 50 nucleotides to alter 5 bases simultaneously.

We emphasize that the presence of a mutation in DNA does not guarantee that all of the protein produced will contain the proper amino acid substitutions. Whenever relatively inactive variants are examined, care must be taken to ensure that the observed function is not due to contaminating protein. Contamination could arise through trivial experimental error, or more insidiously, through misreading of the mutant codon and insertion of the wild-type amino acid into the nascent peptide. This effect has been observed for mutants of trypsin (16) and β-lactamase (17). When large numbers of mutants are being produced, inadvertent substitution of clonal stocks can lead to the misidentification of expressed proteins. As a result, mass spectral analysis of putative mutant proteins may be advisable as a final check for their proper identity (18). It is equally necessary to realize that unwanted mutations may have been introduced during the mutagenesis protocol, and that all mutant DNAs be sequenced to confirm their identity.

B. Mutagenesis Using Denatured Double-Stranded DNA as Template

The production of single-stranded DNA is not always straightforward. The difficulty of this step varies unpredictably depending on subtle properties of the gene of interest. Even when it proceeds smoothly, obtaining single-stranded DNA represents an additional procedure. The production of single-stranded template, however, is not strictly necessary for annealing a primer and initiation of polymerization of the mutant strand. For example, primer annealing and strand polymerization are essential steps in the sequencing of duplex DNA. To mutate double-stranded DNA, the template is denatured with a basic solution containing sodium hydroxide. The solution is then neutralized and the primer allowed to anneal. Extension of the primer yields a second strand containing the mutation of interest.

Selection of mutant DNAs can be achieved by one of several methods. One method involves the use of a second primer in the reaction that reactivates an antibiotic resistance gene or removes a unique restriction site. A second method makes use of multiple unique restriction sites surrounding the site of mutagenesis (19) (Fig. 3). A plasmid containing unique sites A, B, and C is cleaved in separate reactions at A and B and at A and C. The purified products of these reactions are mixed, denatured, and allowed to anneal. In this mixture some "mismatched" annealing will have occurred, yielding double-stranded linear template with long sticky ends containing the site to be mutagenized. The

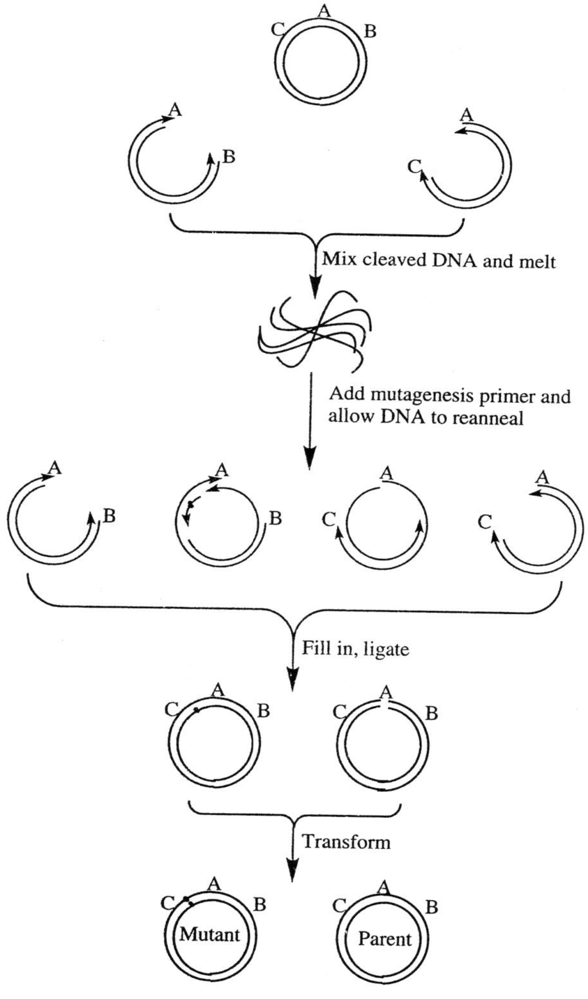

Fig. 3 Mutagenesis of double-stranded DNA using multiple restriction digests and subsequent plasmid reassembly.

mutagenesis primer is added to this template and synthesis and ligation reactions are carried out. Only this mutant DNA should recircularize and be maintained after transformation.

A third strategy takes advantage of the dependence of certain restriction enzymes on the methylation state of their target sequences (20). After synthesis of the mutant strand, the DNA is treated with an enzyme such as *Dpn*I that recognizes a fully methylated four base-pair target sequence. Plasmids that are unmutagenized remain fully methylated and are cleaved. Plasmids containing a mutant strand are hemimethylated and are thus not recognized by the enzyme. After treatment with the enzyme, the DNA is used to transform a *mutS* bacterium that is unable to distinguish between parent and newly replicated strand on the basis of their respective methylation states and repair the introduced mismatch in the mutagenic strand. Obstacles to successful use of this method are that, after transformation, many of the bacterial colonies will contain both the desired mutant plasmid and the wild type. In addition, it is difficult to obtain clear, readable sequencing from DNA isolated from *mutS* strains. Both of these difficulties can be solved by transforming the plasmid into another strain after isolation of plasmid from the *mutS* strain.

A number of difficulties may arise that apply to any method in which a strand is synthesized *in vitro*. Mutations may occasionally occur at secondary sites during *in vitro* synthesis because of imperfect fidelity of replication. For this reason it may be desirable to subclone only that portion of a gene that contains the region to be mutagenized. This reduces the likelihood that mutations will be introduced and decreases the amount of DNA that must be sequenced to confirm the absence of secondary mutations.

C. PCR-Based Methods for Mutagenesis

Polymerase chain reaction (PCR) is the basis for several methods for mutagenesis (Fig. 4). The ability to amplify a segment of DNA from essentially any template virtually eliminates the need for selection or screening and greatly simplifies troubleshooting if problems occur. Because synthesis occurs at high temperatures, PCR methods also eliminate the need for single-stranded template or prior denaturation, and the problems caused by the propensity of some templates to form extensive secondary structure at target sequences.

One of the simplest methods for PCR-based mutagenesis requires that the target site for mutagenesis be directly adjacent to a unique restriction site so that both can be included in a single primer (Fig. 4A). Using this primer and another that contains a nearby unique restriction site, a single fragment can be amplified and subsequently subcloned back into the original vector using the two restriction sites. When a conveniently located unique restriction site does not exist, mutagenesis requires either multiple synthesis steps or synthesis of the entire plasmid followed by recircularization (21).

One method of whole plasmid PCR is accomplished by using complementary primers that each contain the desired mutation and that face in opposite directions (22) (Fig. 4D) Because the primers face in opposite directions, elongation only occurs when they are annealed to the original template. The sequences of the primers thus act as sticky ends that allow recircularization after amplification is completed. Because the amplification product is the same size as the

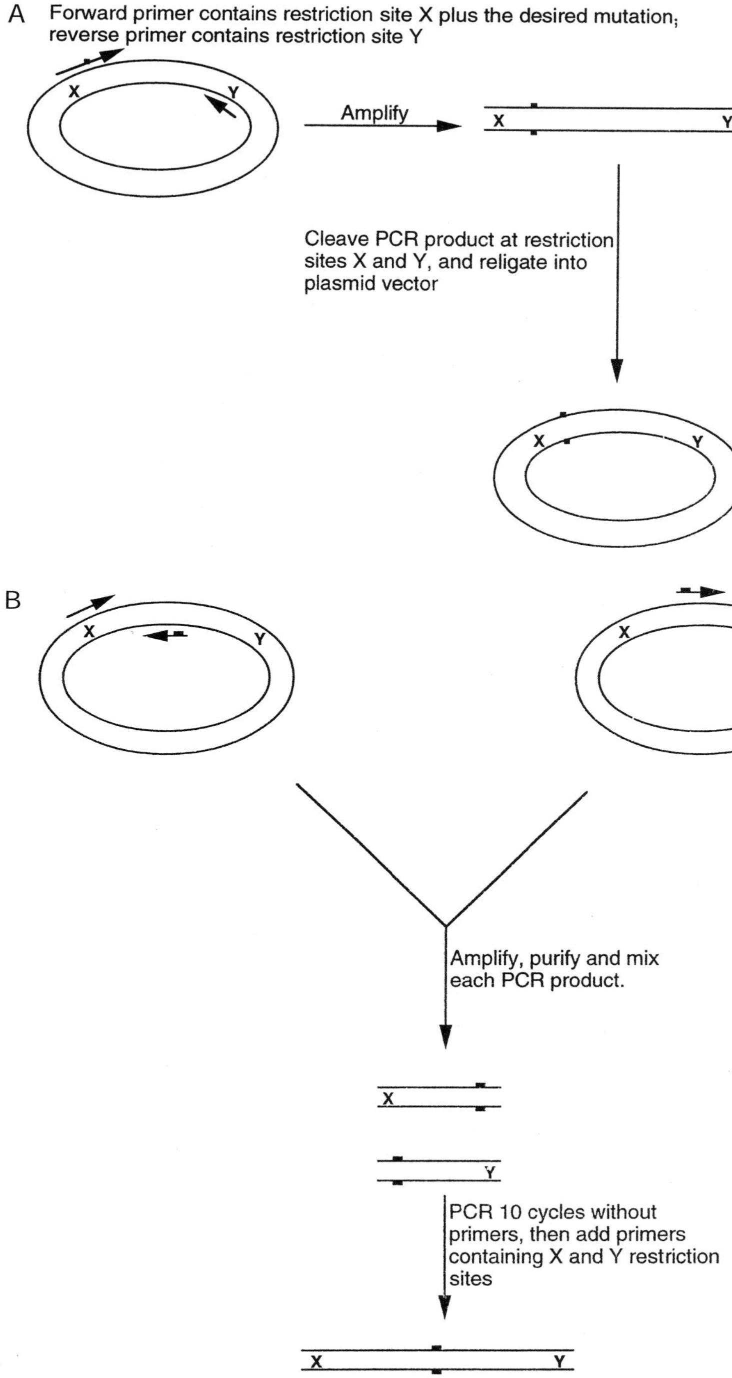

A Forward primer contains restriction site X plus the desired mutation; reverse primer contains restriction site Y

X Y

Amplify

X Y

Cleave PCR product at restriction sites X and Y, and religate into plasmid vector

X Y

B

X Y

X Y

Amplify, purify and mix each PCR product.

X

Y

PCR 10 cycles without primers, then add primers containing X and Y restriction sites

X Y

Fig. 4 Methods for mutagenesis using PCR amplification.

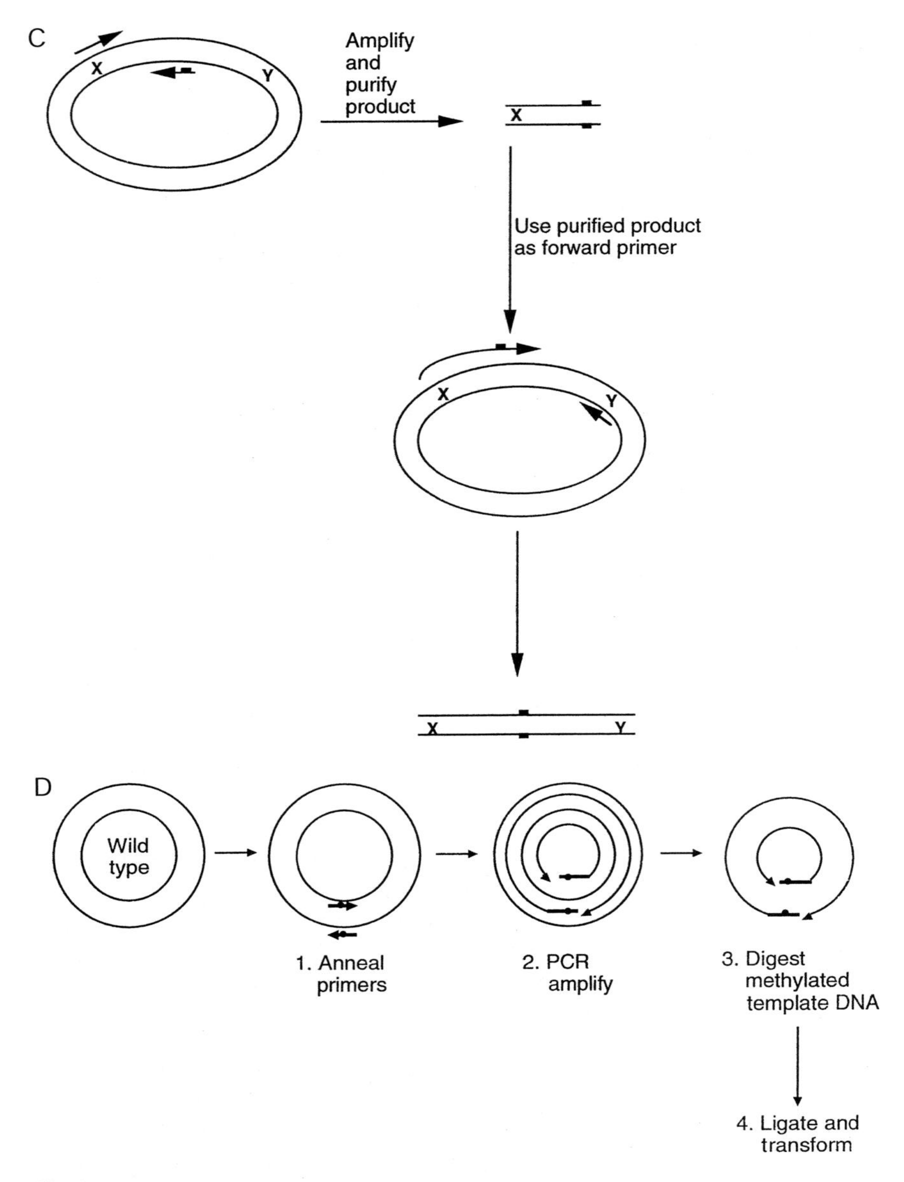

Fig. 4 *(continued)*

original template in this method, some means of selection is required and the PCR reaction products are commonly treated with restriction enzymes that recognize methylated DNA (23). Another method for whole plasmid PCR includes a class IIS restriction enzyme recognition site at the 5′ end of each primer (24) (Fig. 4E). Class IIS enzymes' recognition sites are 5′ to their cleavage sites; thus, this strategy provides another means for providing sticky ends at any site without adding exogenous sequence to the vector.

Two multistep methods for amplifying a segment of the template DNA

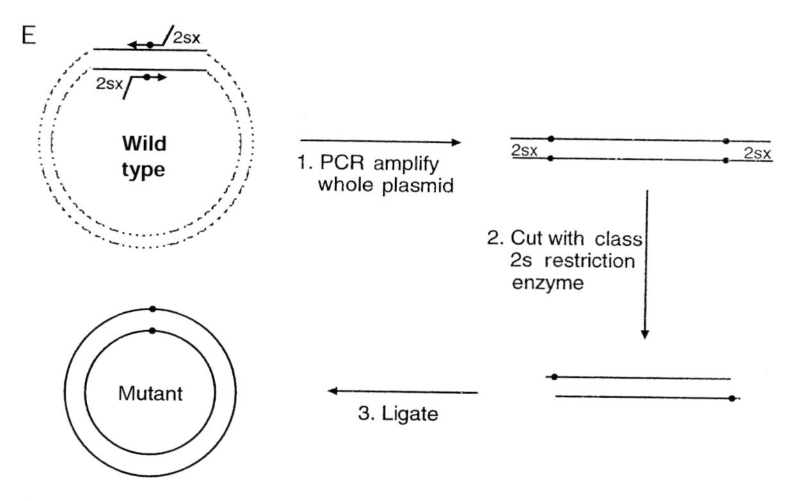

Fig. 4 *(continued)*

that will incorporate the desired mutation are in common use. Both methods require "universal" primers that are situated on either side of the site to be mutagenized and that contain unique restriction sites. The first method then requires two complementary mutagenesis primers (Fig. 4B). In separate reactions, the universal and mutagenic primers are used to produce half-sized PCR products that overlap each other at the site of their mutagenic primers. These fragments are purified and amplified in a third PCR reaction lacking other primers to produce the full-length product (25,26). The second method requires only one mutagenic primer (21) (Fig. 4C). One half-length mutation-containing product is synthesized and purified and then added to a second reaction containing template and the other universal primer. The rate-limiting step in either of these multistep methods is probably the purification of the intermediate fragments. At least one method has been described that simplifies this process by using biotinylated universal primers so that the intermediate products can be isolated with streptavidin-coated magnetic beads (27).

The strength of PCR-based methods for mutagenesis is that they afford a rapid approach for the generation of mutations and do not require production of single-stranded template. As with any PCR-based technique, however, there is the potential for the introduction of secondary mutations, necessitating careful characterization of mutant clones.

D. Cassette Mutagenesis

Mutations can also be introduced by the direct cloning of oligonucleotide duplexes containing the alterations of interest. This cassette mutagenesis is convenient because the mutation can be introduced by automated DNA synthesis rather than by a multistep *in vitro* procedure, facilitating the rapid generation of altered clones. Complementary oligonucleotides are synthesized, annealed, and phosphorylated to create a duplex cassette. The insert is ligated into the vector and the resultant construct is sequenced to ensure that the insert has been introduced uniquely and in the proper orientation. Cassette mutagenesis can be used for the synthesis of libraries of protein variants by randomizing central codons of a template oligonucleotide and annealing two flanking oligonucleotides to it to act as adapter arms for subsequent ligation.

A disadvantage of this technique is that one or more unique restriction sites must be appropriately placed within the target gene to introduce the cassette. This can be a problem if appropriate restriction sites are absent, although these can often be introduced through the mutagenesis techniques discussed previously without alteration of the wild-type amino acid sequence. Cassette mutagenesis is especially useful when experimental plans call for a particular region of the protein to be subjected to extensive and repeated alteration, thus justifying an initial investment in the resources needed to introduce strategically placed restriction sites. The ability to insert small segments of DNA into a gene at a specific site can facilitate attachment of histidine tags for protein purification, insertion of proteolytic target sites for cleavage of fusion proteins, and construction of peptide or protein-based phage display libraries. Cassette mutagenesis is also well suited to the simultaneous introduction of multiple mutations throughout a defined region.

IV. Strategies for Engineering Proteins

Proteins are complex molecular machines, and their alteration must be planned to minimize damage to existing function while maximizing the potential for achieving important insights. These goals are often in conflict, and the choice of protein engineering strategy plays a pivotal role in striking a balance. Each of the engineering strategies described in the following sections has intrinsic advantages and disadvantages that help dictate their feasibility for a given application. Much also depends on preexisting knowledge about the protein. Are particular regions known to be important? Can particular amino acids be identified as candidates for mutagenesis? If so, which amino acids should be substituted for them? The ability to develop rational answers for these questions varies proportionately with available data. The following sections will outline options for macromolecular transformations and address the strengths and weaknesses of each.

We focus on protein engineering through site-directed mutagenesis. We define site-directed mutagenesis as any technique that is capable of selectively modifying a protein at one or more defined positions. These include the introduction of a specific amino acid at a predetermined position, the randomized introduction of varied amino acids at several positions simultaneously, and the deletion or addition of entire protein domains. We do not discuss random mutagenesis through chemical mutagens or UV light because these experimental strategies are not aimed at the engineering of particular protein regions. We include strategies for chemical modification, in addition to genetic techniques, because the end result, site-specific protein modification, is the same, and because the approaches can often be combined. As we noted in Sections I and II, achieving worthwhile results through protein engineering can be challenging, and an experimenter must be aware of every option for manipulating protein chemistry.

A. Chemical Modification

We now associate protein engineering with the alteration of proteins through manipulation of their genes and subsequent expression. This was not always the case. Early examples of protein engineering relied on chemical modification to alter already folded proteins (28). This strategy was necessary because of the absence of cloned genes or techniques of their manipulation. Currently, the strategy is rarely the only available option, but it remains an important route for the introduction of chemistries that go beyond those afforded by the 20 naturally occurring amino acids.

1. *Chemical Mutation*

Selective chemical modification of enzymes requires that an amino acid be uniquely reactive. One such residue is the catalytic serine of serine proteases. This serine is a highly reactive nucleophile because of its membership in the catalytic triad. The serine hydroxyl can react specifically with either *p*-toluenesulfonyl chloride or phenylmethylsulfonyl fluoride to introduce a labile leaving group. Displacement of the leaving group with varied nucleophiles can introduce alanine (29), cysteine (30,31), or selenocysteine (32,33). The loss of serine greatly reduces the hydrolytic activity of these mutant proteins, and anhydrotrypsin or anhydrochymotrypsin are useful affinity ligands for the purification of protease inhibitors. The protease variants containing selenocysteine or cysteine possess relatively unimpaired aminolysis, allowing them to be used for peptide ligation (32,34).

Kaiser and Lawrence (28) performed a more radical modification of protein chemistry by modifying papain with NADH. This alteration was possible because papain contains an active-site cysteine that is highly reactive, providing a unique point of attachment for the cofactor. Modification of this key residue abolished protease activity but did not block substrate binding. As a result, derivatization of the cysteine with cofactor introduced a modest level of reductase activity for substrates binding in the normal hydrophobic substrate binding pocket.

The modification of papain was possible because of the uniquely reactive thiol of cysteine. Many other examples exist of derivatization of thiols to introduce novel chemistries into proteins. Schultz and co-workers have used a modified hapten molecule to introduce a thiol into an antibody by affinity labeling. Once covalently anchored by the affinity label, this thiol acted as a nucleophile and directly increased catalysis by a hydrolytic antibody (35). The unique ability of sulfur to form disulfide linkages also allowed the attachment of reactive groups and spectroscopic probes close to the active site (35,36).

It is also possible to insert thiols through site-directed mutagenesis of residues to cysteine and to modify the introduced cysteines to assay novel chemistries near an active site. Raines and co-workers have introduced a cysteine into RNase A in place of a catalytic lysine (37) (Fig. 5). This alteration reduced the catalytic efficiency of the enzyme, as measured by $kcat/Km$, by 85,000-fold relative to wild type. The coupling of varied haloalkylamines reintroduced positive charge and restored activity to differing extents depending on their structure. These results demonstrated that the strategy can extend the options available for probing the effect on catalysis produced by varying the

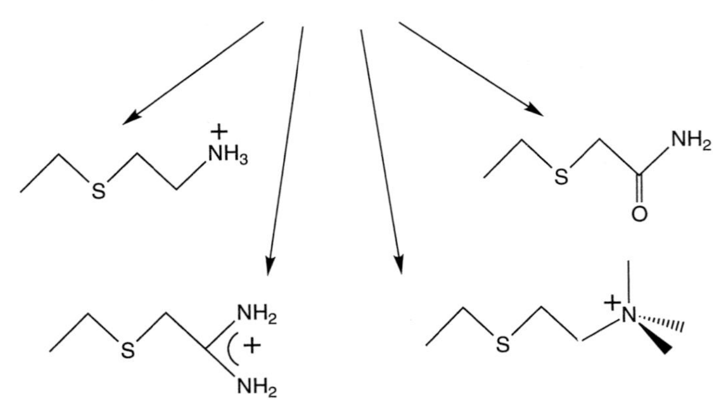

Fig. 5 Derivatization of an introduced cysteine to afford semisynthetic RNase A (37).

nature and placement of charge beyond those provided by the 20 naturally occurring amino acids.

Chemical mutation has been used to construct semi-synthetic DNA-cleaving proteins by coupling small synthetic molecules to proteins that bind DNA. One of these conjugates, between CAP protein and copper-phenanthroline, proved to be an effective agent for selective cleavage of megabase DNA (38). This conjugate overcomes the obstacle of nonselective cleavage by strategic placement of the attached phenanthroline. The phenanthroline is accessible to DNA at the cognate binding site but is blocked by the protein from association with other DNA within the genome.

The chemical mutation of amino acids is often overlooked as a tool for protein engineering because genetic manipulations are usually easier to perform and offer a route to more diverse variants. Why be limited to the handful of amino acids that one can alter chemically, when one can replace any residue genetically via site-directed mutagenesis? Also, many experimenters are more familiar with biological approaches than chemical ones. Chemical mutation will, however, continue to possess significant advantages for introducing novel chemical functionality into already-folded proteins. These advantages are especially apparent for modification of cysteine, which is both experimentally straightforward and highly amenable to the introduction of diverse chemistries.

2. *Domain Conjugation*

The conjugation of functionally autonomous units is another strategy for the synthesis of proteins processing novel function (Fig. 6). Existing properties of the partners are retained and the combination of existing function leads to hybrid properties. One can also conjugate protein domains directly by genetic fusion at the amino or carboxy termini, and this will be the focus of a subsequent section. The fusion of domains by chemical cross-linking *in vitro*, however, allows the attachment of nonprotein moieties and permits attachment of domains to any exposed amino acid residue.

The combination of protein subunits during evolution has generated multifunctional polypeptides (39,40) that possess a wide range of properties. These provide compelling evidence for the effectiveness of domain fusion as a strategy for functional modification of proteins. One example is tissue-type plasminogen activator or t-PA (41–43). t-PA consists of a trypsin-like protease domain, two kringle domains, an epidermal growth factor domain, and a fibronectin-like finger domain. Its physiological role in the fibrinolytic cascade is the cleavage of a single protein substrate, plasminogen. t-PA must be highly specific to avoid systemic degradation of other proteins, and must achieve this specificity in spite of 40% identity of its protease domain with trypsin, a protease that promiscuously digests diverse proteins.

One mechanism that t-PA employs to attain this specificity is the ability to bind a third molecule, fibrin, through its kringle 2 domain. Plasminogen also contains kringle domains and binds to fibrin. This mutual fibrin binding brings protease and substrate into close proximity and reduces the K_m of t-PA for plasminogen by 400-fold. This fibrin-mediated targeting mechanism has allowed the protease domain of t-PA in the absence of fibrin to evolve toward ineffectiveness relative to trypsin. t-PA is therefore less liable to cleave potential substrates encountered randomly. Thus, two engineering lessons can be learned from t-PA: (*i*) one domain can direct the function of an attached domain to a target and (*ii*) the individual function of attached domains can coevolve to optimize the function of the entire complex for its specific physiologic role.

The multifunctional combinations apparent in the natural evolution of multidomain proteins lend themselves to replication in the laboratory. Conjugation of antibodies to enzymes affords highly sensitive and widely used probes. Other applications of the vast potential of the immune repertoire to the synthesis of multifunctional conjugates demonstrate the essentially infinite scope for conjugate synthesis. Antibodies supply an extra fibrin-binding domain to t-

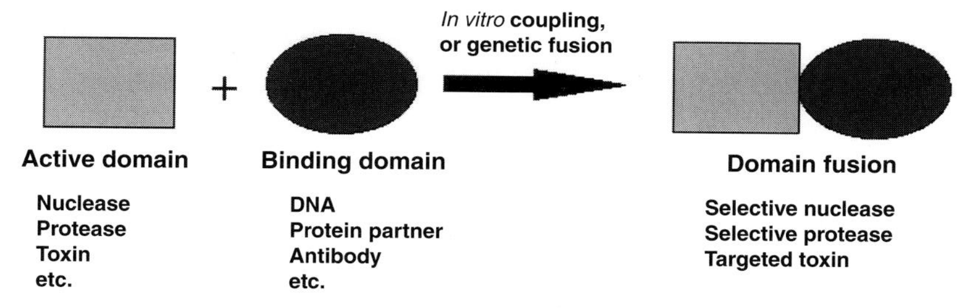

Fig. 6 Engineering by domain conjugation.

PA, and t-PA–antibody conjugates possess a 3 to 10-fold greater affinity for fibrin (44,45). Antibody conjugates may also have important applications for anticancer therapeutics. For example, the antibody domain of antibody–toxin conjugates directs the toxin to selectively expressed antigens on the surface of tumors or tumor vasculature (46–48). These "immunotoxins" selectively kill tumor cells *in vivo*, although the development of resistant tumor cells after initial treatment and tumor regrowth remains an obstacle to effective therapy.

Another example of the engineering of semisynthetic proteins and their subsequent optimization is the synthesis of hybrid proteins that can target sequence-specific DNA cleavage to selected sites within DNA and RNA. Schultz and co-workers have conjugated staphylococcal nuclease, a nonselective nuclease, to either oligonucleotides (49) or DNA binding proteins (50) to adapt it for sequence-specific DNA cleavage. In these experiments, the molecules were coupled through disulfide exchange. Disulfide exchange is a particularly convenient strategy for cross-linking domains because many proteins lack surface cysteines, and introducing cysteine then affords a point of unique reactivity for subsequent linkage. The location for the cysteine that was introduced into staphylococcal nuclease was chosen after close examination of the three-dimensional X-ray crystal structure.

Conjugates between staphylococcal nuclease and either oligonucleotides or proteins cleaved DNA with high selectivity. A drawback to this approach was that the conjugates exhibited a high level of nonspecific hydrolysis due to nonoligonucleotide-directed nuclease activity. To reduce this effect, the nuclease was mutated to reduce its catalytic potential (51). This strategy succeeded because attachment of a DNA binding domain to the nuclease increases the effective concentration of substrate and reduces the need for the nuclease to be an efficient enzyme. Selective cleavage remained efficient, while nonselective cleavage during a given period was greatly reduced. Strictly speaking, this does not alter the specificity of the enzyme, but it does improve the selectivity of the conjugate on a practical level. This combination of a targeting domain and a weakened catalytic domain simulates the process described previously for the evolution of specificity by t-PA and demonstrates that catalytic and binding domains can be joined and subsequently optimized for new roles within the conjugate.

Synthetic molecules can also be used to create hybrid proteins *in situ*. Schreiber, Crabtree, and colleagues have synthesized dimeric analogs of the immunosuppressive drug FK506 (52) to create FK1012. Both halves of the dimer are capable of binding to FK506 binding protein (FKBP12). Multiple copies of FKBP12 were then fused to the intracellular signaling domain of the T-cell receptor, while the native extracellular dimerization domain was removed. Addition of FK1012, but not FK506, activated the receptor, suggesting that binding to two proteins by the dimeric synthetic molecular can bring receptors together and effect cellular signaling. This approach may have wide applications for the effective engineering of macromolecular function *in vivo* since it relies on small molecules to trigger protein association, molecules that will usually prove much more adaptable as potential drugs because of superior pharmocokinetic properties relative to those of proteins. In contrast to standard mutagenesis, this strategy relies on site-directed noncovalent modification of protein chemistry. Similar experiments have also shown that synthetic ligands can also induce heterodimerization (53).

B. Novel Function through Genetic Fusion of Domains

Experimenters can also obtain functional multidomain macromolecules through genetic fusions. As noted previously, genetic fusions are spatially restricted. Protein domains can be linked at the amino or carboxy termini or can be inserted within a host protein, but they cannot be linked through side-chain residues. This restriction is balanced by the simplicity with which gene fusions can be made and multidomain proteins expressed—no *in vitro* synthetic steps or synthesis-related purifications are required. This technique is widely used and, as one would expect from the modular nature of native proteins, has been highly successful. An example of its versatility is the ability to pair DNA binding domains with unrelated activation domains to afford hybrid molecules that retain selective binding of DNA and ligand-induced activation. These fusions have been widely successful and have found particular utility in development of the two-hybrid system used to identify protein–protein interactions *in vivo* (54,55), a system that, as will be discussed later, offers its own potential as a strategy for protein engineering (Section IV,C,3).

The ability to rapidly generate novel function can lead to the development of multidomain proteins possessing potential therapeutic value. An example is the fusion of nucleases to proteins that are associated with the interior of viral particles. Upon assembly of the virus, the viral-derived fusion partner delivers the nuclease to the interior of the virus. Presumbly it will be in close proximity to the viral DNA or RNA. Hydrolysis of DNA or RNA by the nuclease then inactivates the virus and prevents further replication. A fusion between staphylococcal nuclease and Vpx, a protein packaged within the human immunodeficiency virus (HIV) I virion, has been shown to target the nuclease into the interior of the virion (56,57), whereas the ability of fusion proteins to inhibit replication was demonstrated for fusions directed toward packaged yeast Ty transposon (58).

The conjugation of macromolecular domains, either chemically or genetically, has important advantages relative to engineering strategies that attempt to modify function by substituting individual amino acids through oligonucleotide-directed mutagenesis. As will be noted in subsequent sections, the interdependence of key amino acids involved in binding and catalysis makes it difficult to successfully alter protein function through selected mutation of key amino acid residues, as such alterations often disrupt existing function. The conjugation of functional domains avoids this difficulty because the constituents retain their native properties and attain unique properties through their combination. Striking changes in protein function can be achieved without sacrificing the high activity of the native enzymes.

C. Alteration of Function Using Selections and Screens

1. *Genetic Selections and Screens*

New protein functions emerge during evolution through the random occurrence of mutations and the subsequent selection for protein variants that confer an advantage on an organism. Similar strategies can be applied in the laboratory by requiring that either survival or a distinct phenotype of a clonal population of cells be dependent on expression of a variant protein with a desired activity. The need, however, to evolve function on a human time scale, rather than an evolutionary one, demands that great care be exercised in experimental design and execution.

Evolution in the laboratory requires the creation of large numbers of variant genes. Genes can be altered by chemical mutagens, UV light, passage through a mutator host strain, error-prone PCR, or oligonucleotide-directed mutagenesis using primers or cassettes containing randomized nucleotides. The resultant library of modified genes can be examined in the initial host organism or introduced into other strains. Propagation of cells containing the library produces a population of clones, each expressing a variant protein with unique properties.

Generation of a library of clones that express variant proteins is generally straightforward. However, a library is worthless without a method for distinguishing the few clones that express proteins with the target function from the far greater number of clones that do not. This can be done through selections, in which the phenotype confers survival, or through screens, in which the phenotype confers a detectable characteristic (Fig. 7).

Most selections involve plating transformed cells on solid media. The target protein then begins to be expressed and must either destroy a toxic compound or process a precursor molecule to create conditions that favor growth of the clone that expresses it. Clones that express variant protein with the desired activity will grow faster than those lacking the variant, and this can be observed by evaluating colony size or survival. Selections for antibiotic resistance are a familiar example of the power of this approach. Millions of cells, each potentially expressing a different protein variant, can be applied to a single Petri dish, allowing a library of proteins to be rapidly screened for a desired function. Screens are performed similarly, except that plating must be performed at a lower density because all colonies must be identified by some other phenotypic characteristic, such as color.

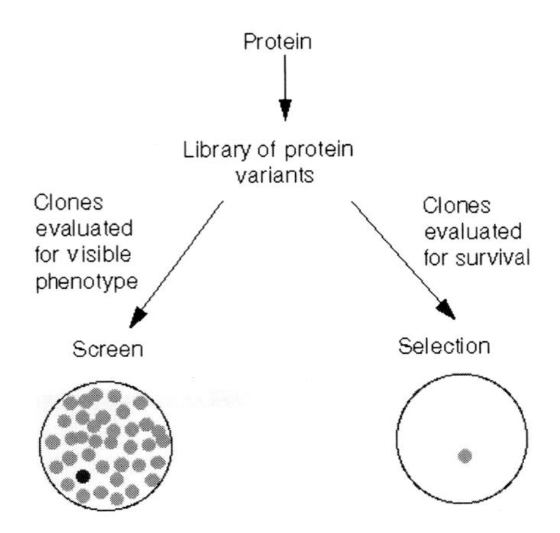

Fig. 7 Design of a genetic selection or screen.

Once a selection or screen is designed, one must devise a mutagenesis strategy to generate a library of expressed proteins. The number of cells that can be transformed with library DNA is physically limited by transformation efficiency of the host strain. The number of cells used can compensate for this limitation to some extent, but the potential number of variants will be limited to hundreds of millions. As a result, alteration of six amino acids is considered the limit for coverage of potential variants, as substitution of all 20 amino acids in each position will create a library of 20^6 or 64,000,000. More residues can be altered, but a decreasing proportion of variants will be sampled in the selection, introducing the potential for the desired variants to be overlooked. It is therefore advantageous for the researcher to restrict random substitutions to as few residues as is possible. This consideration does not preclude random mutagenesis of more than six residues, or even entire proteins or genomes, since a wider scope for mutagenesis may be the only option when no information is available to restrict mutagenesis.

When using oligonucleotides to direct the randomized incorporation of amino acids, one must also consider where the amino acids are coded for in the parent gene. If their codons are close together, oligonucleotide-directed mutagenesis can be an efficient means for randomizing several amino acids simultaneously. If codons are not close, devising an efficient mutagenesis strategy will be limited by the inefficiencies encountered when performing two mutagenesis protocols simultaneously. In addition, even in small libraries, bias in the introduction of mutations, transcription, or translation may lead to some clones being over- or underrepresented.

An example of the use of a selection for protein engineering is the identification of modified trypsin specificities (59). The goal was to probe the ability to alter key residues within the binding pocket of trypsin while retaining primary specificity for lysine or arginine side chains. To accomplish this, Craik and co-workers designed a selection based on an *Escherichia coli* strain that was auxotrophic for arginine. A nonnutritional form of arginine, arginine β-naphthylamide, was present in the plating agar, so that trypsin variants that could hydrolyze it to arginine possessed an advantage for growth. The mutated residues aspartic acid-189 and serine-190 were adjacent, allowing one oligonucleotide primer to be sufficient for library construction. This experiment permitted identification of determinants of the specificity of trypsin for arginine, and suggested that similar strategies should be used to refine protease specificity for other substrates (60).

Trypsin mutants have also been screened by demanding that near wild-type activity be present to generate sufficient nitrogen and carbon from exogenous protein applied to solid media (61). Initial studies of this technique have shown that wild-type revertants of an inactive trypsin mutant can be detected. Other enzymes that have had their specificities or catalytic potency altered by random mutagenesis and subsequent screens or selections include β-lactamase (62), yeast proteinase A (63), *Eco*RV endonuclease (64), *Fok*I, endonuclease (65), and triose-phosphate isomerase (66,67). Enzymes have also been selected for greater thermostability (68,69) and alkaline stability (70), and immunoglobulin libraries have been screened for folding stability (71). Hilvert and colleagues have used selections to probe the electrostatic requirements of catalysis by chorismate mutase (72).

As noted, selections require that the target activity be related to cell growth and survival. Therefore, if an engineered function cannot be related to cell survival, or a least to a marked alteration in the rate of cell growth, selection-based strategies cannot be used. Screens represent an alternative strategy. Screens are fundamentally different from selections in that they do not require that the engineered activity enhance survival. Rather, screens rely on production of a phenotypic change that can be visualized. This affords many more options for experimental design. One recent example of use of a screen to identify proteins with optimized function is the directed evolution of a *p*-nitrobenzyl esterase for improved activity in solutions containing organic solvents (73). This experiment revealed that activity could be enhanced and that, surprisingly, key alterations were distant from the active site. The effect of these alterations could not have been predicted by prior evaluation of available information regarding the protein.

Advantages and disadvantages of selections and screens as tools for protein enngineering must be carefully weighed prior to attempts to incorporate them into an experimental plan. The power to create large numbers of variants can overcome a lack of understanding of how to manipulate the subtleties of protein function. This is important in cases where such knowledge is scarce. Whatever knowledge is available can be used to limit library size. The disadvantage of selections and screens is that they are often difficult or impossible to design. Their feasibility is dependent on the properties of individual proteins, genes, and host organisms, considerations that often make development of a screen or selection an ambitious project in its own right.

2. *Phage Display*

The display of proteins on bacteriophage can avoid some of the limitations of traditional selections and screens as strategies for the identification of phenotypic change [reviewed in (74–78,204]. As with any selection or screen, the key to phage display is linkage of replication with a selectable feature (Fig. 8). DNA from a filamentous bacteriophage is genetically manipulated to express a fusion between a phage coat protein and the protein to be engineered. The fusion protein coded for by the phage genome is then displayed on the surface of the phage. Random mutagenesis of the fusion protein can create a library of proteins with diverse characteristics. If a protein variant is capable of binding to a ligand displayed on solid support, the phage bearing the protein will also bind, whereas phage that lack the ability to bind will be eluted upon washing. Thus, the protein that possesses the desired binding properties will be enriched, and its sequence can be determined by subsequent isolation of bound phage, amplification of DNA, and DNA sequencing of the gene encoding the fusion protein. Candidate proteins can then be expressed and tested for target binding properties using standard assays.

The protein products of either Gene 3 (GpIII) or Gene 8 (GpVIII) are generally utilized as the fusion partners for the protein of interest. GpIII is present in five copies per phage, whereas GpVIII is the major phage coat protein and is present in thousands of copies. GpIII is usually employed for phage display because only one copy of the fusion protein per phage is usually necessary. Additional copies of the fusion protein may even be deleterious, as the binding of multiple copies to support may make the phage difficult to

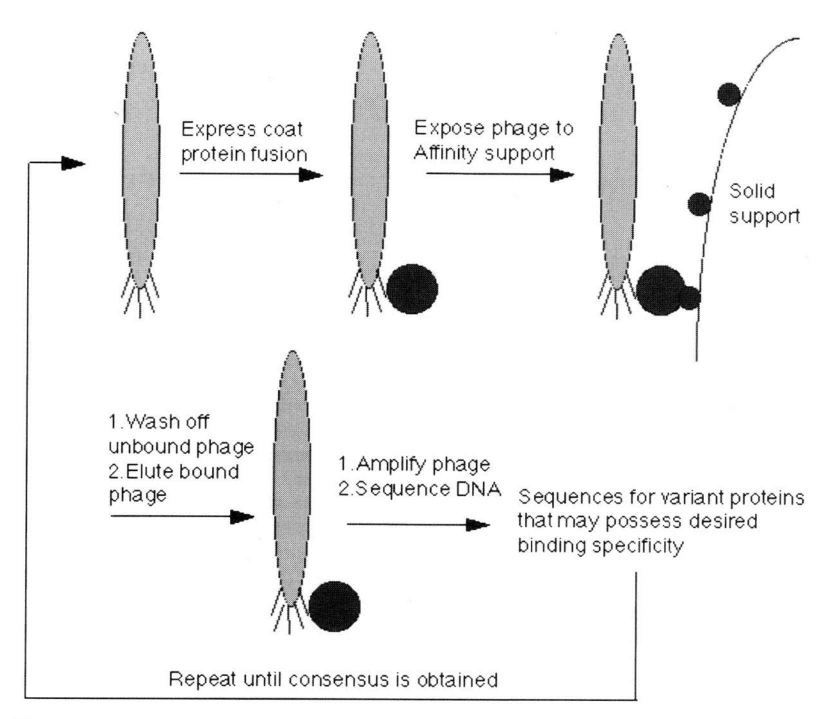

Fig. 8 Phage display of proteins and selection for binding specificity.

remove and make it difficult to discriminate between variant proteins that possess a high affinity for binding from those that possess lower affinities.

The fusion gene can be encoded within the bacteriophage genome or within phagemid DNA. The advantage of the former approach is that it avoids the need to use helper phage. This is a signifiant advantage for some proteins, as an optimal ratio of phagemid to helper phage can be difficult to determine (79). The advantage of the latter approach is that the fusion protein can be contained in a smaller and genetically more stable vector. In addition, vector size is inversely related to transformation efficiency, and smaller vectors allow the creation of larger libraries. For the creation of small libraries and initial troubleshooting, direct cloning of the gene fusion into bacteriophage vectors is the most suitable approach, with the phagemid strategy being held in reserve for applications that demand large library size.

A variety of proteins have been expressed on the surface of bacteriophage. Enzymes include alkaline phosphatase (78), trypsin (79), staphylococcal nuclease (80,81), glutathione transferase (82), and β-lactamase (83). Other proteins include protease inhibitors (77,84), four-helix bundles (85), growth hormone (86), and, most often, antibodies (87). If a protein can be expressed in the bacterial periplasm in active and correctly folded form, and remain active as a C-terminal fusion protein, it is likely that expression through phage display will be possible. It is advisable to first express the wild-type protein on the phage and then assay isolated phage for the appropriate enzyme activity or immunoreactivity to confirm the presence of intact fusion, and in the case of enzymatic ativity, the occurrence of proper folding.

Phage display functions by detecting ligand binding, not catalysis, and is therefore better suited for the engineering of binding specificity than for enhanced catalysis. Theoretically, if binding was dependent on catalysis, phage display could be employed, but only limited studies have been performed in this area. The attachment of a suicide inhibitor of β-lactamase to solid support allows phage that display β-lactamase to be bound through covalent coupling (83,88). It has not yet been shown, however, that the potential for covalent binding can be linked to an alteration in enzyme specificity.

The most striking examples of the modification of function have involved proteins that bind to other proteins but are not catalysts themselves. Bovine pancreatic trypsin inhibitor (BPTI) has been expressed on the surface of phage and screened for binding to immobilized human neutrophil elastase (HNE) (84). The goal of this study was ambitious because BPTI has evolved to inhibit trypsin-like enzymes that have a primary specificity for arginine, whereas HNE has a primary specificity for small apolar amino acids. The selected BPTI variants inhibited HNE with K_d values ranging from 1 to 2.8 p*M*, demonstrating more than six orders of magnitude greater affinity for HNE than native BPTI. Similarly, the inhibition of the relatively nonspecific and low-affinity protease inhibitor ecotin has been made highly selective for urokinase-type plasminogen activator by randomization of the key specificity-determining amino acids (77). A final example is phage display of human growth hormone, which has been utilized to obtain variants with enhanced affinity and specificity for the human growth hormone receptor (86).

Antibodies have attracted more attention as partners for phage display (87,89). In contrast to standard monoclonal or polyclonal methods, phage display can be faster, does not require the use of animals, can be used to generate humanized antibodies, and can afford antibody specificity for DNA (90) and other macromolecules that normally provoke a poor immune response in animals. Normally the putative ligand is immobilized on solid support, but there has also been a report that intact cells can be used as a matrix for eliciting antibody specificities to cell surface proteins (91). Peptide libraries have been screened for binding to different cell types (92) or even to organs *in vivo* (93), and there is no reason that similar experiments cannot be performed with phage displaying antibodies or other proteins.

Antibodies are widely utilized for the generation of ligand specificities through phage display because of their proven ability to adapt to bind different ligands with high affinity and specificity. Once randomized, however, other protein scaffolds can mimic antibody binding sites. Ku and Schultz have expressed a four-helix bundle protein, cytochrome b_{562} on the surface of phage and have shown that mutagenesis of two surface loops yields a protein that binds bovine serum albumin with a K_d of 290 n*M* (85).

Several factors must be considered prior to the decision to employ phage display as a tool for protein engineering. As with any technique that employs mutagenesis, the power to sort vast numbers of variants for selected function is accompanied by complex experimental challenges. Vector preparation and expression and validation of protein-bearing phage can be time consuming. Screening may require the sequencing of dozens of clones before a consensus is obtained. Once consensus sequences are identified, candidate proteins will need to be expressed and assayed for function. These steps necessitate a significant investment in time and require thoughtful experimental design. If an experimental goal is sufficiently important, however, and if structural knowledge suggests that random mutagenesis can yield novel specificity, then the obstacles become relatively less important and the technique more attractive. This analysis is especially important for phage display, as its conceptual simplicity makes it an extremely inviting technology for the engineering of many proteins, including ones where other techniques may be much more likely to yield informative results.

One question that often occurs during the design of phage display experiments is "How many residues should be randomized?" As noted previously, transformation efficiency effectively limits total library size. As a result, randomization of seven residues or greater creates diversity that is too large to be fully assayed, so that some variant proteins will never be screened. If seven residues are equally interesting, this may not be a problem. But if six residues are of exceptional interest, while the seventh is of marginal interest, including the marginal residue in a mutagenesis strategy will dilute the ability to successfully probe combinations at key locations. Furthermore, if a particular substitution at the seventh residue disrupts folding or expression, it can prevent identification of an accompanying favorable set of modifications to key residues.

3. *Two-Hybrid Screening*

A remarkable demonstration of the applicability of domain fusion to the synthesis of proteins of novel function has been the development of two-hybrid screening. This technique relies on the modular nature of dimeric transcriptional activators composed of a DNA binding domain and an activation domain. The DNA binding domain can be fused to one putative protein partner, while the activation domain can be fused to the other. If an interaction occurs, the DNA binding and activation domains will be brought together and transcriptional activation will take place (54,55,94,95).

In contrast to phage display, this technique functions *in vivo* and can be directly employed to monitor or engineer intracellular protein–protein interactions (Fig. 9). The potential of this methodology for the engineering of macromolecular interactions has been demonstrated by Fields and co-workers, who fused a library of randomized peptides to the activation domain of Gal 4 (96). This library of hybrid proteins was then screened for interaction with a hybrid protein consisting of retinoblastoma protein fused to the DNA binding domain of Gal 4. Consensus peptide sequences were obtained, and peptides were shown to bind with affinities ranging from 13 to 23 μM. It is likely that the mutagenesis of protein surfaces can be used to similarly engineer protein–protein interactions.

4. *DNA Shuffling*

Homologous recombination is a major source of genetic variation *in vivo*, and Stemmer has mimicked this phenomenon to develop the technique of DNA shuffling (Fig. 10), which builds similar variation *in vitro* (97–99). In this technique DNase I randomly cleaves the gene of interest. The resultant duplex fragments are denatured and the single strands are allowed to anneal. DNase I

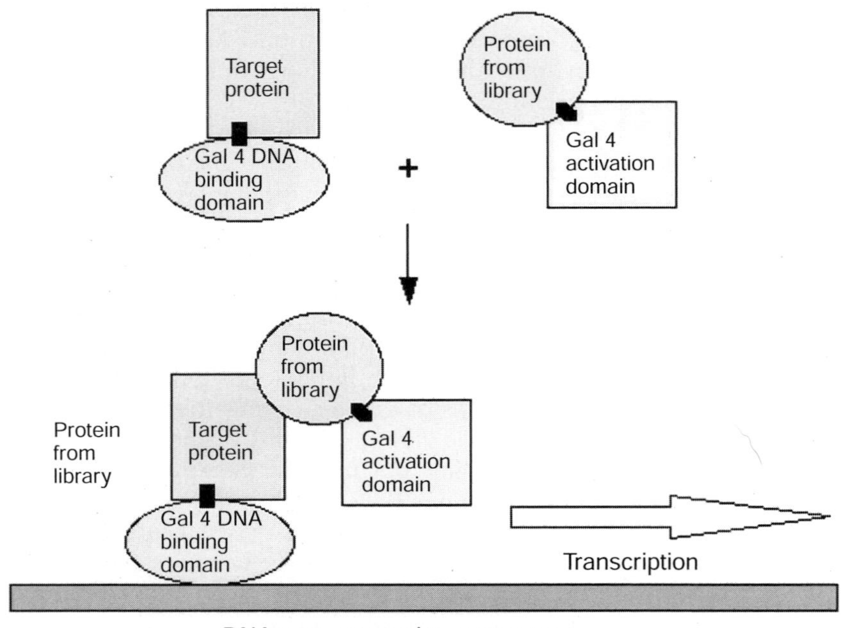

Fig. 9 Two-hybrid screen for engineering protein–protein interactions.

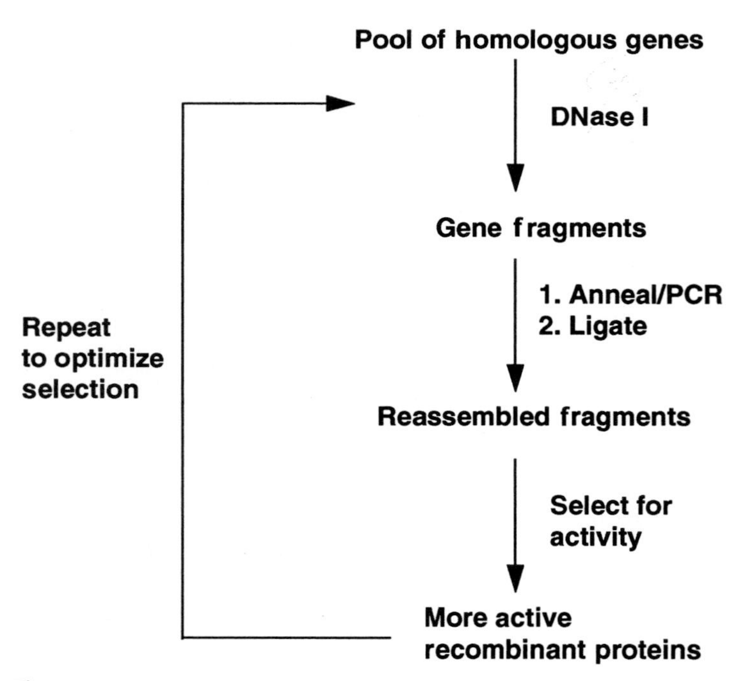

Fig. 10 DNA shuffling.

cleaves DNA relatively nonselectively, and the technique relies on the digestion being tightly controlled so that fragments are relatively large and remain able to reform the entire coding sequence. Addition of *Taq* DNA polymerase fills in the gaps that remain after renaturation and reconstitution of the gene. The error rate for *Taq* polymerase for small fragments is 0.7%, so the pool of reconstituted genes contains various mutations. The gene is then inserted into a plasmid and a second round of DNase I digestion and reconstitution combines various mutations, a process that mimics recombination events *in vivo*. At this point, the pool of recombinant plasmids is introduced into an organism, and a selection or screen is performed for clones that exhibit altered properties. These can then be subjected to further DNA shuffling to permit additional optimization. This technique has been successfully employed to improve function of β-lactamase (99) and green fluorescent protein (97).

5. *Catalytic Antibodies*

Antibodies are the most convenient route for the generation of new binding specificities within a protein scaffold. Their potential for molecular diversity is immense, allowing association with almost any ligand. Application to protein engineering is limited because the immune system has evolved to optimize binding, not catalysis. However, by choosing small molecules that resemble the transition state for chemical reactions, catalytic antibodies can be obtained (100,101). Progress has been substantial and the field has been extensively reviewed (102,103). Many different chemical reactions can be catalyzed and, as strategies for antibody design become more sophisticated, catalytic rates of some catalytic antibodies are approaching those of analogous enzymes.

Results that directly relate to protein engineering include the solution of an X-ray crystal structure for an antibody capable of the hydrolysis of amino acid esters (104). Analysis of the structure revealed that the active site of the antibody contains a serine and a histidine in an alignment that is similar to that found in the catalytic triad of serine proteases. This catalytic antibody was able to cleave amide bonds with labile aromatic leaving groups, but not unactivated peptide bonds. Using the crystal structure as a guide, it may be possible to optimize the active site of the antibody through mutagenesis and achieve catalysis that more closely mimics that of serine proteases. Such structure-guided design experiments should be a test of the potential of antibodies as enzyme-like catalysts and of the ability to combine antibody production with *in vitro* mutagenesis. These experiments will be aided by advances in the heterologous expression of antibodies, which will facilitate the acquisition of protein for detailed structure–function studies (105).

As with other forms of protein engineering, the optimization of antibody catalysis by specific substitutions using site-directed mutagenesis may not be straightforward. It has been shown that during affinity maturation the residues that contact the hapten remain constant, and that residues away from the binding site are responsible for subtle conformational changes that increase catalysis (106). The changes necessary to improve function may, therefore, be difficult to rationally predict. The ideal strategy for optimizing the amidase catalytic antibody described previously may be to retain the histidine and serine and build on their catalytic potential using a combination of techniques. The crystal structure can guide the introduction of residues at positions where they would be most likely to contribute to catalysis. The entire variable region can then be mutagenized and a selection developed for variants that will be able to best take advantage of the potential catalytic machinery within the combining site. Such a selection might employ phage display if a suitably reactive affinity resin could be developed. Hydrolysis of peptide bonds will depend on precise positioning of substrate relative to multiple amino acids.

D. Deletion Mutagenesis

When little is known about the determinants of structure and function of a particular protein, deletion mutagenesis (107) can be an effective option. In this technique the experimenter deletes portions of the gene encoding the protein and expresses the truncated variants. These can often be expressed in active form, and the preservation of function is probably due to the modular division of many proteins. Deletions may cripple one module and affect a given set of functions, while leaving other modules intact and active. Thus, important clues about the location of various features can be ascertained. Examples of the many proteins whose function has been probed by deletion mutagenesis include guanylyl cyclase (108), calreticulin (109), heparin cofactor II (110), Lam B, a sugar-specific porin (111), adenylyl cyclase (112), and Raf-1 kinase (113).

E. Introduction of Selected Amino Acids by Oligonucleotide-Directed Mutagenesis

The most common strategy for the introduction of site-directed mutations into a protein is the substitution of one natural amino acid for another. Technical details for this approach are described in Section III. The approach is simple and can be combined with any of the other strategies described previously. Specific applications are discussed throughout this chapter and will not be addressed here. A general consideration is that, as with other techniques for mutagenesis, many mutations will yield negative results. It is important to realize that these results can still be worth considering and carefully characterizing if they overturn an existing hypothesis. This is especially true because the introduction of selected amino acids by oligonucleotide-directed mutagenesis usually involves relatively few proteins, allowing more time for the study of each, even if the desired function is not present. Understanding how a variant protein can be expressed but not possess a predicted function can be just as important as understanding why another variant succeeded. This is especially true if laboratory experiments reflect a long-term interest in the protein.

F. Scanning Mutagenesis

The large number of amino acids within a protein complicates the analysis of protein function through site-directed mutagenesis. To replace each residue of a particular protein with the 19 other amino acids and study the resulting variants is impossible. What residues should be substituted? What residues should be introduced? Given three-dimensional structural data, the choice of substitution is usually not difficult. However, in the absence of detailed structural data, experiments become less clear. It is necessary, therefore, to use whatever information is available to limit the scope of mutagenesis to those substitutions that are most likely to afford interesting results. Scanning mutagenesis addresses this problem by providing rationales for targeting one or more subsets of amino acids for alteration. A loss of function of a mutant protein then suggests that the altered residue was important for some activity and merits closer study.

Alanine is often introduced because, unlike glycine, it retains normal stereochemistry, while, unlike the other 18 amino acids, its small size and lack of charge makes it unlikely that it will introduce steric or electronic conflicts. If a substitution does not result in a significant change of function, the original residue is probably not critical for wild-type activity. Most residues will fall into this category. The experimenter will usually be more interested in residues that *are* functionally important. Their replacement will cause a measurable change in protein structure or function. Such changes do not indicate that the residue was directly involved in an important functional interaction, because alterations may decrease activity by lowering protein expression or by preventing proper folding. It is essential, therefore, to evaluate the level of expression of variant proteins derived from alanine scanning, because negative results obtained from a mutant that is not well expressed are not informative. If a protein is expressed, it is helpful to possess an assay for a secondary attribute of protein structure or function whose preservation would support the suggestion that the protein is properly folded.

Cunningham and Wells have used alanine scanning mutagenesis to map interactions between human growth hormone (hGH) and its receptor (114–116). Initially, a related technique, homolog scanning mutagenesis, was employed to replace 7- to 30-residue segments of hGH with equivalent segments from noninteracting hGH homologs (117). This approach identified three different regions of human growth hormone that were likely to be involved in receptor binding and reduced the number of amino acids potentially involved in hGH–receptor interactions from more than 200 in full-length growth hormone to 62. Substitution of each residue within the three regions with alanine identified several variants that possessed substantially lower affinity for human growth hormone receptor. The authors infer that lower affinity is due to the removal of protein–protein interactions rather than protein misfolding, because a battery of eight monoclonal antibodies continued to recognize the variants with affinities that are similar to wild-type. Alanine scanning cannot always identify every important residue—20% of the alanine mutants of hGH could not even be purified. Presumably these either were not expressed or were digested by cellular proteases.

The biggest drawback to scanning mutagenesis as a probe for macromolecular function is the effort required to make, purify, and assay large numbers of mutants. Few laboratories can afford the resources to alter every amino acid in a protein, even if each is only changed to alanine. It is necessary, therefore, to use all available information to limit the number of amino acid residues that are candidates for alteration. As noted previously, homolog scanning is one option. Sequence alignments can also be valuable. For example, when scanning for key residues, it is reasonable to assume that they will be conserved in related homologs. Thus, limiting mutagenesis to conserved residues is an initial approach to the practical employment of alanine scanning. Further, particularly with enzymes, the observed properties of a protein will implicate particular amino acids as being functionally important. For example, ornithine decarboxylase is a pyridoxal-dependent enzyme known to convert ornithine to putrescine (118). In similar enzymes, aspartic or glutamic acids are known to play important roles in catalysis. This realization allowed scanning to be confined to conserved acidic residues, limiting experiments to the alteration of fewer than 10 amino acids. Similarly, it is also possible to restrict mutagenesis to charged residues, cysteine, potential glycosylation or phosphorylation sites, or other types of amino acid that might be involved in some measurable attribute.

Of course, the substitution of individual amino acids by mutagenesis is not difficult, and ambitious scanning experiments are appropriate if the goal is especially worthwhile. Ebright and co-workers have used alanine scanning to create 81 variants of TATA binding protein (TBP) (119) to probe protein–protein interactions during eukaryotic transcription initiation. They chose the location for mutations by using the crystal structure of the TBP–DNA complex to identify the solvent-accessible residues of TBP. They mutated the selected residues to alanine and analyzed the binding of TBP to various transcription factors *in vitro*. The mutations that resulted in reductions in binding affinity were used to construct a model of the interactions of TBP with transcription factors IIA, IIB, IIF, and DNA polymerase. This use of alanine scanning is particularly interesting because it affords information about the geometry of multiprotein complexes that are too large or too complex to be readily probed by X-ray crystallography. The crystal structure of the ternary DNA–TBP–TFIIA complex (120) confirmed that Ebright and co-workers had correctly identified the location of interactions between TFIIA and TBP.

Another example of the ability of scanning mutagenesis to provide structural information is the spatial location of selectivity-determining residues of the Shaker potassium channel (121). This protein, like other membrane proteins, has resisted crystallization, necessitating other approaches for the acquisition of structural information. In this case, the three-dimensional structure of a proteinaceous inhibitor of the channel was known. Scanning mutagenesis of both the inhibitor and the channel was performed. Comparison of the binding of sets of variant partners allowed assignment of pairwise interactions and suggested structural relationships. Once again, such information regarding complex macromolecular interactions would have been difficult if not impossible to obtain using any other currently available experimental strategy.

G. Insertion of Unnatural Amino Acids

The chemistry of native proteins is limited by the chemistry of the 20 natural amino acids. Combinations of these amino acids generate an enormous diversity of native protein function and chemistry. However, when mutagenesis is used to probe protein structure, there are questions that cannot be answered through insertion of native amino acids. To address these questions, methods have been developed for the insertion of nonnative amino acids into proteins. One strategy incorporates unnatural amino acids during translation, and a second strategy employs peptide synthesis to produce synthetic or semisynthetic proteins that contain unnatural amino acids.

A biosynthetic method developed by Schultz and colleagues has been extensively reviewed elsewhere (122–124). The method requires the introduction of a stop codon into the gene encoding the protein of interest. This codon is recognized in an *in vitro* transcription and translation system by a tRNA that has been either chemically or genetically modified to contain an anticodon loop that is complementary to a stop codon, and that has been aminoacylated with an unnatural amino acid. The amino acid is then introduced into proteins during translation. The Schultz laboratory has used this strategy to alter diverse proteins, including β-lactamase (125), ras (126), lysozyme (127), and staphylococcal nuclease (128). Lester and colleagues have even demonstrated that it is possible to use this methodology to incorporate unnatural amino acids into proteins within intact cells, and have employed this technique to probe the nicotinic receptor binding site (129).

One drawback of this strategy as a tool for the study of mutant proteins is that relatively little protein can be produced—substantial effort is necessary to isolate as little as 1 mg. Although the quantity of protein that can be economically produced *in vitro* is not large, it has been possible to solve the X-ray crystal structure of a mutant staphylococcal nuclease containing a nitrobutanoic side chain in place of catalytically critical glutamic acid 43 (128). A more substantial obstacle to widespread use of this methodology is its complexity and the requirement for large amounts of enzymes, unnatural amino acids, and specialized reagents. If the technique can be successfully packaged in "kit" form, however, its use should become widespread. A limited number of unnatural amino acids that are similar in structure to natural amino acids have also been introduced using auxotrophic bacterial strains (130).

An alternative method for the introduction of unnatural amino acids relies on chemical synthesis [reviewed in (131)]. For relatively short peptides of fewer than 40 residues, this is a straightforward procedure using modern techniques for manual or machine-based peptide synthesis. The potential for undesired chemical modifications and errors during peptide synthesis, however, increases with length. The yield of full-length peptide decreases and the purification of correctly synthesized long peptides away from damaged or incorrect sequences becomes increasingly difficult. As a result, proteins are synthesized using a modular methodology in which relatively short 30- to 40-residue peptides are synthesized separately and subsequently coupled to yield the full-length polypeptide. This polypeptide can be purified in a relatively straightforward manner because few other potential products are present, and can then be folded *in vitro* and assayed for activity. The key procedure becomes the chemical assembly of full-length protein from mid-sized peptide components.

One approach to the assembly of short peptides into protein-sized polypeptides employs a modified subtilisin variant that has been engineered to shift relative kinetics of peptide bond formation and peptide bond cleavage to favor peptide bond formation (132–134). This variant, termed subtiligase, was developed from an elegant series of protein engineering studies. Classic experiments with semisynthetic thiolsubtilisin, in which the catalytic serine of subtilisin was replaced with cysteine, had shown that the rate of peptide coupling could be increased relative to the rate of peptide bond cleavage (34). This result was confirmed with analogous genetically engineered thiolsubtilisin (132). A drawback in these studies was that the overall enzymatic rate was reduced, decreasing the efficiency of ligation. Further mutagenesis of subtilisin repositioned the cysteine and reduced steric crowding in the active site. This partially restored catalytic activity while maintaining the favorable ratio for peptide bond formation. This engineered enzyme, which was a remarkable achievement in the design of protein activity in its own right, catalyzed the coupling of peptides of 12 to 26 residues to afford full-length active RNase A (133). Wells and co-workers used this synthetic scheme to introduce fluorohistidine at positions 12 and 119 of RNase A. The differing pK_a of fluorohistidine relative to histidine allowed the contributions of histidine at each position to be dissected.

Kent and co-workers have developed another synthetic strategy that relies on the chemical coupling of peptide fragments (135,136). In this case, the carboxy termini of each peptide are chemically activated via an N-terminal α-bromoacyl moiety, while the C terminus contains an α-thiocarboxylic acid (Fig. 11) (137). These chemistries are specifically reactive toward coupling with each other to form a thioester linkage connecting the two peptides. Because of its high selectivity, this technique is straightforward and can be used if the nonamide linkages at the peptide joining sites do not interfere with the experimental question to be addressed. The Kent laboratory has also developed methodology to generate native amide linkages (136). These strategies have been used to synthesize HIV-1 protease and interleukin 8 to address the contribution of individual backbone hydrogen bonds (137–139) and to examine the introduction of a β-turn mimetic (135).

The chemical techniques for protein synthesis are restricted to relatively small proteins, a disadvantage not shared by the method of Schultz and co-

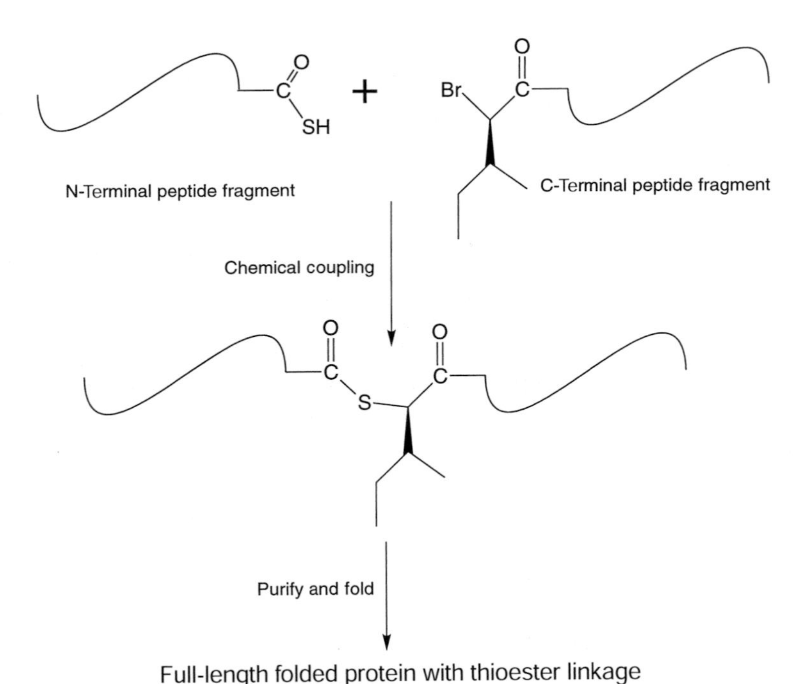

Fig. 11 Chemical peptide ligation (137).

workers. However, synthetic techniques offer the possibility of obtaining milligram quantities of protein, allow ready substitution of more than one unnatural amino acid, provide wider scope for the introduction of chemically novel backbone linkages, and are not dependent on the efficiency of transcription or translation in a particular system. These strategies are experimentally demanding, even for laboratories that specialize in peptide chemistry, and their use should be restricted to particularly important problems that cannot be addressed by standard protocols for site-directed mutagenesis.

V. Case Studies: Introduction of Function by Site-Directed Mutagenesis

A. Structure-Based Mutagenesis

As noted in preceding sections, a sophisticated knowledge of protein structure and mechanism is not always required for protein engineering. Knowledge of protein structure and function can, however, greatly facilitate the design and interpretation of experiments. The investigator can focus on obtaining a relatively small number of variants, each of which possesses a high likelihood of yielding an informative result. This approach can be termed structure-based mutagenesis. Use of the term does not imply that application of phage display, alanine scanning, or other techniques does not benefit from knowledge of the protein structure and mechanism. Rather, the term implies a mutagenesis strategy that requires detailed information about a protein target. The reader may also note that we have avoided referring to "random" vs "rational" strategies for mutagenesis throughout this chapter—such distinctions are meaningless because even a strategy that employs randomized residues needs to be rationally designed using all available information.

The solution of the X-ray crystal structure of a protein, particularly one that is in complex with substrate, ligand, substrate analog inhibitor, or cognate DNA, can suggest many of the key residues involved in binding and catalysis. The number of protein structures that have been solved is increasing rapidly and will be an increasingly important source of information for mutagenesis. Even if a structure of a given protein is not known, it is becoming increasingly likely that the structure of a related homolog will be available. Alternatively, information from sequence alignments, alanine scanning, NMR spectroscopy, or mechanistic studies can be employed to suggest the identity of key amino acids. These residues can then be altered, either individually or in combination, to confirm function, probe the molecular details of function, or alter catalytic activity or specificity.

1. Alteration of Catalytic Residues of Serine Proteases

Catalysis by serine proteases was an early focus for knowledge-based site-directed mutagenesis. Serine proteases were ideal candidates for initial studies because they were among the first enzymes to have been visualized by X-ray crystallography and because extensive mechanistic studies had been performed. As a result, the importance of the catalytic triad of aspartic acid, histidine, and serine was well appreciated. Exact roles of these residues during catalysis, however, were not clear, and the development of oligonucleotide-directed mutagenesis provided an opportunity to obtain important insights.

One study replaced the catalytic aspartic acid of trypsin with asparagine (140). Cysteine proteases normally contain an asparagine as the third member of the triad, so the outcome of this replacement was not obvious. The mutagenesis experiments revealed that the introduction of asparagine greatly reduced catalysis of the variant trypsin. The impact of this result was amplified by solution of the three-dimensional structure of the mutant (141), which revealed that the introduced asparagine was stabilizing the active-site histidine as an inactive tautomer. These experiments set the precedent for the now common and productive interplay between site-directed mutagenesis and crystallography.

Mutagenesis of subtilisin further established the importance of each member of the catalytic triad. Carter and Wells (142) replaced each triad member with alanine, individually and in combination. These experiments revealed that each residue was critical, and that their removal compromised enzyme activity by 1000- to 100,000-fold. Similar experiments with subtilisin also confirmed the importance of an asparagine as an oxyanion hole during formation of the transition state (143). The catalytic triad of trypsin has also been systematically probed and reveals that even relatively conservative replacements of triad members fail to restore catalytic activity (16). Significant enzyme activity can be retained, approximately 1.5% of wild type, by alteration of the position of triad members (144), suggesting that precise placement of charge is not essential for substantial activity. Once again, the X-ray crystal structure of this variant was critical for understanding the structural origins of its relatively proficient catalysis.

2. *Alteration of Primary Substrate Binding Pocket in Serine Proteases*

Serine proteases have served as a paradigm for using site-directed mutagenesis to probe determinants of specificity of enzymes for their substrates [reviewed in (145)]. They are convenient enzymes for this task because their binding and catalytic residues are largely independent. Much of this research has been motivated by the desire to alter protease specificity to generate novel enzymes of practical value. Successful alteration of specificity would also confirm our understanding of the molecular basis for substrate binding—if we know enough about how a protein functions to successfully alter it, then our knowledge is probably well grounded.

An early example of an attempt to alter specificity through protein engineering involved the replacement of aspartic acid-189 at the base of the primary amino acid binding site of trypsin with lysine (146). Aspartic acid-189 normally conveys specificity for arginine or lysine side chains in substrate peptides, so the substitution of lysine was expected to reverse the specificity and allow trypsin to recognize aspartic acid or glutamic acid. This result was not observed. Instead, the enzyme possessed low activity and a chymotrypsin-like specificity for large hydrophobic side chains.

The failure of this experiment to achieve the predicted result is explained by the X-ray crystal structure, which reveals that, rather than pointing out of the binding site toward the substrate, the introduced lysine is buried in the base of the pocket (147). In a similar experiment, mutagenesis of arginine to aspartic acid in aspartate amino transferase also resulted in a large loss in catalysis, although in this case some additional catalysis toward positively charged amino acids was observed (148). These results have been interpreted as evidence that the electrostatic field of these binding pockets is organized specifically for recognition of positively charged amino acids, and that one alteration, albeit a critical one, cannot overcome this preorganization (149). Alternatively, lysine is not a conservative steric substitution, and its potential for different positions with the substrate binding pocket may account for its inability to simply replace aspartic acid. A shorter analog of lysine, such as ornithine, might function better.

The conversion of a serine protease from trypsin-like specificity for positively charged side chains to chymotrypsin-like specificity for large hydrophobic side chains would seem to be more straightforward than reversal of charge preference, since the highly related and well-studied enzymes trypsin and chymotrypsin serve as models for the alterations that would be required. Conceptually, the success of this experiment is assured, as alteration of every amino acid that is different between trypsin and chymotrypsin will certainly convert substrate specificity. The uncertainty lies in the number of mutations required to accomplish the conversion of specificity. Unfortunately for the prospect of simple engineering of specificity, a classic series of experiments (150) revealed that extensive alterations were necessary before trypsin could be adapted for even a low level of catalysis with chymotrypsin specificity.

Hedstrom and co-workers first mutated residues within the primary S1 binding pocket of trypsin to those found in chymotrypsin. The most critical of these alterations was substitution of the aspartic acid at the base of the pocket with serine. This dramatic change greatly reduced native trypsin activity toward substrates containing arginine and lysine, but the removal of negative charge did not enhance activity toward substrates containing hydrophobic residues. Only after an additional nine amino acids in two surface loops were exchanged was significant chymotrypsin-like activity generated (Table II). Mutation of tyrosine 172 to tryptophan further increased activity toward hydrophobic side chains (151).

The extensive changes required to alter specificity toward a well-characterized end point emphasizes the difficulties of introducing functional changes—difficulties that will only be made greater when proteins are altered without homologous models to act as a guide. Moreover, even with homology as a guide, extensive changes can fail to yield observable alteration in specificity, as similar experiments to introduce trypsin-like specificity into chymotrypsin failed to yield desired results (152). Why can specificity be so difficult to engineer? Determinants of substrate specificity can be subtle, complex, and unpredictable, and this unpredictability results from an absence of clear spatial or structural relationships between the key determinants of specificity and the substrate. Similar conclusions have also been inferred from attempts to engineer the specificity of dihydrofolate reductase (153) and cytochrome *c* (73), and from following antibody maturation (106).

The attempts to convert trypsin to chymotrypsin offer discouraging evidence of the difficulty of engineering specificity in proteins. Fortunately, however, engineering other serine proteases has proven more successful. Both α-lytic protease and subtilisin have a broad specificity for small hydrophobic

Table II
Effect on Kinetics of Hydrolysis Suc-Ala-Ala-Pro-Phe-*p*-Nitroanilide of Stepwise Alteration of Trypsin to Chymotrypsin[a]

Enzyme[b]	Catalytic efficiency ($M^{-1}s^{-1}$)	Efficiency relative to chymotrypsin
Ch	560,000	1
Tn	9	0.000014
D189S Tn	21	0.000038
Tn (S1,L1,L2)	1700	0.003
Tn (S1,L1,L2) Y172W	83,000	0.14

[a] From Hedstrom, L., Szilagyi, L., and Rutter, W. J. (1992). Converting trypsin to chymotrypsin: The role of surface loops. *Science* **255,** 1249–1253; and Hedstrom, L., Perona, J. J., and Rutter, W. J. (1994). Converting trypsin to chymotrypsin: Residue 172 is a substrate specificity determinant. *Biochemistry* **33,** 8757–8763.
[b] Ch, Chymotrypsin; Tn, trypsin; S1, mutations introduced within S1 binding pocket; L1, mutations introduced within loop 1 (residues 185–188); L2, mutations introduced within loop 2 (residues 221–225).

amino acids. Specificity for these side chains can be readily modified by mutagenesis within the S1 binding pocket (154,155). Specificity in the S4 pocket has also been altered (156). It has even proven possible to introduce a specificity for dibasic amino acid pairs into a variant subtilisin (157). Why have these two proteases proven easier to engineer than trypsin? One contributing factor is that their S1 binding pockets are apolar and lack the electrostatic complexity of the trypsin pocket. This reduction in complexity means that mutagenesis is less likely to remove subtle determinants necessary for efficient catalysis. In addition, the S1 pockets of both subtilisin and α-lytic protease are largely composed of amino acid side chains, whereas much of the binding pocket of trypsin is defined by the amino acid backbone of the protein. The position of the backbone is difficult to alter predictably, limiting opportunities to modify the S1 pocket of trypsin. It is likely that both reasons contribute to differences in the feasibility of engineering proteases, and that considerations of polarity and side-chain composition will also dictate the potential for successful specificity alterations in other proteins.

B. Introduction of Metal Binding Sites into Proteins

Engineering proteins for altered function is most exciting when the alteration affords a distinctly different substrate specificity or catalysis. Previous sections describe the dramatic changes in substrate specificity that can be achieved via domain fusion, and also the problems that can be encountered during attempts to alter protein specificity by more limited amino acid substitutions. Even more difficult than the modification of specificity, however, is the *de novo* introduction of completely new function into proteins. Novel catalysis or modes of substrate binding will usually require the introduction of multiple amino acids. Each amino acid must be able to assume a defined orientation relative to neighboring residues—a requirement that is particularly difficult to achieve in attempts to mimic enzyme active sites because of their inherent complexity. For example, the catalytic triad of serine proteases has been a popular motif for catalyst design, but this approach requires that three residues be rigidly maintained in properly defined orientations relative to a substrate binding cleft, a daunting task that has yet to be achieved (158). Binding sites for small molecules will be difficult to design for similar reasons, and it may be much more practical to obtain them through phage display or the immune response.

Experimenters have found that the design of metal binding sites is more feasible, since introduced side chains need to associate with a simple ligand that also acts to drive their orientation. In addition, the structures of many native metal binding sites are available as guides. It has proven possible to introduce metal binding sites of varying degrees of sophistication into proteins to introduce a variety of novel properties [reviewed in (159)]. The relative simplicity of metal binding sites makes them ideal initial targets for testing strategies for the introduction of groups of amino acids in defined orientations.

Metal binding sites have been introduced to control enzyme activity. Control of enzyme activity is not only an interesting goal by itself, but also serves as an effective validation of the design concept. An early example was the introduction of a cysteine into the binding pocket of staphylococcal nuclease in place of leucine (160). This cysteine was protected from disulfide exchange between protein monomers by its position within the substrate binding pocket, and its introduction had little effect on catalysis. On addition of mercury or organomercurial compounds, however, the cysteine bound the metal ion, and the additional steric bulk prevented enzyme activity. Activity could then be regenerated by the addition of dithiothreitol.

Introduced metal binding sites can also directly subvert native catalytic machinery of enzymes (Fig. 12). Histidine was introduced near the catalytic histidine of trypsin (161). Computer modeling suggested that chelation of zinc by the two histidines could occur with geometry similar to that observed in thermolysin. Subsequent addition of zinc to the variant enzyme reduced catalytic activity, which could then be restored by addition of EDTA to chelate the zinc. An X-ray crystal structure of the complex confirmed that zinc was bound as predicted between the introduced histidine and the histidine of the catalytic triad (162). This interaction removes the histidine from its normal location adjacent to the catalytic serine and aspartic acid, and this disruption of the catalytic triad accounts for the observed loss of catalysis. Recently this work has been extended to the design of tridentate metal binding sites. Addition of a third histidine ligand into the protein scaffold lowers the K_i for inhibition by copper from 70 μM to 100 nM (163). Similar analysis has also been used to introduce metal binding sites into antibodies (164), offering the possibility of combining the chemistry of metals with the diversity of the immune system.

Metal binding sites can also be introduced into proteins to alter substrate specificity. Craik and colleagues employed molecular modeling to search the surface of trypsin for a pair of amino acids that could form a tridentate binding site for zinc with a substrate containing a third histidine at its P2′ position (165,166). Histidine was introduced at positions 143 and 151, and this variant was able to cleave a peptide containing tyrosine at the P1 position significantly more rapidly than was wild-type trypsin. Interestingly, specificity showed

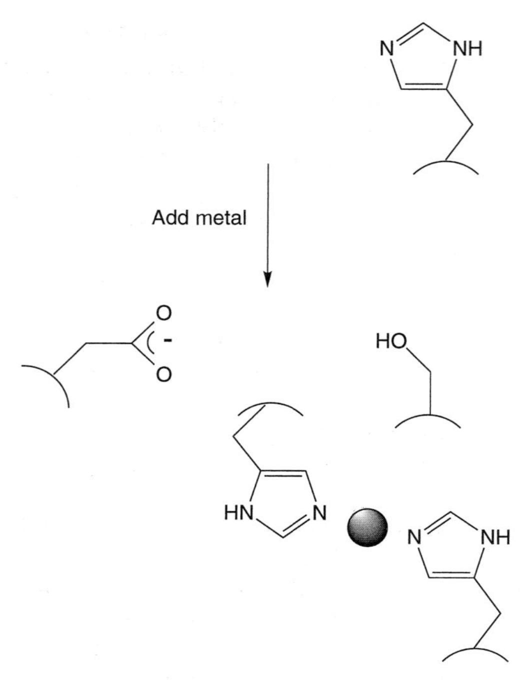

Fig. 12 Formation of a metal complex containing histidine residues, one within the catalytic triad and one nearby, on addition of metal.

metal dependence, as the difference in reactivity was much more pronounced when nickel was added rather than zinc. The structural origins of the introduced substrate specificity have been confirmed by solution of the three-dimensional crystal structure of the trypsin variant in complex with substrate and zinc (167). Most recently, metal-dependent activity has been further enhanced by an additional mutation, D189S, in the S1 pocket of trypsin (166), again revealing that an iterative process of design, testing, and redesign can yield progressively more impressive results.

The specificity of DNA binding proteins has also been regulated by the introduction of metal binding sites. Cuenuod and Schepartz (168) have conjugated a peptide containing a basic DNA binding domain to a tridentate pyridyl complex. On addition of Fe(II) the hexacoordinate system is formed, bringing DNA binding domains into proximity and increasing the cooperativity of binding.

C. Alteration of Selectivity without a Structural Guide

As described previously, the rational mutagenesis of serine proteases has been guided by an exceptionally deep background of high-resolution structural information. Specificity can also be altered in the absence of such data and can confirm the roles of particular amino acids. An example is the reversal of substrate specificity of 12- and 15-lipoxygenase (169). At the time this study was performed, there was no high-resolution three-dimensional structure of any member of this enzyme family. However, the authors identified amino acids at four positions that were conserved within known 12-lipoxygenases, but that were not found in 15-lipoxygenase. The authors were fortunate, given the small number of sequences in the alignment, that they were able to so tightly focus their experiment.

These residues were mutated individually and in combination to examine their importance in determining substrate selectivity. Alteration of three of the four residues within 15-lipoxygenase yielded an enzyme that performed 12- and 15-lipoxygenation equally well. Further mutations adjacent to what appeared to be a key residue for determining specificity boosted the ratio to 15:1, a striking reversal given the wild-type enzymes normal 9-fold preference for 15-lipoxygenation. An X-ray crystal structure of soybean 15-lipoxygenase subsequently confirmed the accuracy of the structural predictions derived from mutagenesis (170,171).

D. Substrate-Assisted Catalysis

As noted previously, the alteration of protease specificity has often proven to be more difficult than expected. Simple alteration of the primary binding pocket for peptide substrates either has worked poorly or has merely yielded enzymes that are mediocre mimics of related naturally occurring proteases. Clearly, other design strategies are required to develop novel substrate specificities. One such strategy has been termed "substrate-assisted catalysis."

In substrate-assisted catalysis, the primary specificity pocket is unaltered. New specificity is generated by altering the preference for substrate amino acids that are adjacent to the primary specificity determinant. This is accomplished by removal of a key catalytic residue, such that restoration of significant enzymatic activity requires that a substrate contain the missing residue in a position that allows it to occupy the vacant position in the enzyme active site. Wells and co-workers pioneered this strategy by replacing the catalytic histidine in subtilisin with alanine (172) to afford H64A subtilisin.

As expected, removal of histidine almost completely inactivated the protein for cleavage of most substrates. However, some activity could be reconstituted if a histidine was present in either the P2 or the P1′ position of the substrate peptide (173). This endows the modified subtilisin with a new and more narrow specificity. Although the wild-type enzyme preferentially hydrolyzes substrates containing hydrophobic residues, cleavage by the variant requires both a hydrophobic residue and an adjacent histidine.

A major practical problem with H64A subtilisin as a tool for protein cleavage was that, even when histidine was present in substrate, activity was still low. This undoubtedly reflects the inability of the substrate histidine to fully mimic the precise positioning of the native histidine in the catalytic triad, as well as the entropic disadvantage of a catalytic triad being dispersed on two macromolecules. To overcome these obstacles, mutagenesis was performed to obtain new subtilisin variants with increased activity (174). Similarly, substrate phage display has been used to identify particularly labile target sequences (173). This combination of a more reactive engineered protease and the identification of labile substrates facilitates application of the variant protease for laboratory procedures. H64A subtilisin has recently become commercially available from New England Biolabs (Beverly, MA).

Substrate-assisted catalysis has also been demonstrated with a variant of trypsin lacking its catalytic histidine (175). This finding suggests that it should be possible to introduce dependence on substrate-assisted catalysis into a wide range of serine proteases to yield a family of restriction proteases. Subtilisin and trypsin are both serine proteases, but are the product of convergent evolution and possess completely different three-dimensional structures. Thus, it is worth noting that the functional conservation between the two enzymes was sufficient to allow the same engineering strategy to succeed equally well in both cases.

E. Engineering Transcription Factors

The control of transcription through sequence-specific interactions between proteins and DNA is one of the central features of modern molecular biology. This recognition directly guides the course of both development and disease, and engineered proteins with novel DNA-binding specificities would offer tools for understanding gene regulation. Zinc finger proteins have been a paradigm for this research (176) and are ideal candidates for protein engineering, because members of the family recognize many different DNA sequences. There is also a large database of their amino acid sequences. Correlation of protein sequence with DNA binding selectivity suggested that zinc fingers exist as largely discrete modules. This hypothesis has been tested by substituting one zinc finger for another with a different DNA recognition preference. Many of these substitutions result in an alteration in specificity without loss of affinity. An example of this approach is the use of sequence alignments by Desjarlais and Berg (177) to create a consensus sequence that differed in key specificity-defining amino acids. Manipulation of these residues yielded trizinc finger proteins with rationally varied specificities.

Phage display of zinc fingers has also been used to elicit novel specificities (178–181). Selected variants have K_d values less than 60 pM (178) and can block transcription *in vivo* (179). Pabo and colleagues have also generated structurally divergent transcription factor motifs that can be linked to generate hybrid proteins (180). Their fusion of the homeodomain from Oct-1 and a two-domain zinc finger peptide bound to the predicted DNA sequence with high affinity and selectivity. Neither the Oct-1 or the zinc finger component was able to bind effectively when assayed separately, and the high specificity was due to a chelating effect that allowed low-affinity interactions to combine to yield high-affinity DNA recognition.

The design of the fusion protein was guided by examination of the crystal structure to ensure optimal alignment of the fusion region, but it is important to note that, given the restriction of fusion to either the N or C terminus, sophisticated analysis is not always necessary. Even in the absence of a high-resolution cocrystal structure, engineering experiments can be done using varied geometries for fusion and by varying the interdomain linker length to find a length that is neither too rigid nor too flexible. The modular nature of zinc fingers can also be used to direct the activity of enzymes. *Fok*I is a type IIS restriction endonuclease, itself a modular protein, that contains a DNA-binding domain and a DNA-cleaving domain. This DNA-cleaving domain has been attached to varied zinc fingers to create artificial Type IIs restriction endonucleases (182).

F. Engineering Proteins for Improved Therapy

Tissue type plasminogen activator (t-PA) plays a key physiological role in the initiation of the fibrinolytic cascade by cleavage of the Arg^{560}-Val^{561} peptide bond of plasminogen. This cleavage yields the active serine protease plasmin, which degrades the fibrin core of blood clots to soluble fragments. t-PA is now in widespread use as a therapeutic agent and significantly enhances the prognosis for patients with acute myocardial infarction. Streptokinase, a less expensive bacterial protein, and urokinase type plasminogen activator, a recombinant human protein like t-PA, have also been used therapeutically. t-PA displays less systemic plasminogen activation and fibrinogen depletion than either of these other treatments (183,184) as a result of an interaction with fibrin, which stimulates t-PA's activity and helps to localize activity to the site of thrombi. Also, unlike streptokinase, t-PA is not antigenic and can be used on multiple occasions.

Despite its efficacy, many have questioned the cost effectiveness of t-PA use because of its greater cost relative to that of streptokinase (185). As a result, much effort has been directed toward the design of improved t-PA variants to reduce the risk of hemorrhage and increase the cost effectiveness of treatment. These studies have focused on increasing fibrin / fibrinogen binding specificity, zymogenicity, circulating half-life, and resistance to inhibitors, and have employed domain deletions, alanine scanning, and the structure-based substitution of key amino acids. These efforts have been the subject of lengthy reviews (43,186), and our aim is to demonstrate briefly how varied engineering strategies can be employed to improve the therapeutic prospects of a protein.

1. Domain Deletions

t-PA is a 527-amino acid protein containing five distinct structural domains (43). Residues 4–50 form a finger domain closely related to the fibrin-binding finger structures of fibronectin. Residues 51–87 form a growth factor domain that is homologous to the precursor of epidermal growth factor. Residues 88–175 and 176–263 each form kringle domains, characterized by their formation of three specifically ordered intradomain disulfide bonds. Residues 276–527 form a serine protease with a primary substrate specificity and sequence that are similar to those of trypsin.

Many of the initial engineering studies of t-PA were desgined to delineate the function of the nonprotease domains of t-PA. Both the finger domain and the epidermal growth factor-like domain appear to carry determinants that play a role in clearance of t-PA from the bloodstream. Deletion of the finger domain extended the circulating half-life of t-PA 20-fold in a rat model (187), and deletion of the epidermal growth factor-like domain resulted in 4- to 10-fold slower clearance compared to wild type in rat, guinea pig, and rabbit models (188–190). Deletion of the finger domain or the second kringle domain has been shown to reduce the fibrin binding affinity of t-PA. A mutant, which contains only the kringle two domain and the protease domain and which has been mutated to remove its two glycosylation sites in clinical trials (191).

Although stimulation by fibrin is reduced four- to fivefold, and inhibition by the circulating plasminogen activator inhibitor 1 (PAI-1) is equally efficient relative to wild-type t-PA, clearance from the bloodstream is significantly reduced. In a rabbit jugular vein thrombosis model, this mutant exhibits 3.9-fold higher thrombolytic potency than wild type. The mutant also shows no greater systemic degradation of fibrinogen than wild type and can be expressed in *E. coli,* which may make it significantly less expensive to produce.

2. *Directed Mutagenesis*

Site-directed mutagenesis has been used to improve the therapeutic efficacy of t-PA. Among the first of these studies were investigations into the effect of glycosylation on the activity of t-PA. t-PA possesses three *N*-linked glycosylation consensus sites that are known to be glycosylated under some circumstances. These sites are at residues 117 (kringle 1), 184 (kringle 2), and 448 (protease domain). Position 184 is only glycosylated on about 50% of expressed t-PA (192,193). Glycosylation at residues 184 and 448 is complex, and that at 117 is of the high mannose type (193). Initial studies involved treatment of cells expressing t-PA with tunicamycin to prevent glycosylation or treatment of the protein with endo β-*N*-acetylglucosaminidase H to remove glycosylation (194). Such experiments indicated that the glycosylation had a negligible effect on t-PA activity but that the rate of clearance of these nonglycosylated forms of t-PA is significantly slower. The mutant N117Q displayed approximately threefold slower clearance than wild type, confirming that at least the high mannose glycosylation site plays a role in clearance (195).

Homology modeling using the trypsin–BPTI complex as a model of the t-PA–PAI-1 interaction indicated that arginine 304 and the loop KHRR 296–299 could interact with a negatively charged surface on PAI-1 (196,197). t-PA variants containing the mutations R304S, R304E and the deletion Δ296–302 retained comparable catalytic activity toward plasminogen in the presence of fibrin, and similar levels of stimulation of activity by fibrin. However, their interaction with PAI-1 was significantly reduced. Arginine-304 apparently forms a salt bridge with glutamate 350 of PAI-1, and the PAI-1 mutant E350R effectively restores inhibitor interaction with R304E t-PA (198). These determinants of PAI-1–t-PA interaction were confirmed by alanine scanning (199). It may be possible to increase cost effectiveness by administering PAI-1–resistant t-PA in lower doses than are required with wild type.

Many chymotrypsin-like serine proteases are secreted in an inactive form termed a "zymogen," which is activated by cleavage at a specific site to form a new N terminus. t-PA, however, is only 5- to 10-fold less active toward plasminogen in its single-chain form than in its cleaved form in the absence of fibrin (200), and in the presence of fibrin the two forms are equally active (199). Increasing t-PA's zymogenicity might help in localizing its activity to the site of a clot and reducing systemic plasminogen activation and fibrinogen depletion. The homologous protease chymotrypsin appears to use a hydrogen-bonded triad of residues (aspartate-194, histidine-40, and serine-32) to stabilize the oxyanion hole and elements of the primary substrate binding pocket in an inactive conformation in the single-chain state of the enzyme. After activation cleavage, chymotrypsin's new N terminus participates in an alternative interaction with aspartate 194, inducing subtle conformational changes that activate the enzyme. t-PA lacks this zymogen stabilizing triad. The mutations A292S and F305H are apparently able to introduce such a triad and reduce the activity of the single-chain form 141-fold without affecting the activity of the cleaved form (200). This activity is measured with a small chromogenic substrate, however, rather than plasminogen, so the true measure of the zymogenicity of the enzyme *in vivo* is unknown.

The dependence of t-PA specificity on fibrin has also been a focus for protein engineering. One study has shown that the role of aspartate-477, a part of the loop forming the oxyanion hole, in switching between zymogen and active states may also be accomplished by the interaction of t-PA with fibrin (201). The two-chain forms of mutants D477N and D477E were compared with their counterparts in a double mutant t-PA that also contained the mutation R275E so that it could not undergo activation cleavage. In the presence of fibrin, none of these mutants displayed significant zymogenicity. However, all of these mutants displayed greater than wild-type stimulation by fibrin compared to their level of activity in its absence, and the single-chain forms displayed 5.3- to 10.9-fold greater stimulation by fibrin than their two-chain counterparts. In addition to elucidating a means for enhancing fibrin dependence, this study suggests that fibrin specificity and zymogenicity are linked functions. In another study, the residues LSPF 420–423 were investigated as a potential site of direct interaction between t-PA and plasminogen because they form a solvent-exposed hydrophobic surface, and its burial in a protein–protein interface would likely be energetically favorable (202). Compared to wild type, two mutants (R275E, S421G and R275E, P422G) in this study displayed 91- to 94-fold greater fibrin dependence. This hydrophobic surface therefore may provide a weak interaction in the absence of fibrin that aids in productive orientation of the enzyme–substrate complex, but that is insignificant in the presence of fibrin.

G. Minimization of Polypeptide Hormone: Emerging Synthetic Chemistry of Macromolecular Design

The synthesis of small molecules in the laboratory is often a multistep process requiring a variety of chemical transformations. It should not be surprising that the adaptation of proteins for new functions benefits from a similar approach. Any single alteration may improve some aspect of the activity of a protein, while diminishing others, and repeated alterations may be necessary to gradually set the stage for a highly active final variant. This approach is becoming more attractive as the number of available techniques for engineering proteins increases, because diverse methodologies can be applied sequentially and in combination to achieve experimental goals.

An example of this methodical approach to macromolecular engineering is seen in work on the minimization of a polypeptide hormone (203) (Fig. 13; see color insert). The intent of this study was to reduce the size of a peptide hormone to facilitate its use as a lead for drug discovery. The hormone, atrial natriuretic peptide (ANP), is a cyclic 28-residue peptide involved in the regulation of salt balance and blood pressure. Earlier attempts to minimize the size of the hormone had yielded analogs that bound with 1000-fold lower affinity than the wild type. This lack of success persuaded Wells and co-workers to adopt a multistep strategy aimed at progressively minimizing size while maximizing activity.

ANP is cyclized by disulfide formation, and the first goal was to introduce a new disulfide linkage to yield a smaller ring. Wells and co-workers employed alanine scanning to identify the amino acid residues that were critical for receptor binding. Most of these residues were clustered, allowing them to design a smaller disulfide ring that still contained this cluster. This dramatic change was necessary before peptide size could be substantially reduced, but it reduced binding affinity by 100-fold. To restore binding affinity, residues within the ring were optimized through phage display for binding to the extracellular domain of the natriuretic peptide receptor A. The phage display screening yielded a peptide that bound with affinity comparable to that of the wild type. This peptide was still long, and 11 C-terminal residues outside of the disulfide-linking core were subsequently deleted. This deletion reduced affinity 300-fold, but another application of phage display screening yielded a new peptide with restored binding. Finally, the additional N-terminal residues that alanine scanning had indicated were not critical were removed to yield a peptide of minimal length that was capable of binding receptor with an affinity that was only 7-fold less than that possessed by full-length ANP.

The combination of alanine scanning, deletion mutagenesis, specific site-directed substitutions, and repeated rounds of phage display screening allowed the achievement of an ambitious experimental outcome, demonstrating that the techniques of protein engineering can be combined to optimize macromolecular function. As with synthetic organic chemistry, optimal results will be achieved if ambitious overall aims can be broken down into sequential procedures, each of which is modest and experimentally straightforward. Also, as with synthetic chemistry, success will require a sound understanding of physical properties of the molecules being manipulated and a sustained effort that focuses on long-term outcomes.

VI. Conclusion

Just as synthetic organic chemistry provides tools for the successful manipulation of small molecules, protein engineering affords techniques for the transformation of macromolecules. As with any branch of synthetic chemistry, new strategies for protein engineering will continue to be discovered while old ones will be refined and optimized. This combination of spectacular breakthroughs and steady advances will supply new options and opportunities for protein modification, allowing previously unattainable goals to be reached.

How will protein engineering advance during the future? Laboratories in every field of biological research will take advantage of progress in the field and will routinely achieve functional transformations that would have seemed impossible only five years previously. This general ability of nonspecialist laboratories to transform proteins will become critically important as large-scale sequencing efforts and advances in protein expression systems make increasing numbers of biologically and mechanistically important recombinant proteins available for study. Specialist laboratories will continue to explore advanced techniques and define the strengths and limitations of the field. This

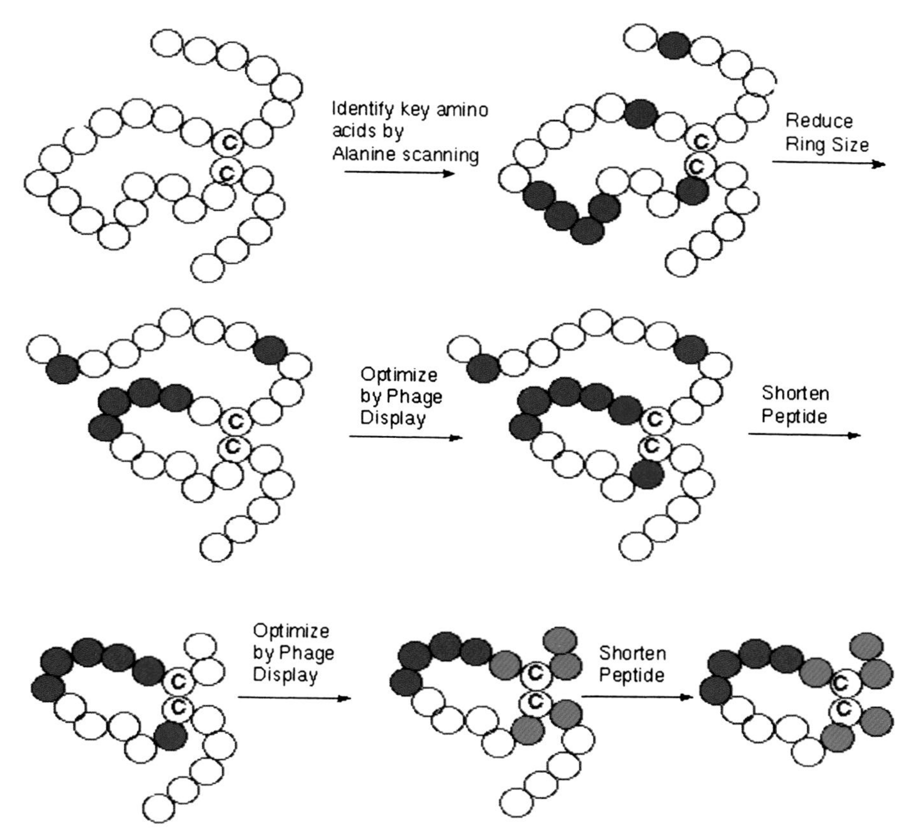

Fig. 13 Stepwise minimization of a polypeptide hormone. Residues in magenta were identified as important during the initial round of alanine scanning. Residues in green were optimized in the first round of phage display. Residues in blue were occupied in the second round of phage display.

should be facilitated by thorough characterization of both positive and negative results, a goal that should be supported by authors and journals alike.

As noted repeatedly in this chapter, this potential is accompanied by the danger that the ease with which protein engineering can be pursued will lead to the pursuit of impractical goals. However, given thoughtful preparation, protein engineering will bridge chemistry and biology to generate unique insights.

Acknowledgments

The authors wish to thank Dr. Donald Doyle and Dr. Susan Hamilton for insightful comments. G. S. C. was supported by National Institutes of Health Predoctoral Training Program Grant T32GMO8203. D. R. C. is an Assistant Investigator with the Howard Hughes Medical Institute.

References

1. "Protein Engineering: Principles and Practice" (1996). (Cleland, J. L., and Craik, C. S., eds.). Wiley-Liss, New York.
2. "Protein Engineering and Design" (1996). (Carey, P. R., ed.). Academic Press, San Diego.
3. Hutchinson, C. A., Phillips, S., Edgell, M. H., Gillam, S., Jahnke, P., and Smith, M. (1978). Mutagenesis at a specific position in a DNA sequence. *J. Biol. Chem.* **253,** 6551–6560.
4. Smith, M. (1985). In vitro mutagenesis. *Annual Rev. Genetics* **19,** 423.
5. McPherson, M. J. (ed.) (1991). "Directed Mutagenesis: A Practical Approach." IRL Press.
6. Trower, M. K. (ed.) (1996). "In Vitro Mutagenesis Protocols." Humana Press.
7. Vernet, T., and Brousseau, R. (1996). *In vitro* mutagenesis. *In:* "Protein Engineering and Design" (1996). (Carey, P. R., ed.). Academic Press, San Diego, pp. 155–179.
8. Sambrook, J., Fritsch, E. F., and Maniatis, T. (1989). "Molecular Cloning." Cold Spring Harbor Laboratory Press, Cold Spring Harbor, New York.
9. Kunkel, T. A. (1985). Rapid and efficient site-specific mutagenesis without phenotypic selection. *Proc. Natl. Acad. Sci.* **82,** 488–492.
10. Kunkel, T. A., Bebenek, K., and McClary (1991). "Efficient site-directed mutagenesis using uracil-containing DNA." *Methods Enzymol.* **204,** 125–139.
11. Sayers, J. R., Krekel, C., and Eckstein, F. (1992). Rapid high-efficiency site-directed mutagenesis by the phosphorothioate approach. *Biotechniques* **13,** 592.
12. Shen, T-J., Zhu, L-Q., and Sun, X. (1991). A marker-coupled method for site-directed mutagenesis. *Gene* **103,** 73–77.
13. Chong, S., and Garcia, G. A. (1994). An oligonucleotide-directed, *in vitro* mutagenesis method using ssDNA and preferential DNA amplification of the mutated strand. *Biotechniques* **17,** 721–723.
14. Kolodziej, P., and Young, R. A. (1989). RNA polymerase II subunit RPB3 is an essential component of the mRNA transcription apparatus. *Mol. Cell. Biol.* **9,** 5387–5394.
15. Swanson, M. E., Carlson, M., and Winston, F. (1990). *SPT6,* an essential gene that affects transcription in *Saccharomyces cerevisiae,* encodes a nuclear protein with an extremely acidic amino terminus. *Mol. Cell. Biol.* **10,** 4935–4941.
16. Corey, D. R., and Craik, C. S. (1992). An investigation into the minimum requirements for peptide hydrolysis by mutation of the catalytic triad of trypsin. *J. Am. Chem. Soc.* **114,** 1784–1790.
17. Schimmel, P. (1989). "Hazards of deducing enzyme structure-activity relationships on the basis of chemical applications of molecular biology." *Acc. Chem. Res.* **22,** 232–233.
18. van Dongen, W. D., van Bommel, J. H., van Wassnaar, P. D., Heerman, W., and Haverkamp, J. (1994). Rapid identification of specific mutation in the sequence of enzyme variant produced by protein engineering using high-performance liquid chromatographic / fast atom bombardment techniques. *Biol. Mass. Spec.* **23,** 675–681.
19. Lai, D., Zhu, X., and Petska, S. (1993). A simple and efficient method for site-directed mutagenesis with double-stranded plasmid DNA. *Nucl. Acids. Res.* **21,** 3977–3981.

20. Instructions for "MORPH: Site-Specific Plasmid DNA Mutagenesis Kit." 5 Prime to 3 Prime Inc., Boulder, CO.
21. Landt, O., Grunert, H.-P., and Hahn, U. (1990). A general method for rapid site-directed mutagenesis using the polymerase chain reaction. *Gene* **96,** 125–128.
22. Papworth, C., Bauer, J. C., and Braman, J. (1996). Site-directed mutagenesis in one day with >80% efficiency. *Strategies* (Stratagene) **9,** 3.
23. Weiner, M. P., Costa, G. L., Schoettlin, W., Cline, J., Mathur, E., and Bauer, J. C. Site-directed mutagenesis of double-stranded DNA by the polymerase chain reaction. *Gene* **151,** 111–123.
24. Stemmer, W. P. C., and Morris, S. K. (1992). Enzymatic inverse PCR: A restriction site independent, single-fragment method for high-efficiency, site-directed mutagenesis. *Biotechniques* **13,** 217–219.
25. Higuchi, R., Krummel, B., and Saiki, R. K. (1988). A general method of *in vitro* preparation and specific mutagenesis of DNA fragments: Study of protein and DNA interactions. *Nucl. Acids. Res.* **16,** 7351–7367.
26. Dulau, L., Cheyrou, A., and Aigle, M. (1989). Directed mutagenesis using PCR. *Nucl. Acids Res.* **17,** 2873.
27. Hall, L., and Emery, D. C. (1991). A rapid and efficient method for site-directed mutagenesis by PCR, using biotinylated universal primers and streptavidin-coated magnetic beads. *Protein Engineering* **4,** 601.
28. Kaiser, E. T., and Lawrence, D. S. (1984). Chemical mutation of enzyme active sites. *Science* **226,** 505–511.
29. Pusztai, A., Grant, G., Stewwart, J. C., and Watt, W. B. (1988). Isolation of soybean trypsin inhibitors by affinity chromatography on anhydro-Sepharose 4B. *Anal. Biochem.* **172,** 108–112.
30. Neet, K., and Koshland, D. E. (1966). The conversion of serine at the active site of subtilisin to cystein: A "chemical mutation." *Biochemistry* **56,** 1606–1611.
31. Polgar, L., and Bender, M. L. (1966). A new enzyme containing a synthetically formed active site. Thiol-subtilisin. *J. Am. Chem. Soc.* **88,** 3153–3154.
32. Wu, Z.-P., and Hilvert, D. (1989). Conversion of a protease into an acyl transferase: Selenosubtilisin. *J. Am. Chem. Soc.* **111,** 4513–4514.
33. O'Connor, M. J., Dunlap, R. B., Odom, J. D., Hilvert, D., Pusztai-Carey, M., Shenoy, B. C., and Carey, P. R. (1996). Probing an acyl enzyme of selenosubtilisin by Raman Spectroscopy. *J. Am. Chem. Soc.* **118,** 239–240.
34. Nakatsuka, T., Sasaki, T., and Kaiser, E. T. (1987). Peptide segment coupling catalyzed by the semisynthetic enzyme thiolsubtilisin. *J. Am. Chem. Soc.* **109,** 3808–3810.
35. Pollack, S. J., Nakayama, G. R., and Schultz, P. G. (1988). Introduction of nucleophiles and spectroscopic probes into antibody combining sites. *Science* **242,** 1038–1040.
36. Pollack, S. J., and Schultz, P. G. (1989). A semisynthetic catalytic antibody. *J. Am. Chem. Soc.* **111,** 1929–1931.
37. Messmore, J. M., Fuchs, D. N., and Raines, R. T. (1995). Ribonuclease A: Revealing structure–function relationships with semisynthesis. *J. Am. Chem. Soc.* **117,** 8057–8060.
38. Pendergrast, P. S., Ebright, Y., and Ebright, R. (1994). High-specificity DNA cleavage agent: Design and application to kilobase and megabase DNA substrates. *Science* **265,** 959–961.
39. Ikeo, K., Takahashi, K., and Gojobori, T. (1995). Different evolutionary histories of kringle and protease domains in serine proteases: A typical example of domain evolution. *J. Mol. Evol.* **40,** 331–336.
40. Campbell, I. D., and Baron, M. (1991). The structure and function of protein modules. *Phil. Trans. R. Soc. Lond. B* **332,** 165–170.
41. van Zonnevld, A-J., Veerman, H., and Pannekoek, H. (1986). Autonomous functions of structural domains on human tissue-type plasminogen activator. *Proc. Natl. Acad. Sci. USA* **83,** 4670–4674.
42. Madison, E. L., Coombs, G. S., and Corey, D. R. (1995). Substrate specificity of tissue type plasminogen activator. *J. Biol. Chem.* **270,** 7558–7562.
43. Madison, E. L. (1994). Probing structure–function relationships of tissue-type plasminogen activator by site-specific mutagenesis. *Fibrinolysis* **8,** 221–236.
44. Haber, E., Quartermous, T., Matsueda, G. R., and Runge, M. S. (1989). Innovative approaches to plasminogen activator therapy. *Science* **243,** 51–56.
45. Runge, M. S., Bode, C., Matsueda, G. R., and Haber, G. R. (1987). Antibody-enhanced thrombolysis: Targeting of tissue plasminogen activator *in vivo*. *Proc. Natl. Acad. Sci. USA* **84,** 7659–7662.

46. Burrows, F. J., and Thorpe, P. E. (1993). Eradication of large solid tumors in mice with an immunotoxin directed against tumor vasculature. *Proc. Natl. Acad. Sci. USA* **90,** 8996–9000.
47. Vitetta, E. S., Thorpe, P. E., and Uhr, J. W. (1993). Immunotoxins: Magic bullets or misguided missiles. *Trends Pharm. Sci.* **14,** 148–154.
48. Siegal, C. B. (1995). Targeted therapy of carcinomas using BR96 SFV-PE40, a single-chain immunotoxin that binds to the LE(Y) antigen. *Seminars in Cancer Biology* **6,** 289–295.
49. Corey, D. R., and Schultz, P. G. (1987). Generation of a hybrid sequence-specific single-stranded deoxyribonuclease. *Science* **238,** 1401–1403.
50. Pei, D., and Schultz, P. G. (1990). Site-specific cleavage of duplex DNA with a λ repressor–staphylococcal nuclease hybrid. *J. Am. Chem. Soc.* **112,** 4579–4580.
51. Corey, D. R., Pei, D., and Schultz, P. G. (1989). Generation of a catalytic sequence-specific hybrid DNase. *Biochemistry* **28,** 8277–8286.
52. Spencer, D. M., Wandless, T. J., Schreiber, S. L., and Crabtree, G. R. (1993). Controlling signal transduction with synthetic ligands. *Science* **262,** 1019–1024.
53. Belshaw, P. J., Ho, S. N., Crabtree, G. R., and Schreiber, S. L. (1996). Controlling protein association and subcellular localization with a synthetic ligand that heterodimerization of proteins. *Proc. Natl. Acad. Sci. USA* **93,** 4604–4607.
54. Fields, S., and Song, O. (1989). A novel genetic system to detect protein–protein interactions. *Nature* **340,** 245–246.
55. Allen, J. B., Walberg, M. W., Edwards, M. C., and Elledge, S. J. (1995). Finding prospective partners in the library—the two hybrid system and phage display find a match. *Trends Biochem. Sci.* **20,** 511–516.
56. Natsoulis, G., and Boeke, J. D. (1991). New antiviral strategy using capsid-nuclease fusion proteins. *Nature* **352,** 632–635.
57. Natsoulis, G., Seshaiah, P., Federspiel, M. J., Rein, A., Hughes, S. H., and Boeke, J. D. (1995). Targeting of a nuclease to murine leukemia virus capsids inhibits viral multiplication. *Proc. Natl. Acad. Sci. USA* **92,** 364–368.
58. Wu, X., Liu, H., Xiao, H., Kim, J., Seshaiah, P., Natsoulis, G., Boeke, J. D., Hahn, B. H., and Kappes, J. C. (1995). Targeting foreign proteins to human immunodeficiency virus particles via fusion with Vpr and VpX. *J. Virol.* **69,** 3389–3398.
59. Evnin, L. B., Vasquez, J. R., and Craik, C. S. (1990). Substrate specificity of trypsin investigated by using a genetic selection. *Proc. Natl. Acad. Sci. USA* **87,** 6659–6663.
60. Perona, J. J., Evnin, L. B., and Craik, C. S. (1993). A genetic selection elucidates structural determinants of arginine versus lysine specificity in trypsin. *Gene* **137,** 121–126.
61. Venekei, I., Hedstrom, L., and Rutter, W. J. (1996). A rapid and effective procedure for screening protease mutants. *Prot. Engineering* **9,** 85–93.
62. Oliphant, A. R., and Struhl, K. (1989). An efficient method for generating proteins with altered enzymatic properties: Application to β-lactamase. *Proc. Natl. Acad. Sci. USA* **86,** 9094–9098.
63. van den hazel, H. B., Kielland-Brandt, M. C., and Winther, J. R. (1995). Random substitution of large parts of the propeptide of yeast proteinase A. *J. Biol. Chem.* **270,** 8602–8609.
64. Vipond, I. B., and Halford, S. E. (1996). Random mutagenesis targeted to the active site of the *EcoRV* restriction endonuclease. *Biochemistry* **35,** 1701–1711.
65. Waugh, D. S., and Sauer, R. T. (1994). A novel class of *FokI* restriction endonuclease mutants that cleave hemi-methylated substrates. *J. Biol. Chem.* **269,** 12298–12303.
66. Hermes, J. D., Blaklow, S. C., and Knowles, J. R. (1990). Searching sequence space by definably random mutagenesis: Improving the catalytic potency of an enzyme. *Proc. Natl. Acad. Sci. USA* **87,** 696–700.
67. Blacklow, S. C., Liu, K. D., and Knowles, J. R. (1991). Stepwise improvements in catalytic effectiveness: Independence and interdependence in combinations of point mutations of a sluggish triosephosphate isomerase. *Biochemistry* **30,** 8470–8476.
68. Liao, H., McKenzie, T., and Hageman, R. (1986). Isolation of a thermostable enzyme variant by cloning and selection in a thermophile. *Proc. Natl. Acad. Sci. USA* **83,** 576–580.
69. Haruki, M., Noguchi, E., Akasako, A., Oobatake, M., Itaya, M., and Kanaya, S. (1994). A novel strategy for stabilization of *Escherichia coli* ribonuclease HI involving a screen for an intragenic suppressor of carboxyl-terminal deletions. *J. Biol. Chem.* **269,** 26904–26911.
70. Cunningham, B. C., and Wells, J. A. (1987). Improvement in the alkaline stability of subtilisin using an efficient random mutagenesis and screening procedure. *Protein Engineering* **1,** 319–325.

71. Kolmar, H., Frisch, C., Gotze, K., and Fritz, H. J. (1995). Immunoglobulin mutant library genetically screened for folding stability exploiting bacterial signal transduction. *J. Mol. Biol.* **252,** 471–476.
72. Kast, P., Asif-Ullah, M., Jiang, N., and Hilvert, D. (1996). Exploring the active site of chorismate mutase by combinatorial mutagenesis and selection: The importance of electrostatic catalysis. *Proc. Natl. Acad. Sci. USA* **93,** 5043–5048.
73. Moore, J. C., and Arnold, F. H. (1996). Directed evolution of a *para*-nitrobenzyl esterase for aqueous-organic solvents. *Nature Biotech.* **14,** 458–467.
74. Smith, G. P., and Scott, J. K. (1993). Libraries of peptides and proteins displayed on filamentous phage. *Meth. Enzymol.* **217,** 228–257.
75. Bradbury, A., and Cattaneo, A. (1995). The use of phage display in neurobiology. *Trends Neurosci.* **18,** 243–249.
76. O'Neil, K. T., and Hoess, R. H. (1995). Phage display: Protein engineering by directed evolution. *Curr. Op. Struc. Biol.* **5,** 443–449.
77. Wang, C-I., Yang, Q., and Craik, C. S. (1996). Phage display of proteases and macromolecular inhibitors. *Meth. Enzymol.* **267,** 52–68.
78. McCafferty, J., Jackson, R. H., and Chiswell, D. J. (1991). Phage-enzymes: Expression and affinity chromatography of functional alkaline phosphatase on the surface of bacteriophage. *Protein Engineering* **4,** 955–961.
79. Corey, D. R., Shiau, A. K., Yang, Q., Janowski, B. A., and Craik, C. S. (1993). Trypsin display on the surface of bacteriophage. *Gene* **128,** 129–134.
80. Ku, J., and Schultz, P. G. (1994). Phage display of catalytically active staphylococcal nuclease. *Biorg. & Med. Chem.* **2,** 1413–1415.
81. Light, J., and Lerner, R. A. (1995). Random mutagenesis of staphylococcal nuclease and phage display selection. *Bioorg. & Med. Chem.* **3,** 955–967.
82. Widerstein, M., and Mannervik (1995). Glutathione transferases with novel active sites isolated by phage display from a library of random mutants. *J. Mol. Biol.* **250,** 115–122.
83. Soumillion, P., Jespers, L., Bouchet, M., Marchand-Brynaert, J., Winter, G., and Fastrex, J. (1994). Selection of β-lactamase on filamentous bacteriophage by catalytic activity. *J. Mol. Biol.* **237,** 415–422.
84. Roberts, B. L., Markland, W., Ley, A. C., Kent, R. B., White, D. W., Guterman, S. K., and Ladner, R. C. (1992). Directed evolution of a protein: Selection of potent neutrophil elastase inhibitors displayed on M13 fusion phage. *Proc. Natl. Acad. Sci. USA* **89,** 2429–2433.
85. Ku, J., and Schultz, P. G. (1995). Alternate protein frameworks for molecular recognition. *Proc. Natl. Acad. Sci. USA* **92,** 6552–6556.
86. Lowman, H. B., Bass, S. B., Simpson, N., and Wells, J. A. (1991). Selecting high-affinity binding proteins by monovalent phage display. *Biochemistry* **30,** 10832–10838.
87. Marks, J. D., Hoogenboom, H. R., Bonnert, T. P., McCafferty, J. R., Griffiths, A. D., and Winter, G. (1991). By-passing immunization. *J. Mol. Biol.* **222,** 581–597.
88. Vanwetswinkel, S., Marchand-Brynaert, J., and Fastrez, J. (1996). Selection of the most active enzymes from a mixture of phage-displayed β-lactamase mutants. *Bioorg. Med. Chem. Lett.* **6,** 789–792.
89. Nilsson, B. (1995). Antibody engineering. *Curr. Op. Struc. Biol.* **5,** 450–456.
90. Barbas, S. H., Ditzel, H. J., Salonen, E. M., Yang, W.-P., Silverman, G. J., and Burton, D. R. (1995). Human autoantibody recognition of DNA. *Proc. Natl. Acad. Sci. USA* **92,** 2529–2533.
91. Bradbury, A., Persic, L., Were, T., and Cattaneo, A. (1993). Use of living columns to select specific phage antibodies. *Biotechnology* **11,** 1565–1569.
92. Barry, M. A., Dower, W. J., and Johnston, S. A. (1996). Toward cell-targeting gene therapy vectors: Selection of cell binding peptides from random peptide-presenting phage libraries. *Nature Med.* **2,** 299–305.
93. Pasqualini, R., and Ruoslahti, E. (1996). Organ targeting *in vivo* using phage display libraries. *Nature* **380,** 364–366.
94. Chien, C. T., Bartel, P. L., Sternglatz, R., and Fields, S. (1989). The two-hybrid system: A method to identify and clone genes for proteins that interact with a protein of interest. *Proc. Natl. Acad. Sci. USA* **88,** 9578–9582.
95. Bartel, P. L., and Fields, S. (1995). Analyzing protein–protein interactions using two hybrid system. *Meth. Enzymol.* **245,** 241–263.
96. Yang, M., Zining, W., and Fields, S. (1995). Protein–peptide interactions analyzed with the yeast two hybrid system. *Nucl. Acids. Res.* **23,** 1152–1156.

97. Crameri, A., Whitehorn, E. A., Tate, E., and Stemmer, W. P. C. (1996). Improved green fluorescent protein by molecular evolution using DNA shuffling. *Nature Biotechnology* **14,** 315–319.
98. Stemmer, W. P. C. (1994). Rapid evolution of a protein *in vitro* by DNA shuffling. *Nature* **370,** 389–391.
99. Stemmer, W. P. C. (1994). DNA shuffling by random fragmentation and reassembly: *In vitro* recombination for molecular evolution. *Proc. Natl. Acad. Sci. USA* **91,** 10747–10751.
100. Pollack, S. J., Jacobs, J. W., and Schultz, P. G. (1986). Selective chemical catalysis by an antibody. *Science* **234,** 1570–1573.
101. Tramantano, A., Janda, K., and Lerner, R. (1986). Catalytic antibodies. *Science* **234,** 1566–1570.
102. Lerner, R. A., Benkovic, S. J., and Schultz, P. G. (1991). At the crossroads of chemistry and immunology: Catalytic antibodies. *Science* **252,** 659–667.
103. Schultz, P. G., and Lerner, R. A. (1995). From molecular diversity to catalysis: Lessons from the immune system. *Science* **269,** 1835–1842.
104. Zhou, G. W., Guo, J., Huang, W., Fletterick, R., and Scanlan, T. S. (1994). Crystal structure of a catalytic antibody with a serine protease active site. *Science* **265,** 1059–1064.
105. Ulrich, H. D., Patten, P. A., Yang, P. L., Romesberg, F. E., and Schultz, P. G. (1995). "Expression Studies of Catalytic Antibodies." *Proc. Natl. Acad. Sci. USA* **92,** 11907–11911.
106. Patten, P. A., Gray, N. S., Yang, P. L., Marks, C. B., Wedemayer, G. J., Boniface, J. J., Stevens, R. C., and Schultz, P. G. (1996). The immunological evolution of catalysis. *Science* **271,** 1086–1092.
107. Ward, H. J., Tims, D., and Ferscht, A. R. (1990). Protein engineering and the study of structure–function relationships in receptors. *Trends Pharmacol. Sci.* **11,** 280–284.
108. Wedel, B., Harteneck, C., Foerster, J., Friebe, A., Schultz, G., and Koesling, D. (1995). Functional domains of soluble guanylyl cyclase. *J. Biol. Chem.* **270,** 24871–24875.
109. Camacho, P., and Lechleiter, J. D. (1995). Calreticulin inhibits repetitive intracellular Ca^{2+} waves. *Cell* **82,** 765–777.
110. Sheffield, W. P., and Blajchman, M. A. (1995). Deletion mutagenesis of heparin cofactor II: Defining the minimum size of a thrombin inhibiting serpin. *FEBS Lett.* **365,** 189–192.
111. Klebba, P. E., Hofnung, M., and Charbit, A. (1994). A model of maltodextrin transport through the sugar specific porin, LamB, based on deletion analysis. *Embo J.* **13,** 4670–4675.
112. Tang, W.-J., Stanzel, M., and Gilman, A. G. (1995). "Truncation and alanine-scanning mutants of type I adenylyl cyclase." *Biochemistry* **34,** 14563–14572.
113. Ghosh, S., and Bell, R. M. (1994). Identification of discrete segments of human Raf-1 kinase critical to high affinity binding to Ha-Ras. *J. Biol. Chem.* **269,** 30785–30788.
114. Bass, S. H., Mulkerrin, M. G., Wells, J. A. (1991). A systematic mutational analysis of hormone binding determinants in the human growth hormone receptor. *Proc. Natl. Acad. Sci. USA* **88,** 4498–4502.
115. Cunningham, B. C., and Wells, J. A. (1989). High resolution epitope mapping of hGH–receptor interactions by alanine scanning mutagenesis. *Science* **244,** 1081–1085.
116. Wells, J. A. (1996). Binding in the growth hormone receptor complex. *Proc. Natl. Acad. Sci. USA* **93,** 1–6.
117. Cunningham, B. C., Jhurani, P., Ng, P., and Wells, J. A. (1989). Receptor and antibody epitopes in human growth hormone identified by homolog-scanning mutagenesis. *Science* **243,** 1330–1336.
118. Ostermann, A. L., Kinch, L. N., Grishin, N. V., and Phillips, M. A. (1995). Acidic residues important for substrate binding and cofactor reactivity in eukaryotic ornithine decarboxylase identified by alanine scanning mutagenesis. *J. Biol. Chem.* **270,** 11797–11802.
119. Tang, H., Sun, X., Reinberg, D., and Ebright, R. H. (1996). Protein–protein interactions in eukaryotic transcription initiation: Structure of the preinitiation complex. *Proc. Natl. Acad. Sci. USA* **93,** 1119–1124.
120. Nikolov, D. B., Chen, H., Halay, E. D., Usheva, A. A., Hisatake, K., Kun Lee, D., Roeder, R. G., and Burley, S. K. (1995). Crystal structure of a TFIIB-TBP-TATA-element ternary complex. *Nature* **377,** 119–128.
121. Ranganathan, R., Lewis, J. H., and MacKinnon, R. (1996). Spatial localization of the K^+ channel selectivity filter by mutant cycle-based structure analysis. *Neuron* **16,** 131–136.
122. Ellman, J., Mendel, D., Anthony-Cahill, S., Noren, C. J., and Schultz, P. G. (1993). Biosynthetic method for introducing unnatural amino acids site-specifically into proteins. *Meth. Enzymol.* **202,** 301–336.

123. Cornish, V. W., and Schultz, P. G. (1994). A new tool for studying protein structure and function. *Curr. Op. Struc. Biol.* **4,** 601–607.
124. Mendel, D., Cornish, V. W., and Schultz, P. G. (1995). Site-directed mutagenesis with an expanded genetic code. *Annu. Rev. Biophys. Biomol. Struc.* **24,** 435–462.
125. Noren, C. J., Anthony-Cahill, S. J., Griffith, M. C., and Schultz, P. G. (1989). A general method for site-specific incorporation of unnatural amino acids into proteins. *Science* **244,** 182–188.
126. Chung, H. H., Benson, D. R., and Schultz, P. G. (1993). Probing the structure and mechanism of ras protein with an expanded genetic code. *Science* **259,** 806–809.
127. Mendel, D., Ellman, J. A., Chang, Z., Veenstra, D. L., Kollman, P. A., and Schultz, P. G. (1992). Probing protein stability with unnatural amino acids. *Science* **256,** 1798–1802.
128. Judice, J. K., Gamble, T. R., Murphy, E. C., de Vos, A. M., and Schultz, P. G. (1993). Probing the mechanism of staphylococcal nuclease with unnatural amino acids: Kinetic and structural studies. *Science* **261,** 1578–1581.
129. Nowak, M. K., Kearney, P. C., Sampson, J. R., Saks, M. E., Labarca, C. G., Silverman, S. K., Zhong, W., Thorson, J., Abelson, J. N., Davidson, N., Schultz, P. G., Dougherty, D. A., and Lester, H. A. (1995). Nicotinic receptor binding site probed with unnatural amino acid incorporation in intact cells. *Science* **268,** 439–442.
130. Beiboer, S. H. W., van den Berg, B., Dekker, N., Cox, R. C., and Verheij, H. M. (1996). Incorporation of an unnatural amino acid in the active site of porcine pancreatic phospholipase A_2. *Protein Engineering* **9,** 345–352.
131. Wallace, C. J. A. (1995). Peptide ligation and semisynthesis. *Curr. Op. Biotech.* **6,** 403–410.
132. Abrahmsen, L., Tom, J., Butcher, K. A., Kossiakoff, A., and Wells, J. A. (1991). Engineering subtilisin and its substrates for efficient ligation of peptide bonds in aqueous solution. *Biochemistry* **30,** 4151–4159.
133. Jackson, D. Y., Burnier, J., Quan, C., Stanley, M., Torn, J., and Wells, J. A. (1994). A designed peptide ligase for total synthesis of ribonuclease A with unnatural catalytic residues. *Science* **266,** 243–247.
134. Chang, T. K., Jackson, D. Y., Burnier, J. P., and Wells, J. A. (1994). Subtiligase: A tool for semisynthesis of proteins. *Proc. Natl. Acad. Sci. USA* **91,** 12544–12548.
135. Baca, M., and Alewood, P. F., and Kent, S. B. (1993). Structural engineering of the HIV-1 protease molecule with a beta-turn mimic of fixed geometry. *Prot. Sci.* **2,** 1085–1091.
136. Dawson, P. E., Muir, T. W., Clark-Lewis, I., and Kent, S. B. (1992). Synthesis of proteins by native chemical ligation. *Science* **266,** 776–779.
137. Baca, M., and Kent, S. B. (1993). Catalytic Contribution of flap-substrate hydrogen bonds in HIV-1 protease explored by chemical synthesis. *Proc. Natl. Acad. Sci. USA* **90,** 11638–11642.
138. Schnolzer, M., and Kent, S. B. (1992). Constructing proteins by dovetailing unprotected synthetic peptides: Backbone engineered HIV protease. *Science* **256,** 221–225.
140. Craik, C. S., Roczniak, S., Largman, C., and Rutter, W. J. (1987). The catalytic role of the active site aspartic acid in serine protease. *Science* **237,** 909–912.
141. Sprang, S., Standing, T., Fletterick, R. J., Stroud, R. M., Finer-Moore, J., Xuong, N.-H., Hamlin, R., Rutter, W. J., and Craik, C. S. (1987). Three-dimensional structure of Asn102 mutant of trypsin: Role of Asp102 in serine protease catalysis. *Science* **237,** 905–909.
142. Carter, P., and Wells, J. A. (1988). Dissecting the catalytic triad of a serine protease. *Nature* **332,** 564–568.
143. Bryan, P., Pantoliano, M. W., Quill, S. G., Hsiao, H.-Y., and Poulos, T. (1986). Site-directed mutagenesis and the role of the oxyanion hole in subtilisin. *Proc. Natl. Acad. Sci. USA* **83,** 3743–3745.
144. Corey, D. R., McGrath, M. E., Vasquez, J. R., Fletterick, R. J., and Craik, C. S. (1992). An alternate geometry for the catalytic triad of serine proteases. *J. Am. Chem. Soc.* **114,** 4905–4907.
145. Perona, J. J., and Craik, C. S. (1995). Structural basis of substrate specificity of the serine proteases. *Protein Science* **4,** 337–360.
146. Graf, L., Craik, C. S., Patthy, A., Roczniak, S., Fletterick, R. J., and Rutter, W. J. (1987). Selective alteration of substrate specificity by replacement of aspartic acid-189 with lysine in the binding pocket of trypsin. *Biochemistry* **26,** 2616–2622.
147. Sprang, S. R., Fletterick, R. J., Graf, L., Rutter, W. J., and Craik, C. S. (1988). Studies of specificity and catalysis in trypsin by structural analysis of site-directed mutants. *CRC Crit. Rev. Biotech.* **8,** 225–236.
148. Cronin, C. N., Malcolm, B. A., and Kirsch, J. F. (1987). Reversal of substrate charge specificity by site-directed mutagenesis of aspartate aminotransferase. *J. Am. Chem. Soc.* **109,** 2222–2223.

149. Hwang, J-K., and Warshel, A. (1988). Why ion pair reversal by protein engineering is unlikely to succeed. *Nature* **334,** 270–272.
150. Hedstrom, L., Szilagyi, L., and Rutter, W. J. (1992). Converting trypsin to chymotrypsin: The role of surface loops. *Science* **255,** 1249–1253.
151. Hedstrom, L., Perona, J. J., and Rutter, W. J. (1994). Converting trypsin to chymotrypsin: Residue 172 is a substrate specificity determinant. *Biochemistry* **33,** 8757–8763.
152. Venekei, I., Szilagyi, L., Graf, L., and Rutter, W. J. (1996). Attempts to convert chymotrypsin into trypsin. *FEBS Lett.* **379,** 143–147.
153. Posner, B. A., Li, L., Bethell, R., Tsuji, T., and Benkovic, S. J. (1996). Engineering specificity for folate into dihydrofolate reductase from *Escherichia coli. Biochemistry* **35,** 1653–1663.
154. Estell, D. A., Graycar, T. P., Miller, J. V., Powers, D. B., Burnier, J. P., Ng, P. G., and Wells, J. A. (1986). Probing steric and hydrophobic effects on enzyme–substrate interactions by protein engineering. *Science* **233,** 659–663.
155. Bone, R., Silen, J. L., and Agard, D. A. (1989). Structural plasticity broadens the specificity of an engineered protease. *Science* **339,** 191–195.
156. Rheinnecker, M., Baker, G., Eder, J., and Fersht, A. R. (1993). Engineering a novel specificity in subtilisin BPN′. *Biochemistry* **32,** 1199–1203.
157. Ballinger, M. C., Tom, J., Wells, J. A. (1995). Designing subtilisin BPN′ to cleave substrates containing dibasic residues. *Biochemistry* **34,** 13312–13319.
158. Matthews, B. W., Craik, C. S., and Neurath, H. (1994). Can small cyclic peptides have the activity and specificity of proteolytic enzymes? *Proc. Natl. Acad. Sci. USA* **91,** 4103–4105.
159. Mathews, D. J. (1995). Interfacial metal binding site design. *Curr. Op. Biotech.* **6,** 419–424.
160. Corey, D. R., and Schultz, P. G. (1989). Introduction of a Metal-Dependent Regulatory Switch into an Enzyme. *J. Biol. Chem.* **264,** 3666–3669.
161. Higaki, J. N., Haymore, B. L., Chen, S., Fletterick, R. J., and Craik, C. S. (1990). Regulation of serine protease activity by an engineered metal switch. *Biochemistry* **29,** 8582–8586.
162. McGrath, M. E., Haymore, B. L., Summers, N. L., Craik, C. S., and Fletterick, R. J. (1993). Structure of an engineered, metal activated switch in trypsin. *Biochemistry* **32,** 1914–1919.
163. Halfon, S., and Craik, C. S. (1996). Regulation of proteolytic activity by engineered tridentate metal binding loops. *J. Am. Chem. Soc.* **118,** 1227–1228.
164. Iverson, B. L., Iverson, S. A., Roberts, V. A., Getzoff, E. D., Tainer, J. A., Benkovic, S. J., and Lerner, R. A. (1990). Metalloantibodies. *Science* **249,** 659–662.
165. Willett, W. S., Gillmor, S. A., Perona, J.J., Fletterick, R. J., and Craik, C. S. (1995). Engineered metal regulation of trypsin specificity. *Biochemistry* **34,** 2172–2180.
166. Willett, W. S., Brinen, L. S., Fletterick, R. J., and Craik, C. S. (1996). Delocalizing trypsin specificity with metal activation. *Biochemistry* **35,** 5992–5998.
167. Brinen, L. S., Willett, W. S., Craik, C. S., and Fletterick, R. J. (1996). X-ray structures of a designed binding site in trypsin show metal-dependent geometry. *Biochemistry* **35,** 5999–6009.
168. Cuenuod, B., and Schepartz, A. (1993). Altered specificity of DNA-binding proteins with transition metal dimerization domains. *Science* **259,** 510–513.
169. Sloane, D. L., Leung, R., Craik, C. S., and Sigal, E. (1991). A primary determinant for lipoxygenase positional specificity. *Nature* **354,** 149–152.
170. Boyington, J. C., Gaffney, B. J., and Amzel, L. M. (1993). The three-dimensional structure of an arachidonic acid 15-lipoxygenase. *Science* **260,** 1482–1486.
171. Nelson, M. J., and Seitz, S. P. (1994). The structure and function of lipoxygenase. *Curr. Op. Struc. Biol.* **4,** 878–884.
172. Carter, P., and Wells, J. A. (1987). Engineering enzyme specificity by "substrate-assisted catalysis." *Science* **237,** 394–399.
173. Matthews, D. J., and Wells, J. A. (1993). Substrate phage: Selection of protease substrate by monovalent phage display. *Science* **260,** 1113–1116.
174. Carter, P., Abrahmsen, L., and Wells, J. A. (1991). Probing the mechanism and improving the rate of substrate-assisted catalysis in subtilisin BPN′. *Biochemistry* **30,** 6142–6148.
175. Corey, D. R., Willett, W. S., Coombs, G. S., and Craik, C. S. (1995). Trypsin specificity increased through substrate-assisted catalysis. *Biochemistry* **34,** 11521–11527.
176. Choo, Y., and Klug, A. (1995). Designing DNA-binding proteins on the surface of filamentous phage. *Curr. Op. Biotech.* **6,** 431–436.
177. Desjarlais, J. R., and Berg, J. M. (1993). Use of a zinc-finger consensus framework and specificity rules to design specific DNA binding proteins. *Proc. Natl. Acad. Sci. USA* **90,** 2256–2260.

178. Rebar, E. J., and Pabo, C. O. (1994). Zinc finger phage: Affinity selection of fingers with new DNA-binding specificities. *Science* **263,** 671–673.
179. Choo, Y., Sanchez-Garcia, I., and Klug, A. (1994). *In vivo* repression by a site-specific DNA-binding protein designed against an oncogenic sequence. *Nature* **372,** 642–645.
180. Pomerantz, J. L., Sharp, P. A., and Pabo, C. O. (1995). Structure-based design of transcription factors. *Science* **267,** 93–96.
181. Wu, H., Yang, W.-P., and Barbas, C. F. (1995). Building zinc fingers by selection: Toward a therapeutic application. *Proc. Natl. Acad. Sci. USA* **92,** 344–348.
182. Kim, Y-G., Cha, J., and Chandrasegaran, S. (1996). Hybrid restriction enzymes: Zinc finger fusions to *FokI* cleavage domain. *Proc. Natl. Acad. Sci. USA* **93,** 1156–1160.
183. Hoylaerts, M., Rijken, D. C., Lijnen, H. R., and Collen, D. (1982). Kinetics of the activation of plasminogen by human tissue plasminogen activator. *J. Biol. Chem.* **257,** 2912–2919.
184. Matsuo, O., Rijken, D. C., and Collen, D. (1981). Comparison of the relative fibrinogenolytic, fibrinolytic, and thrombolytic properties of tissue plasminogen activator and urokinase *in vitro*. *Thromb. Haemost.* **45,** 225–229.
185. Mark, D. B. et al. (1995). Cost effectiveness of thrombolytic therapy with tissue-type plasminogen activator as compared with streptokinase for acute myocardial infarction. *N. Engl. J. Med.* **332,** 1418–1424.
186. Keyt, B. A., Paoni, N. F., Bennett, W. F. (1996). Site-directed mutagenesis of tissue-type plasminogen activator. *In* "Protein Engineering: Principles and Practice" (1996) (Cleland, J. L., and Craik, C. S., eds.). Wiley-Liss, New York, pp. 435–466.
187. Larsen, G. R., Metzger, M., Henson, K., Blue, Y., and Horgan, P. (1989). Pharmacokinetic and distribution analysis of variant forms of tissue-type plasminogen activator with prolonged clearance in rat. *Blood* **73,** 1842–1850.
188. Browne, M. J., Carey, J. E., Chapman, C. G., Tyrrell, A. W. R., Entwisle, C., Lawrence, G. M. P., Reavy, B., Dodd, I., Esmail, A., and Robinson, J. H. (1988). A tissue-type plasminogen activator mutant with prolonged clearance *in vivo*. *J. Biol. Chem.* **263,** 1599–1602.
189. Browne, M. J., Chapman, C. G., Dodd, I., Esmail, A. F., and Robinson, J. H. (1990). Deletion of a tripeptide sequence from the growth-factor domain of tissue-type plasminogen activator prolongs *in vivo* circulation. *Thromb. Res.* **59,** 687–692.
190. Johannessen, M., Diness, V., Pingel, K., Petersen, L. C., Rao, D., Lioubin, P., O'Hara, P., and Mulvihill, E. (1990). Fibrin affinity and clearance of t-PA deletion and substitution analogues. *Thromb. Haemost.* **63,** 54–59.
191. Kohnert, U., Rainer, R., Verheijen, J. H., Weenig-Verhoeff, E. J. D., Stern, A., Opitz, U., Martin, U., Lill, H., Prinz, H., Lechner, M., Kresse, G.-B., Buckel, P., and Fischer, S. (1992). Biochemical properties of the kringle 2 and protease domains are maintained in the refolded t-PA deletion variant BM 06.022. *Prot. Engineering* **5,** 93–100.
192. Bennett, W. F. (1983). Two forms of tissue-type plasminogen activator differ at a single, specific glycosylation site. *Thromb. Haemost.* **50,** 106.
193. Pohl, G., Kallstrom, M., Bergsdorf, N., Wallen, P., and Jornvall, H. (1984). Tissue plasminogen activator: Peptide analyses confirm an indirectly derived amino acid sequence, identify the active site serine residue, establish glycosylation sites, and localize variant differences. *Biochemistry* **23,** 3701–3707.
194. Little, S. P., Bang, N. U., Harms, C. S., Marks, C. A., and Mattler, L. E. (1984). Functional properties of carbohydrate-depleted tissue plasminogen activator. *Biochemistry* **23,** 6191–6195.
195. Hotchkiss, A., Refino, C. J., Leonard, C. K., O'Connor, J. V., Crowley, C., McCabe, J., Tate, K., Nakamura, G., Powers, D., Levinson, A., Mohler, M., and Spellman, M. W. (1988). The influence of carbohydrate structure on the clearance of recombinant tissue-type plasminogen activator. *Thromb. Haemost.* **60,** 255–261.
196. Madison, E. L., Goldsmith, E. J., Gerard, R. D., Gething, M.-J., and Sambrook, J. F. (1989). Serpin resistant mutants of human tissue-type plasminogen activator. *Nature,* **339,** 721–723.
197. Madison, E. L., Goldsmith, E. J., Gerard, R. D., Gething, M.-J. H., Sambrook, J. F., and Bassel-Duby, R. S. (1990). *Proc. Natl. Acad. Sci. USA* **87,** 3530–3533.
198. Madison, E. L., Goldsmith, E. J., Gething, M-J. H., Sambrook, J. F., and Gerard, R. D. (1990). Restoration of serine protease inhibitor interaction by protein engineering. *J. Biol. Chem.* **265,** 21423–21426.
199. Bennet, W. F., Paoni, N. F., Keyt, B. A., Botstein, D., Jones, A. J. S., Presta, L., Wurm, F. M.,

and Zoller, M. J. (1991). High resolution analysis of functional determinants on human tissue-type plasminogen activator. *J. Biol. Chem.* **266,** 5191–5201.

200. Madison, E. L., Kobe, A., Gething, M.-J., Sambrook, J. F., and Goldsmith, E. J. (1993). Converting tissue plasminogen activator to a zymogen: A regulatory triad of Asp His-Ser. *Science* **262,** 419–421.
201. Strandberg, L., and Madison, E. L. (1995). Variants of tissue-type plasminogen activator with substantially enhanced response and selectivity toward fibrin co-factors. *J. Biol. Chem.* **270,** 23444–23449.
202. Ke, S., Tachias, K., Lamba, D., Bode, W., and Madison, E. L. (1997). Identification of a critical site on tissue type plasminogen activator that modulates specificity for plasminogen. *J. Biol. Chem.,* **272,** 1811–1816.
203. Li, B., Tom, J. Y. K., Oare, D., Yen, R., Fairbrother, W. J., Wells, J. A., and Cunningham, B. C. (1995). Minimization of a polypeptide hormone. *Science* **270,** 1657–1660.
204. Lowman, H. B., and Wells, J. A. (1991). Monovalent phage display: A method for selecting variant proteins from random libraries. *Methods* **3,** 205–216.

Chapter 5

De Novo Design of Protein Structure and Function

Michael W. Klemba,*‡ Mary Munson,†‡ and Lynne Regan

I. Introduction and Overview

In recent years there has been much interest in the design of sequences that can fold into specified structures. A major impetus for such efforts is that they address the protein folding problem from a novel perspective. Rather than attempting to predict a tertiary fold of a protein from the amino acid sequence, design approaches aim to identify sequences that are compatible with a given fold. In the course of these studies, it has become apparent that the design of sequences that can reproduce the overall fold of a protein is easier than one might have predicted. What has proven more difficult is the design of sequences that reproduce all the physical properties of a natural protein having a unique

* *Present address:* Department of Microbiology, GBF-National Research Center for Biotechnology, Mascheroder Weg1, D-38124 Braunschweig, Germany.
† *Present address:* Department of Molecular Biology, Princeton University, Princeton, NJ.
‡ Department of Molecular Biophysics and Biochemistry, Yale University, New Haven, CT.
Both authors contributed equally to this work.

and densely packed hydrophobic core. In the first half of this chapter we discuss several specific examples of designed proteins and compare the extent to which they have been characterized both structurally and thermodynamically. In the second half of this chapter we discuss some even bolder designs: those of proteins with novel functions. In this area there has been a focus on the design of novel metal-binding sites because these are well defined, display a range of functions *in vivo*, and may be relatively easy to characterize. Finally, we end with a brief discussion of the success and limitations of the design field to date and consider future directions.

II. *De Novo* Design of Structure

The ultimate goal of protein design is the creation of new proteins with predetermined structures and functions. Progress towards this goal continues apace as the principles underlying the structures of natural proteins become better understood. These principles can be used to formulate guidelines for the design of new structures. The guidelines can be viewed in a modular fashion: first, individual secondary structural elements are designed, α helices or β sheets; the secondary structural elements are then linked together with turns or loops; and finally, the elements of secondary structure are packed against each other to form a stable, tertiary structure. Superimposed upon this scheme is the requirement to sequester hydrophobic residues from solvent and to position polar and charged residues on the surface to enhance solubility and prevent aggregation. This requirement leads to what has been referred to as a "binary patterning" of hydrophobic and hydrophilic residues. This binary patterning is found in natural proteins and has been exploited in minimalist protein design efforts (Kamtekar *et al.*, 1993; Kamtekar and Hecht, 1995). Generally, however, protein design does not stop with binary patterning. Proteins are frequently "overdesigned" for thermodynamic stability, with as many potentially stabilizing interactions as possible incorporated. In addition, residues are sometimes included into a design with the specific aim of destabilizing an alternative, unwanted structure. This aspect of the design process has been referred to as "negative design." Section II,A,1–4 discusses these general guidelines for designing secondary structural elements, turn and loop connections, and hydrophobic core packing. Many of the techniques used to compare the structure, stability and native-like properties of designed proteins to natural proteins are described in Section III. Finally, Section IV discusses both specific approaches to and the results of selected designs.

A. Secondary Structural Elements and Their Connections

1. α *Helices*

Experimental studies of monomeric helix formation and statistical surveys of proteins of known structure have provided a wealth of information that facilitates the design of helices and helical proteins. This information, combined with the regular structure of individual α-helices and the ease of producing stable and soluble helices, makes them the most straightforward secondary structural elements to design. In natural proteins α helices are often amphiphilic:

Polar and charged residues are found on the solvent-exposed face of the protein, and hydrophobic residues are present on the other, buried face. In addition to this binary patterning, a major consideration in the design of stable helices is the intrinsic helical propensities of the amino acids. The comparison of helical propensities derived from both statistical and experimental studies has been reviewed (Bryson *et al.*, 1995; Chakrabartty and Baldwin, 1995). In general, good residues for the hydrophilic surface of helices are arginine, lysine, glutamic acid, and glutamine because they have high helical propensities and allow the incorporation of positive and negative charges and polar residues to promote protein solubility in aqueous solutions. Hydrophobic residues with good helix propensities include alanine, leucine and methionine. Residues to be avoided in the design of stable α helices are Pro and Gly and β-branched residues, such as Val and Thr.

Additional specific residues can be placed in the designed helix to provide stability through capping of the N or C termini, ion-pairing, or interacting favorably with the helix dipole (Bryson *et al.*, 1995; Chakrabartty and Baldwin, 1995; Doig and Baldwin, 1995; Fersht and Serrano, 1993). Asparagine, aspartate, serine and threonine have the highest N-cap preferences, since they can accept a hydrogen bond from the free N terminus. At the C terminus, there is less preference among the amino acids, except that β-branched residues are disfavored and glycine is especially favored in natural proteins. Ion-pairing interactions have been added to designs by using pairs of charged residues, such as glutamine and lysine, at the i and $i + 4$ positions on the surface of an α helix.

2. *β Sheets*

Studies of β-sheet formation and stability have been hampered by several problems. First, unlike the regular α helix, β sheets form more flexible, often quite twisted structures. Also, the side-to-side interactions between stands of β-sheet proteins have tended to cause many model systems to aggregate, making structural and thermodynamic characterization difficult. These difficulties have been overcome in a number of experimental studies reviewed in Hecht (1994), Regan (1994), and Smith and Regan (1997) and discussed next.

β Sheets in natural proteins often show a binary patterning of hydrophobic and hydrophilic residues (West and Hecht, 1993). Because every second residue in a β sheet faces in the same direction, the hydrophobic and hydrophilic residues alternate to form hydrophobic and hydrophilic faces. Propensities for individual residues in both the central and edge strands of β sheets, as well as pairwise interactions between residues in β sheets, have now been studied experimentally (Hecht, 1994; Regan, 1994; Smith and Regan, 1997). These studies show that large hydrophobic residues and β-branched residues, such as tyrosine, threonine, isoleucine, phenylalanine, tryptophan, and valine, have high β-sheet propensities even at solvent-exposed positions. In addition, these residues, when paired appropriately on adjacent β strands, can increase the stability of β sheets beyond the sum of the individual propensities. Many of the charged and polar residues, such as aspartate, glutamate, lysine, and glutamine, have low β-sheet forming propensities. This presents a potential problem for the design of β sheets that are both stable and soluble; however, pairwise studies show that oppositely charged side chains, when present in the cross-strand positions, have a significant interaction energy that contributes to β-sheet stability beyond their intrinsic propensities. β Strands usually begin and end in turns, which are discussed in detail next.

3. *Loops and Turns*

Elements of secondary structure are connected with loops and turns. Generally, "turns" refer to short connections having specific backbone dihedral angles. Other connections, having less well-defined structures and lengths, are referred to as "loops." The composition and positional preferences of turns have been well documented (Hutchinson and Thornton, 1994; Rose *et al.*, 1985). For loops, the sequence requirements are much less severe. The difference in tolerance for amino acid substitutions between turns and loops is well illustrated by two examples. When the type II turn of poplar plastocyanin was replaced with random sequences, most (94%) of the mutant proteins no longer bound copper (Ybe and Hecht, 1996). By comparison, when a loop in cytochrome *b*-562 was replaced with random sequences, most (>91%) of the mutant proteins retained functionality, as assessed by their retention of the ability to bind heme (Brunet *et al.*, 1993). More detailed thermodynamic studies of variants of the turns and loops in natural proteins indicate that the most commonly observed sequences for a particular turn or loop are indeed those that give rise to the most stable proteins.

Many designs have connected elements of secondary structure with loop sequences defined by their low intrinsic propensity for α-helix and β-sheet formation, leading to loop sequences that are typically glycine- or serine-rich. The most important consideration when designing loops is to optimize their length. In general, a shorter loop leads to a more stable protein, because when the protein folds there is less entropic loss associated with constraining the ends of a short loop. A systematic study of loops with varying numbers of glycine residues in the four-helix-bundle protein Rop revealed an inverse correlation between the length of the loop and the stability of the protein (Nagi and Regan, 1997). The loop must not be too short, however, or undesired higher-order associations of the connected secondary structural elements may be energetically more favorable than the desired fold. There are two examples of designed four-helix-bundle proteins in which the loop chosen to connect two helices was too short, and higher-order associations were observed. In both cases, simply increasing the loop length circumvented this problem (Ho and DeGrado, 1987; Predki *et al.*, 1995).

4. *Hydrophobic Packing*

The tertiary fold of a protein is the relative spatial orientation of elements of secondary structure. This orientation may be influenced by the linear connection of the elements of secondary structure by turns or loops. In addition, the way in which elements of secondary structure pack against each other is of key importance. The hydrophobic cores of natural proteins are characterized by tight packing, with packing densities that are comparable to those of small organic molecules in crystals [for reviews, see Richards (1977) and Richards and Lim (1993)]. The difficulty in designing core packing arises from the tremendous complexity of assembling hydrophobic side chains into tightly packed arrays. Unfortunately for protein designers, no simple pairwise preferences or patterns of packing residues have been discerned from surveys of natural globular proteins (Behe *et al.*, 1991). Although exact complementarity of core residues is thought to be required to generate this tight packing, there is an apparent plasticity to the overall fold of natural proteins such that they can accommodate some variation in the size and shape of packing residues (Baldwin *et al.*, 1993; Eriksson *et al.*, 1992; Jackson *et al.*, 1993; Kamtekar *et al.*, 1993; Lesk and Chothia, 1980; Lim and Sauer, 1991; Lim *et al.*, 1992, Munson *et al.*, 1996, 1994; Sandberg and Terwilliger, 1991, 1989; Shortle *et al.*, 1990; Steif *et al.*, 1995). This plasticity comes with an energetic cost: Mutations that perturb core packing usually decrease the stability of the protein, and often induce structural changes that are reflected in a diminution of protein activity. Studies of variants of natural proteins with alternative core packings suggest limits on the amount that the total core volume can be increased or decreased before the native-like properties of the protein are abolished. For example, the hydrophobic core of λ repressor (total core volume of 549 $Å^3$) can tolerate approximately $\pm$1–2 methylene units of volume (1 methylene unit is approximately 20 $Å^3$) without losing function (Lim and Sauer, 1989, 1991). Proteins with core volume increases or decreases of more than 1–2 methylene units suffer partial or complete loss of function. In addition to demonstrating restrictions on allowed total core volumes, studies with natural proteins also emphasize the importance of the shape and position of the side chains within the core. Cores with the same amino acid composition, but with the residues present at different positions, show different stabilities and solution properties.

Because the cores of natural proteins can tolerate several different packing arrangements, it is likely that a number of different *de novo* core designs will be able to reproduce a given fold. There are some examples in which computational methodologies have begun to be used to address the complexity of this problem (Desjarlais and Handel, 1995; Ponder and Richards, 1987). Alternatively, in certain cases designs for the hydrophobic core can be tested experimentally in a systematic fashion to identify the important features of a successful design.

One such systematic mutagenic study was performed on the core of the natural four-helix bundle protein Rop (Fig. 1) (Munson *et al.*, 1994, 1996). The objective was to determine to what extent simple patterns of amino acids with varying sizes and shapes in the hydrophobic core would either reproduce or disrupt Rop's structure and native-like properties (analysis of these properties is discussed later). The amino acid residues present in the "a" and "d" positions of the typical helical wheel diagram (Fig. 2) make up the hydrophobic core of Rop. In the wild-type protein, the "a" positions are generally small or polar residues, such as alanine, and the "d" position residues are generally large, such as leucine. The "a" positions in one layer pack against the "d" positions in the neighboring layers, forming a small–large–small–large packing pattern along the length of the bundle. Repacked Rop mutants with all alanines substituted into the "a" positions and leucines in the "d" positions (called Ala_2Leu_2) reproduced the RNA binding property of the wild-type protein and had increased thermal stability. When the alanine/leucine pattern is completely alternated (Leu_2Ala_2), mutants are created that are native-like and stable, but have lost wild-type RNA binding activity. Mutants that have been repacked with alanine and methionine (Ala_2Met_2), or underpacked with all alanines (Ala_4) or with alanines and valines (Ala_2Val_2), have decreased stability and native-like attributes. Overpacking the core with all leucines (Leu_4) creates mutants with

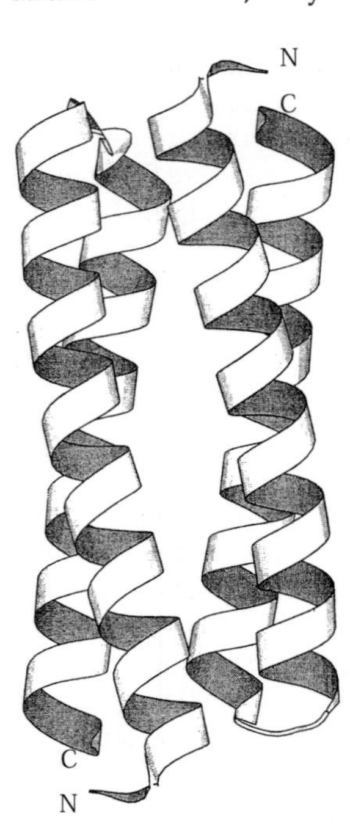

Fig. 1 Ribbon representation of the fold of Rop. N and C termini are indicated. (Munson M., O'Brien, R., Sturtevant, J. M., and Regan, L. (1994). Redesigning the hydrophobic core of a four-helix-bundle protein. *Protein Sci.* **3,** 2015–2022. Reprinted with the permission of Cambridge University Press.)

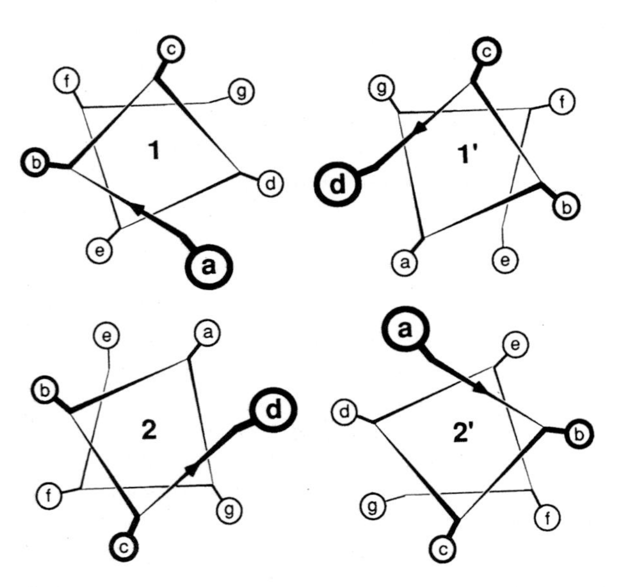

Fig. 2 Schematic illustration of two layers of Rop showing the locations of the a and d positions and the 4 helices (Munson M., O'Brien, R., Sturtevant, J. M., and Regan, L. (1994). Redesigning the hydrophobic core of a four-helix-bundle protein. *Protein Sci.* **3,** 2015–2022. Reprinted with the permission of Cambridge University Press.)

neither native-like nor wild-type-like structure. This work indicates that the use of alanines in the "a" helical positions, and leucine residues in the "d" positions (Fig. 2) in a four-helix-bundle protein will form stable and native-like structures. These results have been used to facilitate the design of other four-helix bundles (Dalal *et al.*, 1997).

III. Characterizing the Design

One aim of protein design is the production of stable proteins with specified secondary and tertiary structure and native-like core packing. This native-like packing is crucial for obtaining a well-structured protein lacking the fluctuating conformations that are seen in nonnative states such as the "molten globule" (Fink, 1995; Haynie and Freire, 1993; Kuwajima, 1989, 1996). The molten globule state is believed to exist as a transient intermediate in the folding pathway of many proteins. Under certain solution conditions, however—for example at low pH, low ionic strength, or in the absence of a stabilizing ligand—the molten globule can be the predominant species at equilibrium (Fink, 1995; Haynie and Freire, 1993). Characterization of designed proteins by several different techniques must be used to compare their solution behavior to that of both natural proteins (native-like) and molten globules (nonnative-like).

If the designed protein is expressed in *Escherichia coli*, resistance to protease digestion can provide an initial indication of folding and stability (Parsell and Sauer, 1989); however, this resistance to degradation is dependent on the sequence and structure of the protein, as well as on the strain of *E. coli* used for expression. Also, proteins with a tendency to aggregate, perhaps due to solvent exposure of hydrophobic residues or improper folding, are often found in inclusion bodies. In this form, overexpression reveals little about the structural integrity of the folded protein. In some cases, renaturation of the inclusion bodies can provide material for further analysis.

Circular dichroism (CD) is the most common technique used to estimate the amount and type of secondary structure (α helix vs β sheet vs. coil) present in a protein. Quantitative measurement of relative amounts of secondary structure is generally possible only when data are available to low wavelength (below 200 nm). An intense CD signal is not alone indicative of a native-like protein: Molten globules often display high levels of secondary structure. CD does, however, provide an initial means by which to assess whether a designed protein is folded at all: If the CD spectrum resembles that of a random coil, it is a clear indication that there are problems with the design!

Thermally induced denaturation transitions, monitored by CD or by differential scanning calorimetry (DSC), are good indicators of both stability and native-like properties. Natural proteins show cooperative denaturation transitions, with significant associated enthalpies of denaturation, whereas nonnative-like proteins show shallow noncooperative denaturation curves with little or no enthalpic component (Yutani *et al.*, 1992). Stability can also be characterized by following chemically induced denaturation transitions using CD and/or fluorimetry. The cooperativity of chemical denaturations is less sensitive than that of thermal denaturations; many proteins with nonnative-like thermal denaturations may still show cooperative chemically induced denaturations.

Perhaps the most powerful tool for establishing the presence of native-like structure in NMR spectroscopy (Alexandrescu *et al.*, 1993; Bai *et al.*, 1993, 1994; Baum *et al.*, 1989; Chyan *et al.*, 1993; Fink, 1995; Hughson *et al.*, 1990; Jeng *et al.*, 1990; Ohgushi and Wada, 1983) The one-dimensional (1D) ^{1}H NMR spectra of small natural proteins show substantial dispersion of aliphatic and aromatic proton chemical shifts away from random coil values. The lack of unique chemical environments in the hydrophobic core of nonnative-like proteins often results in poor dispersion of the aliphatic and aromatic proton chemical shifts. In addition, the rate of backbone amide proton exchange with the solvent is a useful indicator of native-like structure, providing a measure of both the stability and the conformational dynamics of the structure. Because nonnative-like structures are more mobile, less stable, and have a lower activation energy separating them from the denatured state, relatively rapid exchange rates for backbone amide protons are observed.

The affinity for the hydrophobic fluorescent dye 8-anilino-1-naphthalene sulfonate (ANS) can provide insights into the nature of the hydrophobic core in a designed protein. Binding of this dye has been used as an indicator of the molten globule state. Wild-type or fully denaturated proteins display low affinity for this dye, but it interacts with moderately high affinity with the loosely packed hydrophobic core of molten globules. This interaction can be detected by the increase in fluorescence intensity that occurs when the dye is transferred from an aqueous to a hydrophobic environment.

In addition to characterization of the solution thermodynamic properties of designed proteins, determination of their structures at high resolution is an obvious and important goal. To date, there are only a few designed proteins for which high-resolution structural characterizations have been achieved. It is not yet clear to what extent the difficulties that have been encountered are a consequence of the presence of nonnative-like features in the designed proteins. Conversely, crystallization of a protein is not necessarily evidence for native-like properties. It has been suggested that the crystallization process may lock a nonnative designed protein into a single conformation (Betz *et al.*, 1996), and there appear to be at least two examples of this phenomenon (Betz *et al.*, 1996; Schafmeister *et al.*, 1993).

IV. Selected Examples of Designed Proteins

In this section, specific examples of *de novo* protein design are discussed. The rational behind these designs and the characterization of the resulting proteins are outlined, and the successes and failures of the designs are evaluated.

A. All α-Helical Proteins

A number of natural proteins have two, three-, or four-associating helices as part or all of their structure. Several examples are the two-stranded parallel leucine zipper coiled coils, such as GCN4 (O'Shea *et al.*, 1991); the three-stranded coiled-coil hemagglutinin from influenza virus (Wilson *et al.*, 1981; Bullough *et al.*, 1994); and four-helix bundles as isolated folds or domains of larger protein, such as human growth hormone, T4 lysozyme, myohemerythrin, cytochrome *b*-562, and the RNA-binding protein Rop (Harris *et al.*, 1994; Kamtekar and Hecht, 1995; Paliakasis and Kokkinidis, 1992; Banner *et al.*, 1987). (See Fig. 1.) The helices in these proteins often form coiled-coil structures with "ridges into grooves" packing, with a crossing angle of 20° (Chothia *et al.*, 1981; Crick, 1953). The helices can be either parallel or antiparallel, and the connection between the helices can be formed by short, direct loops or by longer "overhanded" connections (Cohen and Parry, 1986, 1990). The four-helix-bundle structural motif is found in many proteins with diverse functions and is a favorite target of protein designers because of its relative simplicity and pseudo-4-fold symmetry [for a review, see Kamtekar and Hecht, (1995)].

The most basic multihelical design is that of the helix-loop-helix hairpin, examples of which have been designed by several groups (Fezoui *et al.*, 1995; Kuroda, 1995; Myszka and Chaiken, 1994; Olofsson *et al.*, 1995; Olofsson and Baltzer, 1996). See Fig. 3 and Table I. Residues for the helices were chosen after consideration of helical propensities, N- and C- capping preferences, stabilization of the helical dipole, and formation of salt bridges or ion pairs. The hydrophobic core packing residues, as well as the loop residues varied with each design. All of these designs gave rise to peptides which, to differing degrees, formed the correct secondary and tertiary structure, but which were not entirely native-like. Their properties are compared later.

The protein ALIN, designed and characterized by Kuroda and colleagues (Kuroda, 1995), was based on helices from the C-terminal helix of cytochrome *c* with several changes to facilitate NMR analyses, and one substitution of a hydrophobic residue that was expected to be solvent-exposed. A loop containing GTSGN was used to join the helices. Two cysteine residues were placed across the hairpin from each other, such that disulfide bond formation would provide a test of the antiparallel hairpin conformation. Disulfide bond formation did occur and plays a critical role in enhancing the stability of the design. ALIN was shown to be helical by CD, and monomeric by gel filtration chromatography. The ^{1}H NMR spectra were reasonably well dispersed, and amide exchange experiments suggested that regions of the protein near the disulfide bond may have native-like structure, because the observed exchange rates were fairly slow. The rest of the amide protons in the protein, however, exchange more rapidly, suggesting a degree of nonnative-like structure. This observation is also consistent with the protein's noncooperative thermal denaturation behavior and low associated enthalpy. The designed protein CCSL (coiled-coils stem loop) (Fig. 3; Table I) differs significantly from ALIN in the lengths of the helices and the turn, and the identity of the hydrophobic core residues (Myszka and Chaiken, 1994). The helices are each 25 residues long (10 residues more than ALIN) and the loop residues are GRGDMP. Only valine is present in the "a" helical positions and only leucine in the "d" positions (Fig. 3). The other residues in the helices were also not based on a sequence from a natural protein, but were designed *de novo*. The "e" and "g" helical positions contain glutamate and lysine, respectively, in order to form favorable ionic interactions. The remaining residues in the "b," "c," and "f" positions were chosen to be serine and alanine, to increase solubility and helical stability. This design also contains cysteine residues, which were used to demonstrate formation of an antiparallel structure. In this case, the cysteines were at the amino and carboxy termini of the protein. Disulfide bond formation occurs, although in this example it is

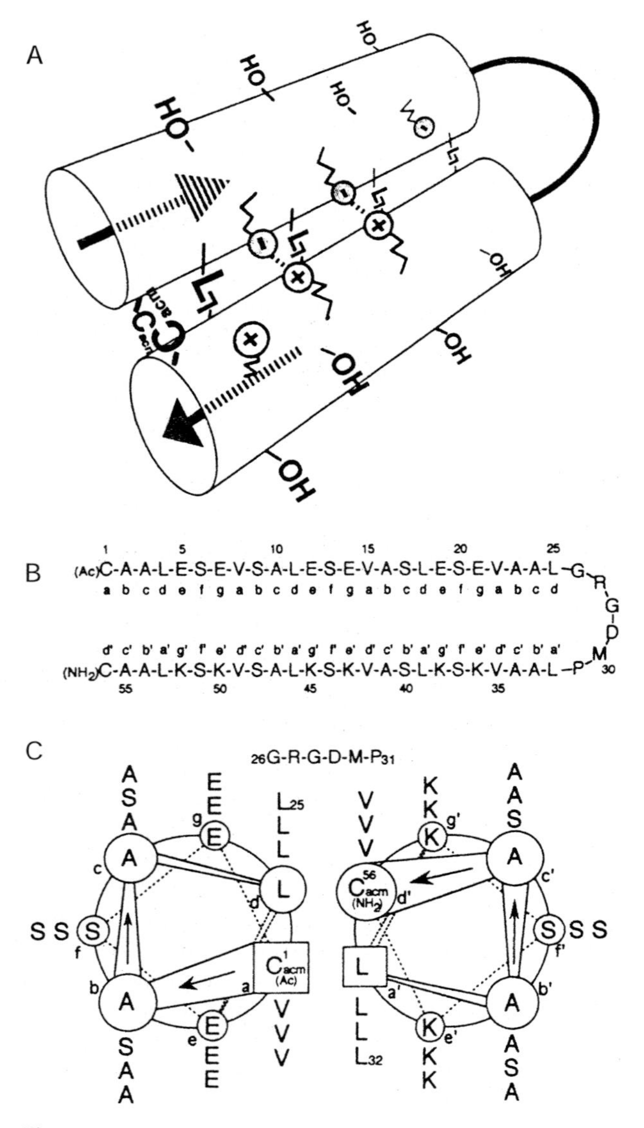

Fig. 3 (A) Proposed structure of an intramolecular coiled coil. (B) The 56-amino acid sequence of the CCSL peptide. (C) Helical wheel representation of the CCSL peptide. (Reprinted with permission from Myszka D., and Chaiken, I. (1994). Design and characterization of an intramolecular antiparallel coiled coil peptide. *Biochemistry* **33,** 2363–2372. Copyright 1994 American Chemical Society.)

not essential for peptide stability. The protein was shown to be helical by CD and monomeric by gel filtration chromatography and analytical ultracentrifugation, and it was reasonably cooperative and stability to both thermal and urea denaturation. This protein appears to show native-like behavior, but this conclusion would be strengthened by further characterization by NMR.

Osterhout and colleagues (Fezoui *et al.*, 1995) designed the protein "$\alpha t \alpha$" with a view to using it as a model system for studying protein folding. This protein has similar lengths of helices and turns to the ALIN protein (Table I). Like CCSL, the sequences were designed *de novo*. An important aim of this design was to use a sequence that was nonsymmetrical and nondegenerate to facilitate assignment and structure determination using NMR. Specific packing interactions were chosen based on those known to stabilize coiled coils. For the surface residues, glutamate, lysine, and arginine were chosen for solubility; alanine for high intrinsic helical propensity; and glutamine and asparagine for solubility and helix capping. The protein was shown to be helical and monomeric; however, the thermal denaturation was noncooperative. A number of NOEs were observed that are consistent with the designed structure, but the protein has not yet been fully assigned.

The final designed helical hairpin to be discussed, SA-42, was created to serve as a scaffold for the design of catalytic activity (Olofsson *et al.*, 1995; Olofsson and Baltzer, 1996); see also Section VI of this chapter. As the sequence in Table I indicates, the artifical amino acids α-aminoisobutyric acid and norleucine were used for extra amino acid variability (to help NMR assignments) and helix stabilization. The loop sequence was GPVD. In addition to the use of leucine and norleucine in the hydrophobic core, two phenylalanines were placed one turn of the helix away from each other in helix 2 to facilitate NMR assignments. The protein was shown to be helical, but rather than assuming the desired monomeric fold, the protein formed dimers in solution. The thermal denaturation does not show a sharp transition, and the NMR spectra were broadened, suggesting exchange between multiple conformations. In the present of low amounts of trifluoroethanol (TFE), the resonances were much sharper, allowing assignments to be made. NMR data collected in TFE allowed modeling of the secondary and tertiary structure of this four-helical-bundle protein.

The first *de novo* four-helix bundle, α_4 (Ho and DeGrado, 1987; Regan and DeGrado, 1988), was designed from a "minimalist" perspective: It contained only glutamic acid and lysine residues on the surface of the bundle, and leucines in the "a," "d," and "e" positions of the hydrophobic core. The design was an iterative process starting with four single, self-associating helices, moving through the intermediate dimeric helix–loop–helix and finally to the four-helix bundle α_4 (Fig. 4). This designed protein was highly helical, monomeric, and extremely stable to guanidine hydrochloride denaturation. However, it showed certain nonnative-like characteristics: a noncooperative thermal denaturation,

Table I
Sequences of Designed Helix–Loop–Helix Proteins[a]

Name	Sequence: Helix 1 -- Loop -- Helix 2
ALIN	YERDELMACLKKATN--GTSGN--TEREDLIACVKRATH
CSSL	CAALESEVSALESEVASLESEVAAL--GRGDMP--LAAVKSKLSAVKSKLASVKSKLAAC
$\alpha t \alpha$	DWLKARVEQELQALEAR--GTDS--NAELRAMEAKLKAEIQK
SA-42	NAADNEAAIKALAEHNAAK--GPVD--AAQNAEQLAKAFEAFARAG

[a] Amino acid identities are given by the standard one-letter code, plus A (isobutyric acid) and N (norleucine).

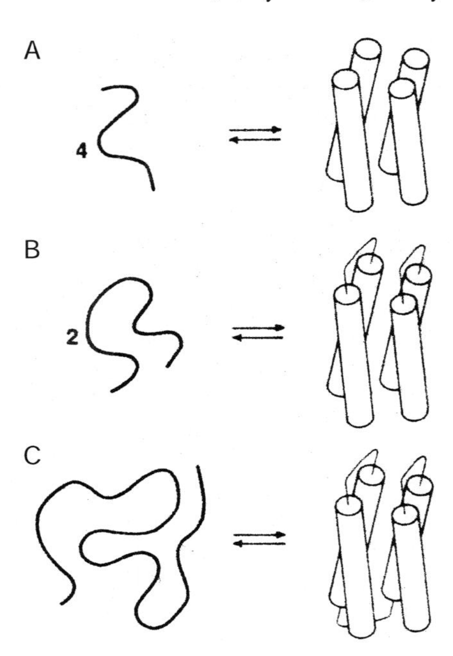

Fig. 4 A schematic illustration of the incremental approach to the design of α_4.

binding of the hydrophobic dye ANS, and poorly dispersed NMR resonances with rates of amide exchange faster than those of natural proteins. Subsequent designs of α_4 have attempted to advance toward a native-like protein by introducing residues other than leucine residues in the core, or by attempting to constrain the protein by introducing histidine residues for specific metal-ion coordination on the surface of the protein (Handel *et al.*, 1993; Raleigh *et al.*, 1995). These redesigns have resulted in more cooperative thermal denaturations, and better dispersed NMR spectra, demonstrating the importance of iterative design and thorough characterization at each step.

The protein Felix (Hecht *et al.*, 1990) was designed as a four-helix bundle using a different design philosophy: Here a goal was to use a sequence that was not stripped to the basics but that better reflected the diversity found in natural proteins. The designed sequence employed 19 of the 20 naturally occurring amino acids and had no sequence homology to any known protein. The overall residue composition of Felix was one-quarter charged (placed on the surface to form salt bridges), one-quarter polar, one-quarter large hydrophobic residues in the core, and one-quarter alanine. Specific placement of residues to form N and C caps and for favorable interaction with the helix dipole were included. An advantage of this approach is that the sequence diversity of the design may facilitate structural characterization by NMR. A potential disadvantage is that if a design is unsuccessful, it is less apparent where to begin to improve it than it is with the minimalist approach.

Gel filtration chromatography confirmed a monomeric structure in solution, and the protein was shown by CD to be helical. The stability of the protein, assessed by GuHCl denaturation, however, was low in comparison to that of natural proteins. A design feature that was incorporated into Felix to aid in characterization was to include two cysteine residues, positioned such that a disulfide bond could form between them if the protein folded with the correct topology. The disulfide bond formed readily in the folded protein, suggesting that although protein stability was low, the overall desired tertiary structure was attainable.

The protein "peptitergent" was designed to form amphiphilic α-helices that would associate with and solubilize membrane proteins to aid their crystallization. When the crystal structure of peptitergent was determined in the absence of a membrane protein, it was found to be a four-helix bundle (Schafmeister *et al.*, 1993). These peptides form an antiparallel four-helix bundle with an alanine-leucine hydrophobic core, reminiscent of the core of the natural protein Rop. There is, however, a significant difference between the structure of peptitergent and that of a natural protein: There is an approximately 60Å^3 cavity at the center of the bundle. The bundle is somewhat helical in solution, but no characterization of its thermodynamic properties has yet been reported. This represents one of the few examples where the structure of a *de novo* protein is known at high resolution. Another example of a structurally characterized designed protein is a three-stranded coiled coil, which was actually designed to be a parallel two-stranded coiled coil (Lovejoy *et al.*, 1993). These examples emphasize the essential importance of high-resolution structural characterization as the ultimate test of the realization of a design.

Several sequences have been designed to form antiparallel four-helix bundles (Betz *et al.*, 1996). The hydrophobic packing was based on the core of Rop, with various residues (valine, alanine, threonine, or leucine) in the "a" positions, and leucine in the "d" positions. The surface residues were a combination of charged and polar residues. The coil-AL (alanine in the "a" positions) and coil-TL were shown by analytical ultracentrifugation and CD to be monomeric but had low helicity. The coil-VL and coil-LL proteins were tetrameric and helical in a concentration-dependent fashion. The thermal denaturation of coil-VL is similar to that of natural proteins, whereas that of coil-LL is broader, suggesting nonnative-like character. Accordingly, the coil-LL protein binds the hydrophobic dye ANS more strongly than does coil-VL, and the NMR resonances are broader for coil-LL than for coil-VL.

B. β Hairpins

Several peptides, either taken from proteins of known structure or of *de novo* design, have been shown to have significant β-hairpin structure in solution (Blanco *et al.*, 1994; Klauser *et al.*, 1991). The production of soluble, monomeric β hairpins was of great interest, because β-sheet systems have been plagued with problems of insolubility. A monomeric β hairpin provides the simplest model system in which to study β-sheet formation.

Two peptides (called peptide 2 and peptide 3) were designed to improve on a peptide from tendamistat (de Alba *et al.*, 1995). Peptide 2 incorporated residues with higher β-sheet propensities (sequence: IYSNPDGTWT); peptide 3 used Ser instead of Pro in the turn (sequence: IYSNSDGTWT). A combination of 1D NMR, fluorescence, and CD was used to demonstrate that the peptides are monomeric. CD studies suggest a low amount of β-sheet formation, but CD of β structure is typically less easy to interpret quantitatively than that of α helixes, as β turns and aromatic residues have a strong influence. NMR structures show that peptide 2 indeed forms a single β-hairpin conformer, whereas peptide 3 has two forms (as yet unassigned) in equilibrium. The maximum extent of β-strand formation is estimated to be about 30%, based on NOE intensities.

Several other peptides have been designed *de novo* to form β-hairpins containing type 1′ β turns (Ranírez-Alvarado *et al.*, 1996). The peptide BH8 (Fig. 5) (sequence: RGITVNGKTYGR) was designed to be a model β-hairpin peptide. The BH8 peptide included the most favorable sequence for a type 1′ β-turn: NG. The residues at the other positions were based on statistical considerations, except for the Arg-Gly sequence at both ends, which was intended to prevent lateral oligomerization by the introduction of charge. A peptide named BH1 was also synthesized. BH1 contains a number of alanine residues, which have low intrinsic β-sheet forming propensity and interact poorly across two β strands. This peptide was synthesized as a negative control and was not expected to form a hairpin (sequence: RGATANGATAGR). Other peptides that are variants of BH8 were synthesized to test the contribution to β-sheet stability of different side-chain interactions (BH3 sequence: RGATANGKTYGR and BH4 sequence: RGITVNGATAGR). Formation of the desired monomeric β turn and the β-hairpin structure by BH8 was suggested by initial CD studies and confirmed by more detailed NMR analysis. By contrast, the "control" peptides BH1, BH3, and BH4 peptides appeared to be unfolded by CD; by NMR, it appeared that the turn conformation was populated to some extent in these peptides, but that the hairpin itself did not form. These results emphasize the importance of the turn in promoting the β-hairpin conformation and suggest that the hairpin may be additionally stabilized by incorporating residues in the strands that have high intrinsic β-sheet forming propensities and that interact favorably.

C. α/β Proteins

A common motif for an α/β fold that is particularly common in enzymes is the α/β TIM barrel (named after the enzyme triose phosphate isomerase). The first set of proteins designed to form α/β barrels (Fig. 6) were constructed from repeating units (Goraj *et al.*, 1990; Beauregard *et al.*, 1991). The proteins contained between 2 and 12 repeats of a single turn-β–turn-α sequence of the form turn1 (DARS), β strand (GLVVYL), turn2 (GKRPDSG), and α helix (TARELLRHLVAEG). The criteria used to choose these sequences included residue frequencies and secondary structure lengths from known β/α-barrel proteins, observed packing of the α helix on the β strand according to the rules of Cohen *et al.* (1982), and the utility of certain residues for NMR and fluorescence studies. The peptides with 4 to 12 repeats were highly expressed in *E. coli*, but formed inclusion bodies. Heptarellin, octarellin, and nonarellin (7, 8, and 9 repeats, respectively) were chosen for further study. In urea gradient gels, purified octarellin, but not heptarellin or nonarellin, showed a cooperative transition. Using Fourier transform infrared (FTIR) spectroscopy, it was calculated that octarellin contained about 35% α helix and 40% β strand; Raman spectroscopic analyses also suggested a fraction of β strand similar to that observed in TIM. The thermodynamic behavior of octarellin is unusual. When secondary structure was monitored by CD at 222 nm, octarellin was found to

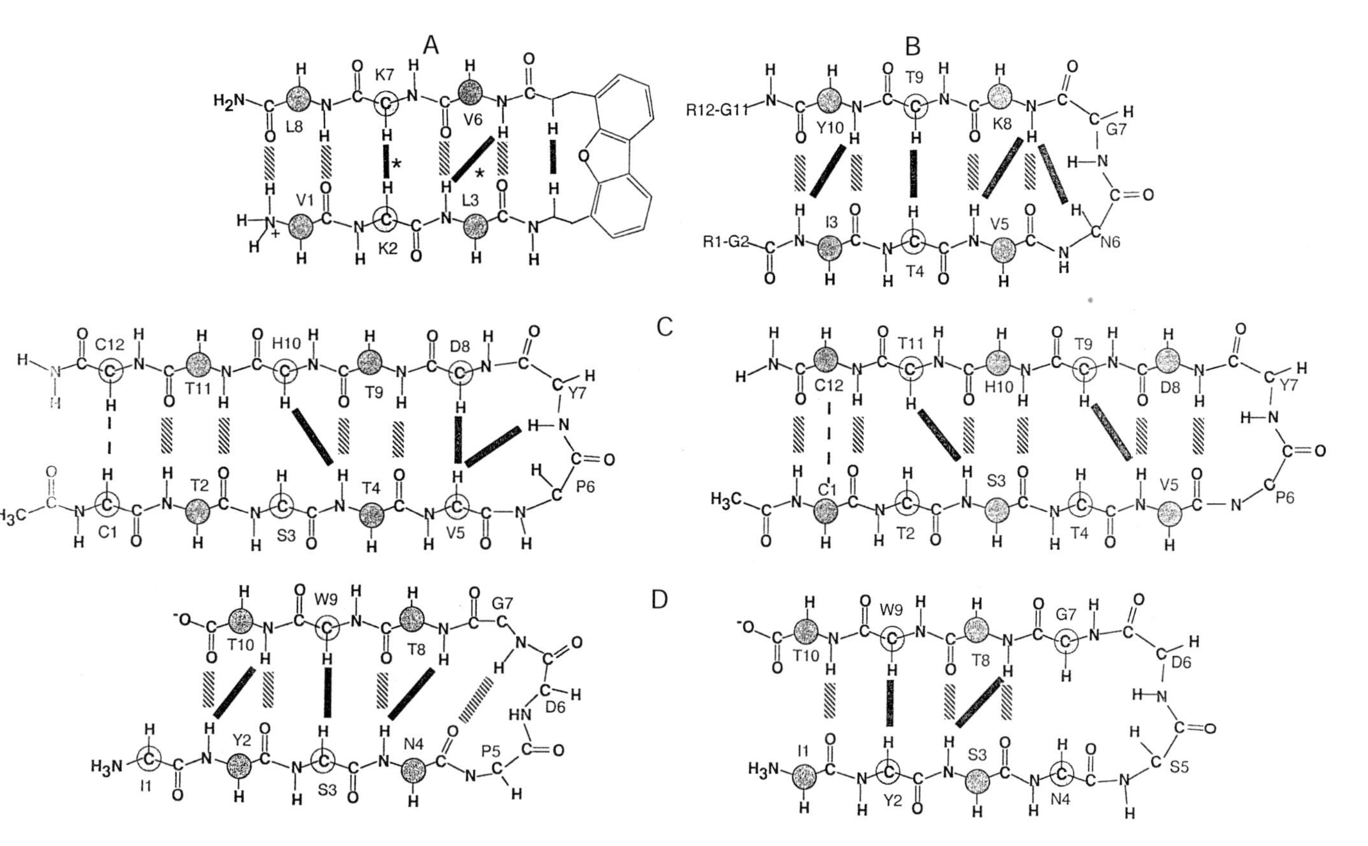

Fig. 5 Schematic representations of several β hairpins. (A) Dibenzofuran-based design. (B) BH8. (C) BB-O BB-R. (D) Peptides 2 and 3. H-bonds and NOEs are indicated.

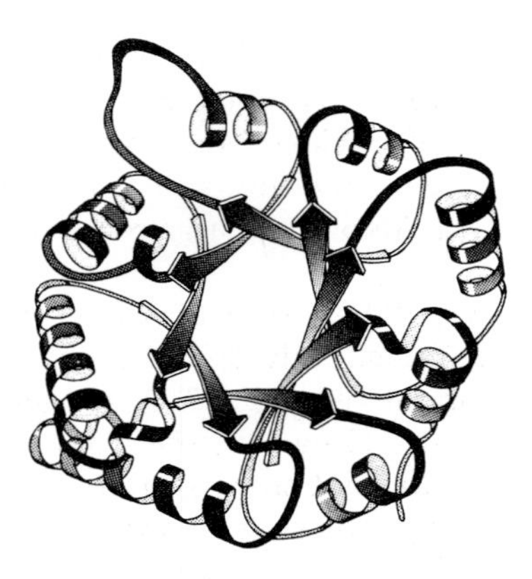

Fig. 6 Ribbon representation of the TIM barrel fold.

be stable above 45°, but irreversibly denatured at significantly higher temperatures. The near-UV region of the CD, derived from aromatic residues, is indicative of tertiary structure formation. This signal is lost irreversibly at temperatures greater than 39°C. The differential loss of secondary and tertiary structure is unusual and, without further characterization, cannot be readily explained. The behavior suggests a structure and stabilization that is different from that of a natural protein.

Several other β/α-barrel proteins have been designed and characterized (Tanaka *et al.*, 1994a, b). In these designs, factors in addition to the design considerations discussed previously for forming α/β-barrel proteins were taken into account. First, the designs used two different alternating β strands, because the orientation of the side chains in a β strand of an α/β barrel changes whether the strand is even or odd numbered. This leads to a fourfold symmetry instead of the eightfold symmetry used in octarellin. The β-strand sequences were chosen from frequencies of amino acids at those positions in natural β/α-barrel proteins. The amino acid sequences for the helices and the loops were based on those principles discussed previously in Sections II,A,1 and II,A,3. Three versions were made (named A-A-A-A, L-L-L-L, and L-βL-L-L, the best characterized). All three proteins were found by gel filtration chromatography to be monomeric. By CD, they exhibited significant secondary structure, with perhaps more β strand than predicted. They also showed cooperative unfolding transitions in guanidine hydrochloride; the L-βL-L-L protein, with a midpoint of denaturation of 2.6 *M*, was the most stable. The denaturation curve for L-βL-L-L was similar whether monitored by CD or fluorescence, indicating that the Trp residue was buried in the core, and that the unfolding was a typical two-state transition characteristic of small, natural proteins. However, the NMR resonances of L-βL-L-L were quite broad, the protein bound the dye ANS, and the thermal denaturations were noncooperative; these results suggested a nonnative-like structure.

To try to increase the native-like features of the β/α-barrels, the next stage of the design process (Tanaka *et al.*, 1994a, b) was to design a protein having only twofold sequence symmetry, with four different helices and four different β strands. The sequences were chosen to improve complementary packing in the core of the protein; the lower symmetry allowed the use of small residues opposite large aromatic ones, and vice versa, as is commonly seen in natural barrels. The redesigned protein was monomeric at pH < 5, but dimeric above this pH. The CD spectrum gave an estimate of 30% helix and 40% strand. The protein showed a cooperative guanidine hydrochloride denaturation transition with a concentration midpoint of 3.9 *M* whether monitored by CD or by fluorescence. This protein, however, still bound ANS and had broad NMR resonances.

A subsequent redesign (Tanaka *et al.*, 1994a, b) increased loop length to fill in "weak points" in the packing, and had slightly different helix and sheet sequences. This protein was monomeric in the range pH 3–7, and also showed a significant amount of secondary structure. However, the NMR resonances were again broad, and the amide protons did not show significant protection against exchange with solvent. Their exchange rates were more similar to those observed in unfolded proteins than in a stably folded structure. Interestingly, in spite of this, the proteins exhibited reasonably cooperative thermal denaturation transitions at low pH (2.8–3.3) with some associated enthalpy. Together these results indicate that the proteins appear to adopt an overall α/β fold and to be reasonably stable, but still have some nonnative-like character. Further redesign and characterization may surmount this final problem.

D. All β Proteins

A novel β-sheet protein, called "minibody," was designed using three β strands from two β sheets from the heavy-chain variable domain of an immunoglobulin molecule with two connections from hypervariable loops (Pessi *et al.*, 1993). In addition, residues were substituted in the loops that could bind metal if the desired tertiary fold was obtained (see Section V,A,2). The minibody was monomeric, but sparingly soluble (about 10 μM). CD analysis showed characteristic β-strand features, which were denatured in a reasonably cooperative fashion in the presence of urea. More recent work has concentrated on redesigns to increase the solubility of the protein to facilitate more detailed structural characterization.

Betabellin 14D and β doublet (Yan and Erickson, 1994; Quinn *et al.*, 1994) are two recent versions of the "betabellin" design. The basic design consists of two β sheets packed against each other to form a β sandwich. Each sheet is composed of four β strands connected by three turns (Fig. 7). The inside face of each sheet, the "filling" of the sandwich, is composed of a series of hydrophobic residues; the outside face is composed of mostly polar residues. The loops were chosen to be D-Lys-D-Ala and D-Ala-D-Lys, as these sequences are favored to form type-1′ β turns. Also, a disulfide bond was engineered to form at the dimer interface (Fig. 8). Formation of this disulfide bond was critical for both the structure and stability of the protein. Without the disulfide bond, the protein was mostly unfolded when monitored by CD and NMR. After disulfide bond formation, the protein was folded and had both well-dispersed NMR resonances and a cooperative thermal denaturation suggestive of native-like structure. It is important to note that although the disulfide bond facilitates structure formation by holding the two sheets together, it imposes severe limitations on the volume of residues that can be used in the "filling" of the sandwich. This limitation on the surface area that is buried may be part of the reason why the sheets do not associate in the absence of the disulfide.

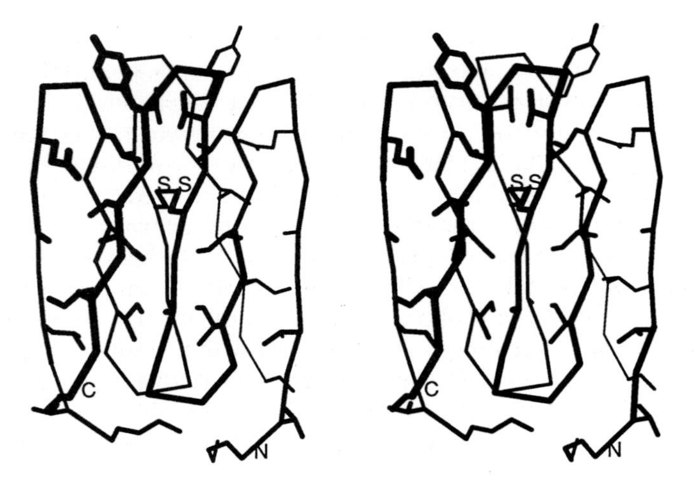

Fig. 7 Schematic illustration of the designed betabellin-based fold of betadoublet and betabellin-14.

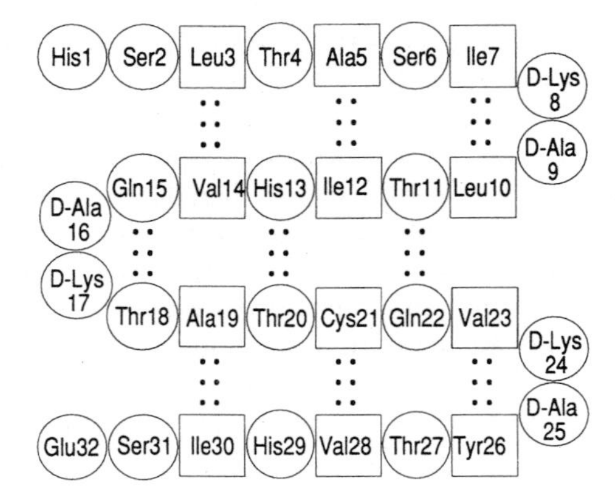

Fig. 8 Sequence of betabellin-14.

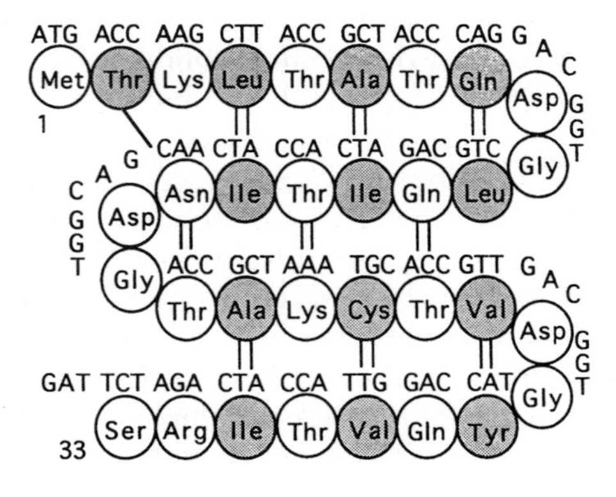

Fig. 9 Sequence of betadoublet.

A similar β-sandwich protein, betadoublet (Figs. 7 and 9) was designed *de novo* using knowledge acquired from the betabellin series of proteins (Quinn *et al.*, 1994). The design shares several similarities with betabellin 14D; a designed disulfide bond between the sheets of the dimer, residues that are likely to form type-1′ β-turns (betadoublet contains Asp-Gly turns), and a hydrophobic dimeric interface. Like betabellin 14D, the disulfide-bonded form of the protein was soluble, dimeric, and showed significant β-sheet formation. In addition, its thermal denaturation was cooperative, and the NMR resonances fairly well dispersed. The protein did bind ANS, however, and its amide exchange rates were faster than those of natural proteins. Both these designs represent significant progress towards creating a β-sheet protein that mimics a natural one. Subsequent redesigns may well achieve completely native-like behavior. Betabellin 14D and betadoublet represent two of the most highly "evolved" β-sheet designs. Their success again serves to illustrate that the iterative approach to design is proving very successful.

V. *De Novo* Design of Function

The creation of specific binding interactions between proteins and small molecules is the first step in designing functional proteins. Metal ions have been the most common targets of binding-site design efforts, both because the structural requirements for high-affinity sites are well known for a variety of metals, and because metal ions play a central role in many biological processes. Binding sites for larger molecules are more difficult to design because of the complexity involved in defining and achieving the appropriate placement of amino acid side chains for electrostatic, hydrogen bonding, and van der Waals interactions with the target ligand. Heme-binding sites have been designed in four-helix-bundle proteins by taking advantage of the two open coordination sites on the heme iron atom. The design of binding-sites for a few nonmetal ligands, such as 1,1,1-trichloro-2,2-bis(*p*-chlorophenyl)ethane (DDT) and opioid peptides, have been attempted by using physical models to identify and evaluate potential specific binding interactions.

A. Designed Metal-Binding Sites

Metal-binding sites are attractive design targets because of their involvement in a host of critical life processes. In addition, the requirements for metal binding by proteins are relatively well understood and the affinities of natural metal sites are high (around 10^{-11} *M* for Zn binding to carbonic anhydrase) (Christianson, 1991; Glusker, 1991). Metal-binding site designs have been attempted for a variety of purposes: facilitation of protein purification, stabilization of protein structure, probing of protein three-dimensional structure, regulation of enzymatic activity, creation of model systems, and the introduction of novel activities into proteins. Different levels of "structural definition" are required for these designs; for example, the creation of a metalloprotein model system requires reproduction of the metal and ligand identity and the specifics of the coordination geometry of the original system. Stabilization of protein structure, on the other hand, may only require that two or three metal ligands be in close spatial proximity without concern for the exact mode of interaction. This discussion is limited to those novel metal sites that are composed of three or four naturally occurring amino acid ligands and have a specific structural component to their design. The introduction of novel metal sites into both natural and *de novo* designed proteins is covered.

1. *Design of Novel Function to Modulate Existing Function*

The rational manipulation of enzyme specificity has long been a goal of protein engineering. Craik and colleagues have developed an intriguing approach to the alteration of substrate specificity in the proteolytic enzyme rat anionic trypsin (Brinen *et al.*, 1996; Willet *et al.*, 1995, 1996). Their approach is centered on the design of a novel metal-binding site composed of two ligands from trypsin and one from the P2′ site of the substrate or inhibitor. In trypsin nomenclature, the P2′ site is the second amino acid following the site of proteolytic cleavage. The presence of metal is therefore expected to favor the binding of substrates or inhibitors having a metal-ligating residue, such as histidine, in the P2′ site, and thus should "delocalize" the specificity of binding (Willett *et al.*, 1996).

Novel three-histidine metal-binding sites were identified by computer modeling of the crystal structure of trypsin D189S (a trypsin variant with aspartate 189 mutated to serine) complexed with BPTI (bovine pancreatic trypsin inhibitor). Potential histidine ligands were introduced into the crystal structure and evaluated for solvent accessibility, favorability of the side-chain rotamer, histidine–histidine α carbon distances, and potentially unfavorable interactions with the rest of the protein. A suitable site was found by substituting histidines at positions 143 and 151 of trypsin and at the P2′ position of the substrate. Proteolytic activity of trypsin N143H, E151H (TnN143H/E151H) toward peptide substrates containing tyrosine (Y) or arginine (R) at the P1 position and histidine (H) at the P2′ position was measured in the absence and presence of metal. Since wild-type trypsin is selective for basic amino acids in the P1 position, cleavage of a substrate having tyrosine at the P1 position would provide a strong indication of altered substrate specificity.

In the absence of metal, TnN143H/E151H had a wild-type level of activity against a typical amide substrate containing an arginine residue in the P1 position, demonstrating that the activity of TnN143H/E151H against normal trypsin substrates was unaffected. In the presence of Zn(II), TnN143H/E151H was able to hydrolyze a peptide that had His in the P2′ position and Tyr in the P1 position (AGPYAHSS) with a 10-fold higher activity than wild-type trypsin. Wild-type trypsin had an unexpectedly high level of activity against this substrate in the presence of Zn(II), possibly due to a bidentate metal complex with E151 and the substrate histidine acting as ligands. Consistent with this hypothesis, the mutation E151Q abolished this activity. Importantly, the authors demonstrated that the catalytic activity of TnN143H/E151H was dependent on the concentration of metal ion.

By incorporating the additional D189S mutation, the catalytic activity of TnN143H/E151H toward the peptide AGPYAHSS in the presence of Zn(II) was increased 150-fold over that of the wild type (Willett *et al.*, 1996). D189 is found in the S1 subsite of trypsin and mediates trypsin specificity by forming an ion pair with the basic residue in the substrate P1 position. The triple mutant

TnN143H/E151H/D189S hydrolyzed arginine-containing peptides in the presence of metal ions with a catalytic efficiency 350-fold greater than that of the TnD189S mutant. Thus, substrate discrimination in TnN143H/E151H/D189S has largely been transferred from the P1 to the P2′ position by simultaneously introducing stabilizing protein–P2′ interactions through metal ligation and disabling favorable protein–P1 interactions.

The crystal structures of N143H/E151H trypsin complexed with A86H ecotin (a polypeptide trypsin inhibitor containing a histidine residue in the P2′ position) in the presence of Zn(II), Cu(II), and Ni(II) verify key points of the design (Brinen *et al.*, 1996) (Fig. 10). In the Zn(II) complex, the metal is coordinated in a distorted tetrahedron with the three designed histidine residues. The fourth ligand is not visible in the electron density map and is presumed to be a water molecule. The geometry of Cu(II) coordination was described as distorted square planar, and that of Ni(II) coordination as square pyramidal, with a water molecule clearly visible as the fourth ligand in both cases. Apparently, the geometric preferences of the individual metals are profoundly influencing the coordination geometry of the designed site, and this influence is probably greater than that observed with natural proteins because of the absence in the designed site of "secondary ligands," those residues that hydrogen-bond to the metal ligands. In this design, the observed geometries likely represent a compromise between the preferred geometry of the metal and the conformational preferences of the ligand side chains.

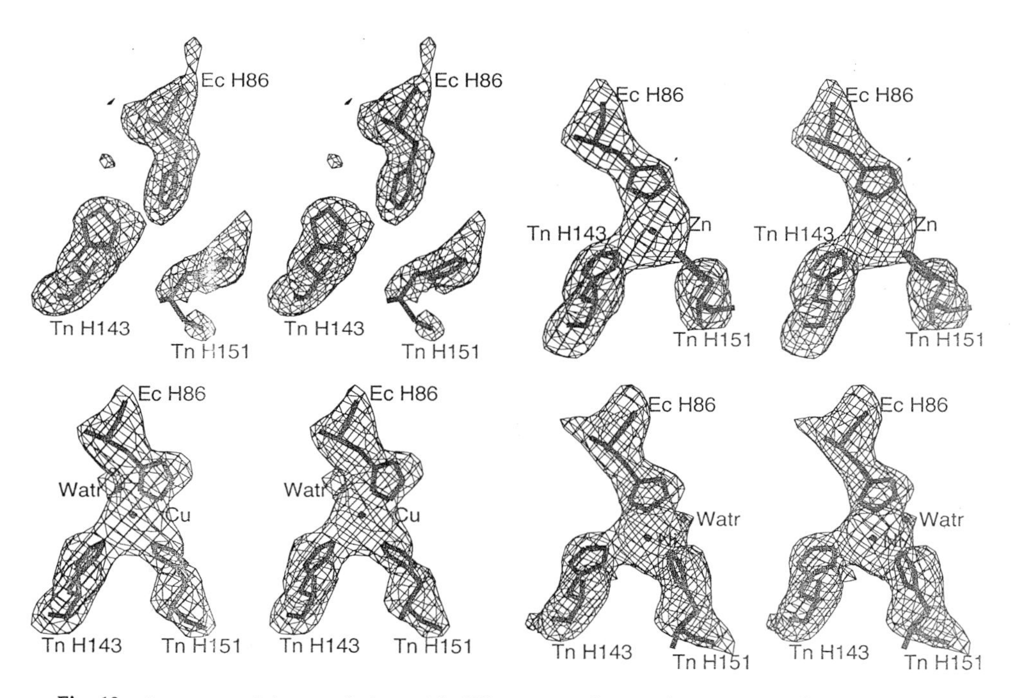

Fig. 10 Structure of the metal sites with different metal-ion substitutions in the trypsin variant. Electron density is shown along with the model.

2. *Carbonic Anhydrase as Template for* de Novo *Metal-Site Design*

All of the known vertebrate carbonic anhydrases (EC 4.2.61, carbonate dehydratase) bind a single Zn(II) ion that is required for enzymatic function (Tashian, 1989). Extensive characterization of this Zn(II) protein, from both structural and functional perspectives, has illuminated the central role that the Zn(II) atom plays in the catalytic mechanism. Carbonic anhydrase has been an enticing model for metal-site design because the Zn(II) site is composed of three histidine residues at positions 94, 96, and 119, which are situated on two adjacent β strands. The relative simplicity of the two-strand Zn(II)-site motif makes it attractive for transfer to other proteins, particularly those with β-sheet structure. This approach requires some means of "matching" the structural context of the target protein with that of the Zn(II) site in carbonic anhydrase, and a structure or high-resolution model of the target protein is required. Underlying this approach is the assumption that any specific interactions between histidine ligands and the surrounding environment in the carbonic anhydrase active site can be stripped away without abolishing the ability to bind metals. Several examples of this approach to metal-site design are discussed next.

Considerable effort has been put toward the introduction of metal-binding sites in antibodies (Iverson *et al.*, 1990; Roberts and Getzoff, 1995; Roberts *et al.*, 1990; Stewart *et al.*, 1994; Wade *et al.*, 1993). The ultimate aim of these designs is the manipulation of catalytic antibody activity through the rational placement of a metal ion. The approach of designing these sites involved first identifying structurally conserved regions in the complementarity-determining region (CDR) of several known antibody structures. The carbonic anhydrase II Zn(II) site was then used as a template for the design of three-histidine metal-binding sites in the antibody-binding pocket. One of these sites was built into a single-chain fluorescein-binding antibody by replacing R34, S89, and S91 with histidine, and the resulting protein (named QM212) was characterized in detail (Iverson *et al.*, 1990; Roberts *et al.*, 1990; Getzoff, 1995; Wade *et al.*, 1993). QM212 bound Cu(II) fairly tightly with a dissociation constant of 3×10^{-7} *M* at pH 6.0, but bound Zn(II) and Co(II) considerably more weakly. Independent mutagenesis of each histidine residue to glutamate resulted in a 50- to 3000-fold drop in Cu(II) affinity, offering indirect evidence for the participation of each histidine in Cu(II) binding (Wade *et al.*, 1993). Fluorescein and Cu(II) appear to bind simultaneously, although the affinity for fluorescein was three orders of magnitude lower in the metal-binding antibody, in part because of the recruitment of two fluorescein-binding residues for metal binding (Iverson *et al.*, 1990; Wade *et al.*, 1993).

Several other three-histidine sites have been introduced into the fluorescein-binding antibody and also into a catalytic antibody designed to hydrolyze esters and amides (Roberts and Getzoff, 1995; Stewart *et al.*, 1994; Wade *et al.*, 1993). These designs bind Zn(II) with micromolar affinity, and it has been suggested that the quenching of antibody tryptophan fluorescence on Zn(II) binding could be exploited in the design of Zn(II) biosensors (Stewart *et al.*, 1994). Although the design of metalloantibodies with reasonably high metal affinity has been demonstrated, no information is yet available on the geometry of metal binding in these sites. This is an important consideration in the design of catalytic metalloantibodies, as the properties of the bound metal are profoundly influenced by the coordination number and geometry.

Skerra and colleagues (Müller and Skerra, 1994) used the structure of thiocyanate-inhibited carbonic anhydrase as the model for the design of two Zn(II) sites in adjacent β strands of the β-barrel retinol-binding protein (RBP). The rationale for using this CA structure, which has a pentacoordinate zinc atom, rather than the uninhibited structure containing a four-coordinate zinc site, was based on the desire for a metal coordination geometry with two "open" coordination sites. This arrangement allows binding of a small-molecule chelator at the open site and permits protein purification by immobilized-metal affinity chromatography (IMAC). The two novel metal-binding sites were identified by searching for regions of protein backbone structural similarity between carbonic anhydrase and RBP. This was achieved by superimposing the Cα atoms of carbonic anhydrase residues 93–97 and 118–120 (the metal-liganding residues are 94, 96, and 119) sequentially along β-strand pairs B/C and D/E of RBP. Promising sites were further evaluated by comparing the Cα and Cβ residues of the carbonic anhydrase histidines to the corresponding atoms in RBP. Two sites, termed RBP/H_3(A) and RBP/H_3(B), were selected for introduction into the protein.

The designed metal site RBP/H_3(A) had a remarkably high affinity for Zn(II), with a dissociation constant of 36 $\pm$ 10 n*M*; the RBP/H_3(B) design bound Zn(II) with approximately 10-fold lower affinity. Interestingly, the authors compared these Zn(II) affinities to those of RBP containing a "His_6 tag" consisting of six sequential histidines at the C terminus of the protein. The protein RBP-His_6 bound two Zn(II) ions with affinities 50- and 100-fold lower than that of RBP/H_3(A), demonstrating that significant increases in metal affinity are possible through the rational placement of metal-binding residues. Affinities for both Ni(II) and Cu(II) were substantially lower than that for Zn(II). The metal preferenes for RBP/H_3(A) differed from those observed with isolated histidine residues and carbonic anhydrase; the structural basis for these differences is at present unclear, but could lie in the particular geometry of coordination in each case. RBP/H_3(A) appears to bind both Zn(II) and Ni(II) in an octahedral fashion: both metals are bound equally tightly in absence and presence of iminodiacetic acid (IDA), a tridentate chelator. The stability of RBP/H_3(A) toward guanidinium hydrochloride was enhanced in the presence of Zn(II) by 8.5 kJ/mol compared to that of the wild-type protein. Single-step IMAC purification of RBP/H_3(A) to homogeneity has been achieved using immobilized Ni(II)–iminodiacetic acid and Ni(II)–pyrrolidone–(*R,S*)-dicarboxylic acid, a novel chealating ligand developed for IMAC purification of RBP/H_3(A) (Schmidt *et al.*, 1996).

Essential for the design of novel functions in proteins is the identification of appropriate frameworks. The suitability of two frameworks, a designed β-sheet protein termed "minibody" and the scorpion protein charybdotoxin, was investigated by attempting the design of carbonic anhydrase-like metal-binding-sites. The "minibody" is a small designed protein that mimics part of the β-sheet structure of an antibody and includes two loops corresponding to the hypervariable loops (see Section IV,D). Rather than grafting a carbonic anhydrase three-histidine site onto the β-strands of the minibody, this group selected the two loops for placement of the metal-binding residues (Pessi *et al.*, 1993). The objective was both to verify the folded structure of the minibody, in which the two loops would be in close proximity, and to demonstrate the utility of this protein as a framework for the design of function. The placement of the metal site was selected by measuring the Cα–Cα distances of the histidine ligands in the crystal structure of carbonic anhydrase II and then searching for a similar set of Cα relationships in the two "hypervariable loops" of the minibody model. Histidine residues placed at positions 13, 30, and 32 appeared to provide a satisfactory arrangement of binding residues. The interaction of metal with the designed protein has been probed in a qualitative fashion. The designed metal-binding protein bound Zn(II) in a monomeric fashion at pH 7.0 as determined by comigration with ^{65}Zn on a gel filtration column. The Zn(II) affinity was determined by a filter binding assay, and the dissociation constant was estimated to be 10^{-6} *M*. Affinities of other metals were determined by competition with ^{65}Zn(II), giving the following order of binding affinities: Cu $>$ Zn $>>$ Cd $>$ Co. Redesigns to enhance the solubility of the original minibody (Bianchi *et al.* 1994) should allow for more detailed structural characterization of the overall protein and the metal-binding site.

The ability of charybdotoxin, a scorpion ion-channel blocking protein, to serve as a framework for the design of novel functional properties was tested by introducing a carbonic anhydrase-like metal-binding site (Vita *et al.*, 1995). The framework of this protein is somewhat unusual, being composed of only 37 amino acids that fold into a single α helix and a three-stranded β sheet. This structure is extremely stable because of the presence of three disulfide bonds. Residues 92–96 and 117–121 of carbonic anhydrase II were superimposed on β-sheet residues 25–29 and 32–36 of the experimentally determined charybdotoxin structure, with an 0.82-Å RMS deviation for the backbone atoms. Histidine substitutions at charybdotoxin positions 27, 29, and 34 were made, along with several other sequence changes: A glutamine and glutamate residue that serve as so-called "secondary ligands" in carbonic anhydrase (Christianson, 1991) were introduced. Secondary ligands are residues that H-bond to the direct metal-binding residues to position them and consequently to enhance metal-binding affinity. Three residues were mutated to eliminate unfavorable steric contacts, and one preexisting histidine residue was changed to phenylalanine to avoid adventitious metal ligation. The resulting sequence was chemically synthesized, and after formation of the three disulfide bonds, appeared to adopt the wild-type structure as determined by CD spectroscopy at pH 7.0 and two-dimensional ^{1}H NMR at pH 3.5. The protein bound Cu(II) with high affinity; a dissociation constant of 40 n*M* at pH 6.5 was measured by following the quenching of tryptophan fluorescence upon the addition of Cu(II). The affinity for Zn(II) was approximately 100-fold lower. Changes in the chemical shifts and line widths of the six histidine residue protons in one-dimensional ^{1}H NMR spectra of the unbound and Zn(II)-bound protein were taken as indirect evidence for the involvement of all three histidines in metal binding. No information is currently available on the geometry of Cu(II) or Zn(II) coordination.

In summary, several designs based explicitly on the structure of the Zn(II) site in carbonic anhydrase have been reported. These designed sites almost always bind metal, although often with affinities that are orders of magnitude lower than that of carbonic anhydrase. A critical assessment of the success of these approaches to metal-site design is difficult until more information on the specific geometry of metal–protein interactions is available.

3. *Computer-Aided Metal-Site Design*

One approach to the design of novel metal-binding sites is the "template" approach described previously, whereby the structural relationship between individual metal-binding residues in a natural protein is grafted onto a new protein framework. A limitation of this approach is the need to find a template for every type of secondary structural context in which a new metal site is desired. Moreover, the primary criterion for selection of sites is usually the superposition of backbone residues, which may not be the best determinant for sites of specific coordination geometry. An alternative approach is to divorce the design process from the database of structurally characterized natural metal sites. Naturally, this increases the level of complexity of the design process enormously, and any attempt to sample the combinatorial possibilities requires the assistance of computational methods. To date, two algorithms have been developed for the design of novel, structurally defined metal-binding sites. One of the algorithms, called Metal Search (Clarke and Yuan, 1995) has been used successfully to design tetrahedral Zn(II) binding sites in two proteins. The program DEZYMER has been developed by Hellinga and colleagues and was used in an attempt to design a blue-copper site in thioredoxin (Hellinga *et al.*, 1991; Hellinga and Richards, 1991). Although a detailed description of these algorithms is beyond the scope of this chapter, a general outline of the approach taken will be given, along with a discussion of the metal sites designed.

The program Metal Search (Clarke and Yuan, 1995) was created to identify potential novel tetrahedral Zn(II)-binding sites in protein structures or models. Metal Search accepts as input a Protein Data Bank (pdb) file of the protein of interest and, keeping the backbone fixed, substitutes each amino acid with histidine and cysteine. The algorithm then searches for groups of four cysteine/histidine substitutions that could bind a single Zn(II) ion with tetrahedral geometry and reasonable metal–ligand distances. The possibility of unfavorable steric interactions between the introduced ligands and the rest of the protein is not addressed by the program and must be evaluated by the designer using computer modeling methods.

High-affinity tetrahedral Zn(II) sites have been successfully designed into two proteins using the Metal Search approach. The first design, consisting of two cysteine and two histidine residues, was introduced into the *de novo* designed four-helix bundle α_4 (Regan and Clarke, 1990). The resulting protein, termed $Z\alpha_4$, bound Zn(II) tightly, with a dissociation constant of about 10^{-8} *M* (Fig. 11). More importantly, spectroscopic characterization of bound Co(II) indicated that metal coordination occurs with tetrahedral geometry, confirming this critical aspect of the design. The protein was significantly stabilized against guanidinium hydrochloride denaturation in the presence of Zn(II), an expected consequence of the high-affinity interaction. This Zn(II) site was characterized in greater detail by independently mutating each of the four metal ligands to alanine (Klemba and Regan, 1995). The goals were both to investigate the effects of removal of each ligand on metal-binding affinity and geometry, and to generate a series of proteins with an open coordination site at the Zn(II) ion for the binding of water and potential catalysis of hydrolytic reactions. Mutation of either histidine to alanine was accompanied by the preservation of tetrahedral

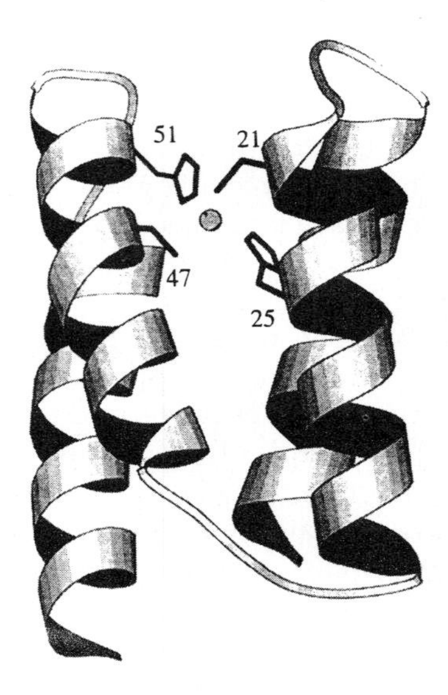

Fig. 11 Ribbon illustration of Zα4.

metal binding and a 10-fold drop in the affinity for Co(II). These two proteins were the first reported examples of tridentate *de novo* metal sites with demonstrated tetrahedral binding geometry and a water or hydroxide molecule as the presumed fourth ligand. The proteins did not display carbonic anhydrase-like catalytic activity against activated esters, such as *p*-nitrophenyl acetate. A plausable explanation for the lack of activity is that the pK_a of the metal-bound water at a His_2Cys site is too high to provide significant concentrations of hydroxide for catalysis at neutral pH: When the Zn ligands of carbonic anhydrase are mutated from His_3 to His_2Cys, the pK_a of the metal-bound water increases to around 9 (Huang *et al.*, 1996).

The second metal-binding site designed using Metal Search was introduced into the B1 domain of streptococcal protein G (Klemba *et al.*, 1995). The B1 protein framework, consisting of a single α helix packed against a four-stranded β sheet, differs substantially from that of α_4 and was used to test the generality of the design approach. In addition, following from the results discussed previously, the composition of the metal site was biased to include at least three histidines so that a three-histidine carbonic anhydrase-like site could eventually be constructed. A His_3Cys metal site was identified by the Metal Search program and modeled into the B1 domain (Fig. 12). In contrast to Zα_4, in which the

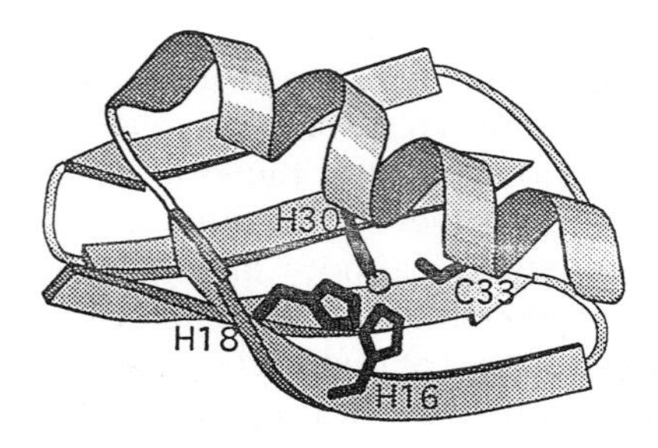

Fig. 12 Ribbon illustration of Zβ1.

metal site could be introduced without unfavorable interactions with the rest of the protein, a steric clash was identified between one of the introduced histidine ligands and a nearby leucine residue at position 5 in the sequence. Modeling of a methionine or alanine in place of Leu-5 suggested that these two substitutions should alleviate the steric clash. Three variants of the metal-binding B1 design were made, having leucine (Zβ1L), methionine (Zβ1M), or alanine (Zβ1A) at position 5. The designed variants all bound Co(II) with high affinity and tetrahedral geometry. The highest affinity was observed for Zβ1M, which had a Co(II) dissociaton constant (4×10^{-6} *M*) approaching those of natural zinc finger peptides. The identity of the position 5 residue had a dramatic impact on the stability of the proteins, with Zβ1A being least stable and Zβ1L being most stable in both the absence and presence of metal. All three variants were substantially stabilized toward thermal denaturation in the presence of metal. Both CD and 2D ^{1}H NMR spectroscopy suggest that the three-dimensional structure of the B1 domain is preserved in Zβ1L.

The computer algorithm DEZYMER was developed for the design of ligand binding-sites in proteins (Hellinga and Richards, 1991). This algorithm utilizes a geometric description of the desired metal site, the three-dimensional structure of a protein, and a library of amino acid side-chain rotamers in order to identify a set of potential binding sites. The potential sites are evaluated for overpacking, underpacking, and satisfaction of hydrogen bonding potential, and if necessary the protein is repacked using an additional algorithm. This approach was tested by attempting to design a $Cys_1Met_1His_2$ blue-copper site into thioredoxin. Introduction of a blue-copper site is a stringent test, as the spectroscopic properties of such sites are highly sensitive to the coordination environment of the copper ion. Characterization of the thioredoxin–blue-copper protein (called chelatin Bα) was complicated by the occurrence of a copper binding site in wild-type thioredoxin; however, an additional copper-binding site, with an affinity of 4×10^{-7} *M* at pH 7.4, was identified (Hellinga *et al.*, 1991). Electron paramagnetic resonance (EPR) characterization of the chelatin Bα site suggested N_2O_2 coordination of the copper, indicating the possibility of copper coordination by the two introduced histidine residues and either water molecules or two nearby carboxylates. Consequently, the characteristic blue-copper spectrum was not observed with this protein. Further redesigns have focused upon removing nearby, competing ligands, and "rotating" the location of the amino acid ligands at the site, to move toward favoring the desired first coordination ligands.

B. Heme-Binding Proteins

The positioning of amino acid side chains for specific, high-affinity heme binding would be a complex combinatorial problem were it not for the two open axial positions in the coordination sphere of the heme-bound iron. Several heme-binding sites have been designed in *de novo* four-helix bundles by introducing histidine residues to specifically coordinate the heme-bound metal. Placing the histidines in the core of the four-helix bundle promoted favorable van der Waals interactions between the porphyrin ring and the hydrophobic amino acids making up the protein core (Fig. 13).

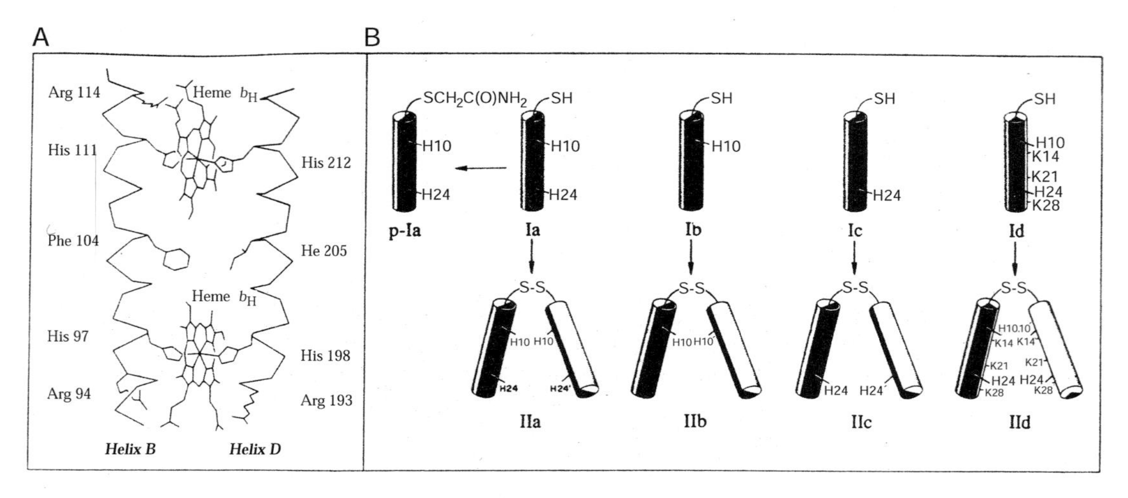

Fig. 13 Schematic illustration of the heme-binding designs.

This strategy was first employed to introduce a heme-binding-site into the *de novo* protein α_2 (Choma *et al.*, 1994). This protein, which forms a dimeric four-helix bundle, has been previously used as a scaffold for designed metal-binding sites (Handel *et al.*, 1993). The heme-binding site was designed by computer modeling of the α_2 bundle with a heme molecule separating the two monomers. Appropriate axial ligands for the heme were selected by changing core leucine residues (one in each monomer) to histidine and evaluating the potential interaction with heme. The L25H mutation appeared to provide satisfactory heme coordination through the histidine Nδ atom. A hydrophobic binding site for the heme moiety was created by truncating three leucine side chains per monomer to alanine where they approached within van der Waals distance of the heme. Finally, a cysteine was introduced at the N terminus to ensure parallel association of the monomers, and thereby orient the Pro-Arg-Arg loops for interaction with the heme propionate groups. The final designed protein was designated $VAVH_{25}$(S-S).

Much of the insight gained from this study lies in the characterization of appropriate control proteins along with $VAVH_{25}$(S-S): α_2(S-S), which has neither the axial histidine ligands nor the hydrophobic binding-site; H_{25}(S-S), which possesses only the two His ligands; and retro(S-S), which corresponds to the reverse sequence of $VAVH_{25}$(S-S). A detailed characterization of the properties of these proteins, including heme-binding stoichiometry and affinity, UV/visible and fluorescence spectral properties, EPR, and redox potentiometry, indicated that only those proteins with designed axial ligands and a hydrophobic binding site [$VAVH_{25}$(S-S) and retro(S-S)] interacted with heme in a specific manner and shielded it from solvent. Interestingly, yet unexpectedly, the reverse sequence, retro(S-S), best reproduced the electronic environment found in natural heme proteins as suggested both by visible wavelength and EPR spectroscopy. Figure 14 shows visible spectra of both retro(S-S) and cytochrome *b*-559, a natural heme protein. The distinct splitting of the α/β region of the reduced cytochrome *b*-559 spectrum is reproduced only by reduced heme–

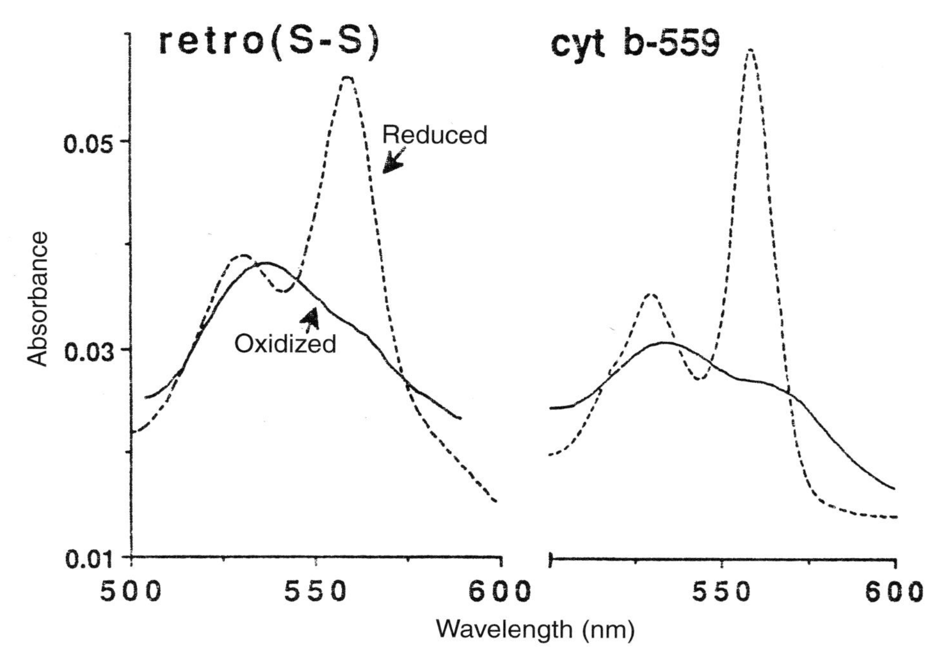

Fig. 14 UV-visible spectra of the α/β region of oxidized (solid line) and reduced (dotted line) retro (S-S) complex (left) and cytochrome b-559 (right).

retro(S-S). Significantly, ferric heme bound by retro(S-S) but not by $VAVH_{25}$ (S-S) was exclusively low-spin at 20 K.

An intriguing explanation for the high affinity and cytochrome *b*-559-like spectra of the retro(S-S) sequence is suggested by the lower stability of this protein compared to that of $VAVH_{25}$(S-S). The reduced stability of retro(S-S), possibly due to the less effective shielding of hydrophobic residues from solvent, may allow more structural flexibility, leading to a more favorable interaction with the heme molecule. If valid, this idea would suggest that the tendency in protein design to maximize the stability of the desired folded structure may not be the most productive general strategy when it comes to designing certain types of functional proteins.

Using a similar approach, multiheme-binding four-helix bundles have been designed and characterized (Farid *et al.*, 1994; Robertson *et al.*, 1994; Gibney *et al.*, 1996). The ultimate goal of these designs is to reproduce the essential features of electron transfer processes occurring during respiration and photosynthesis; an analogy is drawn to maquettes, the small-scale three-dimensional models used by architects and artists. The individual helices of the multiheme-binding four-helix bundles were, like α_2, designed with hydrophobic leucine residues on one face and charged Lys and Glu residues on the other face. In contrast to the α_2-based designs, no attempt was made to explicitly model the interactions between the heme molecule and the leucine residues in the hydrophobic core. The placement of two heme-binding histidines at positions

10 and 24 was guided by the sequence of the cytochrome *b* subunit of cytochrome bc_1. Dimerization of the peptides was effected through oxidation of the N-terminal cysteine residue, giving a two-helix peptide capable of binding two heme molecules through histidine pairs 10,10′ and 24,24′. Also included in the design, and based on the sequence of cytochrome *b*, was a phenylalanine residue intended to separate the two heme-binding sites, and an arginine residue near the 24,24′ site to interact electrostatically with the negatively charged propionic acid side chains of the 24,24′ heme group, thereby raising its redox potential. These two-helix peptides were expected to dimerize in solution, thus providing a four-helix bundle capable of binding four molecules of heme (peptide IIa). Control peptides having only His10 (peptide IIb) or His24 (peptide IIc) were also synthesized (Fig. 13).

The designed peptide forms four-helical bundles in solution, both without and with bound heme, as assessed by CD, gel filtration, and sedimentation equilibrium centrifugation (Robertson *et al.*, 1994). The heme-binding stoichiometries were those expected: four molecules for IIa, and two molecules each for IIb and IIc. The spectra of the heme–protein complexes displayed characteristics similar to those of natural bishistidine ligated heme proteins, including a Soret band at around 412 nm in the oxidized form, and distinctly split α/β absorbancies in the reduced form.

The heme-binding properties of peptides IIb and IIc were studied in detail. Intriguingly, in both cases the first heme molecule bound with considerably higher affinity than the second; for example, the heme dissociation constants for peptide IIb were <0.05 and 0.8 μM for the first and second heme molecules, respectively. The K_d values observed for the first and second heme groups were similar to that found with the α_2-based designs retro(S-S) (0.01 μM) and $VAVH_{25}$(S-S) (0.7 μM), respectively. Redox titrations of diheme–IIb and diheme–IIc were biphasic, indicating redox potentials of -100 and -215 mV for the two heme molecules in peptide IIb. The authors consider two possible explanations for these observations. The first explanation posits that binding of the first heme molecule forces the second heme into a different chemical environment, either through steric interference or through structural organization of the four-helix bundle. A second explanation, and the one preferred by the authors, is that the heme molecules are in similar chemical environments, but that the affinity for and redox potential of the second heme molecule are altered by the charge on the first heme molecule. What experimental evidence might allow discrimination of these two possibilities? Total correlation NMR spectroscopy of tetraheme–IIa shows about 60 Hα–Hβ correlations, approximately half the number expected for the 124-residue peptide. This result suggests a twofold symmetry for the tetraheme four-helix bundle complex; however, *de novo* designed proteins typically have poorly dispersed spectra and broad line widths (Handel *et al.*, 1993), making an accurate count of individual resonances problematic. The environment of a variety of heme analogs in diheme–peptide IIb was probed using resonance Raman spectroscopy (Kalsbeck *et al.*, 1996). All but one of the heme analogs, when bound to protein, gave rise to a single set of ring-skeletal modes; this suggests that the two protein-bound hemes are in a similar chemical environment. In addition, 1-methyl-2-oxomesoheme XIII appears to be bound so that the oxo group is shielded from water. These experiments provide some insight into the orientation of the heme molecule with respect to the four-helix bundle; however, high-resolution structural information might be needed to unambiguously determine the cause of the redox inequivalence in these multiheme four-helix bundle proteins.

The heme-binding four-helix bundle formed from peptide IIa has been the starting point for the introduction of additional functional groups. With the aim of producing a photosynthetic reaction center maquette, coproporphyrin I was covalently linked to the loop region of peptide IIa (Rabanal *et al.*, 1996). Upon dimerization of this modified peptide IIa, a cofacial porphyrin pair was formed with a substantially lower dimer dissociation constant than that found for free coproporphyrin I; however, the quantum yield for electron transfer between the coproporphyrin pair and a bound heme group was expected to be very low (<0.01). In the second example, the loop sequence of peptide IIa was modified to bind a [4Fe-4S] cluster. Simultaneous binding of the [4Fe-4S] cluster and heme was reported (Gibney *et al.*, 1996). The next goal in the design of natural redox protein "maquettes" will be to demonstrate electron transfer.

C. DDT-Binding Design

In an early effort, Gutte and colleagues attempted the design of a 24-residue four-stranded β-sheet structure with hydrophobic faces capable of binding the insecticide 1,1,1-trichloro-2,2-bis(*p*-chlorophenyl)ethane (DDT) (Klauser *et al.*, 1991; Langen *et al.*, 1989; Moser *et al.*, 1983, 1987). The details of the modeling process, including the anticipated van der Waals contacts between the designed sequence and DDT, were not discussed in detail. The selection of appropriate residues was guided by the statistical preference of certain amino acids to appear in β structure, which was the only information available at that time to guide the design.

As might be expected for a β-sheet peptide with two hydrophobic faces, the designed peptide (DDT-binding peptide, or DBP) formed high-order aggregates in aqueous solution (Moser, 1983). β-Like secondary structure was observed by far-UV CD spectroscopy; however, this β structure was not completely abolished in the presence of 8 *M* urea, suggesting that the protein is aggregating under the conditions used for CD analysis. Interactions between the designed peptide and DDT were therefore characterized in 20 m*M* Tris–HCl/50% ethanol, conditions under which DBP appeared to be monomeric.

The designed peptide bound DDT with a dissociation constant of 9×10^{-7} *M* (Klauser *et al.*, 1991). This affinity was approximately 600 times tighter than that observed for a peptide of the same amino acid composition but randomized sequence. Reduction of the hydrophobicity of either DBP or the ligand resulted in a moderate decrease in the binding affinity. Binding of dioxin was weak, suggesting that ligand hydrophobicity alone was not a sufficient criterion for tight binding. Although crystals of the DDT-binding peptide have been grown, no detailed structural information is available, and the exact nature of the DDT–DBP interaction remains to be elucidated. These data demonstrate the feasibility of designing hydrophobic protein–ligand interactions in nonaqueous solvent. The design of such interactions in the absence of organic solvent, while avoiding protein aggregation, is sure to be no small challenge.

D. Design of Opiate Peptide Mimetic Receptor

An attempt to create a peptide model of the opiate receptor family has motivated the design of a 40-residue β-sandwich protein (Kullmann *et al.*, 1984) composed of a pair of face-to-face two-stranded β sheets. The protein framework was built up by incorporating amino acids that have a high potential to form β sheets according to the Chou–Fasman statistical study (Chou and Fasman, 1978). Sheet–sheet packing was designed by placing hydrophobic residues in the proposed interface and polar residues on the solvent-exposed surface. The design of interactions between the receptor mimetic peptide (or RMP) and its ligand was guided by the available experimental data on the structure–activity relationship of various morphine derivatives, particularly the enkephalin family of opioid peptides. Potential receptor–ligand interactions were then modeled using space-filling CPK models.

Structural characterization of the RMP design has been sparse. A far-UV circular dichroism spectrum taken in neutral aqueous solution is suggestive of approximately 50% β structure. No experiments probing the aggregation state of the peptide have been reported; therefore, it is not clear whether this β structure is formed in the context of a monomeric, folded peptide or an aggregate. This distinction is critical particularly when characterizing β-sheet designs, as protein aggregates often appear β-sheet-like by far-UV CD.

The binding of ligands to the receptor mimetic was probed with three different equilibrium techniques, all giving comparable results; only the equilibrium dialysis studies will be discussed here. The affinity of (3-Gly)-[Leu]-enkephalin for RMP was determined to be 5.8×10^{-5} *M* by equilibrium dialysis and appeared to occur with a 1:1 stoichiometry. The specificity of binding, however, appeared to be low. A control [Leu]-enkephalin peptide consisting of all D-amino acids also bound with relatively high affinity. In addition, the binding affinity of the nonpeptide morphine analog naloxone was surprisingly low, suggesting that nonspecific hydrophobic interactions are driving the binding of [Leu]enkephalin and related peptides to RMP.

VI. Design of Catalytic Proteins

The evolutionary process has produced proteins with the ability to catalyze with great efficiency and specificity those chemical reactions necessary for life. The possibility of designing novel enzymes, and thereby expanding the repertoire of protein-catalyzed reactions, has attracted much attention as it would open the door to new research tools, industrial processes, and therapeutic approaches. Means of developing new protein activities include *de novo* design, rational modification of existing enzymes, and selection of catalytic antibodies (Corey and Corey, 1996).

Two main problems impede the *de novo* design of enzymes. The first and foremost problem is the difficulty of producing a stable, native-like structural framework for the introduction of active-site residues. A stable framework is probably essential for the creation of those elements that favor highly efficient protein-mediated catalysis: a substrate binding pocket, desolvation of the substrate, transition-state stabilization, and appropriate positioning of catalytic residues. Those frameworks that have been designed so far are relatively small, and are therefore limited in their ability to accomodate complex arrangements of substrate-binding and catalytic residues. The second limitation to the design of *de novo* enzymes is the inability to systematically search a protein model or structure for potential active sites using computational approaches, as has been done for the identification of new metal-binding sites. This difficulty arises both from the greater complexity of enzyme active sites compared to metal-binding loci, and from the limitations of current computing resources.

Despite these difficulties, several attempts to design *de novo* enzymes have been made. Some of these designs are discussed in detail next.

A. Esterase Activity

A *de novo* four-helix bundle with esterolytic activity has been reported (Hahn *et al.*, 1990). In this design, each of the amino acids of the "catalytic triad" of chymotrypsin (Ser, Asp, and His) was placed at the end of a helix. Both an "oxyanion hole," a region in the active site of chymotrypsin that stabilizes the oxyanion formed in the transition state, and a hydrophobic binding pocket were incorporated into the design. The rest of the helical sequences were designed to create amphipathic helices and thereby promote four-helix-bundle formation. To promote parallel association of the helices and appropriate spatial orientation of the designed catalytic residues, the N termini of all four helices were cross-linked together.

Several enzyme-like properties were reported for the designed protein, termed chymohelizyme-1 or CHZ-1, including saturation kinetics, discrimination between hydrophobic and hydrophilic substrates, and inactivation by an active-site directed chymotrypsin inhibitor. Both activated *p*-nitrophenyl esters and less reactive ethyl esters were reported to be substrates for CHZ-1. Unfortunately, further work suggests that the source of the observed activity is likely not that originally proposed (Corey *et al.*, 1994, Corey and Corey, 1996). At the present time, it appears that CHZ-1 is only active against activated esters and does not require the designed "active site" His and Ser residues for this catalytic activity (Corey and Corey, 1996). The most likely explanation is general-base catalysis of ester hydrolysis, mediated by one or more of the lysine residues introduced on the solvent-exposed surface of the four-helix bundle.

Balzer and colleages have aimed to enhance the acylation rate of a histidine residue in a *de novo* designed dimeric helix–loop–helix protein (see Section IV,A) (Broo *et al.*, 1996). The designed protein, termed SA-42, had been modified to create a "binding site" for the negatively charged *p*-nitrophenyl fumarate ester through the introduction of two positively charged ornithine residues (the resulting protein was named RA-42). In addition, the possibility for stabilization of the negatively charged transition state by interaction with the ornithine residues was cited. RA-42 catalyzed the formation of *p*-nitrophenol from *p*-nitrophenyl fumarate five times faster than SA-42. The authors observed that one of these ornithine residues of RA-42, Orn-15, was acylated in the presence of *p*-nitrophenyl fumarate, and they proposed a two-step reaction starting with the formation of an acylated imidazole intermediate, followed by attack of the Orn-15 side chain on the acyl intermediate, resulting in an acylated ornithine residue. An activity–pH profile suggested that a functional group with a pK_a

close to that of histidine was involved in the rate-limiting step (Broo *et al.*, 1996), and 2D ^{1}H NMR experiments indicated that Orn-15 was selectively acylated; however, the ordering of these two reactions into a "self-functionalization" two-step mechanism has not yet been directly supported. The selective modification of Orn-15 is intriguing, but at present there is no data ruling out the possibility of direct reaction with *p*-nitrophenyl fumarate.

B. Design of an Oxaloacetate-Decarboxylating Four-Helix Bundle

The design of a four-helix-bundle protein that catalyzes a decarboxylation reaction has been described by Johnsson *et al.* (1993). Starting with a well-understood reaction mechanism for the amine-catalyzed decarboxylation of oxaloacetate, the authors have created amine-containing amphiphilic peptides with catalytic power enhanced over that of small-molecule organic amines. The difference between this and most other attempts at the design of *de novo* catalytic peptides is that the "oxaldie" designs do not require the precise alignment of amino acid side chains for catalysis, a goal that is difficult to accomplish given the "molten globule" properties of most current *de novo* frameworks. Rather, Johnsson *et al.* aimed to enhance the reactivity of amine functional groups by taking advantage of electrostatic interactions that occur in the context of a folded four-helix bundle. These interactions are in principle unattainable to the same extent in the unfolded state or by small-molecule amine catalysts.

The mechanism for the decarboxylation of oxaloacetate in the presence of amine proceeds through an unstable carbinolamine intermediate to a Schiff base, which then undergoes decarboxylation to given CO_2, pyruvate, and regenerated amine as the final products. With the oxaldie peptides, both the N-terminal amine and lysine side chains were investigated for their ability to catalyze the decarboxylation of oxaloacetate. A helical wheel diagram of the oxaldie peptides is shown in Fig. 15. Both oxaldie 1 and 2 contain five lysine residues on the hydrophilic face of the helix; the difference between them is the presence of a free (oxaldie 1) or blocked (oxaldie 2) N-terminal amine.

Two strategies were used to enhance the catalytic power of oxaldie at neutral pH: First, as Schiff base formation is the rate-limiting step above pH 4, lowering the pK_a of amine groups in the four-helix bundle should enhance their nucleophilicity, and increase the overall reaction rate. This was achieved in oxaldie 1 by taking advantage of the unfavorable electrostatic interaction between the N-terminal amine in its protonated, positively charged form and the helix dipole, which should suppress the pK_a of the amine. In oxaldie 2, the N terminus was blocked by acylation, leaving the five lysine residues on the solvent-exposed face of each helix as potential catalysts of decarboxylation. The grouping of five lysine residues on the solvent-exposed face of each helix (Fig. 15) was designed to suppress the pK_a of at least one of the side-chain amines by introducing like-charge repulsion. The clustering of lysine residues was also intended to lend electrostatic stabilization to the unstable carbinolamine intermediate, which carries a double negative charge. Although oxaldie 1 contained all of the lysine residues present in oxaldie 2, the N terminus was expected to be the dominant catalytic amine given its significantly lower pK_a.

Peptides oxaldie 1 and oxaldie 2 formed α-helical structures in aqueous

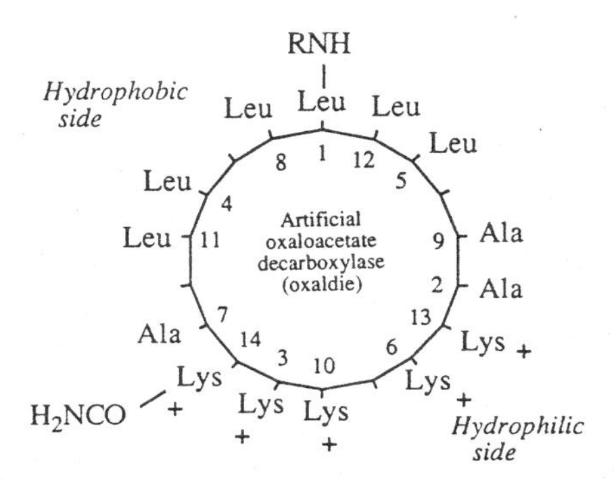

RHN-Leu-Ala-Lys-Leu-Leu-Lys-Ala-Leu-Ala-Lys-Leu-Leu-Lys-Lys-$CONH_2$
1 2 3 4 5 6 7 8 9 10 11 12 13 14

Fig. 15 Sequences of two designed artificial oxaloacetate decarboxylases (oxaidie 1 and oxaldie 2; R for oxaldie 1 is hydrogen, and for oxaldie 2 is an acetyl group together with a representation of these structures on a helical wheel to illustrate the amphiphilic nature of the helices that they build. Five amino groups from the side chains of lysine protrude on one face of the helix. In oxaldie 1, the amino-terminal amine is the principal active site. In oxaldie 2, the ε-lysine amino groups provide the active site. Both peptides were synthesized using standard t-Boc chemistry on polystyrene/1% divinylbenzene resin with the *p*-methyl benzhydrylamine linker. Deprotection of the peptide and release from the support were achieved by treatment with trifluoromethylsulphonic acid in trifluoroacetic acid in the presence of ethanedithiol and thioanisole. Peptides were purified by reversed phase C_{18} HPLC.

solution in a concentration-dependent manner. Oxaldie 1 appeared to form the intended four-helix bundle, whereas oxaldie 2 formed a higher-order aggregate of unspecified size. Analysis of helical structure by ^{1}H NMR indicated that the first four residues of oxaldie 1 were relatively unstable in an α-helical conformation. A decrease of 0.6 pH units was observed for the N-terminal amine in oxaldie 1, despite the likely attenuation of the helix dipole–protonated amine interaction by the fraying of the N termini. Oxaldie 2 was observed to be fully helical by ^{1}H NMR and possessed an amine with a pK_a of 8.9, a value 1.6 pH units lower than the pK_a of a typical lysine side chain (Johnsson *et al.*, 1993).

Both oxaldies 1 and 2 catalyzed the decarboxylation of oxaloacetate with Michaelis–Menten (saturation) kinetics, having K_m values for oxaloacetate of 14 and 48 m*M*, and k_{cat} values of 6.7×10^{-3} and 7.5×10^{-3} sec^{-1}, respectively. These K_m and k_{cat} values, in contrast those of natural enzymes, probably do not reflect the binding of the substrate to a defined binding site followed by interaction with specific catalytic residues; rather, they are likely aggregate values, particularly in the case of oxaldie 2, where more than one of the five amine side chains may react directly with oxaloacetate. To determine whether the observed catalytic activity depends on the folded structure, the authors created a variant of oxaldie 2 that contained two glycine and one proline residues to destabilize helix formation. This peptide, which was limited to a random coil conformation in solution, was sevenfold less effective as a catalyst of decarboxylation than oxaldie 2.

Aspects of the oxaldie-catalyzed reaction mechanism have been verified using chemical trapping and spectroscopic approaches. Imine intermediates were trapped by reduction to amines with $NaBH_3CN$. Formation of an enamine species, which follows imine formation, was monitored spectrophotometrically. The rate of pre-steady-state enamine production in the presence of oxaldie 1 was estimated to be 3–4 orders of magnitude faster than that occuring with butylamine (pK_a 10.6). This rate enhancement was attributed to both the lower pK_a in oxaldie 1 and the presence of side chains capable of stabilizing the transition state through electrostatic interactions. Creation of a random-coil oxaldie 1 analog would provide one test of this model.

VII. Summary and Future Directions

The protein design approach has resulted in the successful reproduction of several protein topologies. This success at the level of "the fold" is contrasted by the difficulty in designing proteins that reproduce the properties of native proteins, rather than those of molten globules. As the importance of internal core packing in the formation of "native-like" proteins is now clearly recognized, detailed thermodynamic analyses should form an integral part of the characterization of designed proteins.

Various attempts to introduce novel functions into proteins have been discussed. Several of the functional designs, particularly the straightforward metal-based systems, have been successful and bind the designed ligands with high affinity. The design of heme-binding proteins has also been successful and will surely lead to further interesting designs. Proteins designed to bind to molecules lacking metal coordination sites have been more difficult to realize and have been less thoroughly characterized. Most difficult of all has been the design of *de novo* catalysts.

Looking to the future, one of the most pressing needs is for high-resolution structures of designed proteins to allow detailed evaluation of the proteins and a truly iterative approach to the design process. Designs aimed at reproducing protein structures should, in the immediate future, focus on producing proteins with well-packed internal cores and native-like physical properties. This is admittedly no small task, but progress is being made. Now that several examples of designed metal-binding sites have been thoroughly characterized, the time is ripe to explore the catalytic potential of these systems. This is an especially tempting avenue, given the ubiquity of metals in nature's catalytic repertoire.

References

Alexandrescu, A. T., Evans, P. A., Pitkeathly, M., Baum, J., and Dobson, C. M. (1993). Structure and dynamics of the acid-denatured molten globule state of α-lactalbumin: A two-dimensional NMR study. *Biochemistry* **32,** 1707–1718.

Bai, Y., Milne, J. S., Mayne, L., and Englander, S. W. (1993). Primary structure effects on peptide group hydrogen exchange. *Proteins Struct. Funct. Genet.* **17,** 75–86.

Bai, Y., Milne, J. S., Mayne, L., and Englander, S. W. (1994). Protein stability parameters measured by hydrogen exchange. *Proteins Struct. Funct. Genet.* **20,** 4–14.

Baldwin, E. P., Hajiseyedjavadi, O., Baase, W. A., and Matthews, B. W. (1993). The role of backbone flexibility in the accommodation of variants that repack the core of T4 lysozyme. *Science* **262,** 1715–1718.

Banner, D. W., Kokkinidis, M., and Tsernoglou, D. (1987). Structure of the ColE1 Rop protein at 1.7 Å resolution. *J. Mol. Biol.* **196,** 657–675.

Baum, J., Dobson, C. M., Evans, P. A., and Hanley, C. (1989). Characterization of a partly folded protein by NMR methods: Studies on the molten globule state of guinea pig α-lactalbumin. *Biochemistry* **28,** 7–13.

Beauregard, M., Goraj, K., Goffin, V., Heremans, K., Goormaghtigh, E., Ruysschaert, J.-M., and Martial, J. A. (1991). Spectroscopic investigation of structure in octarellin (a *de novo* protein designed to adopt the α/β-barrel packing). *Protein Eng.* **4,** 745–749.

Behe, M. J., Lattman, E. G., and Rose, G. D. (1991). The protein-folding problem: The native fold determines packing, but does packing determine the native fold? *Proc. Natl. Acad. Sci. USA* **88,** 4195–4199.

Betz, S. F., Raleigh, D. P., DeGrado, W. F., Lovejoy, B., Anderson, D., Ogihara, N., and Eisenberg, D. (1996). Crystallization of a designed peptide from a molten globule ensemble. *Folding and Design* **1,** 57–64.

Bianchi, E., Venturini, S., Pessi, A., Tramontano, A., and Sollazzo, M. (1994). High level expression and rational mutagenesis of a designed protein, the minibody: From an insoluble to a soluble molecule. *J. Mol. Biol.* **236,** 649–659.

Blanco, F. J., Rivas, G., and Serrano, L. (1994). A stable beta hairpin from protein G. *Nat. Struct. Biol.* **1,** 584–590.

Brinen, L. S., Willett, W. S., Craik, C. S., and Fletterick, R. J. (1996). X-ray structures of a designed binding site in trypsin show metal-dependent geometry. *Biochemistry* **35,** 5999–6009.

Broo, K., Brive, L., Lundh, A.-C., Ahlberg, P., and Baltzer, L. (1996). The mechanism of self-catalyzed site selective functionalization of a designed helix–loop–helix motif. *J. Am. Chem. Soc.* **118,** 8172–8173.

Brunet, A. P., Huang, E. S., Huffine, M. E., Loeb, J. E., Weltman, R. J., and Hecht, M. H. (1993). The role of turns in the structure of an α-helical protein. *Nature* **364,** 355–358.

Bryson, J. W., Betz, S. F., Lu, H. S., Suich, D. J., Zhou, H. X., O'Neil, K. T., and DeGrado, W. F. (1995). Protein design: A hierarchic approach. *Science* **270,** 935–941.

Bullough, P. A., Hughson, F. M., Skehel, J. J., and Wiley, D. C. (1994). Structure of influenza haemagglutinin at the pH of membrane fusion. *Nature* **371,** 37–43.

Chakrabartty, A., and Baldwin, R. L. (1995). Stability of alpha helices. *Adv. Protein Chem.* **49,** 141–177.

Choma, C. T., Lear, J. D., Nelson, M. J., Dutton, P. L., Robertson, D. E., and DeGrado, W. F. (1994). Design of a heme-binding four-helix bundle. *J. Am. Chem. Soci.* **116,** 856–865.

Chothia, C., Levitt, M., and Richardson, D. (1981). Helix to helix packing in proteins. *J. Mol. Biol.* **145,** 215–250.

Chou, P. Y., and Fasman, G. D. (1978). Secondary structure prediction. *Annu. Rev. Biochem.* **47,** 251–276.

Christianson, D. W. (1991). Structural biology of zinc. *Adv. Protein Chem.* **42,** 281–355.

Chyan, C.-L., Wormald, C., Dobson, C. M., Evans, P. A., and Baum, J. (1993). Structure and stability of the molten globule state of guinea-pig α-lactalbumin: A hydrogen exchange study. *Biochemistry* **32,** 5681–5691.

Clarke, N. D., and Yuan, S.-M. (1995). Metal Search: A computer program that helps design tetrahedral metal-binding sites. *Proteins: Structure, Function and Genetics* **23,** 256–263.

Cohen, C., and Parry, D. A. D. (1986). α-Helical coiled coils—a widespread motif in proteins. *Trends Biochem. Sci.* **11,** 245–248.

Cohen, C., and Parry, D. (1990). α-Helical coiled coils and bundles: How to design an α-helical protein. *Proteins Struct. Funct. Genet.* **7,** 1–15.

Cohen, F. E., Sternberg, M. J., and Taylor, W. R. (1982). Analysis and prediction of the packing of α-helices against a β-sheet in the tertiary structure of globular proteins. *J. Mol. Biol.* **156,** 821–862.

Corey, M. J., Hallakova, E., Pugh, K., and Stewart, J. M. (1994). Studies on chymotrypsin-like catalysis by synthetic peptides. *App. Biochem. Biotechnol.* **47,** 199–212.

Crick, F. (1953). The packing of α-helices: Simple coiled-coils. *Acta Cryst.* **6,** 689–697.

Dalal, S., Balasubramanian, S., Regan, L. (1997). Protein alchemy: Changing β-sheet into α-helix. *Nat. Struct. Biol.* **4,** 548–552.

de Alba, E., Blanco, F. J., Jiménez, M., D., Rico, M. Nievo, J. L. (1995). Interaction responsible for the pH dependence of the β-hairpin conformational population formed by a designed linear peptide. *Eur. J. Biochem.* **233,** 283–292.

Desjarlais, J. R., and Handel, T. M. (1995). New strategies in protein design. *Curr. Opin. Biotech.* **6,** 460–466.

Doig, A., and Baldwin, R. (1995). N- and C-capping preferences for all 20 amino acids in α-helical peptides. *Protein Sci.* **4,** 1325–1336.

Eriksson, A. E., Baase, W. A., Wozniak, J. A., and Matthews, B. W. (1992a). A cavity-containing mutant of T4 lysozyme is stabilized by buried benzene. *Nature* **355,** 371–373.

Eriksson, A. E., Baase, W. A., Zhang, X.-J., Heinz, D. W., Blaber, M., Baldwin, E. P., and Matthews, B. W. (1992b). Response of a protein structure to cavity-creating mutations and its relation to the hydrophobic effect. *Science* **255,** 178–183.

Farid, R. S., Robertson, D. E., Moser, C. C., Pilloud, D., DeGrado, W. F., and Dutton, P. L. (1994). Metal–redox centre interactions. *Biochem. Soc. Trans.* **22,** 689–693.

Fersht, A., and Serrano, L. (1993). Principles of protein stability derived from protein engineering experiments. *Curr. Opin. Struct. Biol.* **3,** 75–83.

Fezoui, Y., Weaver, D., and Osterhout, J. (1995). Strategies and rationales for the *de novo* design of a helical hairpin peptide. *Protein Sci.* **4,** 286–295.

Fink, A. L. (1995). Compact intermediate states in protein folding. *Annu. Rev. Biophys. Biomol. Struct.* **24,** 495–522.

Gibney, B. R., Mulholland, S. E., Rabanal, F., and Dutton, P. L. (1996). Ferredoxin and ferredoxin-heme maquettes. *Proc. Natl. Acad. Sci USA* **93,** 15041–15046.

Glusker, J. P. (1991). Structural aspects of metal liganding to functional groups in proteins. *Adv. Protein Chem.* **42,** 1–44.

Goraj, K., Renard, A., and Martial, J. A. (1990). Synthesis, purification and initial structural characterization of octarellin, a *de novo* polypeptide modelled on the α/β barrel proteins. *Protein Eng.* **3,** 259–266.

Hahn, K. W., Klis, W. S., and Stewart, J. M. (1990). Design and synthesis of a peptide having chymotrypsin-like esterase activity. *Science* **248,** 1544–1547.

Handel, T. M., Williams, S. A., and DeGrado, W. F. (1993). Metal ion-dependent modulation of the dynamics of a designed protein. *Science* **261,** 879–885.

Harris, N., Presnell, S., and Cohen, F. (1994). Four helix bundle diversity in globular proteins. *J. Mol. Biol.* **236,** 1356–1368.

Haynie, D. T., and Freire, E. (1993). Structural energetics of the molten globule state. *Proteins Struct. Funct. Genet.* **16,** 115–140.

Hecht, M. H. (1994). *De novo* design of β-sheet proteins. *Proc. Nat. Acad. Sci. USA* **91,** 8729–8730.

Hecht, M. H., Richardson, J. S., Richardson, D. C., and Ogden, R. C. (1990). *De novo* design, expression, and characterization of Felix: A four-helix bundle protein of native-like sequence. *Science* **249,** 884–891.

Hellinga, H. W., and Richards, F. M. (1991). Construction of new ligand binding sites in proteins of known structure I. Computer-aided modeling of sites with predefined geometry. *J. Mol. Biol.* **222,** 763–785.

Hellinga, H. W., Caradonna, J. P., and Richards, F. M. (1991). Construction of new ligand binding sites in proteins of known structure II. Grafting of a buried transition metal binding site into *Escherichia coli* thioredoxin. *J. Mol. Biol.* **222,** 787–803.

Ho, S. P., and DeGrado, W. F. (1987). Design of a 4-helix bundle protein: Synthesis of peptides which self-associate into a helical protein. *J. Am. Chem. Soc.* **109,** 6751–6758.

Huang, C. C., Lesburg, C. A., Kiefer, L. L., Fierke, C. A., and Christianson, D. W. (1996). Reversal of the hydrogen bond to zinc ligand histidine-119 dramatically diminishes catalysis and enhances metal equilibration kinetics in carbonic anhydrase II. *Biochemistry* **35,** 3439–3446.

Hughson, F. M., Wright, P. E., and Baldwin, R. L. (1990). Structural characterization of a partly folded apomyoglobin intermediate. *Science* **249,** 1544–1548.

Hutchinson, E. G., and Thornton, J. M. (1994). A revised set of potentials for β-turn formation in proteins. *Protein Sci.* **3,** 2207–2216.

Iverson, B. L., Iverson, S. A., Roberts, V. A., Getzoff, E. D., Tainer, J. A., Benkovic, S. J., and Lerner, R. A. (1990). Metalloantibodies. *Science* **249,** 659–662.

Jackson, S., elMasry, N., and Fersht, A. (1993a). Structure of the hydrophobic core in the transition state for folding of chymotrypsin inhibitor 2: A critical test of the protein engineering method of analysis. *Biochemistry* **32,** 11270–11278.

Jackson, S., Moracci, M., elMasry, N., Johnson, C., and Fersht, A. (1993b). Effect of cavity-creating mutations in the hydrophobic core of chymotrypsin inhibitor 2. *Biochemistry* **32,** 11259–11269.

Jeng, M.-F., Englander, S. W., Elöve, G. A., Wand, A. J., and Roder, H. (1990). Structural description of acid-denatured cytochrome *c* by hydrogen exchange and 2D NMR. *Biochemistry* **29,** 10433–10437.

Johnson, K., Allemann, R. K., Widmer, H., and Benner, S. A. (1993). Synthesis, structure, and activity of artificial, rationally designed catalytic polypeptides. *Nature* **365,** 530–532.

Kalsbeck, W. A., Robertson, D. E., Pandey, R. K., Smith, K. M., Dutton, P. L., and Bocian, D. F. (1996). Structural and electronic properties of the heme cofactors in a multi-heme synthetic cytochrome. *Biochemistry* **35,** 3429–3438.

Kamtekar, S., and Hecht, M. H. (1995). The four-helix-bundle: What determines a fold? *FASEB J.* **9,** 1013–1022.

Kamtekar, S., Schiffer, J. M., Xiong, H., Babik, J. M., and Hecht, M. H. (1993). Protein design by binary patterning of polar and nonpolar amino acids. *Science* **262,** 1680–1685.

Klauser, S., Gantner, D., Salgam, P., and Gutte, B. (1991). Structure–function studies of designed DDT-binding polypeptides. *Biochem. Biophys. Res. Commun.* **179,** 1212–1219.

Klemba, M., and Regan, L. (1995). Characterization of metal binding by a designed protein: Single ligand substitutions at a tetrahedral Cys2His2 site. *Biochemistry* **34,** 10094–10100.

Klemba, M., Gardner, K. H., Marino, S., Clarke, N. D., and Regan, L. (1995). Novel metal-binding proteins by design. *Nature Struct. Biol.* **2,** 368–373.

Kullmann, W. (1984). Design, synthesis and characterization of an opiate receptor mimetic peptide. *J. Med. Chem.* **27,** 106–115.

Kuroda, Y. (1995). A strategy for the *de novo* design of helical proteins with stable folds. *Protein Eng.* **8,** 97–101.

Kuwajima, K. (1989). The molten globule state as a clue for understanding the folding and cooperativity of globular-protein structure. *Proteins* **6,** 87–103.

Kuwajima, K. (1996). The molten globule state of α-lactalbumin. *FASEB J.* **10,** 102–109.

Langen, H., Epprecht, T., Linden, M., Hehgans, T., Gutte, B., and Buser, H.-R. (1989). Rapid partial degradation of DDT by a cytochrome *p*450 model system. *Eu. J. Biochem.* **182,** 727–735.

Lesk, A., and Chothia, C. (1980). How different amino acid sequences determine similar protein structures: The structure and evolutionary dynamics of the globins. *J. Mol. Biol.* **136,** 225–270.

Lim, W. A., and Sauer, R. T. (1989). Alternative packing arrangements in the hydrophobic core of λ repressor. *Nature* **339,** 31–36.

Lim, W., and Sauer, R. (1991). The role of internal packing interactions in determining the structure and stability of a protein. *J. Mol. Biol.* **219,** 359–376.

Lim, W. A., Farruggio, D. C., and Sauer, R. T. (1992). Structural and energetic consequences of disruptive mutations in a protein core. *Biochemistry* **31,** 4324–4333.

Lovejoy, B., Choe, S., Cascio, D., McRorie, D. K., DeGrado, W. F., and Eisenberg, D. (1993). Crystal structure of a synthetic triple-stranded α-helical bundle. *Science* **259,** 1288–1293.

Mayo, K. H., Ilyina, E., and Park, H. (1996). A recipe for designing water-soluble beta-sheet-forming peptides. *Protein Sci.* **5,** 1301–1315.

Milla, M. E., and Sauer, R. T. (1995). Critical side-chain interactions at a subunit interface in the Arc repressor dimer. *Biochemistry* **34,** 3344–3351.

Moser, R., Thomas, R. M., and Gutte, B. (1983). An artificial crystalline DDT-binding peptide. *FEBS Lett.* **157,** 247–251.

Moser, R., Frey, S., Münger, K., Hehlgans, T., Klauser, S., Langen, H., Winnacker, E.-L., Mertz, R., and Gutte, B. (1987). Expression of the synthetic gene of an artificial DDT-binding polypeptide in *Escherichia coli. Protein Eng.* **1,** 339–343.

Müller, H. N., and Skerra, A. (1994). Grafting of a high-affinity Zn(II)-binding site on the beta-barrel of retinol-binding protein results in enhanced folding stability and enables simplified purification. *Biochemistry* **33,** 14126–14135.

Munson, M., O'Brien, R., Sturtevant, J. M., and Regan, L. (1994). Redesigning the hydrophobic core of a four-helix-bundle protein. *Protein Sci.* **3,** 2015–2022.

Munson, M., Balasubramanian, S., Fleming, K., Nagi, A., O'Brien, R., Sturtevant, J., and Regan, L. (1996). What makes a protein a protein? Hydrophobic core designs that specify stability and structural properties. *Protein Sci.,* **5,** 1584–1595.

Myszka, D., and Chaiken, I. (1994). Design and characterization of an intramolecular antiparallel coiled coil peptide. *Biochemistry* **33,** 2363–2372.

Nagi, A., and Regan, L. (1997). An inverse correlation between loop length and protein stability in a four helix bundle protein. *Folding and Design* **2,** 67–75.

O'Shea, E., Klemm, J., Kim, P., and Alber, T. (1991). X-ray Structure of the GCN4 leucine zipper, a two-stranded, parallel coiled coil. *Science* **254,** 539–544.

Ohgushi, M., and Wada, A. (1983). "Molten-globule state": A compact form of globular proteins with mobile side-chains. *FEBS Letts.* **164,** 21–24.

Olofsson, S., and Baltzer, L. (1996). Structure and dynamics of a designed helix–loop–helix dimer in dilute aqueous trifluoroethanol solution. A strategy for NMR spectroscopic structure determination of molten globules in the rational design of native-like proteins. *Folding and Design* **1,** 347–356.

Olofsson, S., Johansson, G., and Baltzer, L. (1995). Design, synthesis and solution structure of a helix–loop–helix dimer—a template for the rational design of catalytically active polypeptides. *J. Chem. Soc., Perkin Trans.* **2,** 2047–2056.

Paliakasis, C., and Kokkinidis, M. (1992). Relationships between sequences and structure for the four-α-helix bundle tertiary motif in proteins. *Protein Eng.* **5,** 739–748.

Parsell, D., and Sauer, R. (1989). The structural stability of a protein is an important determinant of its proteolytic susceptibility in *Escherichia coli. J. Biol. Chem.* **264,** 7590–7595.

Pessi, A., Bianchi, E., Crameri, A., Venturini, S., Tramontano, A., and Sollazzo, M. (1993). A designed metal-binding protein with a novel fold. *Nature* **362,** 367–369.

Ponder, J. W., and Richards, F. M. (1987). Tertiary templates for proteins: Use of packing criteria in the enumeration of allowed sequences for different structural classes. *J. Mol. Biol.* **193,** 775–791.

Quinn, T. P., Tweedy, N. B., Williams, R. W., Richardson, J. S., and Richardson, D. C. (1994). Betadoublet: *De novo* design, synthesis, and characterization of a β-sandwich protein. *Proc. Natl. Acad. Sci. USA* **91,** 8747–8751.

Rabanal, F., DeGrado, W. F., and Dutton, P. L. (1996). Toward the synthesis of a photosynthetic reaction center maquette: A cofacial porphyrin pair assembled between two subunits of a synthetic four-helix bundle multi-heme protein. *J. Am. Chem. Soc.* **118,** 473–474.

Raleigh, D. P., Betz, S. F., and DeGrado, W. F. (1995). A *de novo* designed protein mimics the native state of natural proteins. *J. Am. Chem. Soc.* **117,** 7758–7759.

Ramírez-Alvarado, M., Blanco, F. J., and Serrano, L. (1996). *De novo* design and structural analysis of a model beta-hairpin peptide system. *Nat. Struct. Biol.* **3,** 304–612.

Regan, L. (1994). Structural biology: Born to be beta. *Curr. Biol.* **7,** 656–658.

Regan, L., and Clarke, N. D. (1990). A tetrahedral zinc(II)-binding site introduced into a designed protein. *Biochemistry* **29,** 10878–10883.

Regan, L., and DeGrado, W. F. (1988). Characterization of a helical protein designed from first principles. *Science* **241,** 976–978.

Richards, F. M. (1977). Areas, volumes, packing, and protein structure. *Ann. Rev. Biophys. Bioeng.* **6,** 151–176.

Richards, F. M., and Lim, W. (1993). An analysis of packing in the protein folding problem. *Q. Rev. Biophys.* **26,** 423–498.

Roberts, V. A., and Getzoff, E. D. (1995). Metalloantibody design. *FASEB J.* **9,** 94–100.

Roberts, V. A., Iverson, B. L., Iverson, S. A., Benkovic, S. J., Lerner, R. A., Getzoff, E. D., and Tainer, J. A. (1990). Antibody remodeling: A general solution to the design of a metal-coordination site in an antibody binding pocket. *Proc. Nat. Acad. Sci. USA* **87,** 6654–6658.

Robertson, D., Farid, R., Moser, C., Urbauer, J., Mulholland, S., Pidikiti, R., Lear, J., Wand, A., DeGrado, W., and Dutton, P. (1994). Design and synthesis of multi-haem proteins. *Nature* **368,** 425–432.

Rose, G., Gierasch, L., and Smith, J. (1985). Turns in peptides and proteins. *Adv. Protein Chem.* **37,** 1–109.

Sandberg, W., and Terwilliger, T. (1989). Influence of interior packing and hydrophobicity on the stability of a protein. *Science* **245,** 54–57.

Sandberg, W., and Terwilliger, T. (1991). Energetics of repacking a protein interior. *Proc. Natl. Acad. Sci. USA* **88,** 1706–1710.

Schafmeister, C. E., Miercke, L. J. W., and Stroud, R. M. (1993). Structure at 2.5 Å of a designed peptide that maintains solubility of membrane proteins. *Science* **262,** 734–738.

Schmidt, A. M., Muller, H. N., and Skerra, A. (1996). A Zn(II)-binding site engineered into retinol-binding protein exhibits metal-ion specificity and allows highly efficient affinity purification. *Chem. Biol.* **3,** 645–653.

Shortle, D., Stites, W., and Meeker, A. (1990). Contributions of the large hydrophobic amino acids to the stability of staphyloccal nuclease. *Biochemistry* **29,** 8033–8041.

Sieber, V., and Moe, G. R. (1996). Interactions contributing to the formation of a beta-hairpin like structure in solution. *Biochemistry* **35,** 181–188.

Smith, C. K., and Regan, L. (1997). The construction and design of beta sheets. *Acc. Chem. Res.,* **30,** 153–161.

Steif, C., Weber, P., Hinz, H.-J., Flossdorf, J., Cesareni, G., and Kokkinidis, M. (1993). Subunit interactions provide a significant contribution to the stability of the dimeric four-α-helical-bundle protein ROP. *Biochemistry* **32,** 3867–3876.

Steif, C., Hinz, H.-J., and Cesareni, G. (1995). Effects of cavity-creating mutations on conformational stability and structure of the dimeric 4-α-helical protein Rop: Thermal unfolding studies. *Proteins Struct. Funct. Genet.* **23,** 83–96.

Stewart, J. D., Roberts, V. A., Crowder, M. W., Getzoff, E. D., and Benkovic, S. J. (1994). Creation of a novel biosensor for Zn(II). *J. Am. Chem. Soc.* **9,** 94–100.

Tanaka, T., Kimura, H., Hayashi, M., Fujiyoshi, Y., Fukuhara, K.-I., and Nakamura, H. (1994a). Characteristics of a *de novo* designed protein. *Protein Sci.* **3,** 419–427.

Tanaka, T., Kuroda, Y., Kimura, H., Kidokoro, S.-I., and Nakamura, H. (1994b). Cooperative deformation of a *de novo* designed protein. *Protein Eng.* **7,** 969–976.

Tashian, R. E. (1989). The carbonic anhydrases: Widening perspectives on their evolution, expression and function. *BioEssays* **10,** 186–192.

Vita, C., Roumestand, C., Toma, F., and Ménez, A (1995). Scorpion toxin as natural scaffolds for protein engineering. *Proc. Nat. Acad. Sci. USA* **92,** 6404–6408.

Wade, W. S., Koh, J. S., Han, N., Hoekstra, D. M., and Lerner, R. J. (1993). Engineering metal coordination sites into the antibody light chain. *J. Am. Chem. Soc.* **115,** 4449–4456.

West, M. W., and Hecht, M. H. (1995). Binary patterning of polar and nonpolar amino acids in the sequences and structures of native proteins. *Protein Sci.* **4,** 2032–2039.

Willett, W. S., Gillmor, S. A., Perona, J. J., Fletterick, R. J., and Craik, C. S. (1995). Engineered metal regulation of trypsin specificity. *Biochemistry* **34,** 2172–2180.

Willett, W. S., Brinen, L. S., Fletterick, R. J., and Craik, C. S. (1996). Delocalizing trypsin specificity with metal activation. *Biochemistry* **35,** 5992–5998.

Wilson, I. A., Skehel, J. J., and Wiley, D. C. (1981). Structure of the haemagglutinin membrane glycoprotein of influenza virus at 3Å resolution. *Nature* **289,** 366–373.

Yan, Y., and Erickson, B. W. (1994). Engineering of betabellin 14D: Disulfide-induced folding of a β-sheet protein. *Protein Sci.* **3,** 1069–1073.

Ybe, J., and Hecht, M. (1996). Sequence replacements in the central β-turn of plastocyanin. *Protein Sci.* **4,** 814–824.

Yutani, K., Ogasahara, K., and Kuwajima, K. (1992). Absence of the thermal transition in apo-α-lactalbumin in the molten globule state. *J. Mol. Biol.* **228,** 347–350.

Index

A

B

C

D

I

L

M

U

W

Y

Z